D0788989

Fractals for the Classroom

Heinz-Otto Peitgen   Hartmut Jürgens   Dietmar Saupe

Fractals for the Classroom

Part Two
Complex Systems and Mandelbrot Set

Evan Maletsky   Terry Perciante   Lee Yunker
NCTM Advisory Board

National Council of
Teachers of Mathematics

Springer-Verlag

Authors

Heinz-Otto Peitgen
Institut für Dynamische Systeme
Universität Bremen
D-2800 Bremen 33
Federal Republic of Germany and
Department of Mathematics
Florida Atlantic University
Boca Raton, FL 33432
USA

Hartmut Jürgens
Institut für Dynamische Systeme
Universität Bremen
D-2800 Bremen 33
Federal Republic of Germany

Dietmar Saupe
Institut für Dynamische Systeme
Universität Bremen
D-2800 Bremen 33
Federal Republic of Germany

Cover design by Claus Hösselbarth.

TI-81 Graphics Calculator is a product of Texas Instruments Inc.
Casio is a registered trademark of Casio Computer Co. Ltd.
Macintosh is a registered trademark of Apple Computer Inc.
Microsoft BASIC is a product of Microsoft Corp.

Library of Congress Cataloging-in-Publication Data
(Revised for volume 2)
Fractals for the classroom.
Vol. 2 published in cooperation with the
National Council of Teachers of Mathematics.
Includes bibliographical references and index.
ISBN 0-387-97041-X (Springer-Verlag New York Berlin Heidelberg : v. 1)
ISBN 3-540-97041-X (Springer-Verlag Berlin Heidelberg New York : v. 1)
ISBN 0-387-97722-8 (New York : v. 2 : alk. paper)
ISBN 3-540-97722-8 (Berlin : v. 2 : alk. paper)
Contents: v. 1. Stragetic activities — v. 2. Complex
systems and mandelbrot set.
1. Fractals. I. Peitgen, Heinz-Otto, 1945-
II. National Council of Teachers of Mathematics.
QA614.86.P45 1991 514'.74 91-11998

Printed on acid-free paper.

Negatives supplied by the authors.
Printed and bound by Hamilton Printing Co., Rensselaer, NY.
Printed in the United States of America.

9 8 7 6 5 4 3 2 1

ISBN 0-387-97722-8 Springer-Verlag New York Berlin Heidelberg
ISBN 3-540-97722-8 Springer-Verlag Berlin Heidelberg New York

To
Hans-Christoph Bömers

Preface

The scientist does not study nature because it is useful; he studies it because he delights in it, and he delights in it because it is beautiful. If nature were not beautiful, it would not be worth knowing, and if nature were not worth knowing, life would not be worth living.

Henri Poincaré

Despite its title, this is not a textbook in the usual sense of the word, nor is it written in a 'popular scientific' style. It has been carefully written for everyone who — even without much knowledge of technical mathematics — wants to know *the details* of chaos theory and fractal geometry. It has been our desire to give the reader a broad view of the underlying notions behind fractals, chaos and dynamics. In addition, we have wanted to show how fractals and chaos relate both to each other and to many other aspects of mathematics as well as to natural phenomena. A third motif in the book is the inherent visual and imaginative beauty in the structures and shapes of fractals and chaos. This book comes in two parts. While Part One is centered more around the geometrical or statical aspects of complexity, i.e., fractal structures and patterns and their generation, this second part addresses the dynamical aspects of complexity: the signs of chaos, the transition from order into chaos, strange attractors, Julia sets and the Mandelbrot set, cellular automata and L-systems. Each of these themes also leaves its own fractal imprints, but it is the evolutionary complexity which is in the foreground here.

Similar to the first volume of *Fractals for the Classroom*, we have designed a computer program for each chapter, the so-called *program of the chapter*. It highlights important constructions and algorithms. All programs were developed under MicroSoft BASIC on Apple Macintosh computers. They were also tested on an IBM-compatible PC.[1] We trust that it will be no problem to get the code running on your machine if only a version of BASIC is available. In addition, most chapters indicate small intriguing computer experiments which allow hands-on experience with the essential messages regarding chaos and fractals.

The entire book has been produced using the TeX and LaTeX typesetting systems, where all figures except for the color images were integrated in the computer files. As with the first volume, we owe our gratitude to many who have assisted us during the writing of this book. Our students Torsten Cordes and Lutz Voigt have produced most of the graphics very skillfully and with unlimited patience. They were joined by two more of our students, Ehler Lange and Wayne Tvedt, during part of the preparation time. Douglas Sperry has read our text very carefully at several stages of its evolution and, in addition to helping to get our English de-Germanized, has served in the broader capacity of copy editor. Friedrich von Haeseler, Guentcho Skordev, Heinrich Niederhausen and Ulrich Krause have read several chapters and provided valuable suggestions. We also thank Przemyslaw Prusinkiewicz and Mitchell J. Feigenbaum for reading parts of the original manuscript and giving valuable advice. Gisela Gründl has helped us with selecting and organizing third-party art work. Claus Hösselbarth

[1] PC users can find some general hints in *Fractals of the Classroom I.*

did an excellent job in designing the cover. Evan M. Maletsky, Terence H. Perciante and Lee E. Yunker read parts of our early manuscripts and gave crucial advice concerning the design of the book.

We were greatly inspired during our efforts by the continued encouragement and enthusiasm we found in many audiences of mathematics and science teachers at national and regional meetings of the NCTM. James D. Gates and Harry Tunis of the NCTM put confidence in our work very early on and did not lose it when we stretched their patience to its limits by not meeting several deadlines.

This book is complemented by strategic classroom activities, which we have co-authored with Evan M. Maletsky, Terence H. Perciante, and Lee E. Yunker. These activity books come in several volumes, the first two of which have already appeared. They directly involve the students in constructing, counting, computing, visualizing and measuring using carefully designed work sheets. These additional volumes focus on the large number of mathematical interrelationships which exist between fractals and the contemporary mathematics curricula found in our schools, colleges and universities.

Finally, we have been very pleased with the excellent cooperation of Springer-Verlag in New York.

Heinz-Otto Peitgen, Hartmut Jürgens, Dietmar Saupe
Bremen, May 1992

Authors

Heinz-Otto Peitgen. *1945 in Bruch (Germany). Dr. rer. nat. 1973, Habilitation 1976, both from the University of Bonn. Since 1977 Professor of Mathematics at the University of Bremen and between 1985 and 1991 also Professor of Mathematics at the University of California at Santa Cruz. Since 1991 also Professor of Mathematics at the Florida Atlantic University in Boca Raton. Visiting Professor in Belgium, Italy, Mexico and USA. Editor of several research journals on chaos and fractals. Co-author of the award winning books *The Beauty of Fractals* (with P. H. Richter) and *The Science of Fractal Images* (with D. Saupe)

Hartmut Jürgens. *1955 in Bremen (Germany). Dr. rer. nat 1983 at the University of Bremen. Employment in the computer industry 1984–85, since 1985 Director of the Dynamical Systems Graphics Laboratory at the University of Bremen. Co-author and co-producer (with H.-O. Peitgen, D. Saupe, and C. Zahlten) of the award winning video *Fractals: An Animated Discussion*

Dietmar Saupe. *1954 in Bremen (Germany). Dr. rer. nat 1982 at the University of Bremen. Visiting Assistant Professor of Mathematics at the University of California at Santa Cruz, 1985–87 and since 1987 Assistant Professor at the University of Bremen. Co-author of the award winning book *The Science of Fractal Images* (with H.-O. Peitgen)

Contents

Introduction

Causality Principle, Deterministic Laws and Chaos

Prediction is difficult, especially of the future.

Niels Bohr

For many, chaos theory already belongs to the greatest achievements in the natural sciences in this century. Indeed, it can be claimed that very few developments in natural science have awakened so much public interest. Here and there, we even hear of changing images of reality or of a revolution in the natural sciences.

Critics of chaos theory have been asking whether this popularity could perhaps only have something to do with the clever choice of catchy terms or the very human need for a theoretical explanation of chaos. Some have prophesied for it exactly the same quick and pathetic death as that of the catastrophe theory, which excited so much attention in the sciences at the end of the 1960's and then suddenly fell from grace even though its mathematical core is counted as one of the most beautiful constructions and creations. The causes of this demise were diverse and did not only have scientific roots. It can certainly be said that catastrophe theory was severely damaged by the almost messianic claims of some apologists.

Chaos theory, too, is occasionally in danger of being overtaxed by being associated with everything that can be even superficially related to the concept of chaos. Unfortunately, a sometimes extravagant popularization through the media is also contributing to this danger; but at the same time this popularization is also an important opportunity to free areas of mathematics from their intellectual ghetto and to show that mathematics is as alive and important as ever.

But what is it that makes chaos theory so fascinating? What do the supposed changes in the image of reality consist of? To these subjects we would like to pose, and to attempt to answer, some questions regarding the philosophy of nature.

Cause and Effect

The main maxim of science is its ability to relate cause and effect. On the basis of the laws of gravitation, for example, astronomical events such as eclipses and the appearances of comets can be predicted thousands of years in advance. Other natural phenomena, however, appear to be much more difficult to predict. Although the movements of the atmosphere, for example, obey the laws of physics just as much as the movements of the planets do, weather prediction is still rather problematic.

Tides Versus Weather

Ian Stewart in his article *Chaos: Does God Play Dice?,* Encyclopædia Britannica, 1990 Yearbook of Science and the Future, makes the following striking comparison:

"Scientists can predict the tides, so why do they have so much trouble predicting the weather? Accurate tables of the time of high or low tide can be worked out months or even years ahead. Weather forecasts often go wrong within a few days, sometimes even within a few hours. People are accustomed to this difference that they are not in the least surprised when the promised heat wave turns out to be a blizzard. In contrast, if the tide table predicted a low tide but the beach was under water, there would probably be a riot. Of course the two systems are different. The weather is extremely complex; it involves dozens of such quantities as temperature, air pressure, humidity, wind speed, and cloud cover. Tides are much simpler. Or are they? Tides are perceived to be simpler because they can be easily predicted. In reality, the system that gives rise to tides involves just as many variables — the shape of the coastline, the temperature of the sea, its salinity, its pressure, the waves on its surface, the position of the Sun and Moon, and so on — as that which gives rise to weather. Somehow, however, those variables interact in a regular and predictable fashion. The tides are a phenomenon of order. Weather, on the other hand, is not. There the variables interact in an irregular and unpredictable way. Weather is, in a word, chaotic."

We speak of the unpredictable aspects of weather just as if we were talking about rolling dice or letting an air balloon loose to observe its erratic path as the air is ejected. Since there is no clear relation between cause and effect, such phenomena are said to have random elements. Yet there was little reason to doubt that precise predictability could, in principle, be achieved. It was assumed that it was only necessary to gather and process greater quantities of more precise information (e.g., through the use of denser networks of weather stations and more powerful computers dedicated solely to weather analysis). Some of the first conclusions of chaos theory, however, have recently altered this viewpoint. Simple deterministic systems with only a few elements (e.g., the quadratic iterator) can generate random behavior, and that randomness is fundamental; gathering more information does not make it disappear. This fundamental randomness has come to be called chaos.[1]

Deterministic Chaos

An apparent paradox is that chaos is deterministic, generated by fixed rules which do not themselves involve any elements of change. We even speak of deterministic chaos. In principle, the future is completely determined by the past; but in practice small uncertainties, much like minute errors of measurement which enter into calculations, are amplified, with the effect that even though the behavior is predictable in the short term, it is unpredictable over the long run.

The discovery of such behavior is one of the important achievements of chaos theory. Another is the methodologies which have been designed for a precise scientific evaluation of the presence of chaotic behavior in mathematical models as well as in real phenomena. Using these methodologies, it is now possible, in principle, to estimate the 'predictability horizon' of a system. This is the mathematical, physical, or time parameter limit within which predictability is ideally possible and beyond which we will never be able to predict with certainty. It has been established, for example, that the predictability horizon in weather forecasting is not more than about two or three weeks. This means that no matter how many more weather stations are included in the observation, no matter how much more accurately weather data are collected and analyzed, we will never be able to predict the weather with any degree of numerical accuracy beyond this horizon of time.

But before we go into an introductory discussion of what chaos theory is trying to accomplish, let us look at some historical aspects of the field. If we look at the development of the sciences on a time-scale on which the efforts of our forbears are visible, we will observe indications of an apparent recapitulation in the present day, even if at a different level. To people during the age of early human history, natural events must have seemed largely to be pure chaos. At first very slowly, then faster and faster, the natural sciences developed (i.e., over the course of thousands of years, the area where chaos reigned seemed to become smaller and smaller. For more and more phenomena, their governing laws were wrung from Nature and their rules were recognized. Simultaneously, mathematics developed hand in hand with the natural

[1] The term chaos was proposed by the American mathematicians James A. Yorke and Tien-Yien Li in their work *Period three implies chaos*, American Mathematical Monthly 82 (1975) 985–992.

sciences, and thus, an understanding of the nature of a phenomenon soon came to also include the discovery of an appropriate mathematization of it. In this way, there was continuous nourishment for the illusion that it was only a matter of time, along with the necessary effort and means, before chaos would be completely banned from human experience.

A landmark accomplishment of tremendous, accelerating effect was made about three hundred years ago with the development of calculus by Sir Isaac Newton (1643–1727) and Gottfried Wilhelm Freiherr von Leibniz (1646–1716). Through the universal mathematical ideas of calculus, the basis was provided with which to apparently successfully model the laws of the movements of planets with as much detail as that in the development of populations, the spread of sound through gases, the conduction of heat in media, the interaction of magnetism and electricity, or even the course of weather events. Also maturing during that time, was the secret belief that the terms determinism and predictability were equivalent.

The Laplace Demon

For the era of determinism, which was mathematically grounded in calculus, the 'Laplace demon' became the symbol. "If we can imagine a consciousness great enough to know the exact locations and velocities of all the objects in the universe at the present instant, as well as all forces, then there could be no secrets from this consciousness. It could calculate anything about the past or future from the laws of cause and effect."[2]

In its core, the deterministic credo means that the universe is comparable to the ordered running of a tremendously precise clock, in which the present state of things is, on the one hand, simply the consequence of its prior state, and on the other hand, the cause of its future state. Present, past, and future are bound together by causal relationships; and according to the views of the determinists, the problem of an exact prognosis is only a matter of the difficulty of recording all the relevant data. The deterministic credo was characteristic of the Newtonian era, which for the natural sciences came to an end, at the latest, through the insights of Werner Heisenberg in the 1927 proclamation of his uncertainty principle,[3] but which for other sciences is still considered valid.

Strict Causality

Heisenberg wrote: "In the strict formulation of the causality law — 'When we know the present precisely, we can calculate the future' — it is not the final clause, but rather the premise, that is

[2]Pièrre Simon de Laplace (1749–1829), a Parisian mathematician and astronomer.

[3]This is also called the indeterminacy principle and states that the position and velocity of an object cannot, even in theory, be exactly measured simultaneously. In fact, the very concept of a concurrence of exact position and exact velocity have no meaning in nature. Ordinary experience, however, provides no evidence of the truth of this principle. It would appear to be easy, for example, to simultaneously measure the position and the velocity of a car; but this is because for objects of ordinary size, the uncertainties implied by this principle are too small to be observable. But the principle becomes really significant for subatomic particles such as electrons.

false. We cannot know the present in all its determining details.

Therefore, all perception is a selection from an abundance of possibilities and a limitation of future possibilities ... Because all experiments are subject to the laws of quantum mechanics, and thereby also to the uncertainty principle, the invalidity of the causality law is definitively established through quantum mechanics."

Classical determinism in its fearful strictness had to be given up — a turning point of enormous importance.

How undiminished the hope in a great victory of determinism still was at the beginning of this century is impressively illustrated in the 1922 book by Lewis F. Richardson entitled *Weather Prediction by Numerical Process,*[4] in which was written: "After so much hard reasoning, may one play with a fantasy? Imagine a large hall like a theater, except that the circles and galleries go right round through the space usually occupied by the stage. The walls of this chamber are painted to form a map of the globe. The ceiling represents the north polar regions, England is the gallery, the tropics in the upper circle, Australia on the dress circle and the antarctic in the pit. A myriad of computers[5] are at work upon the weather of the part of the map where each sits, but each computer attends only to one equation or part of an equation. The work of each region is coordinated by an official of higher rank. Numerous little 'night signs' display the instantaneous values so that neighboring computers can read them. ... From the floor of the pit a tall pillar rises to half the height of the hall. It carries a large pulpit on its top. In this sits the man in charge of the whole theater; he is surrounded by several assistants and messengers. In this respect he is like the conductor of an orchestra in which the instruments are slide-rules and calculating machines. But instead of waving a baton he turns a beam of rosy light upon any region that is running ahead of the rest, and a beam of blue light upon those who are behindhand."

In his book, Richardson first laid down the basis for numerical weather forecasting and then reported on his own initial practical experience with calculation experiments. According to Richardson, the calculations were so long and complex that only by using a 'weather forecasting center' such as the one he fantasized was forecasting conceivable.

Then about the middle of the 1940's, the great John von Neumann actually began to construct the first electronic computer, ENIAC, in order to further pursue Richardson's prophetic program, among others. It was soon recognized, however, that Richardson's

---

[4]Dover Publications, New York, 1965. First published by Cambridge University Press, London, 1922. This book is still considered one of the most important works on numerical weather forecasting.

[5]Richardson uses the word computer here to mean a person who computes.

only mediocre practical success was not simply attributable to his equipment's lack of calculating capacity, but also to the fact that the space and time increments used in his work had not met a computational stability criterion (Courant-Friedrichs-Lewy Criterion), which was only discovered later. With the appropriate corrections, further attempts were soon under way with progressively bigger and faster computers to make Richardson's dream a reality. This development has been uninterrupted since the 1950's, and it has bestowed truly gigantic 'weather theaters' upon us.

Weak Causality

Indeed, the history of numerical weather forecasting illustrates better than anything else the undiminished belief in a deterministic (viz. predictable) world; for in reality, Heisenberg's uncertainty principle did not at all mean the end of determinism. It only modified it, because scientists had never really taken Laplace's credo so completely seriously — as is usual with creeds. The most carefully conducted experiment is, after all, never completely isolated from the influences of the surrounding world, and the state of a system is never precisely known at any point in time. The absolute mathematical precision that Laplace presupposed is not physically realizable; minute imprecision is, as a matter of principle always present. What scientists actually believed was this: from approximately the same causes follow approximately the same effects — in nature as well as in any good experiment. And this is indeed often the case, especially over short time spans. If this were not so, we would not be able to ascertain any natural laws, nor could we build any functioning machines.

The Butterfly Effect

But this apparently very plausible assumption is not universally true. And what is more, it does not do justice to the typical course of natural processes over long periods of time. Around 1960, Ed Lorenz discovered this deficiency in the models used for numerical weather forecasting; and it was he who coined the term 'butterfly effect'. His description of deterministic chaos goes like this:[6] chaos occurs when the error propagation, seen as a signal in a time process, grows to the same size or scale as the original signal.

Thus, Heisenberg's response to deterministic thinking was also incomplete. He concluded that the strong causality principle is wrong because its presumptions are erroneous. Lorenz has now shown that the conclusions are also wrong. Natural laws, and for that matter determinism, do not exclude the possibility of chaos. In other words, determinism and predictability are not equivalent. And what is an even more surprising finding of recent chaos theory

[6]See Peitgen, H.-O. , Jürgens, H., Saupe, D., and Zahlten, C., *Fractals — An Animated Discussion,* Video film, Freeman 1990. Also appeared in German as *Fraktale in Filmen und Gesprächen,* Spektrum der Wissenschaften Videothek, Heidelberg, 1990.

has been the discovery that these effects are observable in many systems which are much simpler than the 'weather'. In fact, they can be observed in very simple feedback systems, even as simple as the quadratic iterator $x \to ax(1-x)$.

Moreover, chaos and order (i.e., causality principle) can be observed in juxtaposition within the same system. There may be a linear progression of errors characterizing a deterministic system which is governed by the causality principle, while (in the same system) there can also be an exponential progression of errors (i.e., butterfly effect) indicating that the causality principle breaks down.

In other words, one of the lessons coming out of chaos theory is that the validity of the causality principle is narrowed by the uncertainty principle from one end as well as by the intrinsic instability properties of the underlying natural laws from the other end.

Chapter 8

Recursive Structures: Growing of Fractals and Plants

The development of an organism may [...] be considered as the execution of a 'developmental program' present in the fertilized egg. The cellularity of higher organisms and their common DNA components force us to consider developing organisms as dynamic collections of appropriately programmed finite automata. A central task of developmental biology is to discover the underlying algorithm from the course of development.

Aristid Lindenmayer and Grzegorz Rozenberg[1]

The historical constructions of fractals by Cantor, Sierpinski, von Koch, Peano, etc., have been labeled as 'mathematical monsters'. Their purpose had been mainly to provide certain counterexamples, for example, showing that there are curves that go through all points in a square. Today a different point of view has emerged due to the ground-breaking achievements of Mandelbrot. Those strange creations from the turn of the century are anything but exceptional counterexamples; their features are in fact typical of nature. Consequently, fractals are becoming essential components in the modeling and simulation of nature. Certainly, there is a great difference between the basic fractals shown in this book and their counterparts in nature: mountains, rivers, trees, etc. Surely, the artificial fractal mountains produced today in computer graphics already look stunningly real. But on the other hand they still lack something we would certainly feel while actually camping in the real mountains. Maybe it is the (intentional) disregarding of

[1]In: *Automata, Languages, Development*, A. Lindenmayer, G. Rozenberg (eds.), North-Holland, 1975.

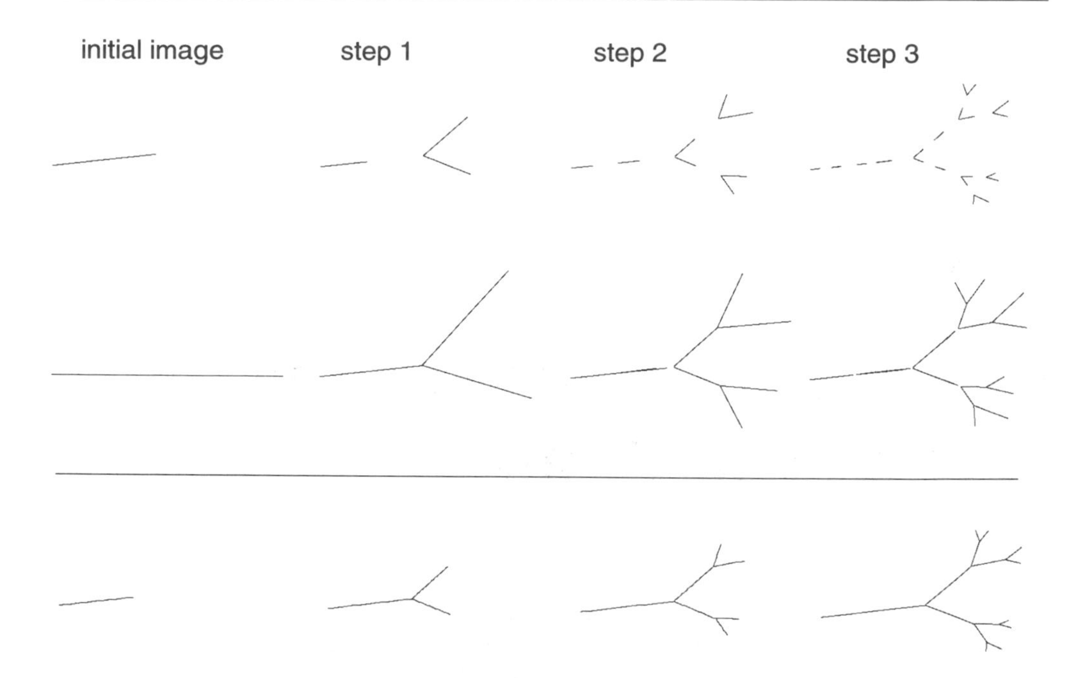

Figure 8.1 : An attempt to simulate growth by an MRCM. The two top rows show the first steps of the iteration of one single MRCM with two different initial images. In the top row a small 'stem', and in the middle row a tall 'stem' is taken as the initial image. Both cases reveal deficiencies. The small stem seems to grow, but unnatural gaps develop between the parts of the structure. The iteration of the tall stem does not produce gaps, however the branches seem to split and do not grow. In contrast, the bottom row shows the development we would like to simulate. Small new twigs are growing from the stem and the other branches. Note that the bottom figures are not just scaled down copies from the middle row.

all developmental processes in the fractal models which is one of the factors responsible for this shortcoming.

Fractals in nature are always a result of some growth process. In contrast, we have regarded fractals as *static*. Indeed, it was a goal to characterize fractals as solutions to equations. And nothing can be more eternal than the solution of an equation. You may argue on the other hand that these fractals have usually been obtained from dynamical processes such as the one in the Multiple Reduction Copy Machine introduced in chapter 1. However, our attention has always been focused on the end result, the attractor of the machine. The process leading to this end product was important merely when it revealed properties of the attractor. In this chapter we pay more attention to the intermediate stages in the pro-

Aristid Lindenmayer

Figure 8.2 : Aristid Lindenmayer, 1925–1989.

duction of a fractal. We begin the discussion with another dialect of the language of fractals which has been created specifically for the description of natural growth processes. This dialect is called *L-systems*.

In figure 8.1 we illustrate the contrast between the way an MRCM may generate a twig from a bush or tree and how one would actually expect a real twig to grow with time.

Development in Time

If we want to include the development in time in our fractal models we have to consider, for example, erosion models for the generation of fractal mountains or growth and evolution models for plants. While erosion models in the context of computer graphics[2] have just started to play a role in current research and will not be discussed in this presentation, growth models for plants have been around for a while; and some of them are also applicable to imaging. The view that growth and form are interrelated actually has a long tradition in biology. In his monumental work *On Growth and Form* D'Arcy Thompson traces its origins back to the late seventeenth century and comments:[3]

[2] K. Musgrave, C. Kolb and R. Mace, *The synthesis and the rendering of eroded fractal terrain*, SIGGRAPH '89, Computer Graphics 24 (1988).

[3] D'Arcy Thompson, *On Growth and Form*, New Edition, Cambridge University Press, 1942.

> *The rate of growth deserves to be studied as a necessary preliminary to the theoretical study of form, and organic form itself is found, mathematically speaking, to be a function of time. [...] We might call the form of an organism an event in space-time, and not merely a configuration in space.*

In 1968 the biologist Aristid Lindenmayer invented a formalization of the description of plant growth that is also very suitable in computer implementations. These formal descriptions are now known as *parallel rewriting systems* or *L-systems*.[4] A recent account of the state of the art of L-systems in the context of pattern formation in botany is presented in the beautiful book *The Algorithmic Beauty of Plants* by Przemyslaw Prusinkiewicz and Aristid Lindenmayer.[5]

But the modeling of growth for plants by L-systems is only one theme of this chapter. We also reexamine many of the fractals we have seen in the preceding chapters and demonstrate how we can describe 'natural' developmental processes which finally create these fractals. We will use two approaches towards this issue: our familiar concept of MRCMs (carefully set up for this problem) and the language of L-systems. In other words, we will see the growth of fractals.

[4]A. Lindenmayer, *Mathematical models for cellular interaction in development, Parts I and II,* Journal of Theoretical Biology 18 (1968) 280–315.

[5]P. Prusinkiewicz, A. Lindenmayer, *The Algorithmic Beauty of Plants*, Springer-Verlag, New York, 1990.

8.1 L-Systems: A Language For Modeling Growth

To begin, let us consider a plant, namely a blue-green alga, a species called *anabaena catenula*. The alga grows in filaments of strings of cells. There are two types of cells: specialized cells, which do not divide and which are called *heterocysts*, and unspecialized cells, which divide and are responsible for the entire growth of the alga. In a laboratory experiment[6] cells divided about every fourteen hours. The cell division was asymmetric in the sense that one of the offspring was generally smaller than the other at the time of division. Moreover, division was governed by a simple rule: if a given cell arose as the left offspring of a division, then its left progeny would be the smaller one in the next generation. Correspondingly, for a right offspring, the right progeny would be the smaller one in the next generation. In a figure illustrating the procedure (figure 8.3) we indicate the reproductive history of a cell by an arrow. The arrow points to the left, when the cell has emerged as a left daughter cell, and it points to the right for a right daughter cell. The smaller cells need about 20% more time to mature. This difference, however, is ignored in our first model as pictured in the figure. Therefore, in each following stage there is no difference between the children of large and small cells. A second model that takes these differences into account will be derived further on.

First Model of Cell Division

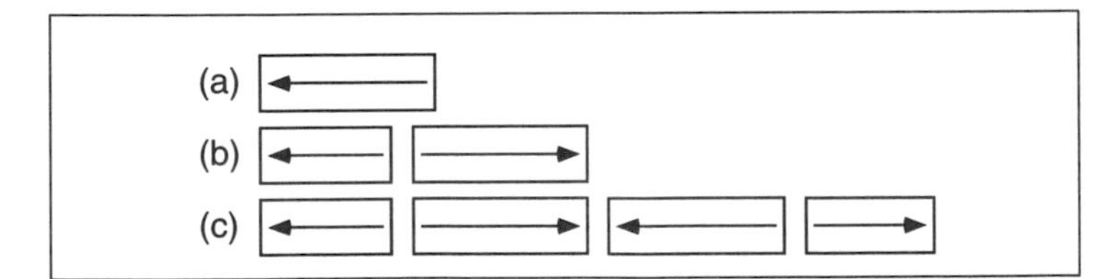

Figure 8.3 : In (a) one large cell is shown with an arrow to the left, indicating that the cell is a left daughter cell. In (b) the cell has subdivided into two cells, a small one on the left and a big one on the right. The arrows again indicate whether a cell has been created on the left or on the right. In the next division, (c), there are a total of four cells.

Reproduction Rules

Let us now write down the reproduction rules in a more formal manner. For that purpose we denote a string of cells by a string of symbols, each symbol denoting an individual cell. Let $\overleftarrow{A}$ stand for a small cell which was created as a left daughter cell (i.e., in the diagram it would appear with an arrow pointing to the left). Likewise, $\overrightarrow{A}$ denotes a small right daughter cell, and $\overleftarrow{B}$ and $\overrightarrow{B}$

[6] G. J. Mitchison and M. Wilcox, *Rule governing cell division in anabaena*, Nature 239 (1972) 110–111.

are the large left and right cells. Now the observed rule for $\overleftarrow{A}$ states that this cell produces a small cell on the left and a large cell on the right. Thus, we can signify the reproduction rule for a small left cell with

$$\overleftarrow{A} \rightarrow \overleftarrow{A}\overrightarrow{B} . \tag{8.1}$$

Using the same principles of notation, we write the reproduction rules for the other cells as

$$\begin{aligned} \overrightarrow{A} &\rightarrow \overleftarrow{B}\overrightarrow{A} \\ \overleftarrow{B} &\rightarrow \overleftarrow{A}\overrightarrow{B} \\ \overrightarrow{B} &\rightarrow \overleftarrow{B}\overrightarrow{A} . \end{aligned} \tag{8.2}$$

With this notation it is now very easy to predict the sequence of cells as they appear. For example, if we start with $\overleftarrow{A}$, we look up its reproduction in the above table, and we find $\overleftarrow{A}\overrightarrow{B}$. Now we apply the same procedure to $\overleftarrow{A}$ and $\overrightarrow{B}$ simultaneously to obtain four cells

$$\overleftarrow{A}\overrightarrow{B}\overleftarrow{B}\overrightarrow{A} .$$

In the next cell division we have eight cells, namely

$$\overleftarrow{A}\overrightarrow{B}\overleftarrow{B}\overrightarrow{A}\overleftarrow{A}\overrightarrow{B}\overleftarrow{B}\overrightarrow{A},$$

then 16 cells

$$\overleftarrow{A}\overrightarrow{B}\overleftarrow{B}\overrightarrow{A}\overleftarrow{A}\overrightarrow{B}\overleftarrow{B}\overrightarrow{A}\overleftarrow{A}\overrightarrow{B}\overleftarrow{B}\overrightarrow{A}\overleftarrow{A}\overrightarrow{B}\overleftarrow{B}\overrightarrow{A},$$

and so forth. Obviously, there is a repeating behavior in this sequence. The cell sequence $\overleftarrow{A}\overrightarrow{B}\overleftarrow{B}\overrightarrow{A}$ is a periodic cycle. This is the case because $\overleftarrow{A}\overrightarrow{B}$, its first half, produces the same sequence as the second half $\overleftarrow{B}\overrightarrow{A}$.

L-Systems: Another Example of an Iterator

What we have described here in (8.1) and (8.2) is a first very simple L-system for the vegetative part of the filament. It consists of the four (re)*production rules* and the definition of an initial state, which we have chosen to be $\overleftarrow{A}$. This initial state in L-systems is called the *axiom*. The concept is just another concrete form of an iterator as pictured in figure 8.6. It is a feedback machine which operates on strings. The production (or *rewriting*) rules determine how a given input string is transformed into an output string.

In fact, string rewriting feedback machines lie at the heart of what is known as formal languages and formal grammars in computer science. On these topics an enormous amount of work has

Development of Anabaena Catenula

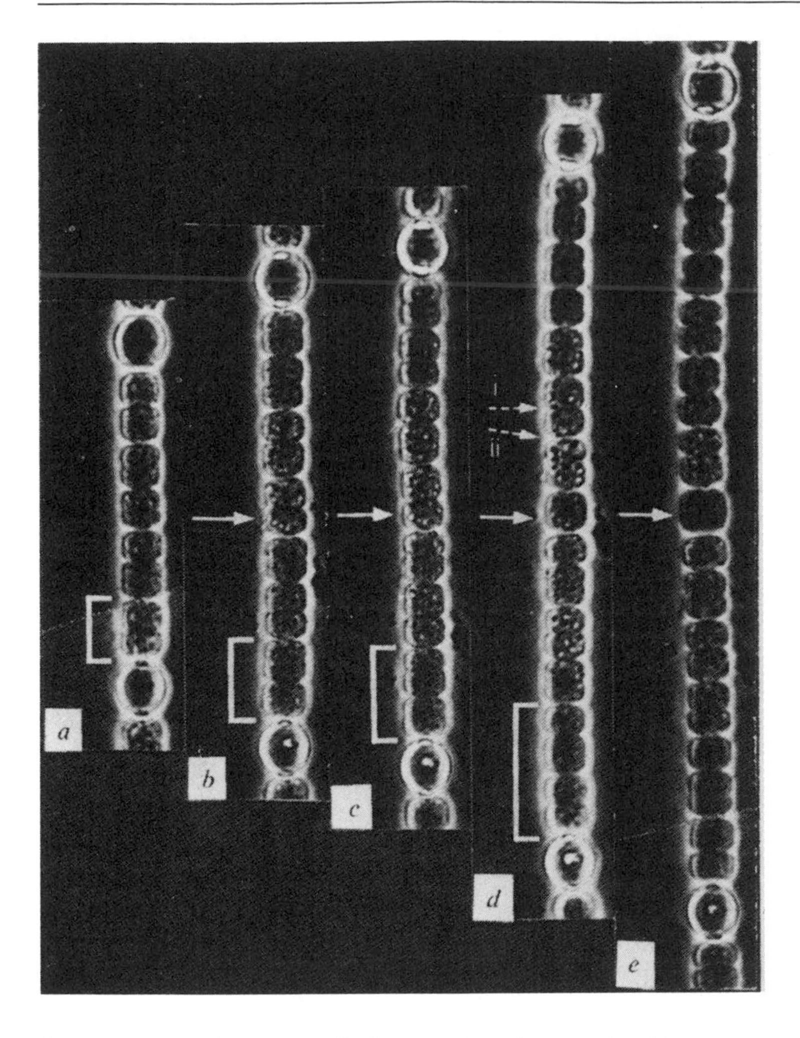

Figure 8.4 : Sequence of photographs of a growing blue-green alga (anabaena catenula) filament, demonstrating the asymmetric division rule and the appearance of a new heterocyst (specialized) cell at arrow. Photographs (a) to (e) were taken at 0, 6, 10, 15 and 21 hours. The division of the cell bracketed in (a) can be followed by referring to the bracketed cells in (b), (c) and (d).

been done. The concept has been explored in many directions and several classes (or types) of feedback machines have been identified. L-systems are string rewriting machines which are characterized by the fact that the production rules are applied simultaneously to all symbols of the input string[7] This property reflects the biological origin of L-systems. Lindenmayer intended to capture, for

[7] This is in contrast to sequential application as is typical for Chomsky grammars.

Modeling Anabaena Catenula with Ages

As a first step to improve the model for cell division for anabaena catenula, we may consider the different time spans needed by small and large cells to mature and to divide. The laboratory observation has been that small cells take about 20% more time on the average between successive divisions. Thus, whereas an initially large cell takes five time steps to divide, an initially small cell takes a total of six to grow and divide. A large cell B might develop in four stages into states denoted by C, D, E and F. Then, a division takes place producing either two cells AB or BA as before, depending on whether the initial cell B was a left daughter or a right one. A small cell A, on the other hand, needs one additional time step for the whole process. This can be simply modeled by introducing a transition from cell type A to type B. Thus, for small cells A the development is in six stages either

$$A \to B \to C \to D \to E \to F \to AB$$

or

$$A \to B \to C \to D \to E \to F \to BA$$

again depending on whether the initial cell was a left or right daughter. As before, let us use arrows to denote left and right daughter cells and summarize the production rules of the model:

$$\begin{aligned} &\overleftarrow{A} \to \overleftarrow{B},\ \overleftarrow{B} \to \overleftarrow{C},\ \overleftarrow{C} \to \overleftarrow{D},\ \overleftarrow{D} \to \overleftarrow{E},\ \overleftarrow{E} \to \overleftarrow{F},\\ &\overrightarrow{A} \to \overrightarrow{B},\ \overrightarrow{B} \to \overrightarrow{C},\ \overrightarrow{C} \to \overrightarrow{D},\ \overrightarrow{D} \to \overrightarrow{E},\ \overrightarrow{E} \to \overrightarrow{F}, \end{aligned}$$

and

$$\overleftarrow{F} \to \overleftarrow{A}\overrightarrow{B},\ \ \overrightarrow{F} \to \overleftarrow{B}\overrightarrow{A}.$$

Starting with $\overleftarrow{A}$ the following transitions will take place:

$$\begin{aligned} \overleftarrow{A} &\to \overleftarrow{B} \to \overleftarrow{C} \to \overleftarrow{D} \to \overleftarrow{E} \to \overleftarrow{F}\\ &\to \overleftarrow{A}\overrightarrow{B} \to \overleftarrow{B}\overrightarrow{C} \to \overleftarrow{C}\overrightarrow{D} \to \overleftarrow{D}\overrightarrow{E} \to \overleftarrow{E}\overrightarrow{F}\\ &\to \overleftarrow{F}\overleftarrow{B}\overrightarrow{A} \to \overleftarrow{A}\overrightarrow{B}\overleftarrow{C}\overrightarrow{B} \to \overleftarrow{B}\overrightarrow{C}\overleftarrow{D}\overrightarrow{C} \to \overleftarrow{C}\overrightarrow{D}\overleftarrow{E}\overrightarrow{D}\\ &\to \overleftarrow{D}\overrightarrow{E}\overleftarrow{F}\overrightarrow{E} \to \overleftarrow{E}\overrightarrow{F}\overleftarrow{A}\overrightarrow{B}\overrightarrow{F} \to \overleftarrow{F}\overleftarrow{B}\overrightarrow{A}\overleftarrow{B}\overrightarrow{C}\overleftarrow{B}\overrightarrow{A}\,. \end{aligned}$$

Obviously, the behavior of the cells is more complicated than in the first crude model. Cells in different cytological states, of different sizes, and even from different generations, coexist (see figure 8.5). The photograph from the original laboratory experiment confirms that this model captures the actual process more accurately (see figure 8.4, which is from G. J. Mitchison and M. Wilcox, *Rule governing cell division in anabaena*, Nature 239 (1972) 110–111.).

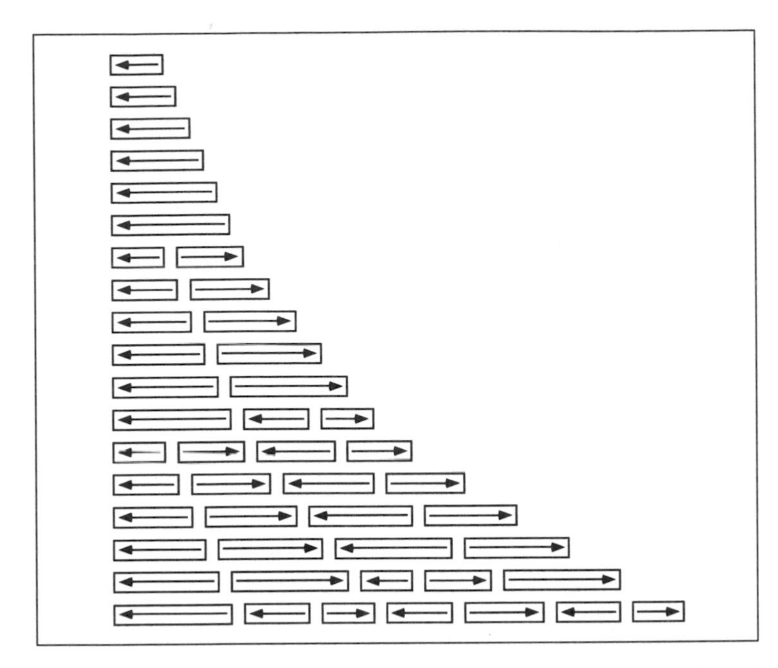

Figure 8.5 : Action of the modified L-system for cell division of blue-green alga anabaena catenula as given in the long transition formula for $\overleftarrow{A}$. Note that cells in different cytological states and of different sizes coexist.

The L-System Machine

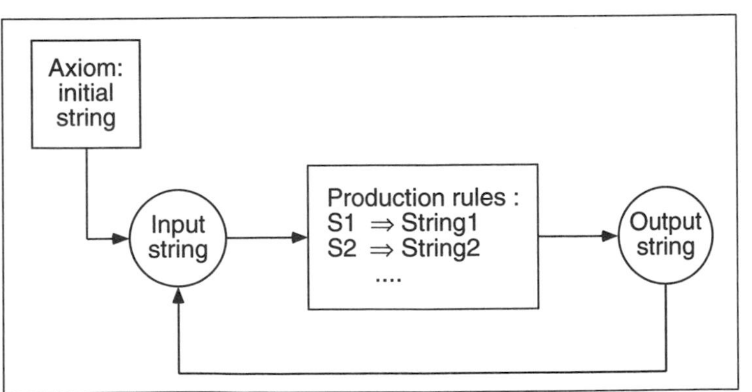

Figure 8.6 : L-system as an iterator. The internal operation of the machine is given by the production rules. Input and output consist of strings of symbols. The initial input string is called the axiom.

example, cell divisions in multicellular organisms, where many divisions may occur at the same time.

In the following we will use so called deterministic context-free L-systems. This means that the rewriting rules depend only on single, isolated symbols (i.e., the substitution of a particular symbol of an input string depends only on this symbol and not on its neighbors). Moreover, for each symbol the machine can work on, there is exactly one rewriting rule.[8]

Deterministic L-Systems

In this chapter we only use context-free L-systems. Such systems are formally defined by an *alphabet*

$$V = \{a_1, a_2, ..., a_n\},$$

the *production map*

$$\begin{aligned} P : V &\to V^* \\ a &\to P(a)\ , \end{aligned}$$

where V^* is the set of all strings formed by symbols from V, and an *axiom*

$$\alpha^{(0)} \in V^*,$$

the initial string.

Note that for all symbols of the alphabet $a \in V$ there is exactly one production (or rewriting) rule $P(a)$. Starting with the axiom $\alpha^{(0)}$, the L-system generates a sequence of strings: $\alpha^{(0)}, \alpha^{(1)}, \alpha^{(2)}, ...$ where the string $\alpha^{(i+1)}$ is obtained from the preceding string $\alpha^{(i)}$ by applying the production rules to all (e.g. m) symbols $\alpha_1^{(i)}, ..., \alpha_m^{(i)}$ of the string simultaneously:

$$\alpha^{(i+1)} = P(\alpha_1^{(i)})P(\alpha_2^{(i)})P(\alpha_3^{(i)}) \cdots P(\alpha_m^{(i)})\ .$$

Formally, we write an L-system using the following notation (exemplified by the basic anabaena model):

L-system: Anabaena (basic model)

Axiom: $\overleftarrow{A}$

Production rules:

$$\begin{aligned} \overleftarrow{A} &\to \overleftarrow{A}\,\overrightarrow{B} \\ \overrightarrow{A} &\to \overleftarrow{B}\,\overrightarrow{A} \\ \overleftarrow{B} &\to \overleftarrow{A}\,\overrightarrow{B} \\ \overrightarrow{B} &\to \overleftarrow{B}\,\overrightarrow{A} \end{aligned}$$

It seems apparent that the approach of L-systems can be very useful in describing growth phenomena in a short and precise manner. The axiom and the production rules constitute the system. These rules are typically derived from research, in our example from controlled laboratory experiments with a strain of the alga

[8] In this special case it does not matter whether the symbols in a string are sequentially rewritten one by one or simultaneously all at once.

organism. Once the rules have been set, they can easily be implemented on a computer in order to check in a simulation if they really capture the essential phases of the development of the plant. This verification would hardly be possible without the tool of a modern computer, at least for L-systems of such complexity as can be expected for real living plants that are somewhat more complicated than the simple string of alga cells.

There is, however, one point we have not yet discussed. Namely, it would be hopeless to carry out the verification process if we had to look at long, confusing strings of symbols. Each symbol might have a precise and understandable interpretation, but the whole string would be too long and too complicated to be understood. This is already very obvious from looking at the results of our above calculations in our very simple model.

Visualization of L-Systems

Again, the computer can help. A visual translation of the generated strings is required, and here it is literally true that a picture is worth a thousand words. On the other hand, the graphical interpretation should reflect the meaning of symbols. In this sense it should be closely coupled to the underlying problem. In the above case of alga, we may picture the cells $\overleftarrow{A}$ and $\overleftarrow{B}$ (and also $\overleftarrow{C}$, $\overleftarrow{D}$, $\overleftarrow{E}$ and $\overleftarrow{F}$) as cylinders or rectangles of varying lengths enclosing an arrow to the left or to the right as in figure 8.5. In this way a long string is interpreted graphically, and a visual inspection of the result will readily reveal important qualities of the underlying L-system model.

The graphical interpretation of strings of symbols is not predetermined in any way. This is one important strong point of the L-system approach to developmental models. The quality of the pictorial representation is completely independent of the generation process of the symbol string and can be adjusted to the available graphics environment and to the intentions of the experiment. For example, the interpretation may be as simple as a straightforward line drawing, or it may be the outcome of an elaborate ray-tracing computation yielding highly realistic 3D images.

Let us summarize. To develop an L-system for a particular biological species we can proceed in the following steps:

- analyze the object in nature (observation) and/or in a laboratory
- set up the rules in an informal way
- formulate rules and the initial state as an L-system
- run a simulation on a computer, the L-system will generate a long string of symbols
- translate the result into a graphical output
- compare the picture (or several pictures for different developmental stages) with the behavior of the real object

Finally, corrections can be made in the model, if necessary, and the steps may be repeated.

In 1968, Lindenmayer was interested in the development of filamentous plants on the cellular level. He considered two classes of filaments, simple (consisting of sequences of cells) and branching. The L-system symbols that he used corresponded to individual cells, thus only organisms consisting of relatively small numbers of cells (of the order of hundreds) could be handled in practice. He chose two species of algae, which are multicellular organisms with a small enough number of cells, to illustrate his concepts. Later on (in the seventies) Lindenmayer extended the interpretation of L-systems, so that L-system symbols were applied to represent entire plant modules, such as an internode, a leaf, or a bud. Using this interpretation, he analyzed the structure of compound inflorescences. This work provided a basis for the realistic visualization of the models of herbaceous (non-woody) plants. This was the first realistic visualization of plant models generated using L-systems.[9] Toward the end of this chapter we will sketch a few of the more elementary of these recent results. But before we turn to these applications of L-systems, let us start the second theme of this chapter: the growth of fractal patterns.

[9]See P. Prusinkiewicz, A. Lindenmayer, J. Hanan, *Developmental models of herbaceous plants for computer imagery purposes,* Computer Graphics 22, 4 (1988) 141–150. Previous application of L-systems to image synthesis, pioneered by Alvy Ray Smith in 1978 used L-systems to generate abstract branching structures that did not correspond to the existing species. See A. R. Smith, *Plants, fractals, and formal languages,* Computer Graphics 18, 3 (1984) 1–10.

8.2 Growing Classical Fractals with MRCMs

Classical fractals like the Cantor set, the Koch curve, the Sierpinski gasket, the Peano curve, etc., were very strange objects at the time they were introduced. Their creators were extremely careful in their definitions and exactly described the construction processes. We will demonstrate that L-systems provide a new language to efficiently and precisely define such constructions. On the other hand, this approach is in some respects rather formalistic and it is a good idea to start in a more visual and more familiar manner. Thus, before we let fractals grow using L-systems, let us find out what we can do with Multiple Reduction Copy Machines and iterated function systems.[10]

Encoding Images ...

We introduced the concept of the MRCM as a tool to encode images. We described several ways of decoding; the basic method is the iteration of the feedback machine, the MRCM. Starting with an arbitrary image one application of the machine produces a new image, a collage of contracted copies of the first image. Applying the machine to this new image and iteration of the process leads to a final image, the attractor of the machine, which is the decoded image.

... Versus Pattern Generation

The fact that the outcome of the process, the attractor of the machine, is completely independent of the starting image has been stressed in the preceding chapters on MRCMs. However, the intermediate stages in the iteration very much depend on the initial image. In this sense, the iteration of an MRCM is an example of recursive pattern generation. In this section we take a closer look at these patterns and observe that by choosing an appropriate initial pattern, we may very well reproduce the historical construction process given for many classical fractals.

The Cantor Set

As a first example, we compare the MRCM encoding of the Cantor set with the original construction process (divide a line segment into three equal parts, remove the middle part, repeat this process ad infinitum). Figure 8.7 shows the blueprint of the MRCM. We use the word 'Cantor' as initial image. Figure 8.8 shows the iteration (left) leading to the Cantor set as final image and compares this with the classical construction (right). Observe, that the classical construction seems to somehow capture the Cantor set C more accurately in every stage: the Cantor set is a subset of all the intermediate stages C_0, C_1, C_2, ... This is not true for the MRCM representation when started with an arbitrary pattern

[10]The relation between L-systems and IFSs was first studied in T. Bedford, *Dynamics and dimension for fractal recurrent sets,* J. London Math. Soc. 33 (1986) 89–100. Another discussion more oriented toward formal languages is presented in P. Prusinkiewicz and M. Hammel, *Automata, languages, and iterated function systems,* in: *Fractals Modeling in 3-D Computer Graphics and Imaging,* ACM SIGGRAPH '91 Course Notes C14 (J. C. Hart, K. Musgrave, eds.), 1991.

MRCM Blueprint of Cantor Set

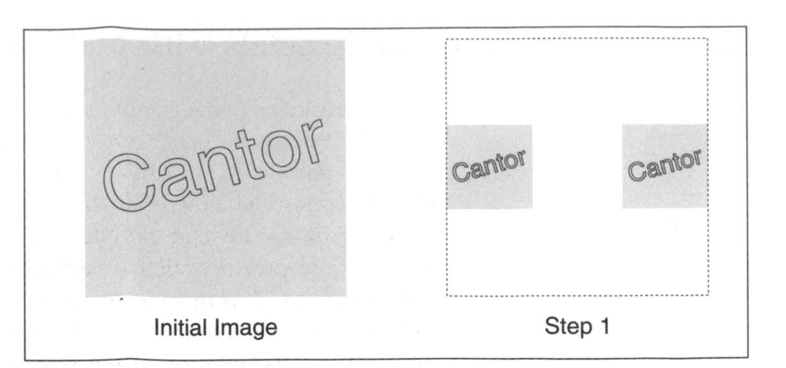

Figure 8.7 : MRCM encoding of the Cantor set by two affine transformations, which reduce by a factor of 1/3 and position as indicated by the squares.

Cantor Set

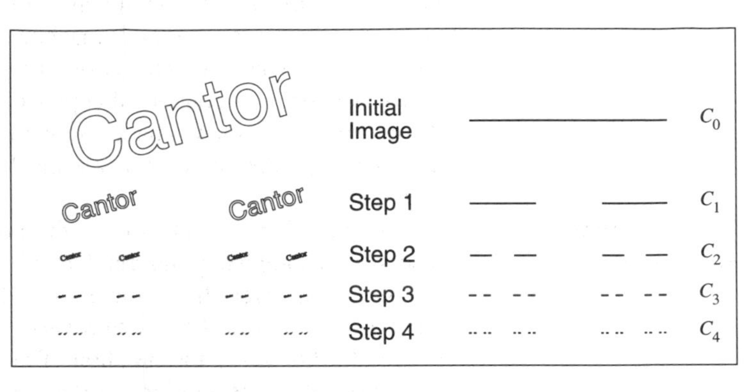

Figure 8.8 : MRCM iteration (left) is compared with the original construction process (right). In the course of the iteration of the MRCM, the initial pattern vanishes and the shape of the Cantor set becomes visible. The classical construction of the Cantor set proceeds by repeatedly removing the middle thirds of all line segments.

as shown in the left part of the figure.

In some sense the MRCM for the Cantor set seems to grasp only the structure of the final image, namely its self-similarity. But we can do more. In the case of the Cantor set this is not a difficult observation. We simply have to do two things:

- use the line segment C_0 as initial image,
- set up the two transformations w_1, w_2 of the MRCM such that C_0 is mapped onto the two line segments of C_1, i.e., the left and right thirds of the initial segment C_0.

Then the iteration of the machine exactly generates the sets of the classical construction as shown on the right in figure 8.8.

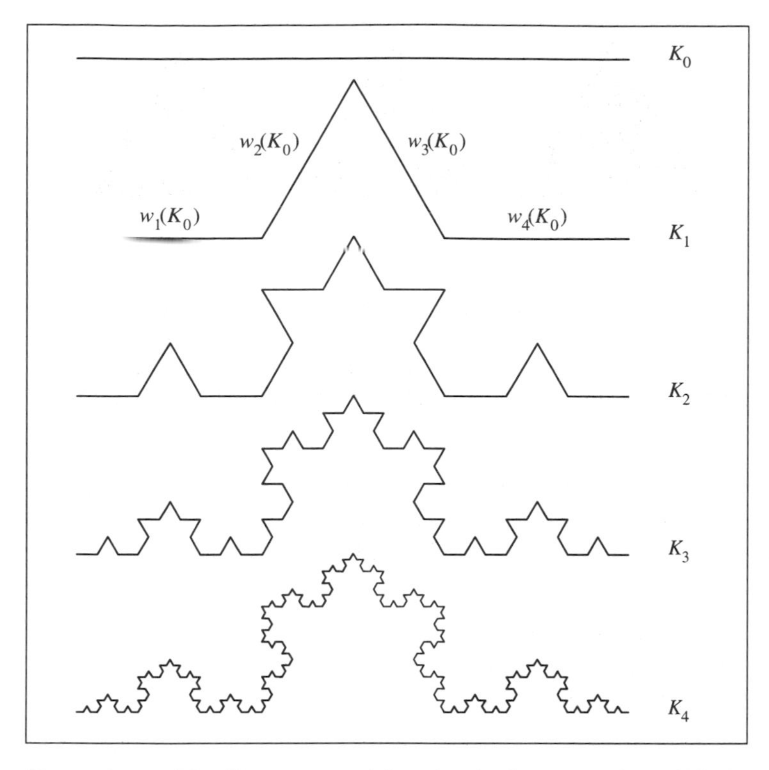

Figure 8.9 : The first stages of the classical construction of Koch. All end points of the generated line segments are part of the final curve.

The Koch Curve

The Koch Curve

Next let us examine some further examples. A close relative of the Cantor set is the Koch curve. Figure 8.9 shows the classical construction of the Koch curve, which is just a variation of the Cantor set construction. Here in each line segment the middle third is not omitted but replaced by two segments forming a tent-like angle. In chapter 2 we noted that self-similarity is built into the construction process: the curve K_{i+1} is composed from four contracted similar copies of K_i. This observation leads directly to the correct set up of an MRCM:

- use K_0, a line segment, as initial image,
- set up four contractions $w_1, ..., w_4$ such that the transformed copies $w_i(K_0)$ are exactly mapped onto the line segments of K_1.

The iteration of this machine exactly generates the sequence K_1, K_2, K_3, ... of the classical construction.

The Sierpinski Gasket

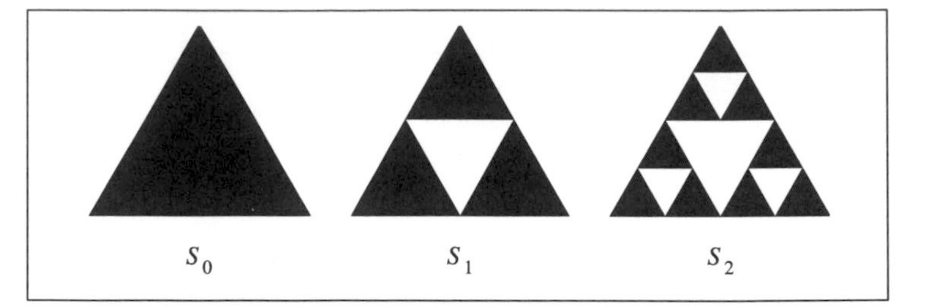

Figure 8.10 : The first stages of the classical construction of the Sierpinski gasket. Starting with a shaded triangle S_0 we repeatedly remove the middle parts of the remaining triangles. Note that in this construction the final set is a subset of all the preceding stages.

The Sierpinski Gasket

Let us now turn to our favorite example, the Sierpinski gasket, and illuminate some further aspects. Figure 8.10 shows its classical construction. The set up of an appropriate MRCM is straightforward:

- use S_0, a shaded equilateral triangle, as the initial image,
- choose three transformations w_1, w_2, and w_3 such that the copies $w_1(S_0), ..., w_3(S_0)$ are aligned with the three triangles in S_1.

Nonunique Transformations ...

You may have noted that the selection of the transformations w_i is not completely determined by S_1. Indeed, even if we assume that $w_1(S_0)$ is the left, $w_2(S_0)$ is the top and $w_3(S_0)$ is the right triangle of S_1, there are still $6^3 = 216$ possible choices because each of the transformations may include reflection and rotations of 120 degrees, 240 degrees or 0 degrees (which may be the only case you may have thought of). But for all possible choices the iteration of the machine generates the desired sequence $S_0, S_1, S_2, ...$ of the classical construction.

The same kind of ambiguity is true for the selection of transformations for the MRCMs which generate the Cantor set and the Koch curve. But in these cases they do not matter either. So why is it interesting to discuss this at all? Let us demonstrate this point for the MRCMs which generate the Sierpinski construction.

... Uncovered by Choice of Initial Pattern

Figure 8.11 shows the iteration of a particular MRCM which encodes the Sierpinski gasket. Taking the triangle S_0 as the initial image would plainly generate the usual classical sequence $S_0, S_1, S_2, ...$ as shown in figure 8.10. However, choosing a somewhat different initial pattern in figure 8.11, a 'rose', we uncover the details of the MRCM. From the top two images (the blueprint of the machine) we can conclude that the lower left transformation involves a 120 degree rotation and a horizontal reflection, the top transformation is just a reduction and the lower right. transformation is a horizontal reflection followed by a 120 degree rotation.

Sierpinski's Rose Garden

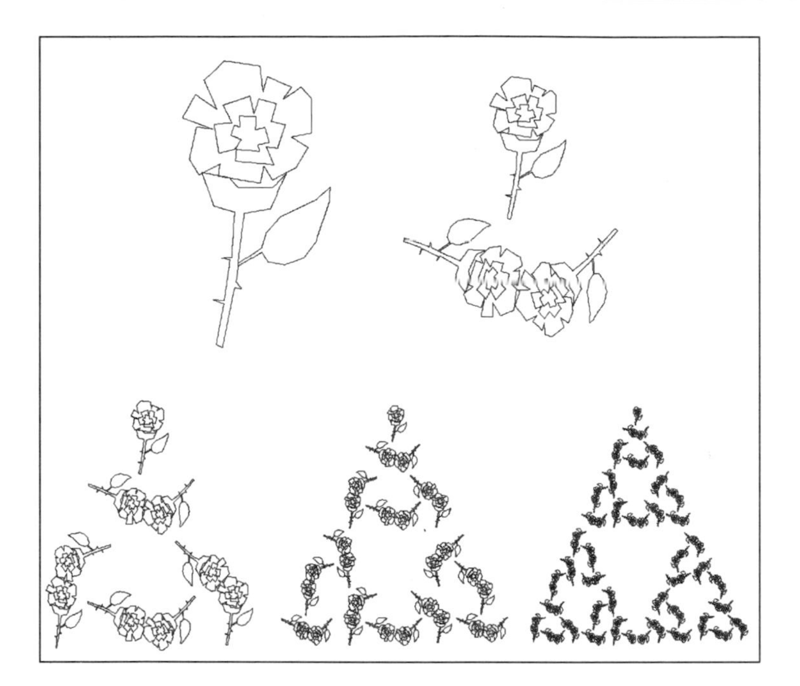

Figure 8.11 : Rose gardens obtained by the iteration of another MRCM which encodes the Sierpinski gasket.

Because of the unusual initial image it is nearly impossible to see from the blueprint or the first iteration that this machine encodes a Sierpinski gasket. Rather, it may remind you of some computer artistic attempt. But there is much more hidden, something which makes this a very special rose garden.

Figure 8.15 shows the iteration of precisely the same MRCM yielding some kind of interesting maze. In contrast to the rose garden, each generated pattern is a subset of the Sierpinski gasket. However, is it possible to find an initial image such that in each stage all line segments join up to form a single path that visits all subtriangles of that stage? The answer is yes. The path always starts at the left corner and ends at the right corner. In the limit it visits all points of the Sierpinski gasket. Figure 8.16 reveals this very surprising construction called the *Sierpinski arrowhead.*

In the figure we choose an arrow pointing from left to right as the initial image. It is positioned at the bottom of S_0. For reference we show S_0 in a dotted pattern to provide a better orientation. You will observe that in the first step the MRCM generates a curve which visits all subtriangles of S_1 from this initial image. In the next step we obtain a curve which visits all subtriangles of S_2, and so on. In the limit this defines a curve which visits all points of the Sierpinski gasket.

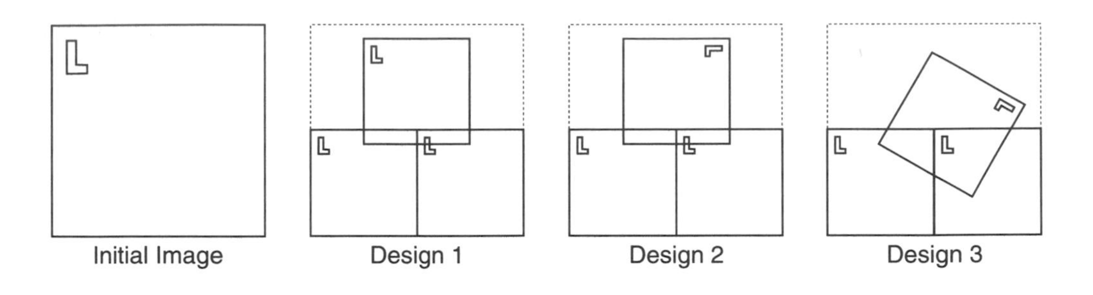

Figure 8.12 : The blueprint for three choices of transformations, the first one of which is the standard configuration for the Sierpinski gasket. Note, that the overlap is required to obtain an equilateral Sierpinski gasket.

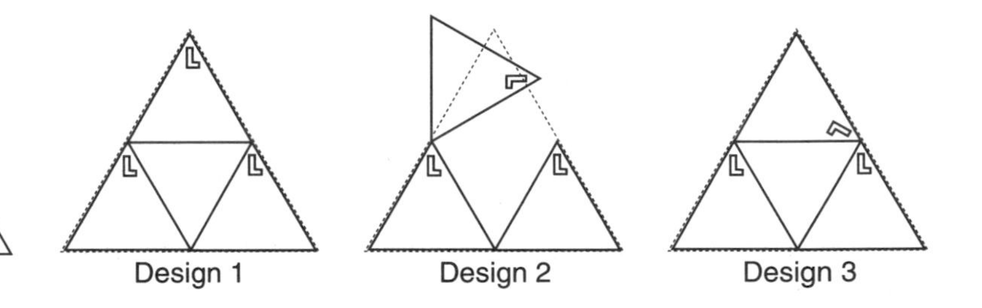

Figure 8.13 : The same choices of transformations as in 8.12, however, using an equilateral triangle as initial image.

MRCM Blueprint Pitfalls There are 216 possible choices of transformations to encode the equilateral Sierpinski gasket with the left vertex at $(0,0)$, the right vertex at $(1,0)$, and the top vertex at $(1/2, \sqrt{3}/2)$. Our standard method for visualizing the MRCM blueprint uses the unit square with the inscribed letter 'L' as an indication of the orientation as the initial image. The 216 different possible choices of transformations are not at all obvious. In fact, one can easily be mislead.

Figure 8.12 shows three different design choices. Guess which of these three settings will generate the Sierpinski gasket! Without much experience we would be inclined to bet on the designs 1 and 2, but certainly not on design 3. Figure 8.14 reveals the surprising answers. Designs 3 yields the Sierpinski gasket while design 2 is as far from the gasket as it can be. What's going on here? Observe that in design 2 the upper transformation involves a 90 degree (clockwise) rotation which does not appear to be dramatically different from design 1 when we just look at the first iteration. In contrast, design 3 involves a 120 degree (clockwise) rotation and the first iteration certainly does not immediately suggest that a Sierpinski gasket will result. But as figure 8.14 shows, it does, and the

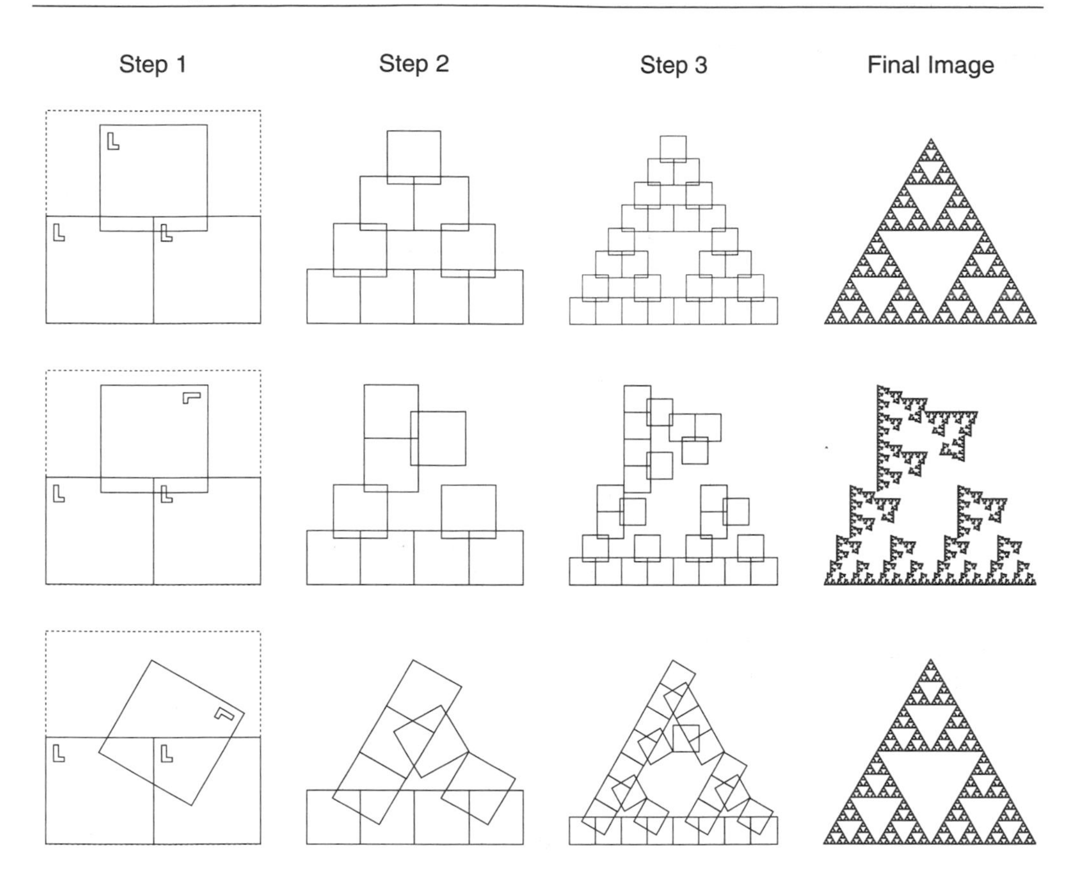

Figure 8.14 : Running the MRCM with the transformations specified in figure 8.12 reveals that only one of the modifications of the standard design produces the Sierpinski gasket.

reason is obvious after the experiment.

The Sierpinski gasket decomposes into three pieces, each of which is a reduced copy of the whole. Moreover, each piece is invariant under multiples of 120 degree rotations and reflections, as we remarked above. Thus, in other words, symmetries which may or may not be suggested by the first iteration and the arbitrary choice of an initial image — like the square in design 2 — of an MRCM are irrelevant. In fact, as we have seen they can be quite misleading. The only symmetries which provide us with choices for the transformations of an MRCM are those which are valid for the limit image. For the design of the Sierpinski gasket an equilateral triangle is appropriate since it has suitable symmetries (see figure 8.13). Note, that such a triangle is used to obtain an MRCM which generates the classical construction as shown in figure 8.10.

The Sierpinski Maze

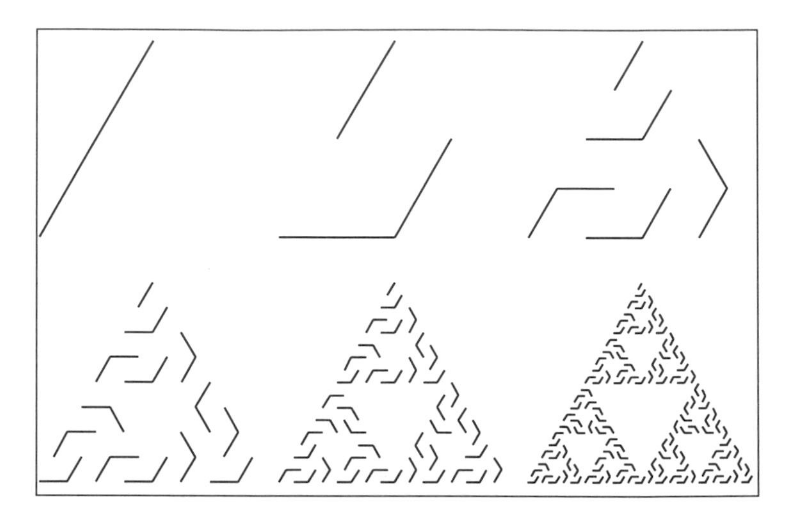

Figure 8.15 : The iteration of the same MRCM as in figure 8.11 started with a different initial image.

The Sierpinski Arrowhead

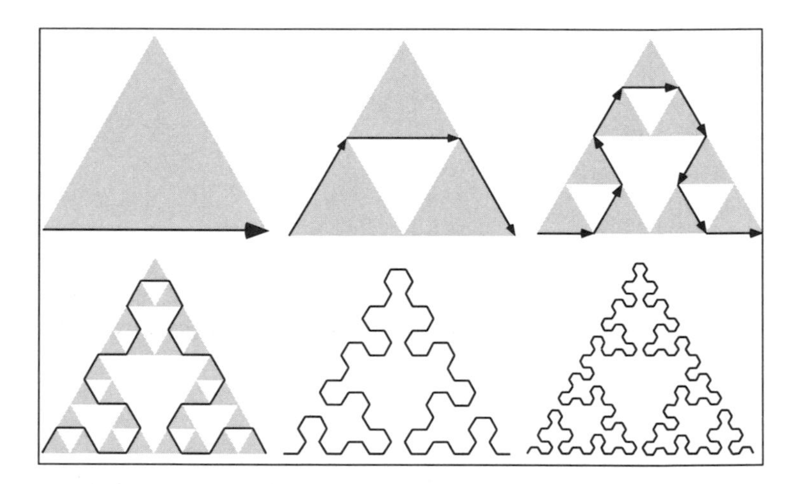

Figure 8.16 : The iteration of the same MRCM as in the previous two figures started with an arrow as initial image generates a sequence of curves which visit all subtriangles of the classical construction and finally visits all points of the Sierpinski gasket. The resulting structure is called the *Sierpinski arrowhead* by Mandelbrot.

Sierpinski Gasket as a Curve

The Sierpinski gasket can be interpreted as a parametrized curve, i.e., there exists a continuous transformation of an interval to the gasket which covers all of its points. This fact can be understood using an addressing scheme which is induced by the transformations w_0, w_1 and w_2 of the MRCM which generates the path sequence of the Sierpinski arrowhead from figure 8.16 (to make things more obvious we use the indices 0, 1 and 2 instead of 1, 2 and 3). Figure 8.17 shows the first two stages of the hierarchy of addresses, which we obtain in this way.

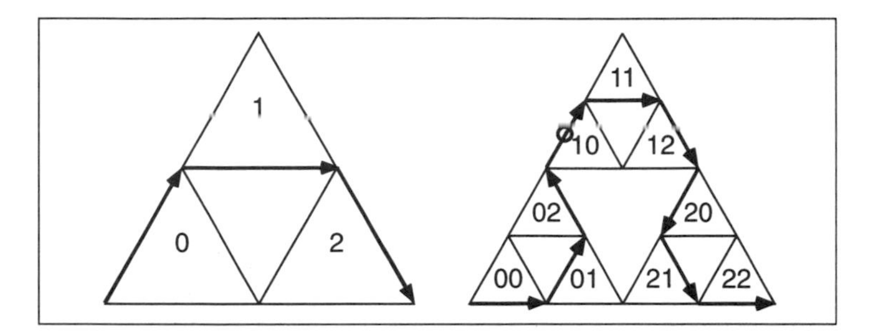

Figure 8.17 : A modified address hierarchy for the Sierpinski gasket.

In the first stage we label the subtriangles of S_0 according to $w_0(S_0)$ (label 0), $w_1(S_0)$ (label 1), $w_2(S_0)$ (label 2). In the next stage we label according to $w_0w_0(S_0)$, $w_0w_1(S_0)$, $w_0w_2(S_0)$, $w_1w_0(S_0)$, ... $w_2w_2(S_0)$. The corresponding labels are $00, 01, 02, 10, ..., 22$. This labeling method is then continued to all further stages. Observe that the arrangement of the transformations implies that in all stages the arrows form a directed curve. The starting and end points of curves obtained from a preceding stage are matched to build up a new curve. This has an important consequence for our addressing scheme, namely that the neighborhood relations in the address space carry over to geometric neighborhood relations of corresponding subtriangles. For example, in the sequence $(0, 1, 2)$ of subtriangles of the first stage each subtriangle shares a vertex with the subtriangle whose address follows in the list. This is not different from the traditional addressing scheme. But the same property also holds for the next stage and all following ones. For example, neighbors in the sorted lists $(00, 01, 02, 10, 11, 12, 20, 21, 22)$ and $(000, 001, 002, 100, ..., 222)$ are also neighbors geometrically, and this is new. Recall from chapters 2 and 6 that the points of the Sierpinski gasket can be identified by infinitely long addresses of three symbols. These addresses were not to be confused with triadic numbers. But it is a good idea to use this type of identification in the case we are discussing here. Identifying the triadic numbers of the unit interval with point addresses as derived in this section defines a continuous transformation of the unit interval onto the Sierpinski gasket. For example the triadic number 0.101 is identified with the address $101000...$ which represents the marked point in figure 8.17 (the circle within triangle 10). Also, note that 0.101 is equal to $0.100222...$ in the triadic system, and that the corresponding address $100222...$ identifies the same point in the Sierpinski gasket.

Peano Curve Construction

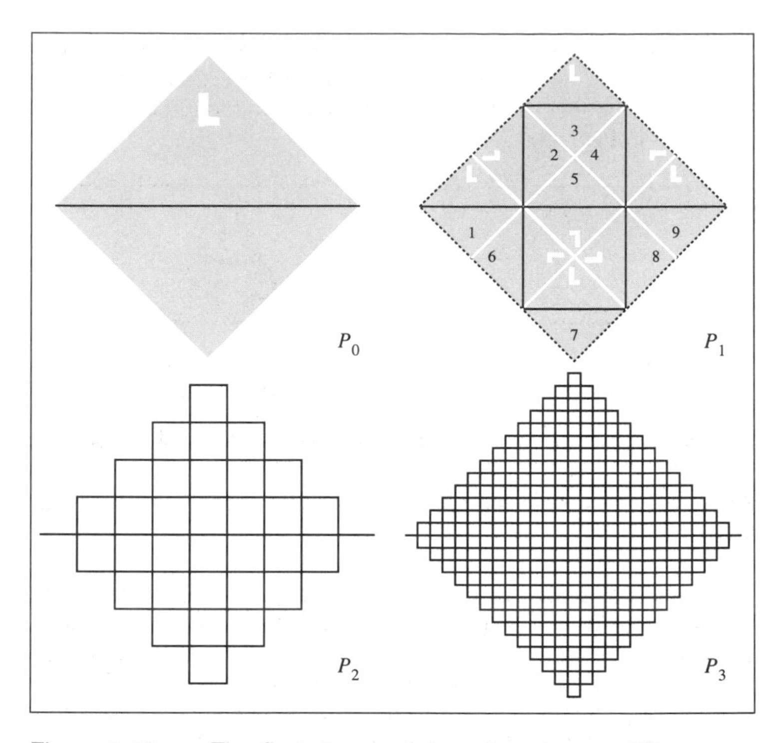

Figure 8.18 : The first stages of the original space-filling curve construction of Peano. The grey background indicates the square which is finally filled by the curve. Furthermore it shows the resulting tiling associated with the first construction step (including rotations) which is used for the design of the MRCM.

The Peano Curve

Our next example is another classical and important curve construction: the original Peano curve. It was proposed by Giuseppe Peano in 1890 as a curve which fills a square and therefore had a strong impact on the discussion of the concept of dimension. Although Peano did not provide any illustrations of his construction let us describe the approach in terms of collections of lines. He started with a straight line P_0, the diagonal of the square which the curve is going to fill (see figure 8.18). For the next step he reduced P_0 by the factor 1/3 and fitted 9 copies into the square in the order shown. The second and eighth copy is rotated by 90 degrees counter-clockwise, the fourth and sixth copy is rotated clockwise and the fifth copy is rotated 180 degrees. All segments together form the new curve P_1. In the following steps this process is repeated: the curve P_{n-1} is scaled down by the factor 1/3 and 9 copies are fitted as in the first step to form the new curve P_n. You will observe, that already the curve P_1 has two points of self-intersection (or touching points). P_2 has 32 such points and in the

Self-Avoiding Space-Filling Curve

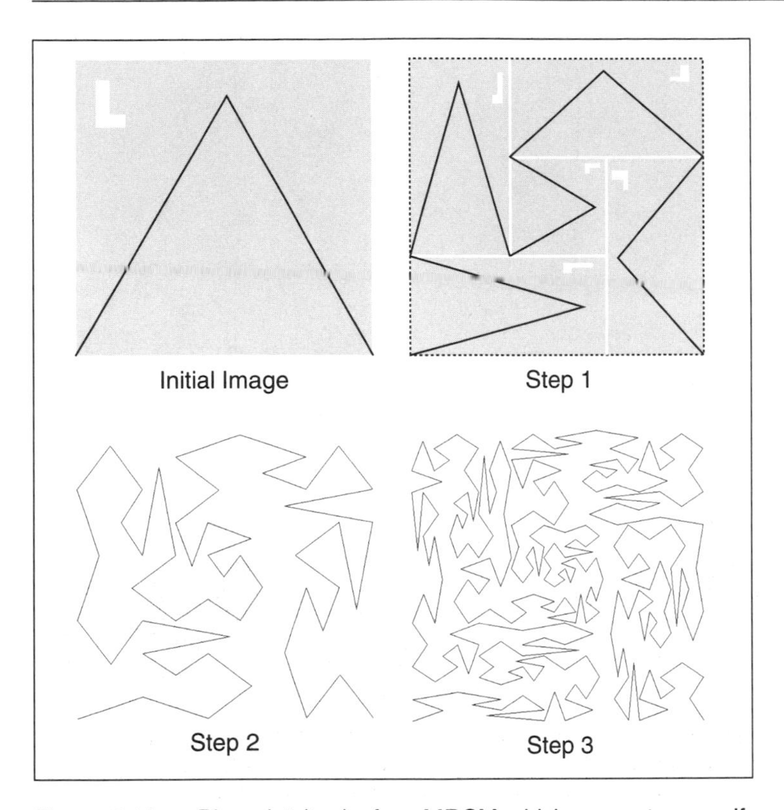

Figure 8.19 : Blueprint (top) of an MRCM which generates a self-avoiding space-filling curve. Already for stage three (bottom right) it is hard to believe that this construction is really self-avoiding.

limit the number increases by the factor 9 from step to step. In this respect Peano's construction is a bit unsatisfying. It seems that the great mathematician David Hilbert shared this feeling: one year later he presented a space-filling curve construction which is self-avoiding. We will discuss his ideas a bit later. First let us return to the original Peano curve and set up an appropriate MRCM. This is a simple task:

- use the curve P_0 as the initial image,
- choose transformations $w_1, ..., w_9$ such that the square is mapped onto the subsquares as indicated in figure 8.18. Thereby we map P_0 onto the line segments of P_1.

Iterating this machine produces exactly the sequence of curves $P_0, P_1, P_2, ...$ which lead to the Peano curve in the limit. The fact that this limit object actually is a curve was already discussed in chapter 2.[11] The fact that the limit curve fills a complete square

[11]See page 112 in *Fractals for the Classroom, Part One.*

The Dragon Curve

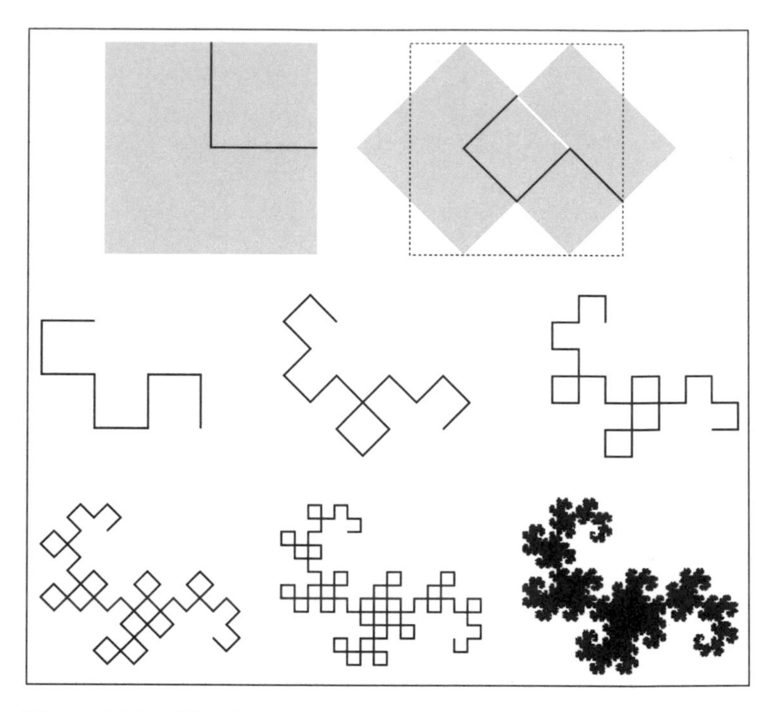

Figure 8.20 : The dragon curve is another example of a space-filling curve. Its boundary is a fractal. In the bottom right hand corner the limit curve of this construction is shown. The starting curve, shown on a grey square, and the first stage (top) are used as the blueprint of an appropriate MRCM.

follows from the MRCM interpretation. Indeed, just observe that the shaded square in figure 8.18 remains invariant under the 9 contractions of the MRCM. Thus, it is the attractor.

Curves That Fill a Cube

In passing, let us add that there even exist curves that cover a full cube of three dimensions. The discovery that such seemingly impossible objects really exist shocked the world of mathematics around the turn of the century. Such excitement can be experienced today not only by first rate mathematicians but by anybody with access to a computer. It takes only a matter of seconds to produce recursive geometric patterns and, with a suitable program on hand, one may join the early pioneers in their hunt for new curve constructions. An implementation of our MRCM should be a first step in that direction. However, with a plain MRCM it is not possible to create a self-avoiding and space-filling curve that is aesthetically pleasing. Figure 8.19 shows one attempt. There is a 'chaotic' ensemble of lines which make it hard to believe that the construction is in fact space-filling and self-avoiding.[12]

[12]It is an interesting and open question whether one can construct such a curve with fewer transformations than

The Dragon Curve

Let us conclude this section with another more recent example, the Harter-Heighway dragon.[13] This is a construction of a space-filling curve which finally fills an area with a fractal boundary. We start with a right angle as shown in figure 8.20. The first stage of the construction is obtained by fitting two copies of this curve (each one reduced by the factor $1/\sqrt{2}$ and rotated clockwise by 45 or 135 degrees respectively) to form a new curve. Again this procedure is repeated stage by stage ad infinitum and can be described by a simple MRCM:[14]

- use the right angle curve as initial image,
- choose transformations w_1, w_2 as indicated by the blueprint shown in figure 8.20.

We should note that the final image shown was obtained by the chaos game. Several relatives of this curve are known. This one (and you will see this in the next section) is especially well suitable for an L-system approach.

five as used in figure 8.19. Note that the space-filling property again follows from the invariance of the initial square under the MRCM.

[13]See C. Davis and D. E. Knuth, *Number Representations and Dragon Curves,* Journal of Recreational Mathematics 3 (1970) 66–81 and 133–149. Also page 66 in B. B. Mandelbrot, *The Fractal Geometry of Nature,* Freeman, 1982.

[14]Note that the curves in the first two stages are self-avoiding while those of all subsequent stages are not. The self-similarity dimension is $\log 2/\log\sqrt{2} = 2$.

8.3 Turtle Graphics: Graphical Interpretation of L-Systems

In order to grow the classical fractals from the previous section by using L-systems the resulting symbol strings must be interpreted graphically. This interpretation is independent of the string generation. We have already pointed out that this feature is of special value. Data generation and data visualization can be separated into independent modules.

In this section we present a very simple graphical interface for symbol strings.[15] It is based on Seymour Papert's concept of *turtle graphics*[16] and especially suited for curves in the plane. With such an interpretation we can formulate the original constructions of the classical fractals in a very concise and compact form, namely as L-systems with just a few production rules. Of course, these production rules must be very carefully set up in order to correctly interface with the graphical interpretation. In other words, all aspects of the graphical interpretation have to be firmly defined.

Let us imagine a turtle sitting on a sheet of paper facing in a certain direction. The tail of the turtle is a bit dirty. Thus, it leaves a trace on the paper as soon as the turtle starts to move. The turtle is thoroughly trained, it understands several commands, which we submit by remote control. The commands are given in the form of symbols (from the list of symbols used in the L-system). These symbols will be just ordinary letters from the alphabet and some special symbols such as $+$ or $-$. Here is a first set of instructions to the turtle:

F move forward by a certain fixed step length l and draw a line from the old to the new position

f move forward as above for F but do not draw the line (raise the tail)

$+$ turn left (counterclockwise) by a fixed angle δ

$-$ turn right (clockwise) by the angle δ

Figure 8.21 shows a short string of symbols and the corresponding graphical interpretation. In this case the angle δ for left and right turns is 90 degrees. Thus the turtle starts to move one step to the right (thereby drawing a line of length l). Then it turns left and continues to draw a line upward. At this point it raises its tail and makes another step upward (without drawing). Then it turns right and draws a line, and so on. With such few and simple directives we can already let the turtle draw amazingly complex images.

---

[15]It was introduced in P. Prusinkiewicz, *Graphical applications of L-systems,* Proceedings of Graphics Interface '86, Kaufmann, 1986, 247–253. See also P. Prusinkiewicz, J. Hanan, *Lindenmayer Systems, Fractals and Plants,* Vol. 79 of Lecture Notes on Biomathematics, Springer-Verlag, New York, 1989.

[16]S. Papert, *Mindstorms: Children, Computers, and Powerful Ideas,* Basic Books, New York, 1980.

Turtle Interpretation

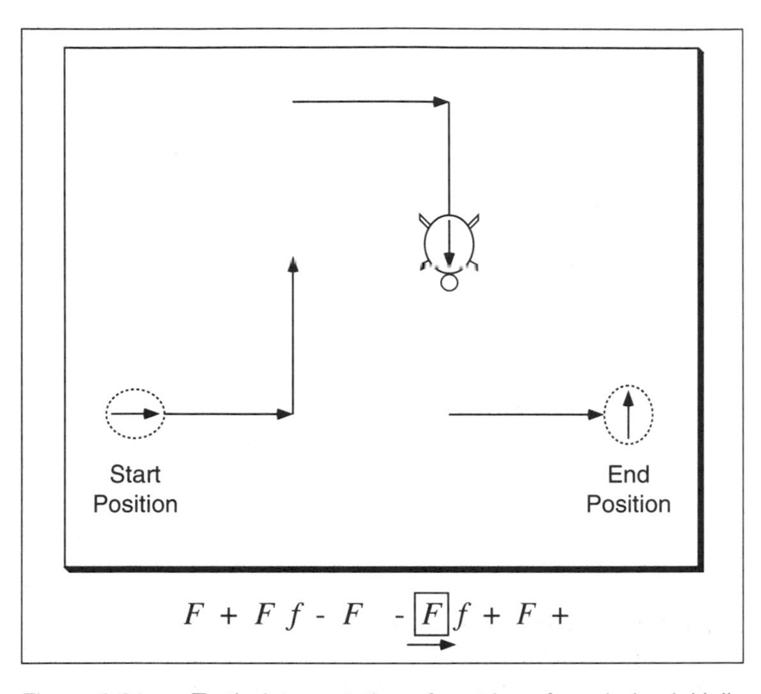

Figure 8.21 : Turtle interpretation of a string of symbols. Initially the turtle is headed left to right. When the interpretation of all 12 symbols is completed the turtle will be headed up.

State Changes Must be Defined Firmly

Instead of drawing a line the turtle could draw an arrow (as shown in 8.21), a dashed line or even a thin cylinder. This is some of the freedom we have in the graphical interpretation. On the other hand, after each command the turtle may have a new position and direction. This change must be defined firmly.

Especially the size of the step length l and the angle δ must be specified before the interpretation can be started. In fact, the size of the angle has an important impact on the shape of the resulting graphics, while the step length only influences the overall size of the image. Let us demonstrate this fact. Figure 8.22 shows the graphical interpretation of two strings:

Top: $F + F + F - F - F$
Bottom: $FF + FF + F + F - F - F + F + FF + F$.

We show the interpretation for three different angles (60, 90 and 120 degrees). The results are quite different. Some are even surprising. In particular this is true for the bottom right hand curve which shows only nine lines although there are 12 symbols F in the string. But if you follow the path of the turtle you will observe that some of the lines are drawn twice.

Two Strings — Three Angles

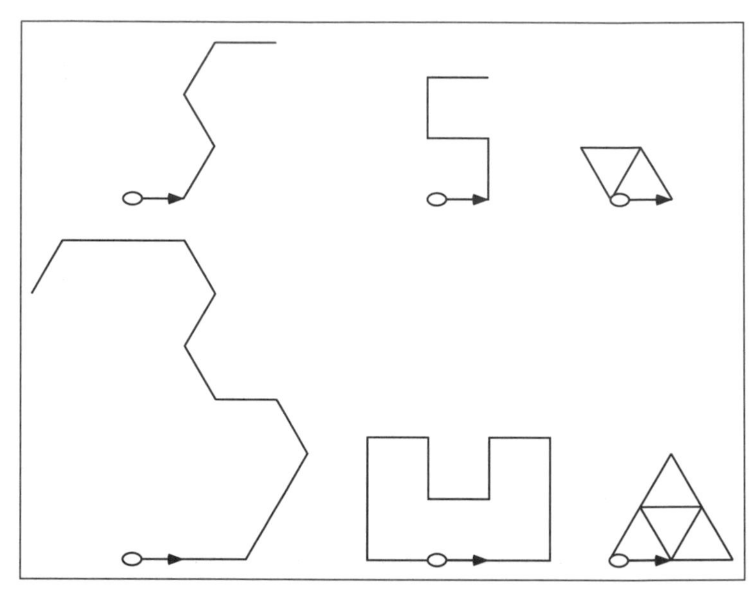

Figure 8.22 : Graphical interpretation of two strings, $F+F+F-F-F$ (upper images) and $FF+FF+F+F-F-F+F+FF+F$ (lower images), and for three different angles δ: 60 degrees (left), 90 degrees (middle) and 120 degrees (right).

State of the Turtle

In mathematical terms (and more useful for programming) we say that the turtle has a state consisting of a current position, given by two coordinates x and y, and a current heading, specified by an angle α. This is written as a triplet (x, y, α). The state of the turtle is changed every time a command is interpreted. Using elementary trigonometry the following table can be deduced:

command	state (x, y, α) is changed to
F	$(x + l\cos\alpha, y + l\sin\alpha, \alpha)$
f	$(x + l\cos\alpha, y + l\sin\alpha, \alpha)$
$+$	$(x, y, \alpha - \delta)$
$-$	$(x, y, \alpha + \delta)$

Here l denotes the step length (by which the turtle moves forward) and δ the angle (by which the turtle turns left or right). We always start with the state $(0, 0, 0)$, i.e., the turtle is headed to the right.

Symbols with Complex Interpretation

It is very convenient to assume that our turtle can also accept commands which represent composite movements. In other words, it should be able to interpret a single symbol as a certain defined sequence of commands like F, f, $+$ and $-$. In the following we

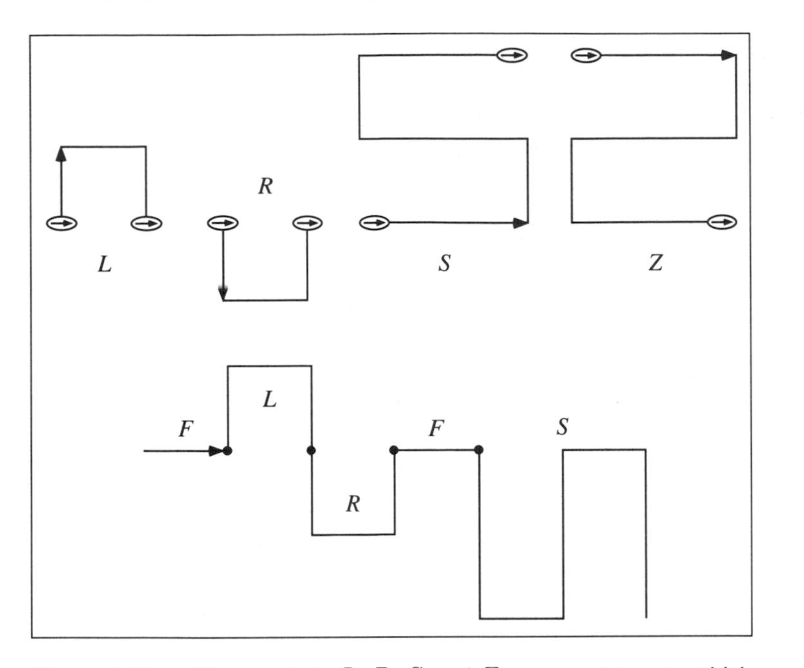

Figure 8.23 : The symbols L, R, S and Z represent curves which are composed from the commands F, $+$ and $-$. The bottom curve is the result of the string $FLRF - S$.

More Symbols

will use the symbols L, R, S and Z. They all represent small curves and you will find the choice of these symbols and their graphical interpretation quite obvious. The turtle will interpret these symbols in the following way:

symbol	interpretation
L	$+F - F - F+$
R	$-F + F + F-$
S	$FF + F + FF - F - FF$
Z	$FF - F - FF + F + FF$

Figure 8.23 shows the graphical interpretation of these symbols for the angle $\delta = 90$ degrees. In all cases the start and end positions of the turtle are marked by an arrow which indicates the direction the turtle is headed to. The symbol L represents a kind of small detour from a direct step forward (it starts with a left turn). The symbol R represents a small detour in the other direction (it starts with a right turn). The S-shaped curve is represented (you would have guessed this) by the symbol S and the reflected S-shaped curve is represented by the symbol Z.

8.4 Growing Classical Fractals with L-Systems

Let us now explore how we can use L-systems for constructing fractals. As a first example we will use the Koch curve. Its description by an L-system is pleasingly simple. In the classical construction we repeatedly replace a straight line segment by a sequence of four lines, as shown in figure 8.24 (right). This sequence can be described by the string $F+F--F+F$ (where we choose the angle $\delta = 60$ degrees). This is a straight line segment forward (F), a left turn by 60 degrees ($+$), another line forward (F), then two turns to the right by 60 degrees each ($--$) (i.e., a total turn of 120 degrees), a line forward (F), a left turn ($+$) and another line forward (F).

Elements of Koch Construction

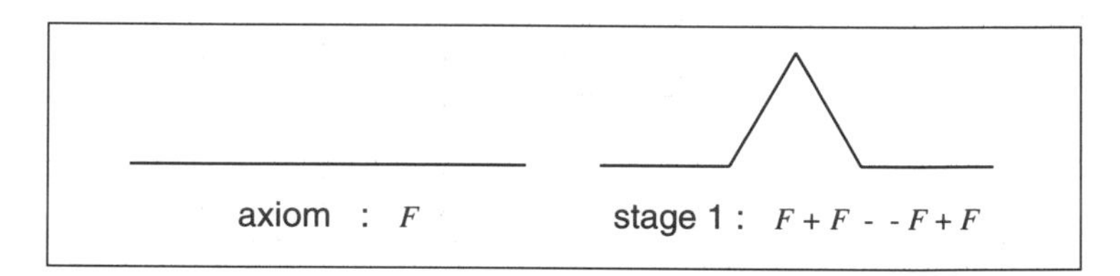

Figure 8.24 : The elements of the classical Koch construction for the curve can be represented by the symbol F (left) and the string $F+F--F+F$ (right). But note that F represents different step lengths l in the two stages shown (see page 39).

Thus, the process of the Koch construction can easily be described by the following L-system:

L-system: Koch Curve
Axiom: F
Production rules: $F \rightarrow F+F--F+F$
$+ \rightarrow +$
$- \rightarrow -$
Parameter: $\delta = 60$ degrees

The second and third production rule simply states that the symbols $+$ and $-$ should be replaced by the same symbols (i.e., they should not change in the substitution process). For now we can assume that the step length l is reduced to one third of the previous step length in each stage of the derivation (we will return to this issue in a moment). Elaborating the L-system, we obtain the following strings of symbols for the first three stages:

Axiom: F
Stage 1: $F+F--F+F$
Stage 2: $F+F--F+F+F+F--F+F--F+F--F+$
$F+F+F--F+F$

Stage 3: $F+F--F+F+F+F--F+F--F+F--F+$
$F+F+F--F+F-$
$F+F--F+F+F+F--F+F--F+F--F+$
$F+F+F--F+F++$
$F+F--F+F+F+F--F+F--F+F--F+$
$F+F+F--F+F-$
$F+F--F+F+F+F--F+F--F+F--F+$
$F+F+F--F+F.$

The L-system prescribes replacing an F (a straight line segment) by a sequence of lines $F+F--F+F$. If we interpret this geometrically, it is exactly the classical replacement. Thus, it is not surprising that the L-system works in exactly the same fashion as the original Koch construction. The advantage of the L-system is that it is directly applicable as a method for (graphical) computer generation of the curve. The graphical interpretation according to the defined rules is the same as in figure 8.9.

Determining the Step Length

In the case of the Koch snowflake curve, it is quite obvious how to choose the step lengths for the turtle for any given stage. For each stage we have to reduce the length by a factor of three. Then the size of the total curve is always the same since a line segment will be replaced by four segments, the first and last of which come to lie on the two ends of the original segment. The situation may be much more complex in many other cases. There is no general theory that permits us to compute the scale of the generated curves.[17] Thus, the only practical way to proceed is to first assume an arbitrary step length, for example equal to 1.0, to compute the whole graphical interpretation of the turtle command string, and then to rescale the result so that it will fit conveniently on the monitor for viewing or on the page for printing.

Our next example is the generation of the Cantor set. Again we start with a straight line. The line is divided into three equal parts and the middle segment is removed. This process is repeated ad infinitum. For an L-system we can formalize this by the production rule $F \to FfF$ (i.e., we replace a line segment by three line segments, of which the middle one is not drawn). This suggests the following L-system:

L-system:	Cantor Set (First Try)
Axiom:	F
Production rules:	$F \to FfF$
	$f \to f$

Figure 8.25 shows the first three stages of this system and its graphical interpretation. Obviously this is not the desired result.

[17] For a solution to this problem for a restricted class of L-systems see F. M. Dekking, *Recurrent Sets,* Advances in Mathematics 44, 1 (1982) 78–104.

Cantor Set by L-System — First Try

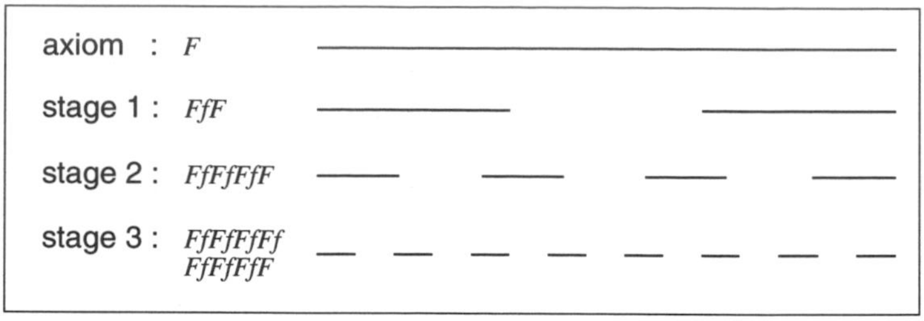

axiom :	F	
stage 1 :	FfF	
stage 2 :	$FfFfFfF$	
stage 3 :	$FfFfFfFf$ $FfFfFfF$	

Figure 8.25 : The first stages of L-system $F \to FfF$ for the Cantor set.

Cantor Set by L-System — Second Try

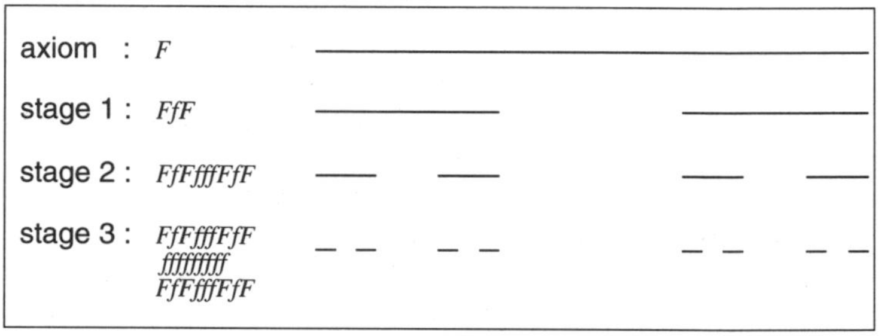

axiom :	F	
stage 1 :	FfF	
stage 2 :	$FfFfffFfF$	
stage 3 :	$FfFfffFfF$ $fffffffff$ $FfFfffFfF$	

Figure 8.26 : The first stages of the correct L-system for the Cantor set.

But what is wrong? Where is the mistake in our L-system?

Let us look at the second stage. We observe that the size of the middle gap is too small. It should be three times as large as it is. The gap is represented by a single symbol f. This particular symbol derived from the f which was already generated in stage one. Obviously it should have been replaced by a sequence of three symbols fff (which represents a gap three times as long as a single f). Thus the L-system for the Cantor set should be:

L-system:	Cantor Set (Second Try)
Axiom:	F
Production rules:	$F \to FfF$
	$f \to fff$

Graphics Uncover Hidden Pitfalls

Indeed, figure 8.26 shows the desired result. This simple example already demonstrates that there are many hidden pitfalls in the derivation of an appropriate L-system for a concrete example. But the graphical interpretation of the generated strings will uncover most mistakes immediately.

Let us now return to the Sierpinski arrowhead and try to find out how we can describe the curve construction (which we introduced using an MRCM) in terms of an L-system. Figure 8.27 shows the first two stages. First, we obviously have to choose the angle

The Sierpinski Arrowhead Generator

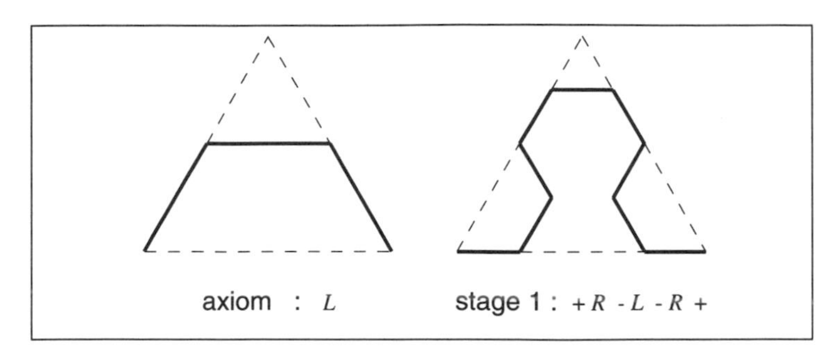

Figure 8.27 : The first two stages of an L-system Sierpinski curve.

$\delta = 60$ degrees. Then we could describe the curve on the left by $+F-F-F+$. The first and the last symbol $+$ are needed to have the turtle facing in the appropriate direction for its first or next movement. But wait, let us recall the symbol L. By definition the turtle interprets this symbol just as the string $+F-F-F+$. Thus, the left curve is simply described by the symbol L. Also using the symbol R (which is interpreted as $-F+F+F-$) we can describe the right curve simply by $+R-L-R+$.

Now imagine flipping the curves of figure 8.27 upside down. Then the left curve would be represented by the symbol R and the right curve would be described by $-L+R+L-$. Combining both observations suggests the following L-system:

L-system:	Sierpinski Arrowhead
Axiom:	L
Production rules:	$L \rightarrow +R-L-R+$
	$R \rightarrow -L+R+L-$
	$+ \rightarrow +$
	$- \rightarrow -$
Parameter:	$\delta = 60$ degrees

For the first three stages this L-system gives:

Axiom: L
Stage 1: $+R-L-R+$
Stage 2: $+-L+R+L--+R-L-R+--L+R+L-+$
Stage 3: $+-+R-L-R++-L+R+L-++R-L-R+-$
$-+-L+R+L--+R-L-R+--L+R+L-+$
$--+R-L-R++-L+R+L-++R-L-R+-+.$

These strings exactly describe the curves which we have already seen in figure 8.16.

Multiple Choices for Correct Production Rules

Our next example demonstrates a minor complication, which is rather typical for L-systems. Figure 8.28 once again shows the original space-filling Peano curve. Its construction starts with just

The Peano Curve

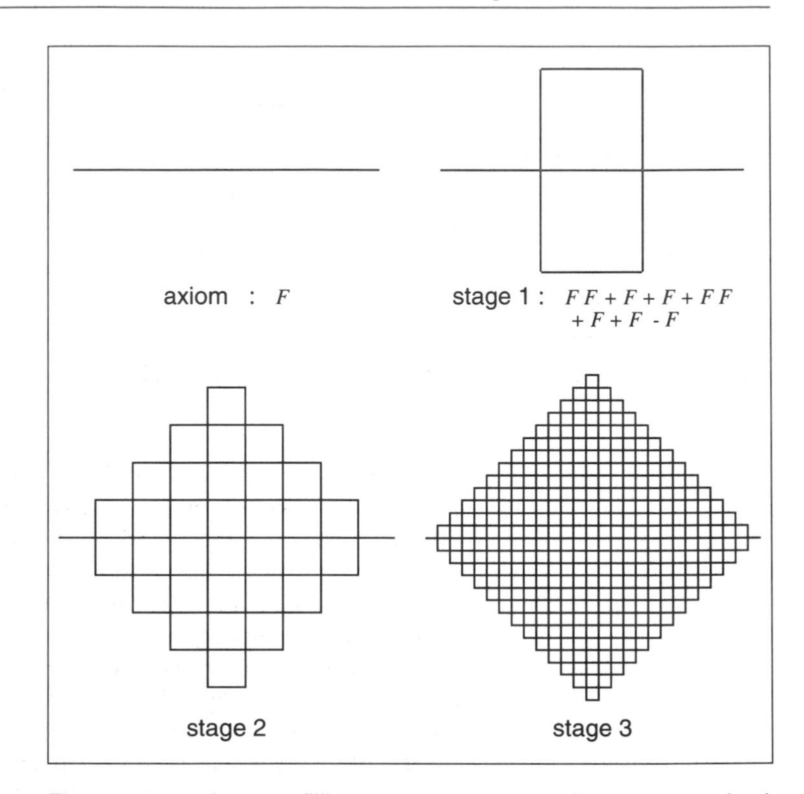

Figure 8.28 : A space-filling construction — the Peano curve: (top) axiom and production, (bottom) the second and third stage.

one line segment. We certainly will not hesitate to encode this by a simple F command. But the curve in the upper right part might cause some doubt. This curve (which we are going to substitute for plain line segments) can be encoded in several ways. After a forward step, the turtle can be told to make a left or a right turn, or to go straight ahead. For example, after a right turn there must be two left turns with line segments in between all three turns. Then, the turtle has again arrived at a point where the curve meets itself, and there is a choice, namely to go straight ahead or to make a left turn. In either case the turtle will trace the top loop of the generator counterclockwise or clockwise, and then it just has to finish up the last line segment to the end of the curve. In terms of turtle commands, these two alternatives can be described as follows (in any case we choose $\delta = 90$ degrees):

$$F - F + F + F + F - F - F - F + F \text{ and}$$
$$F - F + F + FF + F + F + FF.$$

The options for an initial left turn at the first bifurcation of the curve are:

$$F + F - F - F - F + F + F + F - F \text{ and}$$
$$F + F - F - FF - F - F - FF.$$

If the turtle goes straight at the first decision point, we have

$$FF - F - F - FF - F - F + F \text{ or}$$
$$FF + F + F + FF + F + F - F.$$

The complete L-system may work with any of the above choices, for example, we can set:

L-system:	Peano Curve
Axiom:	F
Production rules:	$F \to FF + F + F + FF + F + F - F$
	$+ \to +$
	$- \to -$
Parameter:	$\delta = 90$ degrees

Our last example, which we already discussed from the viewpoint of MRCMs is the dragon curve. The construction starts with an L-shaped curve. Let us call this K_0. The next stage K_1 is obtained by fitting two copies of K_0 rotated clockwise — one rotated by 45 degrees, the second rotated by 135 degrees (see figure 8.29).

First we note that we obviously should use an angle $\delta = 45$ degrees. Then K_0 can be encoded by $--F{+}{+}F$. It is convenient to define for this curve a new symbol: D (i.e., the turtle will interpret D as $- - F + +F$). Let us now analyse K_1. The first part of the curve can be encoded directly (i.e., by the symbol D), but we have to be careful with the second part. Loosely speaking the end points of the two copies of K_0 are fitted together to make up the new curve. Therefore, we have to trace the second copy of K_0 in reversed order: $F - -F + +$. For this we introduce the new symbol E. We arrive at the encoding $-D{+}{+}E$ for the curve K_1 (the minus sign rotates everything by 45 degrees) which also gives us the first rewriting rule: $D \to -D + +E$. The rewriting rule for E is symmetric and reflects the reversed order we found in the second part of the curve. Thus we obtain the L-system:

L-system:	Dragon Curve
Axiom:	D
Production rules:	$D \to -D + +E$
	$E \to D - -E+$
	$+ \to +$
	$- \to -$
Parameter:	$\delta = 45$ degrees

We check the first three stages:

The Dragon Construction

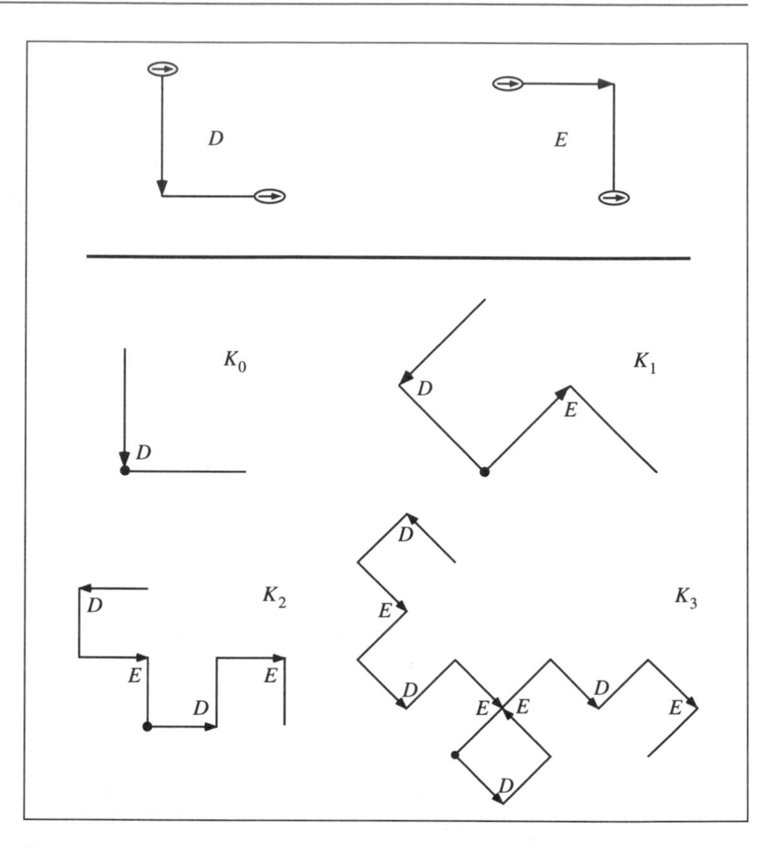

Figure 8.29 : L-system construction of the Dragon curve. The new symbols D and E (top) are the main building blocks.

Axiom: D
Stage 1: $-D++E$
Stage 2: $--D++E++D--E+$
Stage 3: $---D++E++D--E+$
$++-D++E--D--E++$.

The graphical interpretation of these strings indeed generates the curves K_0, K_1, K_2 and K_3 (see figure 8.29).

Not Suitable For MRCMs

Let us now turn to fractal curves which we have not seen in this chapter so far because we cannot obtain them using an MRCM.

Our first example is the quadratic Koch island shown in figure 8.30. The construction starts with a closed polygon of four sides (a square). If we choose an angle $\delta = 90$ degrees this square can be described by the string $F+F+F+F$. In the following stages all line segments are replaced by the zig-zag curve which is shown in the upper right hand corner of figure 8.30. We can encode this curve by $F+F-F-FF+F+F-F$. In summary, we obviously

The Koch Island

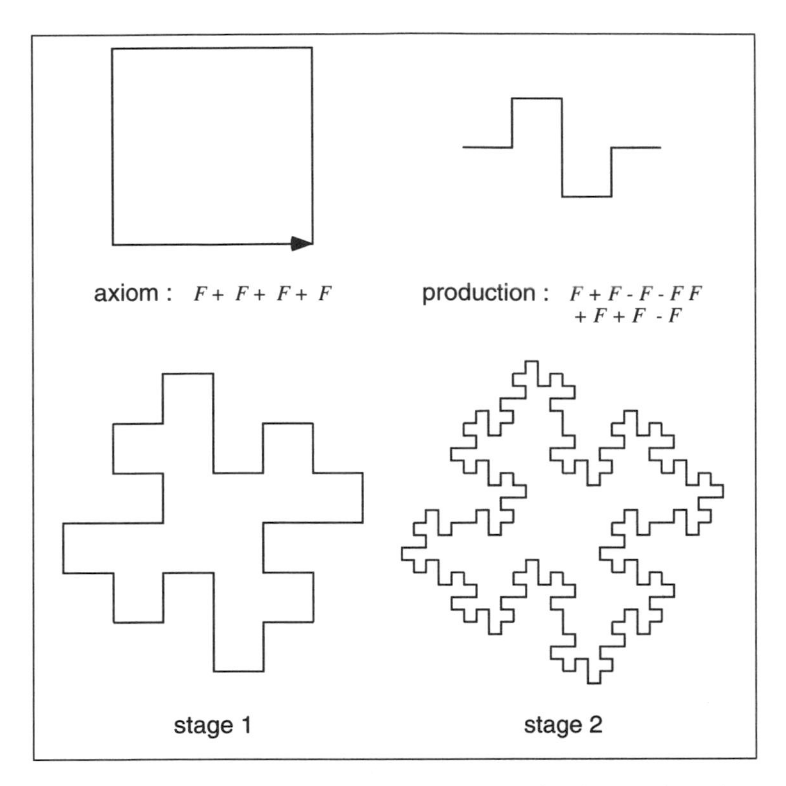

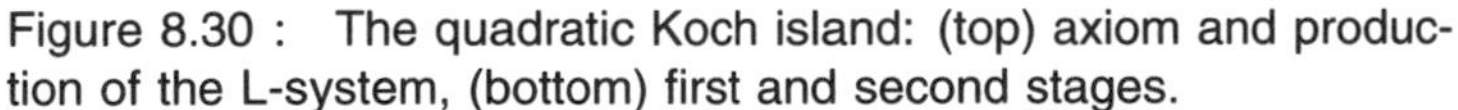
Figure 8.30 : The quadratic Koch island: (top) axiom and production of the L-system, (bottom) first and second stages.

arrive at the following L-system:

L-system:	Quadratic Koch Island
Axiom:	$F + F + F + F$
Production rules:	$F \to F + F - F - FF + F + F - F$
	$+ \to +$
	$- \to -$
Parameter:	$\delta = 90$ degrees

At the bottom of figure 8.30 we show the first stages of this curve construction. Now, why is it not possible to describe this construction by an MRCM?

The answer is simple. This construction does not produce a self-similar image in the sense of an MRCM. On the other hand, an MRCM always generates a self-similar final image, and this property is tightly built into the feedback operation of the machine. In other words, the curve produced in a certain stage is made up of complete (although transformed) copies of the curve which was produced at the preceding stage. The construction process of the quadratic Koch island violates this principle. Thus, it cannot be described by an MRCM.

But restricting the construction to just one side of the Koch island (e.g., the bottom segment, which stems from the first F of the axiom) changes the situation completely. Each of the four sides is made of a curve construction which is self-similar (in the limit) and which can be described by an MRCM. In other words, the only reason why the property of self-similarity is violated is the axiom $F+F+F+F$. Because of this axiom the limit curve is not simply self-similar but a composition of four self-similar curves.

Let us now discuss two curve constructions which are really not feasible for a simple MRCM. Both examples are space-filling curves like the original Peano curve. But whereas the Peano construction creates curves which have self-intersections the following constructions are self-avoiding. This means that each curve (at all stages) has neither self-intersections nor touching points. The first of the two examples is attributed to David Hilbert (1862–1943) and was published in 1891 just one year later than Peano's.[18]

Hilbert's Construction

Hilbert's construction is extraordinarily elegant, especially when we take into consideration that it was created decades before tools like L-systems, MRCMs or recursive computer programming were invented. Let us try a description which follows Hilbert's ideas.[19] In figure 8.31 the dotted square shows the area which we are going to fill by the curve. We divide this square into four quarters. The construction starts with a curve H_0 which connects the centers of the quadrants by three line segments. Assume the size of the segments to be 1. In the next step we produce four copies (reduced by 1/2) of this initial stage and place the copies into the quarters as shown. Thereby we rotate the first copy clockwise and the last one counterclockwise by 90 degrees. Then we connect the start and end points of these four curves using three line segments (of size 1/2) as shown and call the resulting curve H_1. In the second step we scale H_1 by 1/2 and place four copies into the quadrants of the square as in step one. Again we connect using three line segments (now of size 1/4) and obtain H_2. This curve contains 16 copies of H_0, each of size 1/4. As a general rule, in step n we obtain H_n from four copies of H_{n-1} which are connected by three line segments of length $1/2^n$ and this curve contains $4n$ copies of H_0 (scaled down by $1/2^n$).

Inner Versus Outer Replacements

What we have just demonstrated is a geometric construction which is similar to the curve generation using an MRCM. We generate the curve at a certain stage from copies of the curve in the previous stage. This could be called a macroscopic (or outer) re-

[18] D. Hilbert, *Über die stetige Abbildung einer Linie auf ein Flächenstück*, Mathematische Annalen 38 (1891) 459–460.

[19] The original paper of Hilbert is reproduced in the figures 2.35 and 2.36 on pages 110–111 in chapter 2.

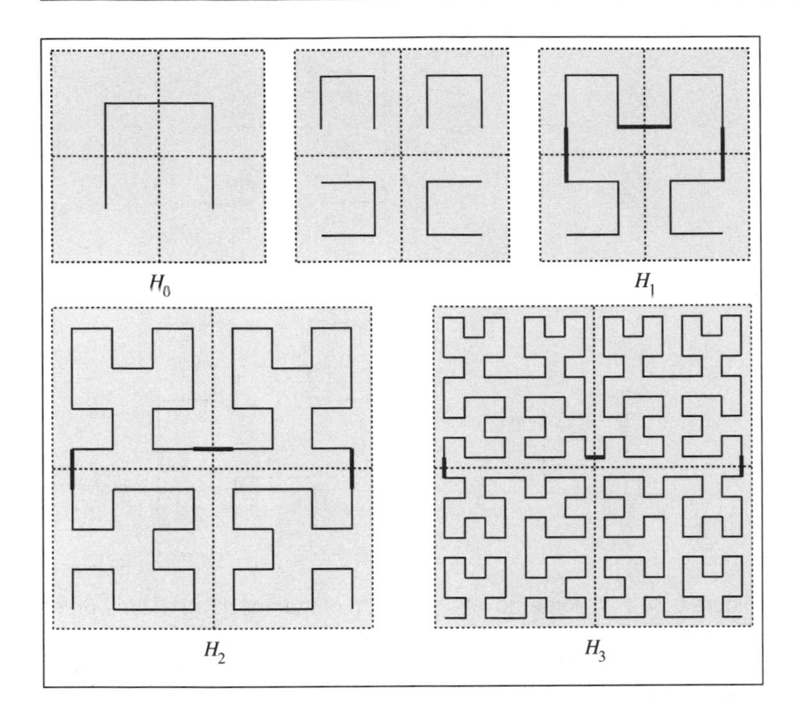

Figure 8.31 : The first stages of the construction of a Hilbert curve.

Four Stages of the Hilbert Curve

placement. In this respect the L-system construction is different. It is restricted to the replacements of the smallest elements (symbols), and we could call this a microscopic (or inner) replacement. Macroscopic replacements are easy to visualize and therefore easy to understand whereas microscopic replacements seem to be much harder to comprehend. Let us explore this point of view. How should we define the second stage H_2 of the Hilbert curve suitable for the L-system approach? Let us look at H_1 again. There are four copies of H_0 in H_1. Loosely speaking each of these copies has to be replaced by a copy of H_1 to obtain H_2. However, this is not quite what we want because the connecting line segments must also be modified. L-systems provide a solution to this problem. With L-systems we only use formal substitutions, and the appropriate size of the line segments is adjusted automatically. Let us demonstrate this in detail.

First we have to choose the angle $\delta = 90$ degrees. Then H_0 can be encoded by the symbol L (which is interpreted by the turtle as $+F-F-F+$) and H_1 can be encoded by $+RF-LFL-FR+$ (see figure 8.32). Please note that in both cases the turtle is headed right at the start and end positions. Since we obtain H_1 from H_0 this determines the first production rule of the L-system. For the

L-System for Hilbert Curve

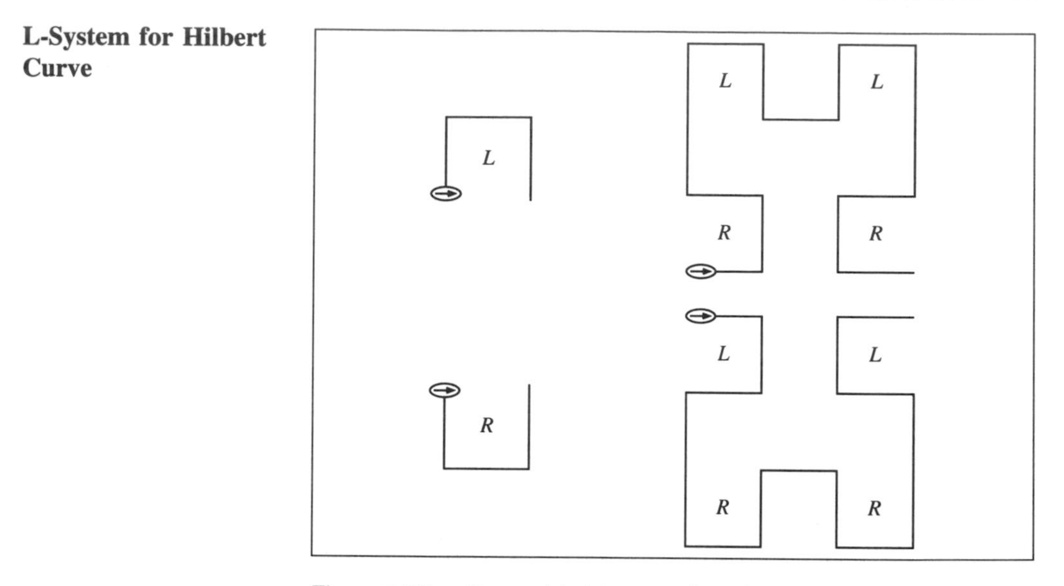

Figure 8.32 : Geometric interpretation of the encoding of the curves H_0 and H_1 using the symbols L and R (left). When starting with axiom R in place of L, we obtain mirror images of the corresponding stages of the Hilbert curve (right).

next step, the generation of H_2, we also need a rewriting rule for R. But the only difference between the mirrored symbols L and R is their orientation. The turtle traces the L-curve in clockwise direction — starting with a left turn — whereas the R-curve is traced counterclockwise, starting with a right turn. Therefore, the rewriting rule for R is a mirror of the rule for L. Instead of right turns, make left turns and vice versa. Moreover, we must exchange the symbols R and L. Thus we arrive at the following L-system:

L-system: Hilbert Curve
Axiom: L
Production rules: $L \to +RF - LFL - FR+$
$R \to -LF + RFR + FL-$
$F \to F$
$+ \to +$
$- \to -$
Parameter: $\delta = 90$ degrees

For the first three stages this L-system gives:

Axiom: L
Stage 1: $+RF - LFL - FR+$
Stage 2: $+ - LF + RFR + FL - F - +RF - LFL - FR + F +$
$RF - LFL - FR + -F - LF + RFR + FL - +$

Another Space-Filling Peano Curve

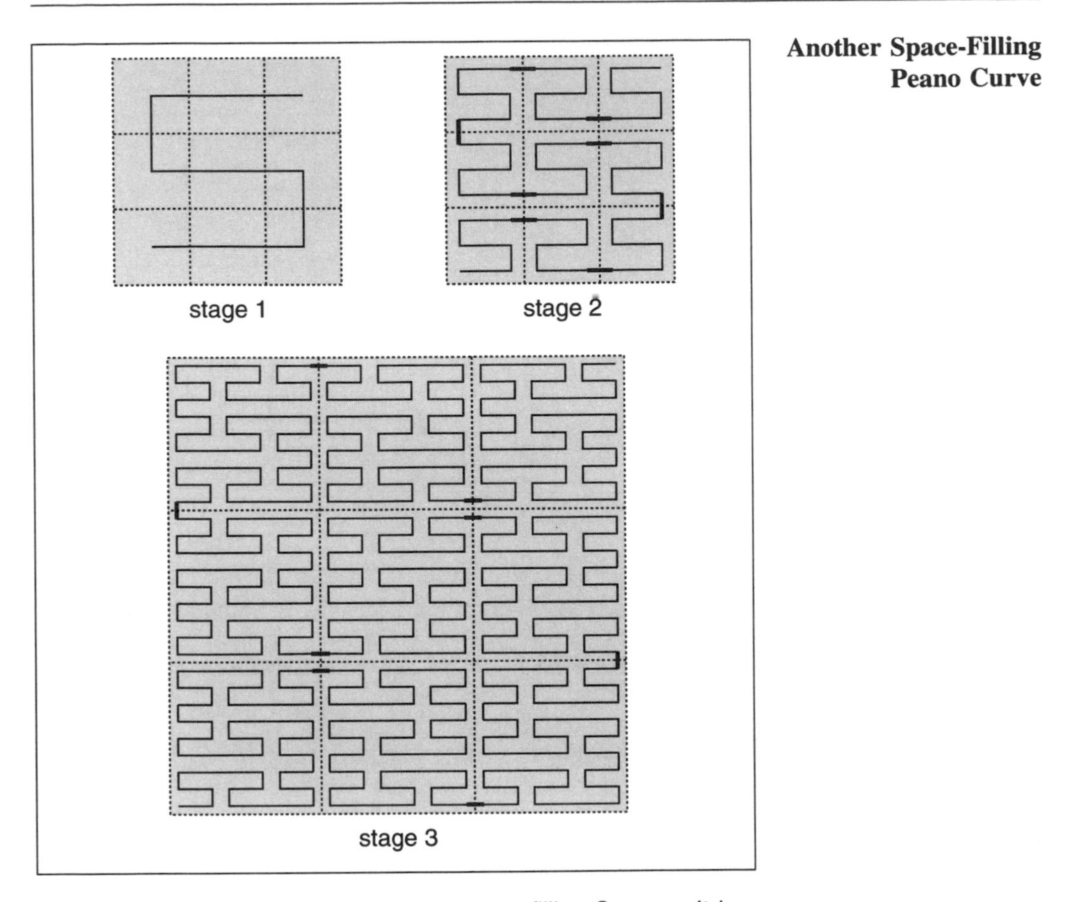

Figure 8.33 : The first stages of the space-filling S-curve. It is a collage of S- and Z-shaped curves.

Stage 3: $+-+RF-LFL-FR+F+-LF+RFR+FL-F-$
$LF+RFR+FL-+F+RF-LFL-FR+-F-$
$+-LF+RFR+FL-F-+RF-LFL-FR+F+$
$RF-LFL-FR+-F-LF+RFR+FL-+F+$
$-LF+RFR+FL-F-+RF-LFL-FR+F+$
$RF-LFL-FR+-F-LF+RFR+FL-+-F-$
$+RF-LFL-FR+F+-LF+RFR+FL-F-$
$LF+RFR+FL-+F+RF-LFL-FR+-+$.

which is a valid description of H_0, H_1, H_2 and H_3 (see figure 8.31).

The SZ-Curve

Let us complete this section with a close relative of the Peano curve. The first stages of this space-filling and self-avoiding construction are shown in figure 8.33. It starts with an S-shaped curve and continues by joining small S- and Z-shaped curves together. Setting the angle $\delta = 90$ degrees we can use our symbols S and

L-System for SZ-Curve

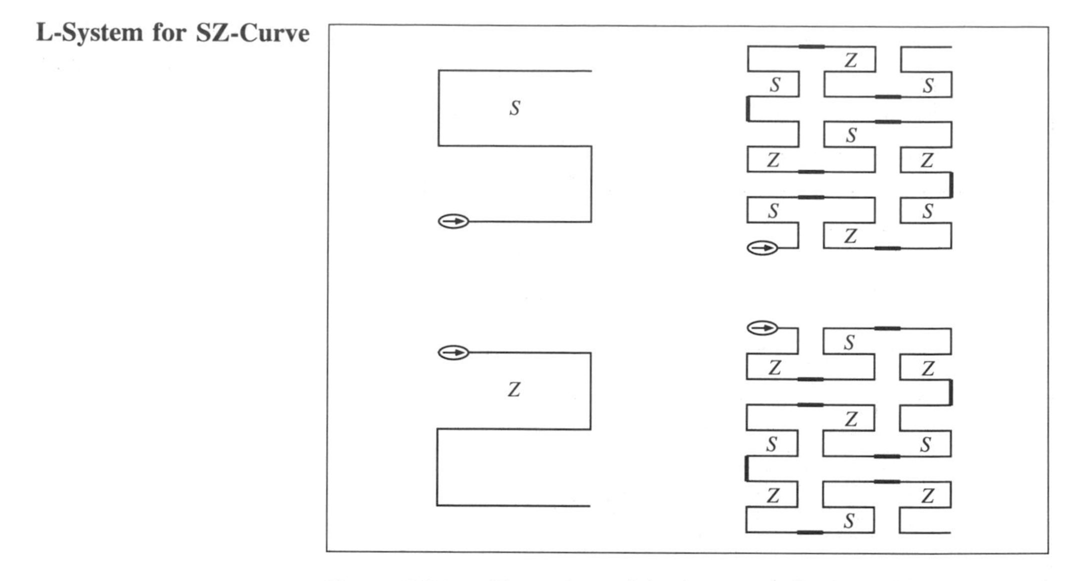

Figure 8.34 : The axiom of the L-system ‘S-shaped Peano curve’ is given by the symbol S. The production rules can be interpreted graphically as the arrangement of S- and Z-shaped curves shown.

Z to encode these curves. The initial curve is simply encoded by the symbol S. The curve of the second stage is given by $SFZFS + F + ZFSFZ - F - SFZFS$ (see figure 8.34). You should observe that the turtle traces the middle part ($ZFSFZ$) of the curve from right to left, thus, this part is a vertical reflection of the first part (we have tried to indicate this by the placement of the letters S and Z in the graphics). The complete L-system is given by:

L-system:	S-Shaped Peano Curve
Axiom:	S
Production rules:	$S \to SFZFS + F + ZFSFZ - F - SFZFS$
	$Z \to FSFZ - F - SFZFS + F + ZFSFZ$
	$F \to F$
	$+ \to +$
	$- \to -$
Parameter:	$\delta = 90$ degrees

8.5 Growing Fractals with Networked MRCMs

From the discussion in the last section it is clear that it requires some (maybe even a bit more) experience to derive appropriate L-systems for fractals that are not strictly self-similar. Therefore it would be a pity if the more geometric approach using MRCMs really could not be extended to such constructions. But you already know the solution to this problem from chapter 5: networked MRCMs or hierarchical iterated function systems. Let us find out how we can use this concept in the present context.

Composing Images by Networked MRCMs

You will recall that all machines of a network of MRCMs work in parallel and all machines produce (in each step) their own image. The transformations of a particular machine may operate not only on its own image, but also on all images of the other machines in the network. In other words, each of the machines may compose its image by transforming the images of all the other machines. We will see immediately that this is exactly what we need.

First Stages of Hilbert Curve — Review

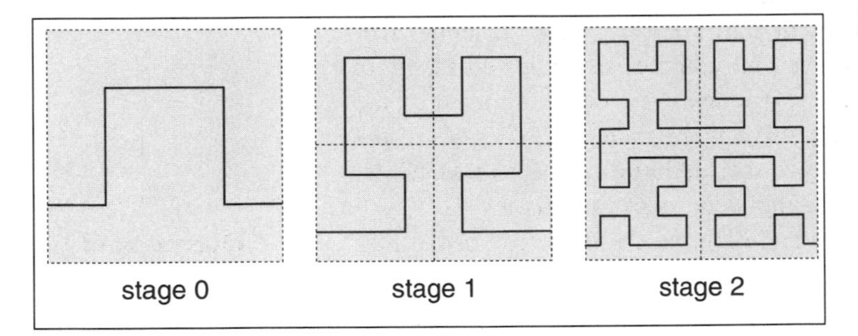

Figure 8.35 : The first stages of a Hilbert curve construction: from stage to stage the size of the line segments is reduced by the factor 1/2.

Let us start with the Hilbert curve construction. Figure 8.35 shows the first stages. However, note that the curves are extended slightly. In contrast to the original construction the curve reaches from the right to the left boundary of the grey square. The purpose of this extension is to simplify the discussion. Now consider the curve of the first stage, which is subdivided into four parts — one in each quadrant (see the center curve in figure 8.35). These curve segments are obviously not small copies of the initial image (the stage 0 curve). Rather, they look a bit like four copies of a P-shaped curve. Some of these are rotated or reflected.

In figure 8.36 we show a close-up of this P-shaped curve segment (from the upper left quadrant) and compare it with the corresponding close-up of the second stage. We have also divided the close-up of the second stage curve into four parts as indicated. This time we find three copies of the P-shaped curve and one copy

Close-Up of One Quadrant

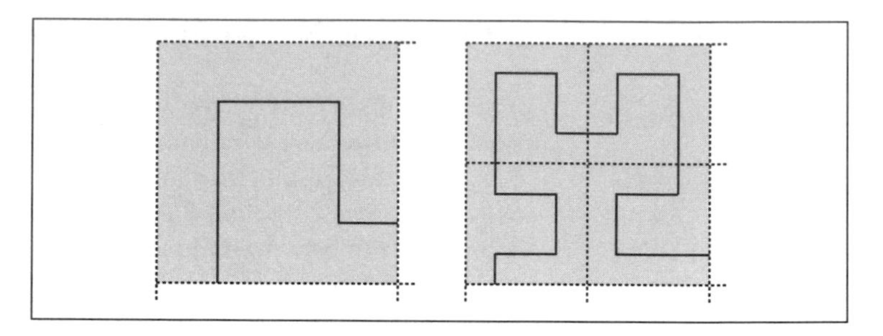

Figure 8.36 : Close-up of the upper left quadrant of the first and second stage curves as indicated in figure 8.35. The close-up of the second stage curve (right) is divided into four parts again.

of the stage 0 curve rotated 90 degrees in the lower left hand quadrant. Based on these observations we are able to describe two networked MRCMs. Let us call them the H- and the P-machines. The H-machine will produce the images which show the curves of the Hilbert construction (as in figure 8.35). The P-machine will start with the P-shaped curve and continue as indicated in figure 8.36. It will operate on its own images and on the product of the H-machine. The H-machine will only operate on the images of the P-machine. Figure 8.37 shows the blueprint of the two MRCMs. Note that both machines use the same transformations.

Operation of the Hilbert MRCMs

In symbolic notation we can describe the operation of the machines by the equations:

$$
\begin{aligned}
H_{n+1} &= w_1(P_n) \cup w_2(P_n) \cup w_3(P_n) \cup w_4(P_n) \\
P_{n+1} &= w_1(H_n) \cup w_2(P_n) \cup w_3(P_n) \cup w_4(P_n) \; .
\end{aligned}
$$

The transformations $w_1, ..., w_4$ are completely determined by the blueprint. They all reduce by the factor 1/2. The design is set up such that H_1, H_2, ... is the sequence of Hilbert curves provided that H_0 is chosen to be equal to the first curve of figure 8.35. P_0, P_1, P_2,... is the sequence of supplementary curves which starts with the P-shaped curve P_0 as indicated on the left of figure 8.36. In the definition of H_{n+1} all transformations are applied to P_n. This reflects that all inputs for the H-machine are connected to the output of P-machine. On the other hand, in the definition of P_{n+1} observe that only w_1 is applied to H_n. So only one input line of the P-machine is connected to the output of the H-machine. All other input is feedback from the output of the P-machine.

A Small Difference

Obviously there is a minor difference between the classical construction and the one just presented, namely those small line segments which extend to the boundary. In fact, those line segments indicate the neighborhood relation of quadrants in the construction (i.e., how the curve leaves and enters the adjacent quadrants). In

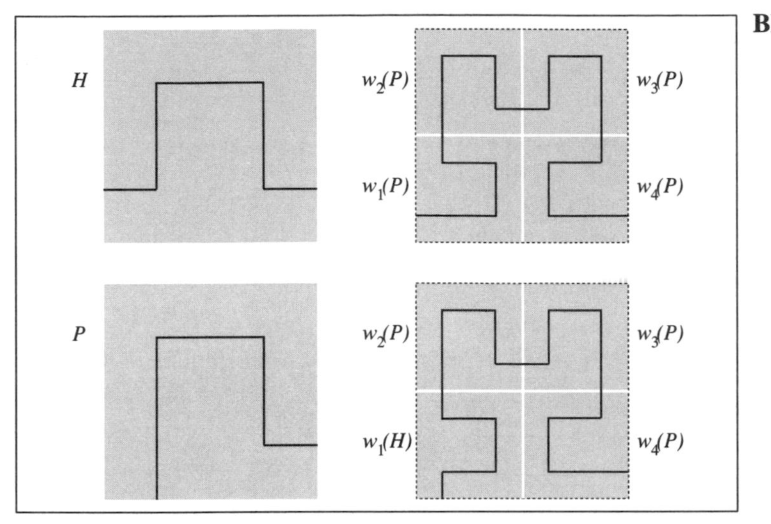

Blueprint of Networked Hilbert MRCMs

Figure 8.37 : Blueprint of the two MRCMs which are required to generate the Hilbert construction. The H-machine is shown above, the P-machine is at the bottom.

the original construction this is hidden by the process of adding the three short line segments in each stage. Finally, let us remark that the MRCM construction can be extended to precisely generate the classical construction but this would require two more MRCMs. These again use the same transformations but start with different initial images. We omit these technical details.

Let us briefly discuss the other example of a self-avoiding Peano curve that we have generated using an L-system. The same kind of analysis that we have demonstrated for the Hilbert curve leads in this case to the blueprint of two networked MRCMs as shown in figure 8.38.

Operation of the Network for Peano's Curve

Let us call the first of these two machines the S-machine. Because of the short vertical line at the top of the S-shaped curve in the blueprint of the other machine, we call that one the \$-machine. The S-machine has four input lines which are connected to the \$-machine (the transformation w_3, w_4, w_6 and w_7 are applied to the product of this machine). The \$-machine also has four input lines which are connected to the other machine (this involves the transformation w_1, w_2, w_5 and w_8). Thus, the operation of this network of MRCMs can be described by the equations:

$$\begin{aligned} S_{n+1} &= w_1(S_n) \cup w_2(S_n) \cup w_3(\$_n) \cup w_4(\$_n) \cup w_5(S_n) \cup \\ &\quad w_6(\$_n) \cup w_7(\$_n) \cup w_8(S_n) \cup w_9(S_n) \\ \$_{n+1} &= w_1(S_n) \cup w_2(S_n) \cup w_3(\$_n) \cup w_4(\$_n) \cup w_5(S_n) \cup \\ &\quad w_6(\$_n) \cup w_7(\$_n) \cup w_8(S_n) \cup w_9(\$_n) \ . \end{aligned}$$

Blueprint of Networked Peano MRCMs

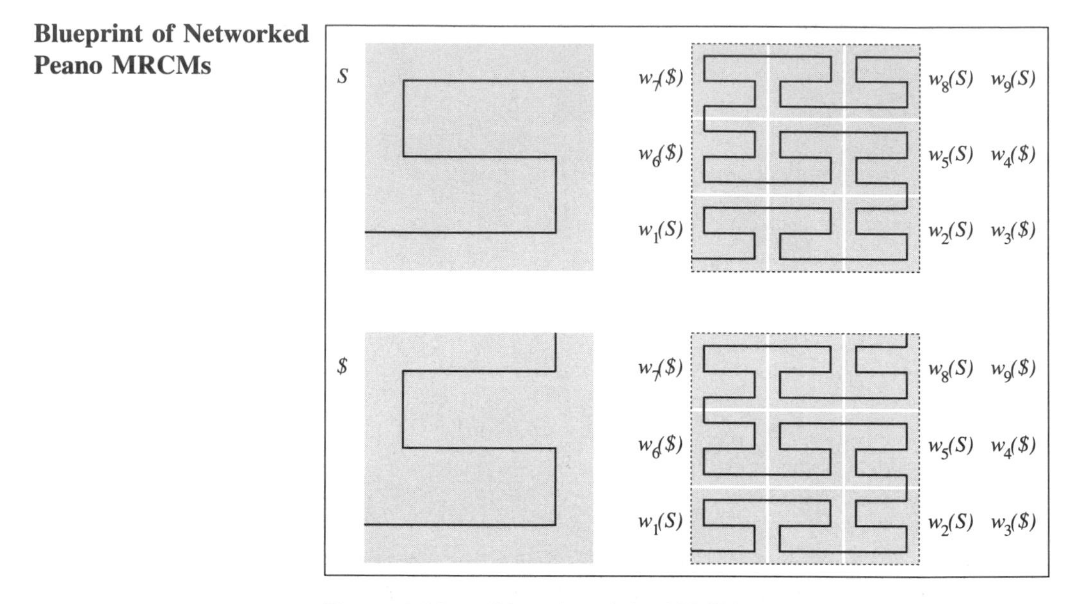

Figure 8.38 : Blueprint of the MRCMs which are required to generate the Peano curve from figure 8.33.

The transformations $w_1, ..., w_9$ are chosen according to the blueprint. They all reduce by the factor 1/3. The initial images S_0 and $\$_0$ have to be selected exactly as indicated by the left side images for the S- and \$-curves in figure 8.38.

Growing of a Twig

Let us conclude this discussion with a network of MRCMs which can grow a twig. Figure 8.39 shows the blueprint of two networked MRCMs which can be used as an example. The attractor of the first machine (L) is a straight line. Here we use two transformations which reduce by a factor of 1/2. Since in this set-up we take just this line as the initial image the machine simply provides this image as a constant output which we use for the stem of the twig. The other MRCM (T) is set up to let the twig grow. Its initial image represents the stem of the twig. The transformation w_3 is applied to the output of the first MRCM. It superimposes the line exactly onto the stem of the twig, thereby fixing the stem once and for all. The other two transformations w_4 and w_5 are chosen in such a way that they fit small copies of the initial image to the end of the stem.

The iteration of the two MRCMs can be described by the equations

$$\begin{aligned} L_{n+1} &= w_1(L_n) \cup w_2(L_n) \\ T_{n+1} &= w_3(L_n) \cup w_4(T_n) \cup w_5(T_n) \ . \end{aligned}$$

Using the attractor L as the initial image L_0 we obtain $L_n = L$. If

Blueprint of Networked Growing-Twig MRCMs

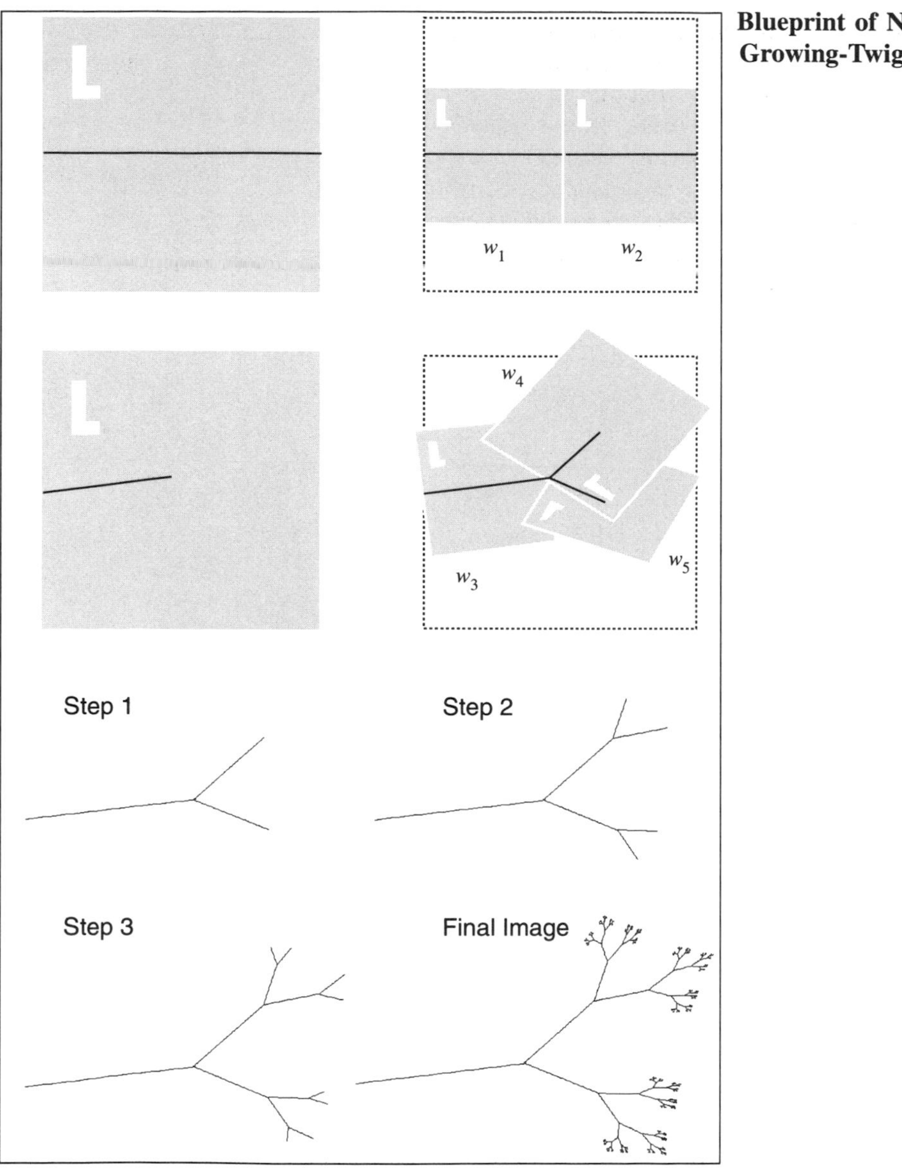

Figure 8.39 : Blueprint of two networked MRCMs which geometrically simulate the growing of a twig (bottom).

we define $S = w_3(L)$ (which we also use as initial image T_0) we obtain a single equation for the production of the twig-machine:

$$T_{n+1} = S \cup w_4(T_n) \cup w_5(T_n) \ .$$

L-Systems Are More Appropriate

Figure 8.39 (bottom) shows the result of the iteration of this network (T_2, T_3, T_4). But it is obvious from the construction that this machinery does not really simulate growth. It rather creates a geometric illusion of this process. In this sense L-systems are much more appropriate. They are designed to capture realistic growth processes. This is an important aspect which we have already mentioned in the introduction to this chapter. So let us now extend this concept and see how we can describe the growing of more complex structures than just strings of algae.

8.6 L-System Trees and Bushes

The representation of a string of cells by a string of symbols is straightforward. But how can we represent a branching structure (as is typical for trees) by a string of symbols which is just a linear sequential list of characters? For a moment this seems to be a problem. But there is a simple solution: we introduce a new symbol which indicates a branch point (we will denote this by a left square bracket '['). This left bracket will be matched by a corresponding closing bracket ']' indicating that at that point the definition of the branch is completed. Take as an example the string

$$ABA[BBAA][CCBB]ABA[AABB]ABA\ .$$

Removing the bracketed portions of the string we obtain

$$ABAABAABA$$

which represents the segments of the stem of a plant from which three branches spring off (at one point of the stem the branches $BBAA$ and $CCBB$ and a bit further up one branch $AABB$).

Turtle Interpretation of Branch Commands

Now what should our turtle do when it encounters these new symbols? First, upon receipt of the [-command it should remember its current position and direction. Technically speaking the turtle state has to be stored and saved. Then the branch can be drawn by the usual interpretation of the subsequent commands. Termination of the branch is triggered by the]-command. The turtle must then return to the location of the branch point, which it remembers.

Stacking of Turtle States

A computer implementation is most conveniently carried out by stacking the turtle states. This allows us to keep track of even complex branching hierarchies. When encountering a [-command, the current state of the turtle is saved at the top of the stack. The]-command, on the other hand, pops the top state from the stack and puts the turtle into this state. Technically the state S of the turtle is described by three numbers, its position (x, y) and its direction, given by the angle α. Thus, $S = \{x, y, \alpha\}$. Initially the stack is empty. Whenever the turtle encounters a [-command we save the current state $S_{n+1} = S$ and increment by 1 the counter n for the number of states on the stack. On the other hand when it receives a]-command we restore its state ($S = S_n$) and decrease the stack counter by 1.

A Weedlike Plant

Now let us examine some simple examples. The first one looks like some weed-like plant. There is a stem with three segments and two main branches. Each segment and branch looks the same

A Regular Weed

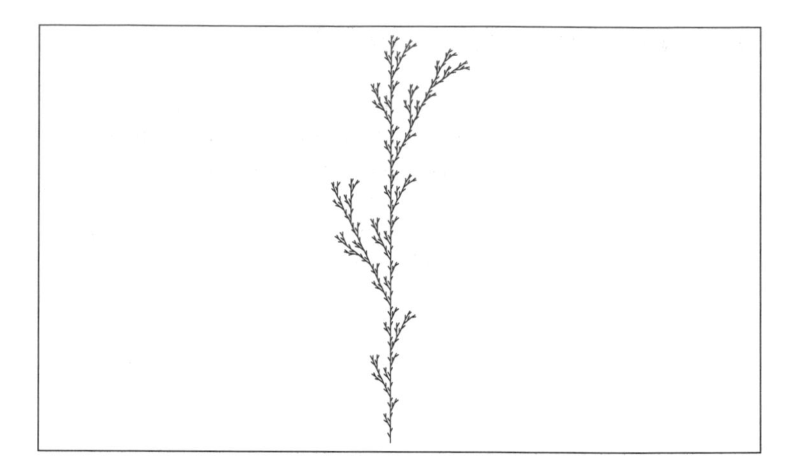

Figure 8.40 : The first stages of a weedlike plant.

(i.e., it also has three smaller segments and two branches). Here is the L-system, again using the symbols which we introduced for turtle graphics. For compactness we omit those production rules from the list which do not change any symbols such as $F \to F$. We do this also in all the remaining L-systems.

L-system: Weedlike Plant I
Axiom: F
Production rules: $F \to F[+F]F[-F]F$
Parameter: $\delta = 25.7$ degrees

Figure 8.40 shows the first five stages. The first two stages are given by:

Axiom: F
Stage 1: $F[+F]F[-F]F$
Stage 2: $F[+F]F[-F]F[+F[+F]F[-F]F]F$
$[+F]F[-F]F[-F[+F]F[-F]F]F[+F]F[-F]F$

Figure 8.41 shows one more example, a two-dimensional bush. The complexity of this figure increases dramatically from step to step. Indeed the image of stage 4 is represented by 160,000 symbols.

And Grass

The next example involves two important production rules, one for F and another one for B. B is an auxiliary symbol which we introduce to model this kind of branching. The turtle simply ignores this symbol. Here is the L-system:

L-system: Weedlike Plant II
Axiom: B

A Simple Bush

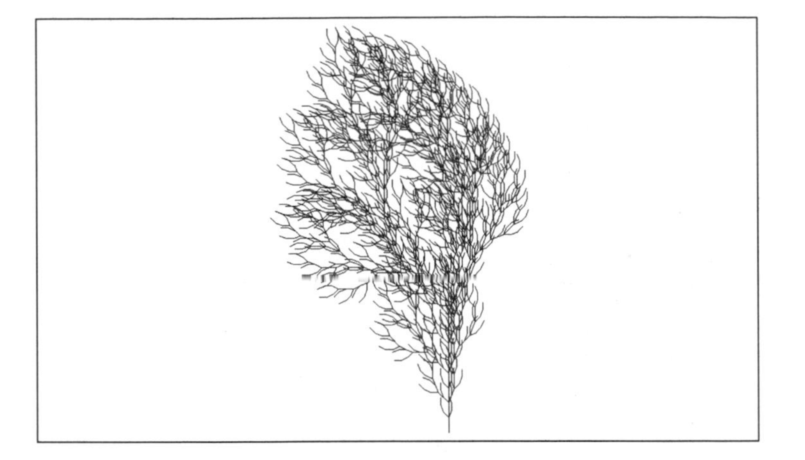

Figure 8.41 : This simple bush is produced by an L-system with axiom F, production $F \to FF + [+F - F - F] - [-F + F + F]$ and angle $\delta = 25$ degrees.

A Different Weed

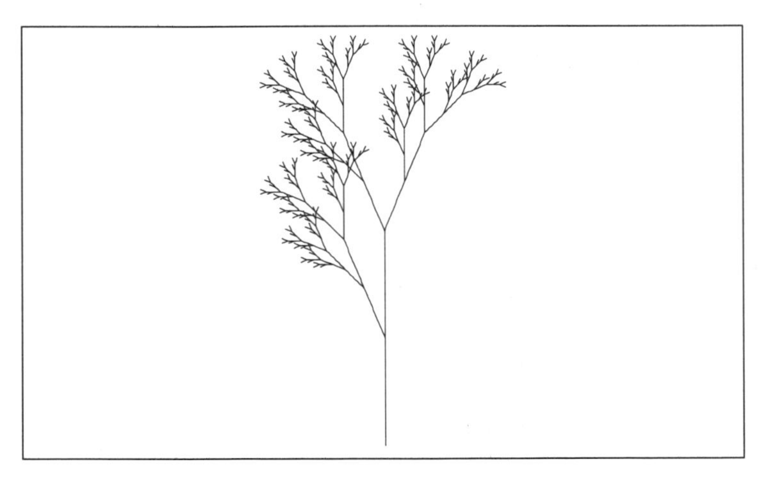

Figure 8.42 : Note the two different types of branching.

Production rules: $F \to FF$

$B \to F[+B]F[-B] + B$

Parameter: $\delta = 20$ degrees

Figure 8.42 shows the first stages. You should be able to easily identify two kinds of branching in this plant. They correspond to the strings $[+B]$ and $[-B] + B$ in the production rule for B.

Adding Randomness to L-systems

So far we only presented L-systems for the generation of images of single plants. But when making pictures of fields of plants, we need to include variations for the same plant so that we are not merely copying one plant image many times. This might lead us

Random Weeds

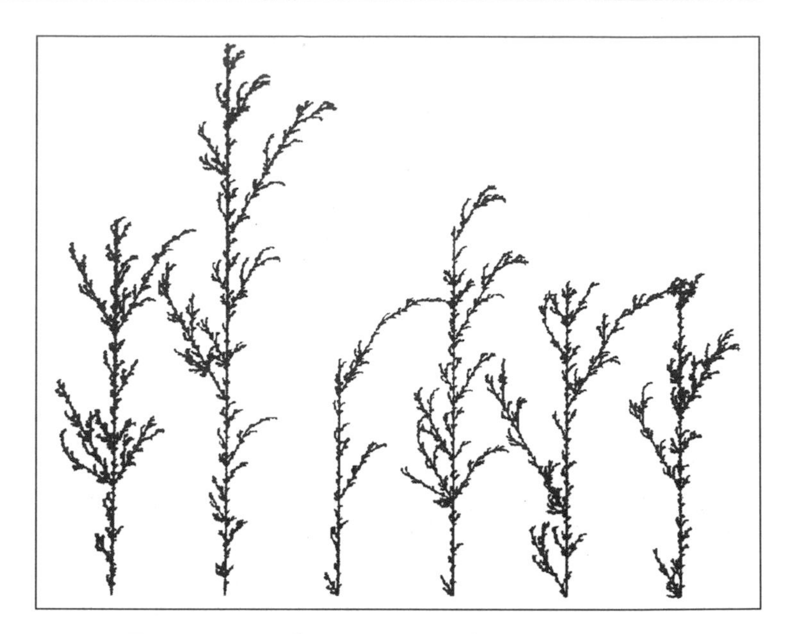

Figure 8.43 : Some samples of 'random weed'.

from deterministic rewriting systems, which we have considered so far, to non-deterministic L-systems. These systems are characterized by multiple production rules for the same symbol (e.g., there may be several production rules for the symbol F). The decision as to which one of the rules to apply in a particular instance is determined by a random process. Most conveniently, we can attach certain probabilities to the individual rules and then choose the rule according to these probabilities. For example, if there are six production rules for one symbol, each one with the same probability (1/6), then the choice could be made rolling a die. Such L-systems are said to be *stochastic*. Figure 8.43 shows an example for the first weedlike structure from above.

L-system:	Stochastic Weedlike Plant III
Axiom:	F
Production rules:	$F \rightarrow F[+F]F[-F]F$ (probability 1/3)
	$F \rightarrow F[+F]F$ (probability 1/3)
	$F \rightarrow F[-F]F$ (probability 1/3)
Parameter:	$\delta = 25.7$ degrees

The random Koch island from chapter 7 also may be described by a stochastic L-system.

L-system:	Random Koch Curve
Axiom:	F

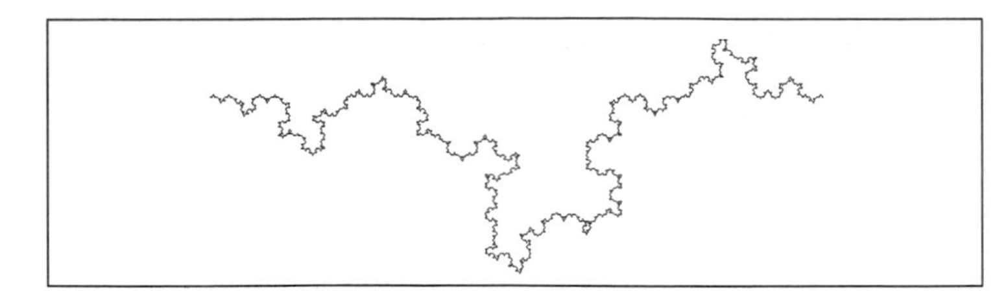

Figure 8.44 : A random Koch curve generated by an L-system.

Random Koch Curve

Production rules: $F \to F - F + +F - F$ with probability 0.5
$f \to F + F - -F + F$ with probability 0.5
Parameters: $\delta = 60$ degrees

For the random Koch island, just replace the axiom by $F - -F - -F$.

The method of stochastic L-systems has in fact been used to produce stunningly realistic images of plants.[20] Of course, the full developmental models for such structures are much more complicated than the simple weeds presented here.

Further Extensions

There are many possible ways to extend the concept of L-systems. For example, the strategy for the expansion of characters can be modified so that one or another rule may apply according to preceding and succeeding symbols. In addition, the turtle which interprets the expanded strings may be allowed to learn a wider vocabulary of commands. For example, parentheses can be used to group drawing commands which define the boundary of a polygon to be filled. Naturally, line-style and color are also parameters of interest, and the turtle may be instructed to move and draw in three dimensions. Curved surfaces may be considered in addition to polygons. Another restriction is the fixed step length associated for example with the symbol F. One can create parameters attached to symbols that specify their interpretation also with precise numerical values such as step length or angle increment. This feature present in so-called parametric L-systems, is very powerful and helps enormously to create more realistic biological models. These extensions are beyond the scope of this chapter, but they can be found along with further references and applications in the book by Prusinkiewicz and Lindenmayer, from which we drew many of the examples given in our presentation.

[20]See P. Prusinkiewicz and A. Lindenmayer, *The Algorithmic Beauty of Plants*, Springer-Verlag, New York, 1990. The figure 8.43 is also from this book.

8.7 Program of the Chapter: L-systems

The program of this chapter enables you to experiment with L-systems. We have implemented the following symbols in small subroutines located at the labels as indicated.

Label 100: F draw a line
Label 300: $+$ turn left
Label 400: $-$ turn right
Label 500: [save the turtle state (for branching)
Label 600:] restore the saved turtle state

For graphical output, this program and most of the other Programs of the Chapter use a square region of the computer screen. This region starts near the upper left corner at the point `(left,left)` and is `w` pixels wide. These parameters are set to `left = 30` and `w = 300` but can be easily changed to fit better to your computer screen, if necessary.

Screen Image of Program

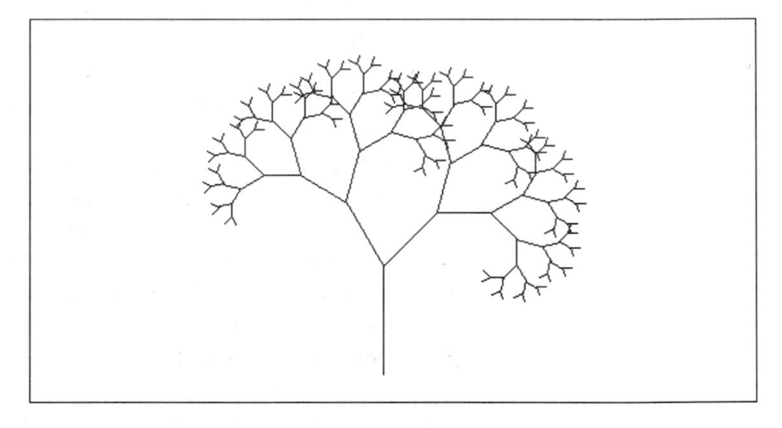

Figure 8.45 : Output of the program 'L-Systems'.

When executed, the program draws a sequence of stages of the development of an L-system of a tree. First, only the stem is drawn. Then the picture is cleared, and the stem with two branches appears. Then subsequent pictures with the tree carrying more and more branches are drawn. The L-system is given by

Axiom: F
Production rule: $F \rightarrow F[+F][-F]$

Observe that the program does not explicitly generate strings of symbols for the stages, which eventually are interpreted by the graphics turtle. Rather, the production rule is integrated into the code for the axiom F as a set of recursive calls to the corresponding

BASIC Programm **L-Systems**
Title Drawing of L-systems using recursion

```
DIM xs(15), dxs(15), ys(15), dys(15)
left = 30 : h = 300
stack = 0
FOR stage = 0 TO 6
    xturtle = left+h/2 : yturtle = left + h
    dx = 0 : dy = -10*(stage+1)
    CLS
    GOSUB 100
NEXT stage
END

REM "F" Drawing and string replacement
100 IF stage > 0 GOTO 150
        LINE (xturtle,yturtle) - STEP (dx,dy)
        xturtle = xturtle+dx : yturtle = yturtle+dy
        RETURN
150 stage = stage -1
    GOSUB 100 : GOSUB 500 : GOSUB 300 : GOSUB 100 : GOSUB 600
    GOSUB 500 : GOSUB 400 : GOSUB 100 : GOSUB 600
    stage = stage + 1
    RETURN

REM "+" Turn right and "-" Turn left
300 dx1 = dx
    dx = .707 * dx1 - .707 * dy
    dy = .707 * dx1 + .707 * dy
    RETURN
400 dx1 = dx
    dx = .866 * dx1 + .500 * dy
    dy = - .500 * dx1 + .866 * dy
    RETURN

REM "[ " push and "]" pop of stack data for branching
500 stack = stack + 1
    xs(stack) = xturtle : ys(stack) = yturtle
    dxs(stack) = dx : dys(stack) = dy
    dx = dx * .7 : dy = dy * .7
    RETURN
600 xturtle = xs(stack) : yturtle = ys(stack)
    dx = dxs(stack) : dy = dys(stack)
    stack = stack - 1
    RETURN
```

parts of the program which represent the symbols that replace F.[21] Let us now discuss some details.

The program draws the L-system for stages 0 through 6. This is controlled by the corresponding `FOR` and `NEXT` statements. For the drawing of each stage, we set up the graphical representation of the axiom F by defining the starting point of the turtle `(xturtle,yturtle)` and its initial step size and direction `(dx,dy)`).[22] Then we go to label 100 which represents the axiom F. The code for F has two parts. If the recursion has already reached the final stage that we want to draw, then a line is drawn and the position of the turtle `(xturtle,yturtle)` is changed. Otherwise, we evaluate the production rule. This is done by `GOSUB` calls. Observe that the sequence

```
GOSUB 100
GOSUB 500
GOSUB 300
GOSUB 100
GOSUB 600
GOSUB 500
GOSUB 400
GOSUB 100
GOSUB 600
```

implements the production rule

$$F[+F][-F] \ .$$

Look up the subroutines to verify this fact. The program parts for symbols $+$ and $-$ change the direction `(dx,dy)` of the turtle. Here we have chosen a counter-clockwise rotation of 45 degrees for $+$ and a clockwise rotation of 30 degrees for $-$. If you want to change these settings you can chose different angles a and use

```
dx = cos(a) * dx1 - sin(a) * dy
dy = sin(a) * dx1 + cos(a) * dy
```

A positive angle leads to a rotation in the counter-clockwise sense, a negative angle in the clockwise sense.

At label 500 the code for the push and pop operations begins. We save the state of the the turtle in the corresponding arrays `xs`, `ys`, `dxs`, and `dys`. Note that the code for [also includes a scaling of the variables `dx` and `dy`. Thus, this is the push operation combined with a little trick appropriate for generating properly dimensioned branches of a tree. The scaling reduces the size of the branches that

[21] Such an approach is possible when the L-system is context-free, i.e., when the production rule for each symbol depends only on the symbol and not on its neighboring symbols. In a context-sensitive system there may be several rules for different contexts.

[22] Note that the length depends on the variable `stage`. This simulates the effect of a growing stem.

appear at higher stages. However, you probably need to remove these statements for experiments with different L-systems.

There are several interesting ways to modify the program. For example, you can change the scaling factor, the angles for the rotations $+$ and $-$, or the initial size of the stem. But you can also generate graphical representations of different L-systems. As an example, let us discuss the changes that are required to generate a Koch curve. The corresponding L-system is

Axiom: F
Production rule: $F \rightarrow F + F - -F + F$

Thus, the corresponding code in the program part of F should be modified to

```
GOSUB 100: GOSUB 300: GOSUB 100: GOSUB 400
GOSUB 400: GOSUB 100: GOSUB 300: GOSUB 100
```

The angles for $+$ and $-$ must both be changed to 60 degrees, and for the initial graphical representation of the axiom you should use

```
xturtle = left : yturtle = left + h/2
dx = 300*.333^stage : dy = 0
```

This allows us to generate different stages of the Koch curve which all fit onto the computer screen.

In further experiments you could try to generate bushes or the weedlike plant of figure 8.40. These L-systems again use the symbol [, however as an indication of a pure push operation without scaling. Thus, do not forget to remove the scaling of dx and dy currently included in the code for this symbol.

Chapter 9

Pascal's Triangle: Cellular Automata and Attractors

Mathematics is often defined as the science of space and number [...] it was not until the recent resonance of computers and mathematics that a more apt definition became fully evident: mathematics is the science of patterns.

Lynn Arthur Steen, 1988

Being introduced to the Pascal triangle for the first time, one might think that this mathematical object is a rather innocent one. Surprisingly it has attracted the attention of innumerable scientists and amateur scientists over many centuries. One of the earliest mentions (long before Pascal's name became associated with it) is in a Chinese document from around 1303.[1] Boris A. Bondarenko,[2] in his beautiful recently published book, counts several hundred publications which have been devoted to the Pascal triangle and related problems just over the last two hundred years. Prominent mathematicians as well as popular science writers such as Ian Stewart,[3] Evgeni B. Dynkin and Wladimir A. Uspenski,[4] and Stephen Wolfram[5] have devoted articles to the marvelous relationship between elementary number theory and the geometrical patterns found in the Pascal triangle. In chapter 2 of *Fractals for*

---

[1]See figure 2.24 in *Fractals for the Classroom, Part One.*

[2]B. Bondarenko, *Generalized Pascal Triangles and Pyramids, Their Fractals, Graphs and Applications,* Tashkent, Fan, 1990, in Russian.

[3]I. Stewart, *Game, Set, and Math*, Basil Blackwell, Oxford, 1989.

[4]E. B. Dynkin and W. Uspenski: *Mathematische Unterhaltungen II*, VEB Deutscher Verlag der Wissenschaften, Berlin, 1968.

[5]S. Wolfram, *Geometry of binomial coefficients*, Amer. Math. Month. 91 (1984) 566–571.

Capturing Pascal's Triangle

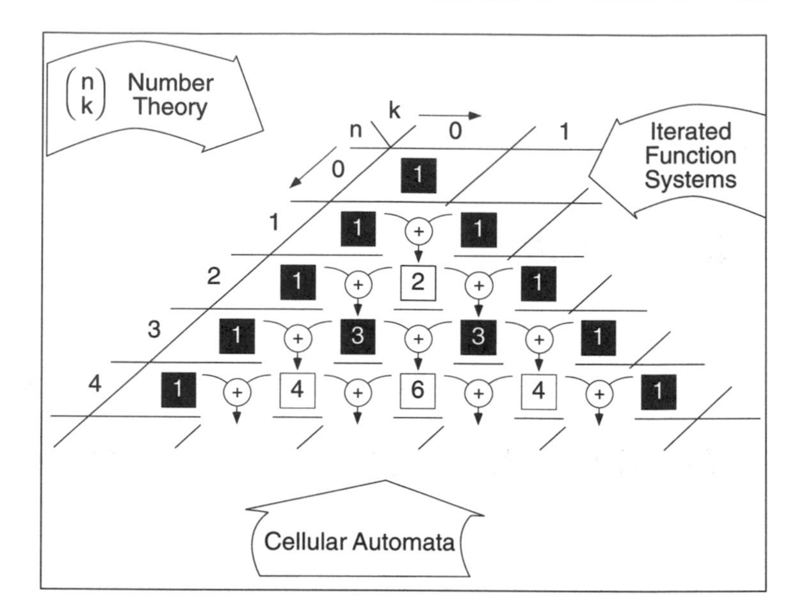

Figure 9.1 : Three approaches to the patterns in Pascal's triangle.

the Classroom, Part One we introduced one example: the relation between the Pascal triangle and the Sierpinski gasket.

One Theme — Many Faces ...

This relationship is indeed a wonderful marvel, and we want to take this opportunity to demonstrate how approaching one mathematical question from totally different angles can beautifully lead to a thorough understanding of that matter. Let us restate the problem.[6] Look at the Pascal's triangle in figure 9.1. It has long been observed that by coloring

- all odd entries black, and
- all even entries white,

we obtain a geometrical pattern which is very closely related to the Sierpinski gasket. Figure 9.1 shows the first 5 rows and the beginning of this pattern formation (first, the black cells only outline a triangle) and figure 9.2 shows the first 128 rows.[7] In fact, the more rows we take into account (e.g. 256, 512, etc.), the more details of the Sierpinski gasket become visible in the geometric pattern.

But we can also use different coloring rules. This leads to all kinds of amazing fractal structures in the triangle. Thus, it is a very interesting question whether there is a way to describe these global pattern formations and how we can find their mathematical

[6]For a more complete discussion see also M. Sved, *Divisibility — With Visibility*, Mathematical Intelligencer 10, 2 (1988) 56–64.

[7]See also figure 2.26 in *Fractals for the Classroom, Part One*.

Modulo 2 Pattern

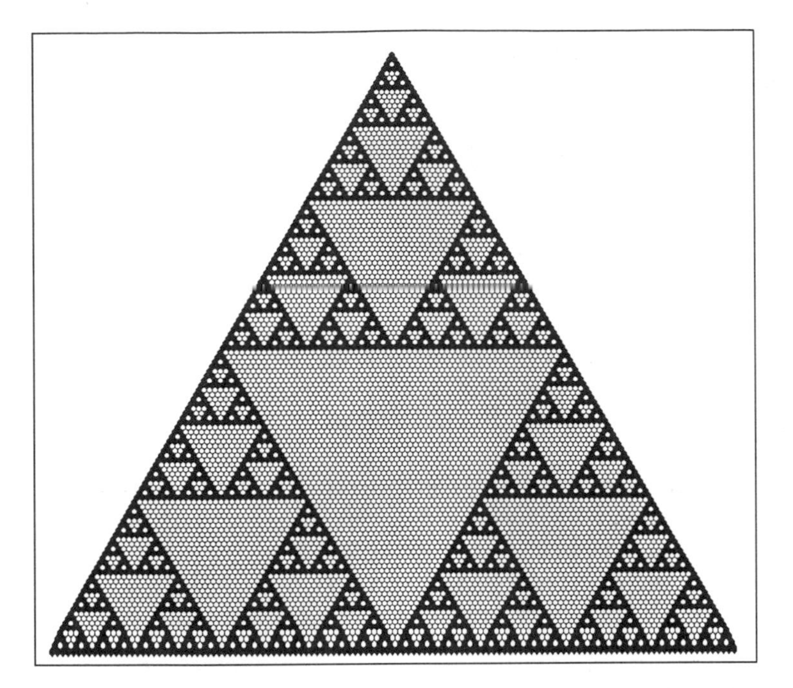

Figure 9.2 : The first 128 rows in Pascal's triangle (each cell carries one entry) colored according to divisibility by 2.

foundations for them.

The most important mathematical interpretation of Pascal's triangle is through binomial coefficients, i.e., the coefficients of the polynomials:

$$\begin{aligned} (1+x)^0 &= 1 \\ (1+x)^1 &= 1+1x \\ (1+x)^2 &= 1+2x+1x^2 \\ &\vdots \\ (1+x)^n &= a_0+a_1x+\cdots+a_nx^n \ . \end{aligned}$$

These coefficients[8] are explicitly given by

$$a_k = \binom{n}{k} = \frac{n!}{(n-k)!k!}, \quad 0 \le k \le n \tag{9.1}$$

where, as usual, factorial n is defined as

$$n! = 1 \cdot 2 \cdot 3 \cdots n$$

[8]The notation $\binom{n}{k}$ was introduced by Andreas von Ettingshausen in his book *Die kombinatorische Analysis*, Wien, 1826.

for $n \geq 1$, and $0! = 1$.[9] Here are some particular cases, which directly follow from these definitions.

$$\binom{n}{0} = 1, \quad \binom{n}{1} = n, \quad \binom{n}{n-1} = n, \quad \binom{n}{n} = 1 \;.$$

Moreover,

$$\binom{n}{k} = \binom{n}{n-k} \;.$$

In other words, introducing a coordinate system for the cells in the triangular array as in in figure 9.1, where $n = 0, 1, 2, ...$ is the row index and $k = 0, 1, 2, ...$ is the column index then the entry in cell with coordinates (n, k) is $\binom{n}{k}$.

Divisibility of Binomial Coefficients

Thus, one approach to the patterns in Pascal's triangle would be to understand the divisibility properties of binomial coefficients. However, computing the entries a_k according to eqn. (9.1) for figures like 9.2 does not lead very far. The reason is that factorials grow extremely rapidly.

n	$n!$	n	$n!$
1	1	6	720
2	2	7	5,040
3	6	8	40,320
4	24	9	363,880
5	120	10	3,628,800

The number 100! has 158 digits,

$$\begin{aligned} 100! = \; & 9332621544\ 3944152681\ 6992388562\ 6670049071 \\ & 5968264381\ 6214685929\ 6389521759\ 9993229915 \\ & 6089414639\ 7615651828\ 6253697920\ 8272237582 \\ & 5118521091\ 6864000000\ 0000000000\ 00000000 \;, \end{aligned}$$

and 1000! about 2568 digits, which surely is beyond the range of common computer arithmetic.[10]

As a first step to overcome these difficulties we use the recursive definition of Pascal's triangle (as indicated in figure 9.1) which is obtained from the addition rule[11]

$$\binom{n+1}{k} = \binom{n}{k-1} + \binom{n}{k} \;. \tag{9.2}$$

[9] For later consideration in the context of cellular automata, we also adopt the convention $\binom{n}{k} = 0$ for $k < 0$ and $k > n$.

[10] The estimate of 2568 digits is obtained by a famous formula developed by James Stirling in 1730 which approximates $n! \approx \sqrt{2\pi n}(n/e)^n$, where $e = 2.71828...$ denotes Euler's number.

[11] See section 2.3 in *Fractals for the Classroom, Part One*.

This fundamental relation avoids the computation of large factorials. However, the binomial coefficients themselves also grow rapidly as the row index n increases. Already in row $n = 34$ we find an entry

$$\binom{34}{17} = 2,333,606,220 > 2^{31} - 1 = 2,147,483,647$$

which cannot be represented exactly in normal computer arithmetic. Fortunately, we do not need the actual numerical values of the binomial coefficients when testing for divisibility. For example, the decision whether a binomial coefficient is odd or even follows directly from the addition rule. Observe that $\binom{n+1}{k}$ is odd provided $\binom{n}{k-1}$ is odd and $\binom{n}{k}$ is even, or vice versa. Systematically we have:

$\binom{n}{k-1}$	$\binom{n}{k}$	$\binom{n+1}{k}$
even	even	even
odd	even	odd
even	odd	odd
odd	odd	even

Cellular Automata ...

This elementary observation is not only of computational importance; it also provides the link to cellular automata. This is another approach to Pascal's triangle which will be explored in the following. We will see that there is a whole class of cellular automata which are closely related to the evolution of divisibility patterns in Pascal's triangle.

... and IFS

However, running a cellular automaton and testing divisibility properties of binominal coefficients have a common property, namely that they are local (or microscopic) procedures. They allow the generation of a geometric pattern but do not at all explain the global (or macroscopic) appearance of the pattern. For example, why do we begin to see the Sierpinski gasket when coloring the odd entries in Pascal's triangle?

To address this problem we once again will bring iterated function systems (IFS) into play. If you recall section 5.3 of *Fractals for the Classroom, Part One*, this does not come as a total surprise and you might have a initial vague idea how this approach to Pascal's triangle could look. We will guide you to this point and explore its relation to divisibility properties and cellular automata, and you will watch most of the pieces of the puzzles falling into place.

9.1 Cellular Automata

Cellular automata have starting points far back in the sciences. In some sense we might say that Pascal's triangle is the first cellular automaton. Their recent development is rooted in the work of Konrad Zuse, Stanislaw Ulam and John von Neumann and is closely related to the first computing machines. During the 1970's and 1980's cellular automata had a strong revival through the work of Stephen Wolfram, who published an interesting survey.[12] Today cellular automata have become a very important modeling and simulation tool in science and technology, from physics, chemistry, and biology, to computational fluid dynamics in airplane and ship design, to philosophy and sociology.

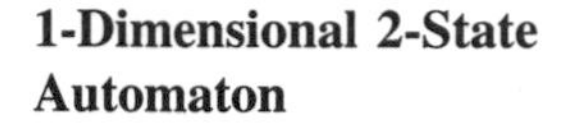

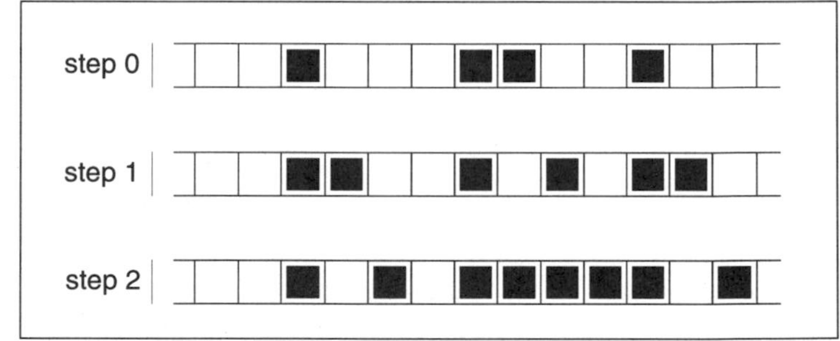

Figure 9.3 : The first steps of a one-dimensional cellular automaton with two states (black and white cells).

Feedback Machines

Cellular automata are perfect feedback machines. More precisely, they are mathematical finite state machines which change the state of their cells step by step. Each cell has one out of p possible states represented by the numbers $0, 1, ..., p-1$. Sometimes we speak of a *p-state cellular automaton.* The automaton can be one-dimensional where its cells are simply lined up like a chain or two-dimensional where cells are arranged in an array covering the plane.[13] Figure 9.3 shows the first steps of a one-dimensional two-state automaton. Sometimes we like to draw the succeeding steps of a one-dimensional cellular automaton one below the other and call the steps *layers.* When running the machine it grows layer by layer as shown in figure 9.4.

To run a cellular automaton we need two entities of information: an initial state of its cells (i.e., an initial layer) and a set of rules or laws. These rules describe how the state of a cell in a new layer (in the next step) is determined from the states of a

[12] S. Wolfram (ed.), *Theory and Application of Cellular Automata,* World Scientific, Singapore, 1986.

[13] In fact, the automaton can have any dimension m, where m is a natural number.

Growing of Layers of a 1-Dimensional Automaton

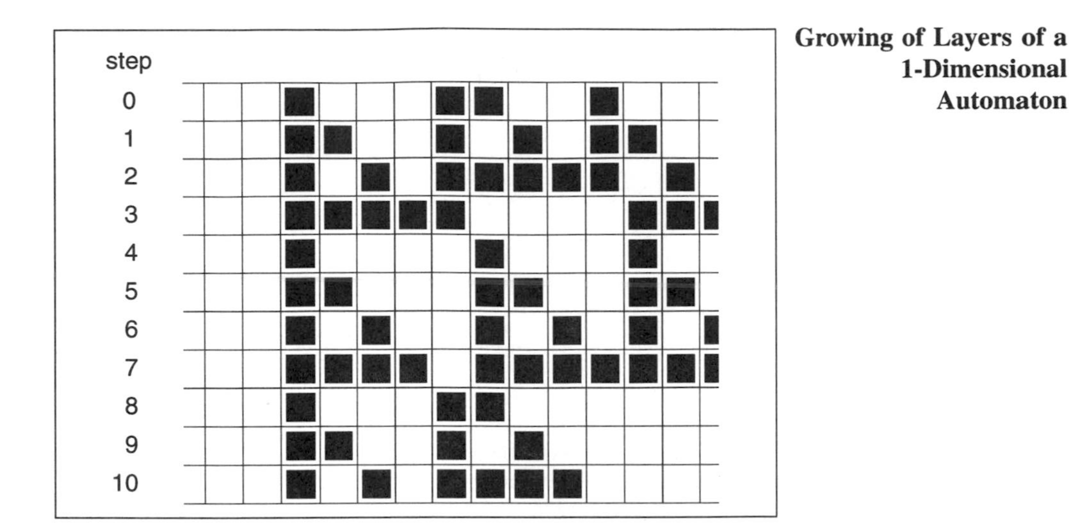

Figure 9.4 : The first layers of the one-dimensional cellular automaton from figure 9.3. The first 10 steps of the iteration are drawn from top to bottom.

Automata Rules

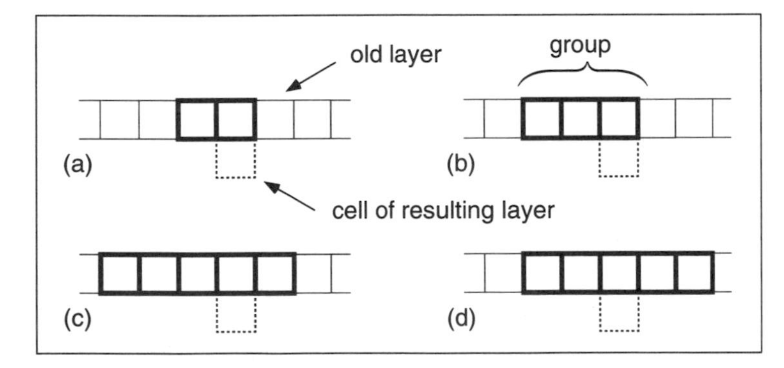

Figure 9.5 : There are several ways a rule may determine the state of a cell in the succeeding layers. In (a) the state of a new cell is determined by the states of two cells, in (b) by the state of three cells. In (c) and (d) the states of five cells determine the state of a new cell, but note that the position of the new cell with respect to the group is different in (c) and (d).

group of cells from the preceding layer. The rules should *not depend* on the position of the group within the layer. Thus, it can be specified by a look-up table or if possible by a formula. Figure 9.6 shows look-up tables for two-state cellular automata which are given by configurations as in (a) and (b) of figure 9.5. These are just two examples of rules for one-dimensional cellular automata. The look-up table (a) was used in figures 9.3 and 9.4.

Automata Look-Up Tables

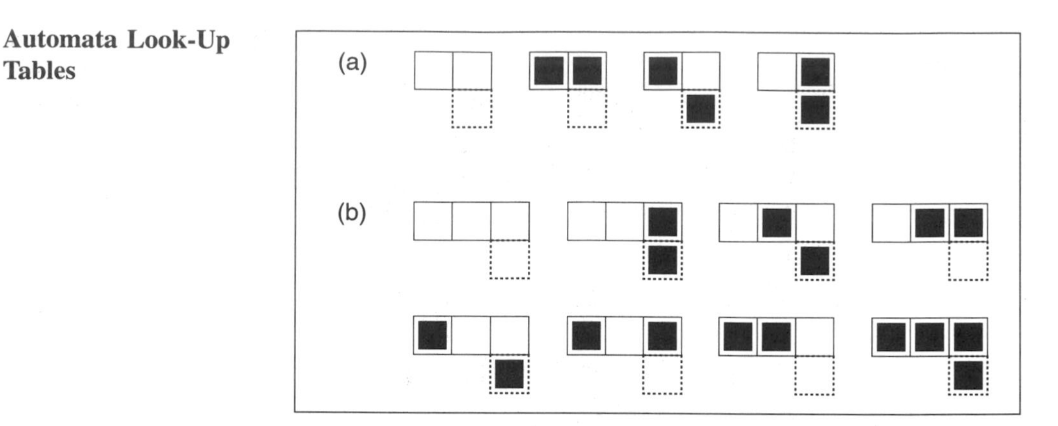

Figure 9.6 : Two examples of look-up tables. (a) four rules for a configuration based on two cells and two states. (b) eight rules for a configuration based on three cells and two states.

Neighborhood in 2-Dimensional Automata

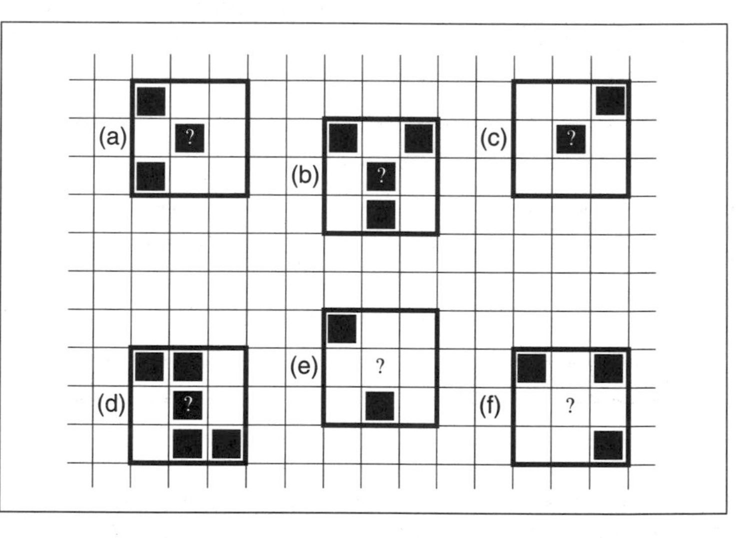

Figure 9.7 : In the Game of Life the neighborhood of a cell decides over life or death. Cell (a) and (b) will stay alive but cell (c) and (d) will die. At (f) a cell will come into life but not at (e).

Game of Life

Particular two-dimensional cellular automata became very popular as the *Game of Life* through the work of John Horton Conway in the 1970's. In the Game of Life each cell is either dead (0) or alive (1) and changes its state according to the states in its immediate neighborhood, including its own state. More precisely, a cell that is alive (symbolized as a black cell) at one time step will stay alive in the next step when precisely two or three cells among its eight neighbors (see figure 9.7) in a square lattice are alive. If more

The Game of Life

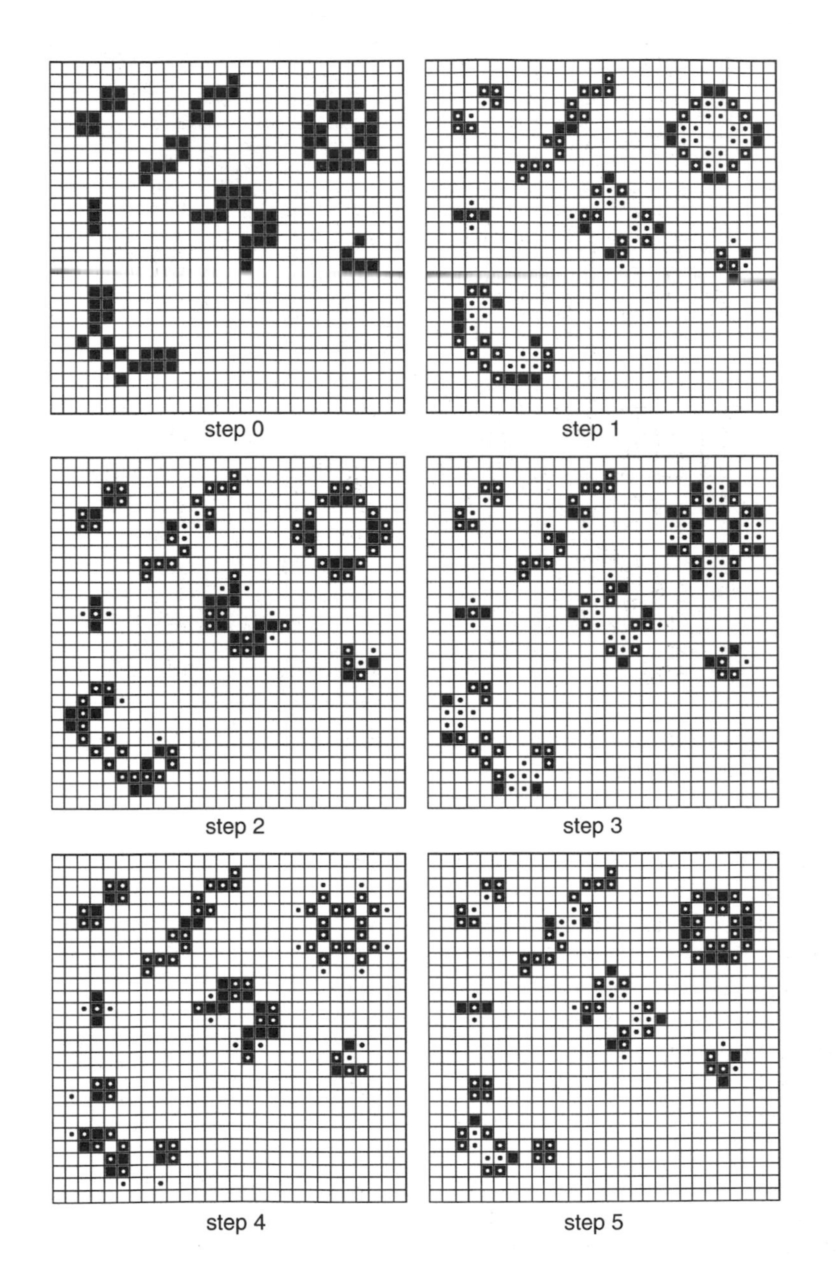

Figure 9.8 : Six successive steps of the Game of Life. Dots indicate the position of living cells of the previous step. Observe that some of the cell clusters shown exhibit a periodic behavior. The center right one is a so-called gliderglider which slowly moves to the left as long as it does not hit another cell cluster. It takes 4 steps to move one cell to the left.

One-Out-Of-Eight Rule

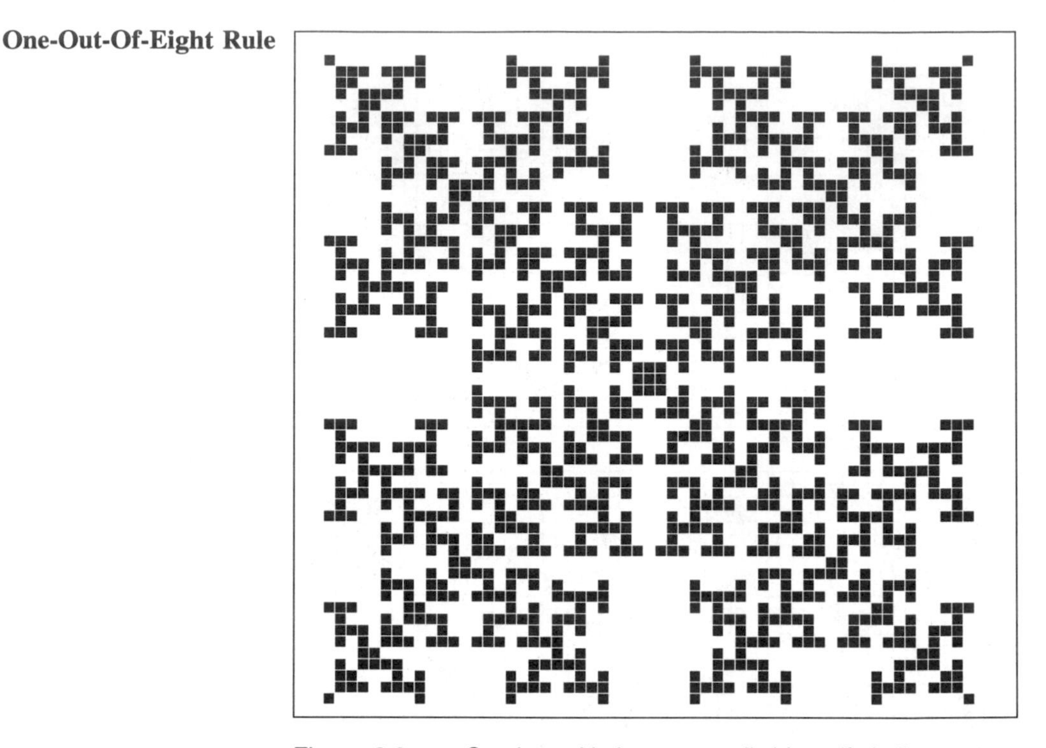

Figure 9.9 : Starting with just one cell this self-similar pattern evolves after 29 steps.

than three neighbors are alive, the cell will die from overcrowdedness. If fewer than two neighbors are alive, the cell will die from loneliness. A dead cell will come to life when surrounded by exactly three live neighbors. Figure 9.8 shows the evolution of the Game of Life in some steps. One of the challenges of the game is to design cell clusters which exhibit a particularly interesting behavior. For example, there are clusters, called *blinkers*, which reproduce themselves after some steps, *gliders* move in a certain direction, *star ships* leave a trace of blinkers, and *guns* periodically eject gliders.

The Number of Games

The rules of the Game of Life are only one choice out of many imaginable sets of rules. For the two possible states and a neighborhood of eight cells generating a new center cell there are $2^{2^9} \approx 10^{154}$ different possible sets.

Let us briefly touch some variants of the Game of Life. The *one-out-of-eight rule* is given by the following set of rules: a cell becomes alive if exactly one of its neighbors is alive; otherwise it remains unchanged. Figure 9.9 shows the resulting pattern which evolves after 29 steps starting with just one living cell. Apparently some self-similarity is built into the formation of this pattern.

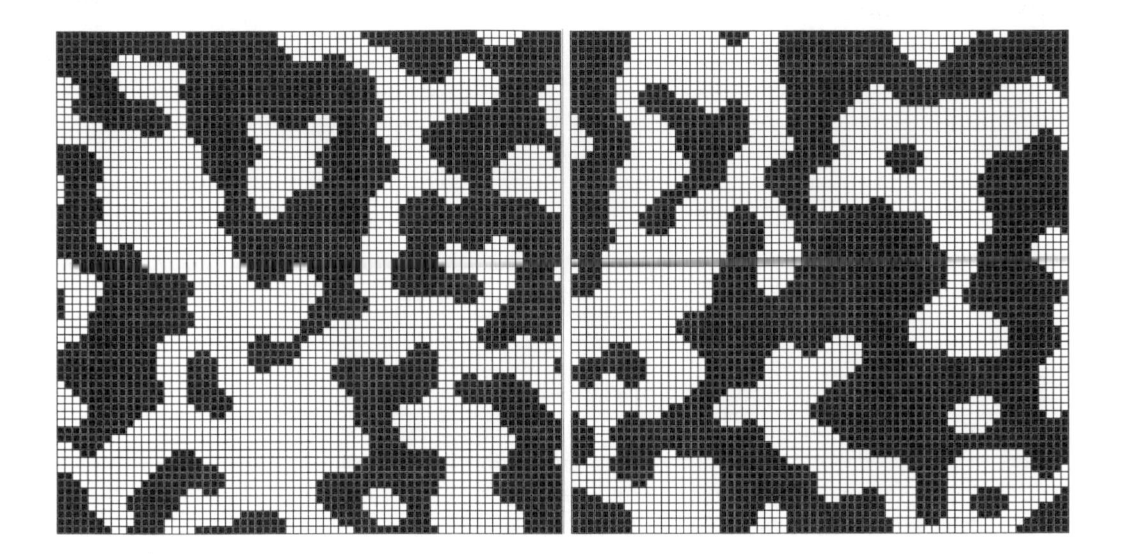

Figure 9.10 : Two examples for a game with the majority rule. Starting from a random distribution of black cells the game settles down (i.e., further iterations do not change the state of cells) to the patterns shown. For the two images, two different initial distributions were used.

Another example, the *majority rule*, is obtained by these conventions: if five or more of the neighborhood of nine cells (including the cell itself) are alive, then this cell will also become or remain alive. Otherwise it will die or remain dead. In other words, the center cell adjusts to the majority in the neighborhood. The resulting patterns resemble some phenomena in statistical physics such as percolation[14] or Ising spin systems. Figure 9.10 shows some experiments which evolve as stable pattern after some 30 steps starting in each case with a random initial distribution of living cells.

Finally, we consider rules which only take four neighbors (again in a two-dimensional square lattice) into account (see figure 9.11). Following Tommaso Toffoli and Norman Margolus[15] we label the center cell by C = Center, and the four neighbors are labelled E = East, W = West, S = South, and N = North. If we allow two states for each cell of this configuration of five cells ($CSWNE$) then the state of $CSWNE$ will be given by five binary digits. For example, $CSWNE = 11010$ indicates that

[14]See section 7.2 in Chapter 7 of *Fractals for the Classroom, Volume One.*

[15]T. Toffoli, N. Margolus, *Cellular Automata Machines: A New Environment For Modelling,* MIT Press, Cambridge, Mass., 1987.

NWSE-Neighbors

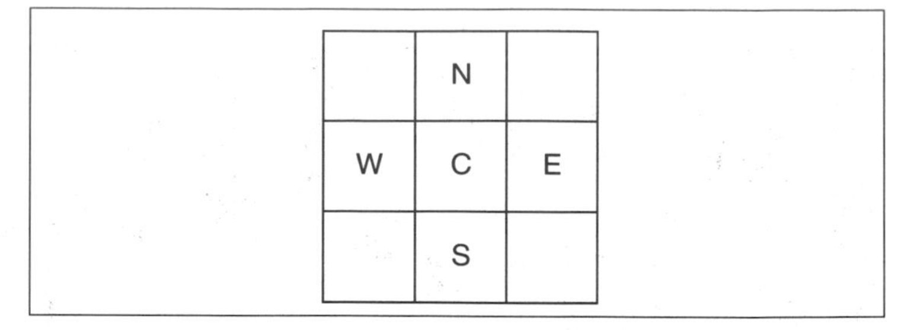

Figure 9.11 : The north, west, south, and east neighbors.

cells C, S, and N are alive, while the other two are dead. A complete set of rules can be given by a table of the 32 possible states of $CSWNE$ and the subsequent values of the center cell C. Note that for such a configuration there are $2^{32} \approx 4 \cdot 10^9$, i.e. 4 billion, possible different tables.

CSWNE	C	CSWNE	C	CSWNE	C	CSWNE	C
00000	0	01000	1	10000	1	11000	1
00001	0	01001	1	10001	1	11001	1
00010	0	01010	1	10010	1	11010	1
00011	0	01011	1	10011	1	11011	1
00100	1	01100	0	10100	1	11100	1
00101	1	01101	0	10101	1	11101	1
00110	1	01110	0	10110	1	11110	1
00111	1	01111	0	10111	1	11111	1

Applying the rules from the above table we obtain a familiar pattern. Figure 9.12 shows the 5th and 15th step starting with just one live cell near the lower left corner. Studying the table you can find a rather simple rule for producing the entries. Note that if the center cell C was dead (0) then the new value depends only on cells W and S. On the other hand, if the cell was alive (1) then it will remain alive in the next step. In fact, we have just seen an example of how we can construct a two-dimensional automaton with the behavior of a one-dimensional one.[16] In other words, cells grow layer by layer like the layers of a one-dimensional cellular automaton (although in this example the layers are diagonals and the pattern grows from bottom left to top right).

Many interesting rules can be expressed by a simple formula. For example, the *parity rule* is given simply by

$$C_{new} = C_{old} + S_{old} + W_{old} + N_{old} + E_{old} \bmod 2 \ . \quad (9.3)$$

[16] In fact, for any given one-dimensional automaton, it is possible to create a corresponding two-dimensional automaton using the same basic idea as in our example.

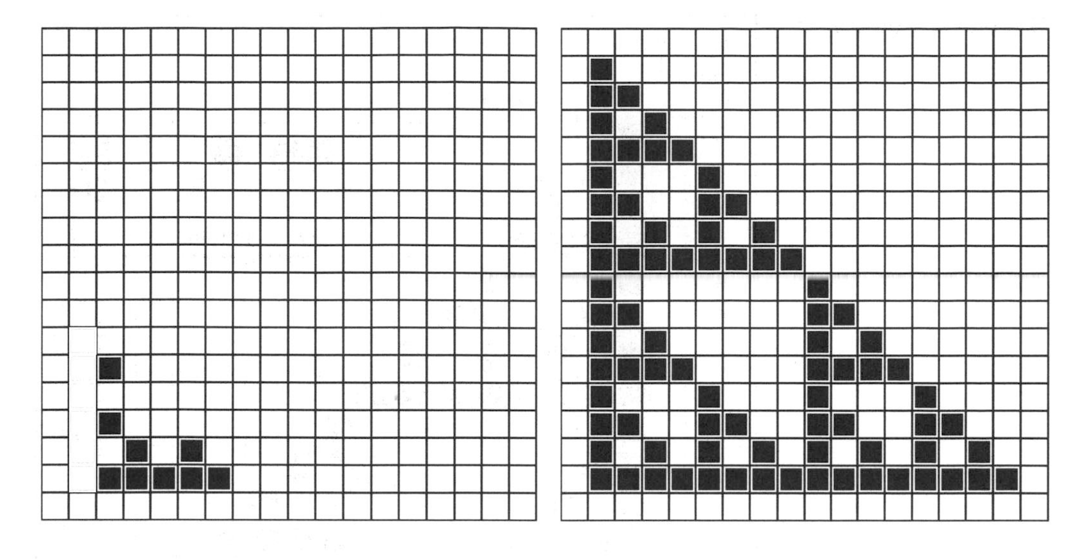

Figure 9.12 : Starting with just one cell this two-dimensional automaton behaves like a one-dimensional one, but the layers grow diagonally from the bottom left corner.

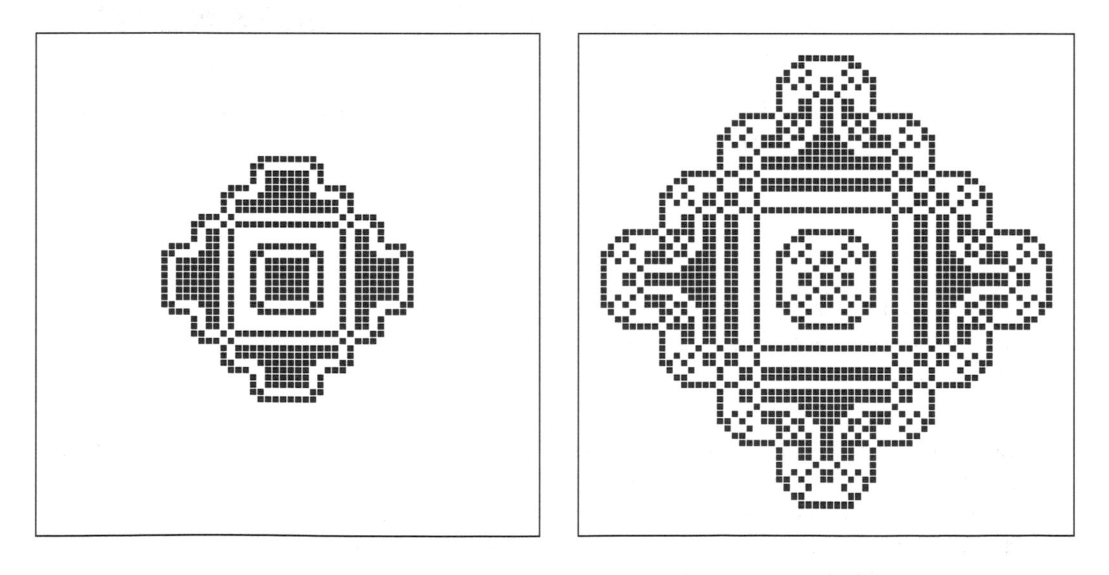

Figure 9.13 : Starting with an 8 by 8 cell cluster these patterns evolve after 13 (left) and 27 (right) steps using the parity rule (9.3).

which means that C_{new} is 0 or 1 if the sum on the right hand side is even or odd, respectively. Here E, W, S, N and C represent the old and new cell states as indicated by the index old and new. Thus for $CSWNE = 11010$ we obtain $C_{new} = 1$, for $CSWNE = 11011$

Sierpinski Automaton

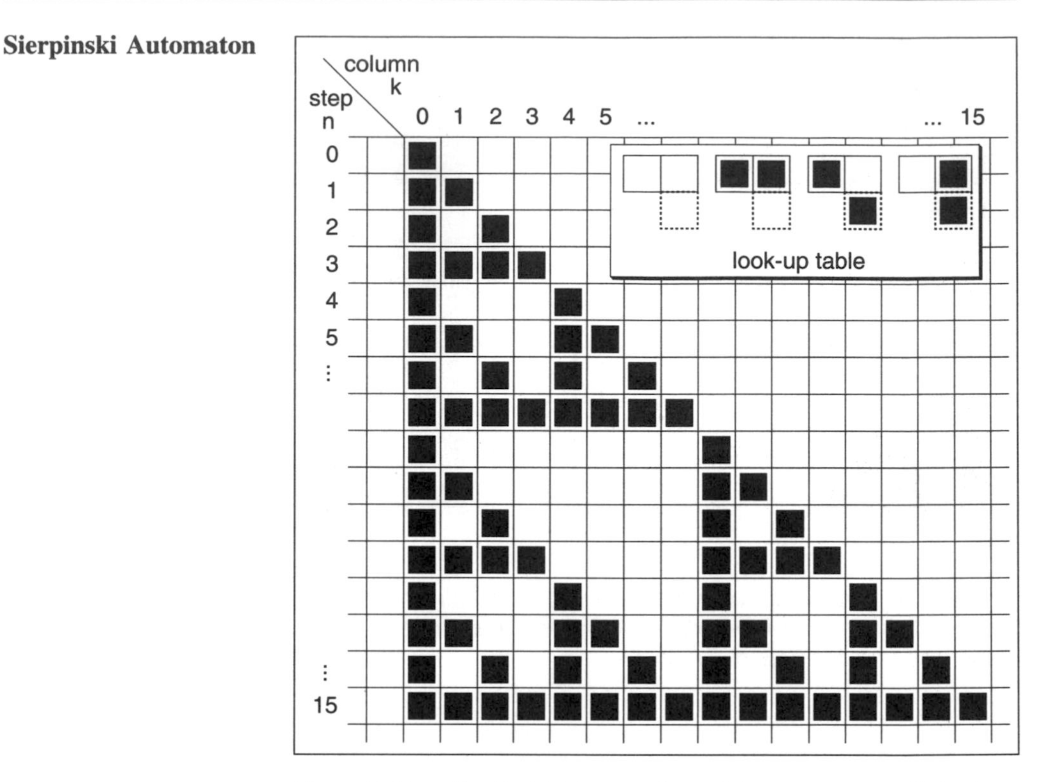

Figure 9.14 : The first 16 layers of a cellular automaton with look-up table displayed in the upper right.

$C_{new} = 0$, and so on. Figure 9.13 shows the evolution of this cellular automaton after 13 and 27 steps starting with a square block of 8×8 black cells.

Pascal's Triangle ...

Let us return to one-dimensional automata. The look-up table used in figure 9.14 reflects the addition of even and odd binomial coefficients. An odd entry is colored black. That is, the evolution of the corresponding cellular automaton will produce the pattern which is obtained from the Pascal triangle when we color cells with odd entries black and cells with even entries white and start with an appropriate initial layer. This is seen in the figure, where we follow the evolution of the first 16 layers starting with the initial layer, which has one black cell.

... and Polynomials

Let us explore the connection between cellular automata and coefficients of polynomials a bit further. First we look at an example involving the powers of the polynomial $r(x) = 1 + x$:

$$\begin{aligned}
(r(x))^0 &= 1 \\
(r(x))^1 &= 1 + x \\
(r(x))^2 &= 1 + 2x + x^2 \\
(r(x))^3 &= 1 + 3x + 3x^2 + x^3 \\
&\vdots \\
(r(x))^n &= a_0(n) + a_1(n)x + a_2(n)x^2 + \cdots + a_n(n)x^n .
\end{aligned}$$

Now let $a_k(n)$ for all integers k and $n \geq 0$ denote the state of cell number k of the n-th layer of a one-dimensional automaton.[17] Starting with $a_0(0) = 1$ and $a_k(0) = 0$ for $k \neq 0$ the rule

$$a_k(n) = a_{k-1}(n-1) + a_k(n-1) \tag{9.4}$$

generates the coefficients of $(r(x))^n$. Equation (9.4) is nothing else but the addition rule in eqn. (9.2) for binomial coefficients.

Now we want to look at the divisibility properties of $a_k(n)$ with respect to an integer p. We write

$$a \equiv b \pmod{p}$$

provided $a - b$ is a multiple of p.[18] Using this language, our test for odd and even binominal coefficients $\binom{n}{k}$ or cells $a_k(n)$ is simply to check whether

$$a_k(n) \equiv 0 \pmod{2} \quad \text{or} \quad a_k(n) \equiv 1 \pmod{2} .$$

Moreover, our addition rules (9.2) and (9.4) imply in mod 2 arithmetic

$a_{k-1}(n)$	$a_k(n)$	$a_k(n+1)$
0	0	0
1	0	1
0	1	1
1	1	0

This is just the look-up table of the 2-state automaton shown in figure 9.14 where black corresponds to 1 and white to 0. Thus, this figure shows the coefficients of the powers of $r(x) = 1 + x$ modulo 2.

[17] Strictly speaking, this is not a finite state automaton because the numbers $a_k(n)$ grow beyond all bounds. However, we will arrive at a finite state machine when restricting our attention to the divisibility properties.

[18] In other words, $a \equiv b \pmod{p}$ provided a and b differ by a multiple of p. For $p = 2$ this means that $a - b$ is even. Furthermore, $a \equiv 0 \pmod{2}$ means that a is even and $a \equiv 1 \pmod{2}$ means that a is odd. The notation was introduced by Carl Friedrich Gauss.

Generalizations

Now we can generalize in two ways: we can look at coefficients modulo integers other than 2 and we can look at arbitrary polynomials. Let us take the example: $r(x) = 1 + x + x^2$:

$$\begin{aligned}
(r(x))^0 &= 1 \\
(r(x))^1 &= 1 + x + x^2 \\
(r(x))^2 &= 1 + 2x + 3x^2 + 2x^3 + x^4 \\
(r(x))^3 &= 1 + 3x + 6x^2 + 7x^3 + 6x^4 + 3x^5 + x^6 \\
&\vdots \\
(r(x))^n &= a_0(n) + a_1(n)x + a_2(n)x^2 + \cdots + a_{2n}(n)x^{2n}
\end{aligned}$$

Do you see an extension of the addition rule for binomial coefficients, eqn. (9.4)? You can check in the first few lines that the law

$$a_k(n) = a_{k-2}(n-1) + a_{k-1}(n) + a_k(n) \tag{9.5}$$

holds. A proof of this relation would proceed by induction. When looking at the divisibility properties with respect to $p = 3$ we obtain the following coefficients:

$$\begin{aligned}
(r(x))^0 &\to 1 \\
(r(x))^1 &\to 1\ 1\ 1 \\
(r(x))^2 &\to 1\ 2\ 0\ 2\ 1 \\
(r(x))^3 &\to 1\ 0\ 0\ 1\ 0\ 0\ 1 \\
&\vdots
\end{aligned}$$

Figure 9.15 shows the evolution of the corresponding 3-state automaton.

Linear Cellular Automata

In a similar way we could start with any polynomial $r(x) = a_0 + a_1 x + \cdots + a_d x^d$ of degree d and integer coefficients a_i, and then look at the coefficients of $(r(x))^n$ modulo some positive integer p for $n = 0, 1, 2, \ldots$ and the result would be that the k-th coefficient of $(r(x))^{n+1}$ is obtained by an addition formula involving $d+1$ coefficients from $(r(x))^n$. In other words, given a polynomial with integer coefficients and a positive integer p there is an associated cellular automaton which generates the coefficients modulo p of the powers $(r(x))^n, n = 0, 1, 2, \ldots$ Since the look-up table is generated by an addition formula these automata are called *linear cellular automata* (LCA).

States and Colors

The choice of the positive integer p determines the number of states of the automaton. If $p = 2$, i.e., we are considering arithmetic modulo 2, then we have an automaton which can be graphically represented in black and white. For $p > 2$ we would need colors to adequately represent the evolution of an automaton. We can often simplify a p-state to a 2-state automaton by the following modification:

Mod 3 Automaton

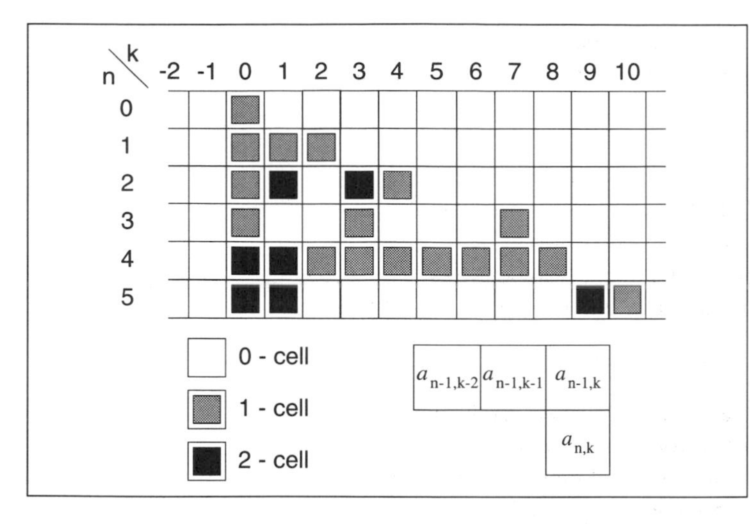

Figure 9.15 : Cellular automaton generated by the polynomial $r(x) = 1 + x + x^2$ and $p = 3$. There are two types of indices, a row index n which runs $n = 0, 1, 2, ...$, and a column index k which runs in the integers $..., -2, -1, 0, 1, 2, ...$ The initial layer, $n = 0$, consists of cells of state 0 except for the cell at $k = 0$, which is a cell with state 1. The rule for the cells of the new layer is: $a_{n+1,k} = a_{n,k-2} + a_{n,k-1} + a_{n,k} \bmod 3$

- Cells representing a nonzero coefficient are colored black.
- Cells representing a zero coefficient are colored white.

With this background of linear cellular automata, we can state a number of very interesting problems:

- *Pattern Formation.* Given a polynomial with integer coefficients and a positive integer p, discuss the global pattern formation which evolves when the automaton has produced for a long time.
- *Colors.* What is the relationship between the global patterns which are obtained for different choices of p and a fixed, given polynomial?
- *Fractal Dimension.* What is the fractal dimension of the global pattern?
- *Higher Dimensions.* Polynomials in one variable generate one-dimensional linear cellular automata. A polynomial in m variables determines a linear cellular automaton in m dimensions. How can we generalize the results to m-dimensional automata?
- *Factorization.* If a polynomial $r(x)$ is the product of two polynomials $s(x)$ and $t(x)$, how is the pattern determined by $r(x)$ related to the patterns determined by $s(x)$ and $t(x)$, and how are the dimensions related?

Actually, the last problem critically depends on the choice of p, the number of states, because what actually counts is whether

$$r(x) \equiv s(x)t(x) \pmod{p} .$$

For example, the polynomial $r(x) = 1 + x$ is irreducible with respect to the integers, i.e., if $r(x) = s(x)t(x)$, then the factorization must be trivial, i.e., $s(x) = 1$ and $t(x) = 1 + x$. If we use arithmetic modulo p, and p is not a prime number, however, then $r(x) = 1 + x$ admits non-trivial factorizations like, for example,

$$1 + x \equiv (1 + 3x)(1 + 4x) \pmod{6} .$$

Several of these problems are still wide open while others have been understood only recently through new tools provided by fractal geometry (namely hierarchical iterated function systems) which stresses again that fractals are more than pretty images.

9.2 Binomial Coefficients and Divisibility

In the remaining part of this chapter we will discuss some of the problems listed at the end of the last section for the particular choice $r(x) = 1 + x$ and positive integers p. Thus, in the following we will only look at the divisibility properties of binomial coefficients,[19] although a similar discussion can be done for general polynomials.[20]

In our discussion we will primarily address the question of whether a binominal coefficient is divisible by p or not. In other words, we consider the black and white coloring of the Pascal triangle interpreted modulo p (see figure 9.16). The question of divisibility can be approached with the aid of prime number factorization. Below we will see that in order to understand the patterns in Pascal's triangle formed by the coefficients divisible by an integer p, we should build on the patterns generated by the prime factors of p.

Pattern in Pascal's Triangle Mod 3

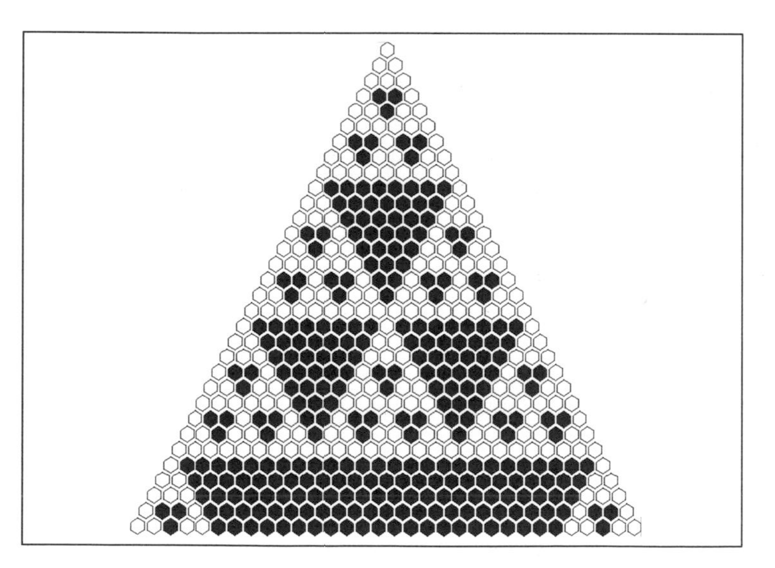

Figure 9.16 : Coefficients in the Pascal triangle which are divisible by 3 are shown in black.

We have seen that we can answer the question of divisibility by recursively computing binominal coefficients using an addition rule like eqn. (9.2) modulo p with a subsequent test as to whether the result is 0 or not. On the other hand, we know very well how

[19] F. v. Haeseler, H.-O. Peitgen, G. Skordev, *Pascal's triangle, dynamical systems and attractors,* to appear in Ergodic Theory and Dynamical Systems.

[20] F. v. Haeseler, H.-O. Peitgen, G. Skordev, *On the hierarchical and global structure of cellular automata and attractors of dynamical systems,* to appear.

Coordinate Systems

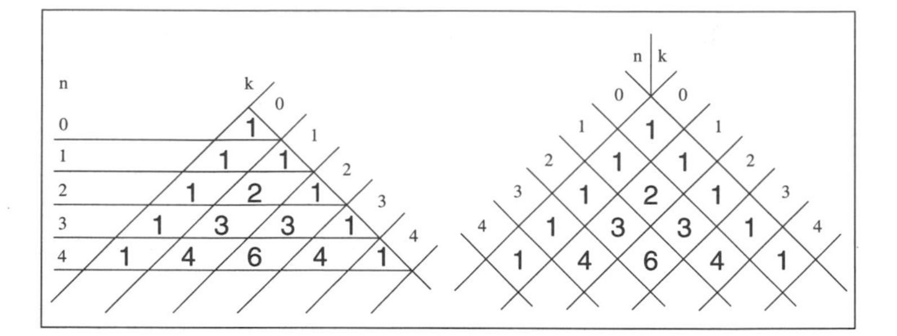

Figure 9.17 : Two coordinate systems for the presentation of binomial coefficients. The new modified system is on the right.

to describe the coefficients without a recursion, namely by

$$\binom{n}{k} = \frac{n!}{(n-k)!k!} .$$

The major question for now will be to understand whether or not these coefficients are divisible by p also by means of a direct, non-recursive computation. It turns out that this problem was solved in a most elegant manner some 150 years ago by the German mathematician Ernst Eduard Kummer.[21] The careful development of Kummer's criterion, which is local in nature, will build a solid foundation for the next step towards understanding the global pattern formation in Pascal's triangle.

It turns out that for the following it will be more convenient to work in a new (n, k)-coordinate system (see figure 9.17). The connection between the old and new representation is easy. In the old system we find at position (n, k) the coefficient $\binom{n}{k}$, while in the new system we have at position (n, k) the binomial coefficient

$$\binom{n+k}{k} = \frac{(n+k)!}{n!k!}$$

Figure 9.18 shows the array in the new system, however, rotated and right angled, together with the usual coloration corresponding to even and odd entries.

Divisibility Sets $P(r)$

We will now describe our problem more formally. We define the following set:

$$P(r) = \left\{ (n,k) \;\middle|\; \binom{n+k}{k} \text{ is not divisible by } r \right\} ,$$

where r is some integer. Thus, figure 9.18 is a graphical representation of a part of $P(2)$.

[21] E. E. Kummer, *Über Ergänzungssätze zu den allgemeinen Reziprozitätsgesetzen,* Journal für die reine und angewandte Mathematik 44 (1852) 93–146. For the result relevant to our discussion see pages 115–116.

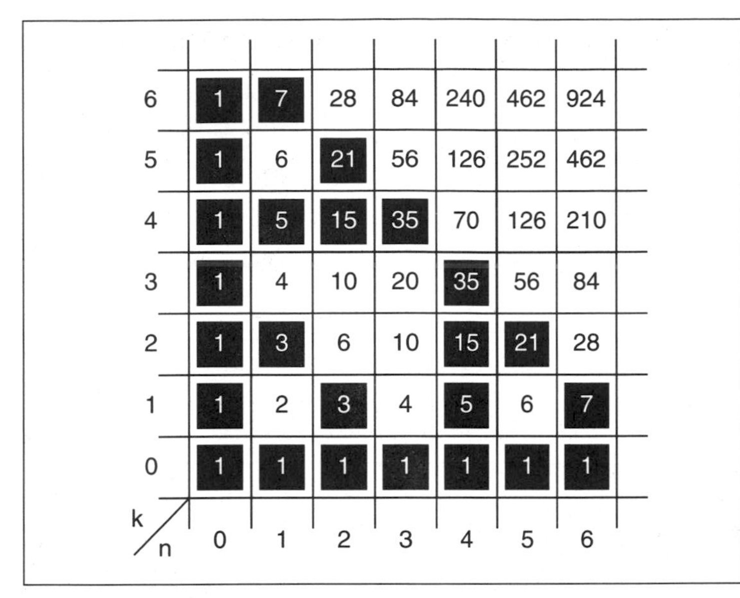

Figure 9.18 : Presentation of Pascal's triangle in the new (n, k)-coordinate system. Position (n, k) shows $\binom{n+k}{k}$. Odd numbers are colored black.

Observe that if p and q are two different prime numbers and a given integer is not divisible by $p \cdot q$, then it is also not divisible by p or q alone. Thus,

$$P(pq) = P(p) \cup P(q), \quad \text{if } p \neq q, \quad p, q \text{ prime} .$$

For example, $P(6) = P(2) \cup P(3)$ (see figure 9.19). This observation can be generalized to the factorization in prime powers. If we consider the prime factorization of an integer r,

$$r = p_1^{\tau_1} \cdots p_s^{\tau_s} ,$$

where the prime numbers p_k are different and the exponents τ_k are natural numbers, then

$$P(r) = P({p_1}^{\tau_1}) \cup \cdots \cup P({p_s}^{\tau_s}) .$$

Thus, to understand the pattern formation for $P(r)$ it suffices to understand $P(p^\tau)$,

$$P(p^\tau) = \left\{ (n, k) \;\middle|\; \binom{n+k}{k} \text{ is not divisible by } p^\tau \right\} ,$$

for a prime number p and some positive integer τ.

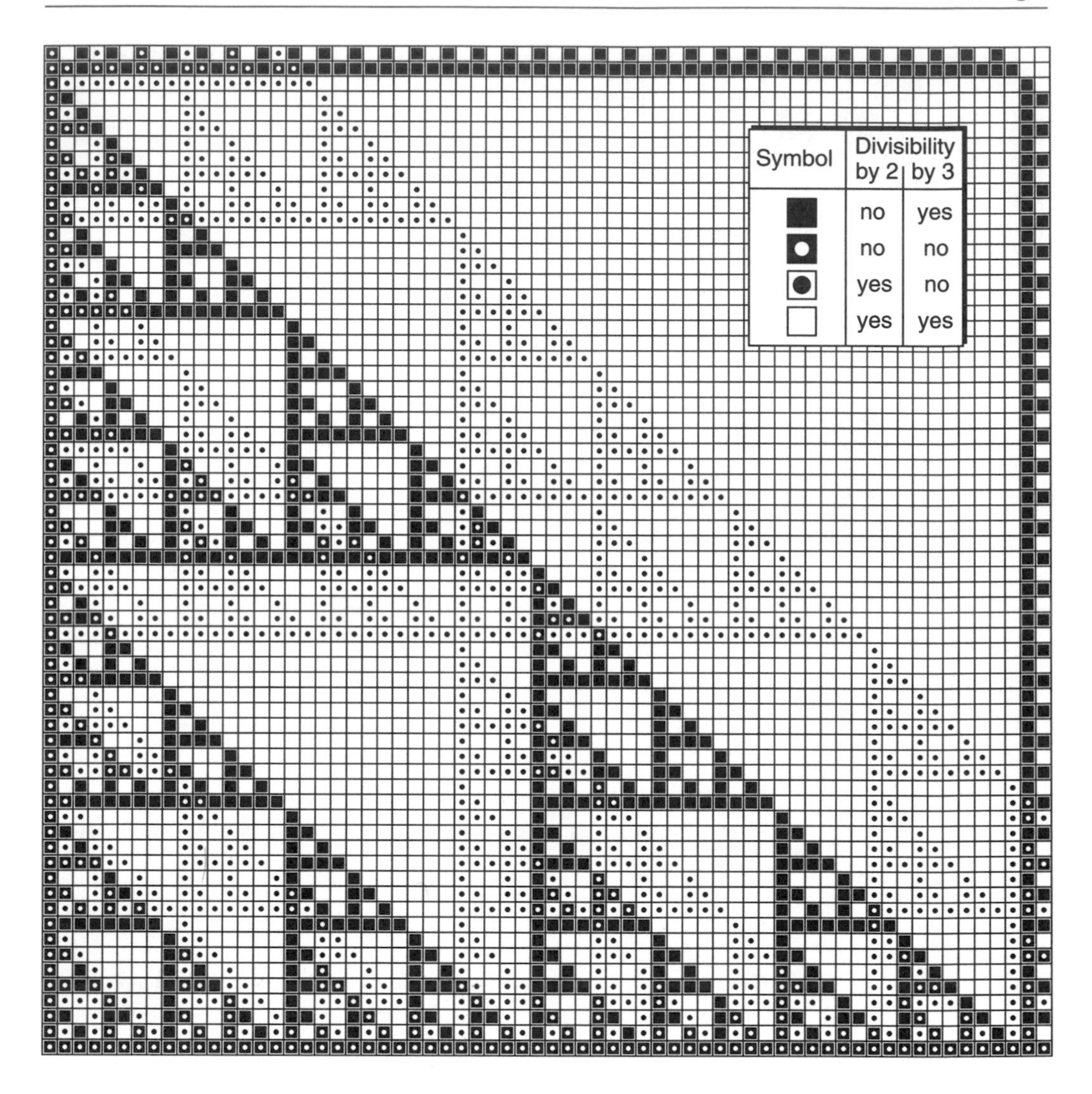

Figure 9.19 : The first 66 rows of Pascal's triangle and its mod-6 pattern generated by a cellular automaton using the rule: $a_{n,k} = a_{n-1,k} + a_{n,k-1} \bmod 6$.

The p-adic Approach

Now let us discuss some gems from elementary number theory attributable to Adrien Marie Legendre (1808), Ernst Eduard Kummer (1852), and Edouard Lucas (1877). These results, together with hierarchical iterated function systems, will completely decipher the patterns in $P(p^\tau)$ for any prime number p and positive integer τ.

We would like to have a direct method to check whether $\binom{n+k}{k}$ is divisible by p^τ. Kummer observed that this information is encoded in the p-adic representation of n and k. You are familiar with decimal expansions like

$$n = d_0 + d_1 \cdot 10 + d_2 \cdot 10^2 + ... + d_m \cdot 10^m,$$

where the numbers $d_i \in \{0, ..., 9\}$ are the decimal digits. Thus n can be represented as the decimal number

$$n = d_m d_{m-1} ... d_1 d_0 \ .$$

Now the p-adic expansion of an integer n is given analogously by

$$n = d_0 + d_1 p + d_2 p^2 + \cdots + d_m p^m$$

where the digits are now $d_i \in \{0, ..., p-1\}$ and m may be different from the m used above in the decimal number. Corresponding to decimal numbers, we introduce the p-adic representation

$$n = (a_m a_{m-1} ... a_1 a_0)_p \ .$$

For example, $n = 17$ and $k = 8$ have the expansions

$$\begin{aligned} n &= (17)_{10} = (10001)_2 = (122)_3 = (32)_5 = (23)_7 = (16)_{11} \\ k &= (08)_{10} = (01000)_2 = (022)_3 = (13)_5 = (11)_7 = (08)_{11}. \end{aligned}$$

Kummer's observation concerns the number of carries which occur when we add n and k in the p-adic representation when p is a prime. For example, let us add the triadic representations:

$$\begin{array}{r} 1\ 2\ 2 \\ +0_1 2_1 2 \\ \hline 2\ 2\ 1 \end{array}$$

Observe that when adding the right most digits we obtain a carry to the next digit. The same is true when adding the second digits of the two numbers. Thus, we obtain two carries. On the other hand, if we add the corresponding binary representations:

$$\begin{array}{r} 10001 \\ +01000 \\ \hline 11001 \end{array}$$

we obtain no carry at all. In other words, if we define

$$\begin{aligned} c_p(n, k) = &\text{ number of carries in the} \\ &\ p\text{-adic addition of } n \text{ and } k \ , \end{aligned}$$

we have demonstrated that $c_2(17, 8) = 0$ and $c_3(17, 8) = 2$. Now we can state Kummer's result.

Kummer's Result

Let $\tau = c_p(n,k)$. $\binom{n+k}{k}$ is divisible by the prime power p^τ but not by $p^{\tau+1}$.

In other words, the prime factorization of $\binom{n+k}{k}$ contains exactly $c_p(n,k)$ factors of p. Applied to our example, $n = 17$ and $k = 8$, we should have that $\binom{n+k}{k}$ has no factors of 2, because $c_2(17,8) = 0$, and exactly two factors of 3, because $c_3(17,8) = 2$. Thus, it is an odd number and divisible by 9, but not by 27. In fact, we compute that

$$\binom{n+k}{k} = \binom{25}{8} = 3^2 \cdot 5^2 \cdot 11 \cdot 19 \cdot 23 \ . \qquad (9.6)$$

which confirms our conclusions.

Interpreting Kummer's result the other way around, we conclude from the factorization in eqn. (9.6), that

$$c_p(17,8) = \begin{cases} 2 \text{ for } p = 3,5 \\ 1 \text{ for } p = 11,19,23 \\ 0 \text{ otherwise} \end{cases}$$

In fact, we check, for example, $c_{11}(17,8) = 1$ by adding 17 and 8 in modulo 11 arithmetic,

$$\begin{array}{r} 1\ 6 \\ +\ {}_1 8 \\ \hline 2\ 3 \end{array}$$

obtaining one carry — as expected.

Lucas' Criterion

To determine whether $\binom{n}{k}$ is odd or even we can use Lucas' criterion[22] as follows. We compute the binary form of n and k, say, $n = 23 = (10111)_2$ and $k = 17 = (10001)_2$. Then we write them one over the other.

10111
10001

Now $\binom{n}{k}$ is odd, if and only if every digit of the bottom number k is less than or equal to the digit of n above. This is the case for our example, $n = 23$ and $k = 17$. In fact, $\binom{23}{17} = 5313$ is odd.

Let us see how Lucas' criterion follows directly from Kummer's result. Let the binary expansions of n and k be

$$n = a_m 2^m + a_{m-1} 2^{m-1} + \cdots + a_0$$
$$k = b_m 2^m + b_{m-1} 2^{m-1} + \cdots + b_0$$

[22] It is related to several published criteria, like the one in I. Stewart, *Game, Set, and Math*, Basil Blackwell, Oxford, 1989, which Stewart attributes to Edouard Lucas following Gregory J. Chaitin's book *Algorithmic Information Theory*, Cambridge University Press, 1987.

with binary digits $a_i, b_i \in \{0, 1\}$. Since $k \leq n$ some of the leading binary digits of k may be 0. First observe that we know from Kummer's result that $\binom{n}{k} = \binom{n-k+k}{k}$ is not divisible by 2 if and only if we have for the 2-adic expansion

$$n - k = c_m 2^m + c_{m-1} 2^{m-1} + \cdots + c_0$$

the property

$$c_i + b_i \leq 1, \quad i = 0, ..., m \ .$$

In other words, in the p-adic addition of $n - k$ and k there will be no carry.

To complete the argument, we have to show two implications. First, if $\binom{n}{k}$ is odd, then Lucas' criterion follows, i.e., $a_i \geq b_i$, for all i. Second, if Lucas' criterion is satisfied, then it follows that $\binom{n}{k}$ is odd. Let i denote an arbitrary index $i \in \{0, ..., m\}$. Now we start with the first part, assuming that $\binom{n}{k}$ is odd. Kummer's result above states that $c_i + b_i \leq 1$. And this implies that $a_i = b_i + c_i$ and, in conclusion, also $a_i \geq b_i$, which is what was to be shown. Now we do the second part. We assume $a_i \geq b_i$. Then $c_i = a_i - b_i$, which implies that $c_i + b_i = a_i \leq 1$. Thus, according to Kummer's result $\binom{n}{k}$ is odd. This finishes the second part and completes the proof of Lucas' criterion.

Mod-p Condition

As a particular case we obtain from Kummer's criterion that $\binom{n+k}{k}$ is not divisible by the prime number p provided $c_p(n, k) = 0$. In other words, we have

$$P(p) = \{(n, k) \mid c_p(n, k) = 0\} \ .$$

Moreover, the number of carries $c_p(n, k)$ is 0 if and only if

$$a_i + b_i \leq p - 1, \quad i - 0, ..., m$$

where a_i and b_i denote the p-adic digits of n and k, i.e.,

$$\begin{aligned} n &= a_0 + a_1 p + a_2 p^2 + \cdots + a_m p^m \\ k &= b_0 + b_1 p + b_2 p^2 + \cdots + b_m p^m \ . \end{aligned}$$

This we will call the *mod-p condition.*

For prime powers, Kummer's result implies that

$$P(p^\tau) = \{(n, k) \mid c_p(n, k) < \tau\}$$

The proof of Kummer's observation can be based on a formula by Legendre dating from 1808 which determines the largest exponent μ of the prime power p^μ which divides $n!$.

Kummer's Result and Legendre's Identity

We recall Kummer's theorem of 1852.

- Let $c_p(n,k)$ be the number of carries in the p-adic addition of n and k and $\tau = c_p(n,k)$. Then $\binom{n+k}{k}$ is divisible by the prime power p^τ but not by $p^{\tau+1}$.

In order to derive this beautiful result we will use a formula by Legendre from 1808 which deals with the divisibility of $n!$ by a prime power. The formula is as follows.

- Let $\mu(n)$ be the largest integer exponent of the prime power $p^{\mu(n)}$ which divides $n! = 1 \cdot 2 \cdots n$. Thus, $n!$ is divisible by $p^{\mu(n)}$ but not by $p^{\mu(n)+1}$. Then

$$\mu(n) = \frac{n-\sigma}{p-1} \,. \tag{9.7}$$

where σ is the sum of the p-adic coefficients $a_i \in \{0,1,...,p-1\}$ of n,

$$\begin{aligned} n &= a_0 + a_1 p + a_2 p^2 + \cdots + a_m p^m \\ \sigma &= a_0 + a_1 + \cdots + a_m \,. \end{aligned} \tag{9.8}$$

To show Legendre's formula we first establish the useful identity

$$\mu(n) = \sum_{i=1}^{\infty} \left\lfloor \frac{n}{p^i} \right\rfloor \,, \tag{9.9}$$

where the brackets $\lfloor\rfloor$ denote the greatest integer less than or equal to the enclosed quantity. For example, $\lfloor 6.1 \rfloor = 6$, $\lfloor 5.9 \rfloor = 5$, $\lfloor \pi \rfloor = 3$.

Note that $\lfloor n/p^i \rfloor = 0$ for large i. Thus, the sum in eqn. (9.9) is a finite sum. Let us first prove this identity (9.9). We observe that the term $\lfloor n/p^i \rfloor$ is the number of elements from $\{1,2,...,n\}$ which are divisible by p^i. For example, if $p=2$, $i=3$ and $n=17$, then $\lfloor n/p^i \rfloor = 2$. In other words, there are two integers less than or equal to 17 which are divisible by 2^3, namely 8 and 16. Next we observe that the sum in eqn. (9.9) counts any factor in the product

$$1 \cdot 2 \cdot 3 \cdots (n-1) \cdot n$$

which is divisible by p^i but not by p^{i+1} exactly i times, namely once in $\lfloor n/p \rfloor$, once in $\lfloor n/p^2 \rfloor$, ..., and once in $\lfloor n/p^i \rfloor$. This accounts for all occurrences of p as a factor of $n!$, i.e., identity (9.9) is established.

Here is an example: $n = 10$ and $p = 2$. Thus, we consider factors of 2 in the product $10! = 2 \cdot 3 \cdot 4 \cdot 5 \cdot 6 \cdot 7 \cdot 8 \cdot 9 \cdot 10$. Indeed, 2 is divisible by 2^1, 4 by 2^2, 6 by 2^1, 8 by 2^3 and 10 by 2^1. Thus, 10! is divisible by 2^8. This is shown in the following representation,

$$\begin{array}{cccccccccccccccccccc} 1 & \cdot & 2 & \cdot & 3 & \cdot & 4 & \cdot & 5 & \cdot & 6 & \cdot & 7 & \cdot & 8 & \cdot & 9 & \cdot & 10 & \\ & & 1 & & & & 1 & & & & 1 & & & & 1 & & & & 1 & \lfloor 10/2 \rfloor = 5 \\ & & & & & & 1 & & & & & & & & 1 & & & & & \lfloor 10/4 \rfloor = 2 \\ & & & & & & & & & & & & & & 1 & & & & & \lfloor 10/8 \rfloor = 1 \\ & & p^1 & & & & p^2 & & & & p^1 & & & & p^3 & & & & p^1 & \end{array}$$

showing the 8 occurrences of the factor $p = 2$. Thus, $\mu(10) = 8$. Observe that the sum

$$\sum_{i=1}^{\infty} \left\lfloor \frac{10}{2^i} \right\rfloor = \left\lfloor \frac{10}{2} \right\rfloor + \left\lfloor \frac{10}{4} \right\rfloor + \left\lfloor \frac{10}{8} \right\rfloor = 5 + 2 + 1 = 8$$

is just the number of these occurrences. Here is another example: $n = 1000$ and $p = 3$. $1000!$ is divisible by 3^{498} but not by 3^{499}, since

$$\begin{aligned}\mu(1000) &= \left\lfloor \frac{1000}{3} \right\rfloor + \left\lfloor \frac{1000}{9} \right\rfloor + \left\lfloor \frac{1000}{27} \right\rfloor \\ &\quad + \left\lfloor \frac{1000}{81} \right\rfloor + \left\lfloor \frac{1000}{243} \right\rfloor + \left\lfloor \frac{1000}{729} \right\rfloor \\ &= 333 + 111 + 37 + 12 + 4 + 1 = 498\ .\end{aligned}$$

Let us now establish Legendre's identity. By means of eqn. (9.9) Legendre's formula (9.7) is equivalent to

$$\sum_{i=1}^{\infty} \left\lfloor \frac{n}{p^i} \right\rfloor (p-1) = n - \sigma\ . \tag{9.10}$$

We proceed by showing this relation. Using the p-adic representation of n in (9.8) and the definition of the brackets $\lfloor\rfloor$ we get

$$\left\lfloor \frac{n}{p^i} \right\rfloor = a_i + a_{i+1}p + \cdots + a_m p^{m-i}\ , \quad i \le m\ .$$

With that we compute the two sums

$$\begin{aligned}\sum_{i=1}^{\infty} \left\lfloor \frac{n}{p^i} \right\rfloor p &= \sum_{i=1}^{m} \left(a_i p + a_{i+1}p^2 + \cdots + a_m p^{m-i+1}\right) \\ &= a_1 p + a_2 p^2 + a_3 p^3 + \cdots + a_m p^m \\ &\qquad + a_2 p + a_3 p^2 + \cdots + a_m p^{m-1} \\ &\qquad\qquad + a_3 p + \cdots + a_m p^{m-2} \\ &\qquad\qquad\qquad \vdots \\ &\qquad\qquad\qquad + a_m p\end{aligned}$$

$$\begin{aligned}\sum_{i=1}^{\infty} \left\lfloor \frac{n}{p^i} \right\rfloor &= \sum_{i=1}^{m} \left(a_i + a_{i+1}p + \cdots + a_m p^{m-i}\right) \\ &= a_1 + a_2 p + a_3 p^2 + \cdots + a_m p^{m-1} \\ &\qquad + a_2 + a_3 p + \cdots + a_m p^{m-2} \\ &\qquad\qquad + a_3 + \cdots + a_m p^{m-3} \\ &\qquad\qquad\qquad \vdots \\ &\qquad\qquad\qquad + a_m\ .\end{aligned}$$

The difference between the two sums is to be computed,

$$\begin{aligned}\sum_{i=1}^{\infty}\left\lfloor\frac{n}{p^i}\right\rfloor(p-1) &= (a_1p+a_2p^2+\cdots+a_mp^m)\\ &\quad -(a_1+a_2+\cdots+a_m)\\ &=(n-a_0)-(\sigma-a_0)\\ &=n-\sigma\ .\end{aligned}$$

This establishes Legendre's identity (9.7).

Now we derive Kummer's criterion from Legendre's identity. Thus, let the p-adic expansions of n and k be

$$\begin{aligned}n&=a_0+a_1p+a_2p^2+\cdots+a_mp^m,\\ k&=b_0+b_1p+b_2p^2+\cdots+b_mp^m\ .\end{aligned}$$

where $a_i, b_i \in \{0,1,...,p-1\}$. Now, if p^ν is the largest prime power of p which divides $\binom{n+k}{k}$ then

$$\nu=\mu(n+k)-\mu(n)-\mu(k),$$

since

$$\binom{n+k}{k}=\frac{(n+k)!}{n!k!}\ .$$

In other words, we have to show that

$$c_p(n,k)=\mu(n+k)-\mu(n)-\mu(k)\ , \tag{9.11}$$

where $c_p(n,k)$ is the number of carries in the p-adic addition of

$$\begin{aligned}n&=(a_ma_{m-1}...a_1a_0)_p \quad \text{and}\\ k&=(b_mb_{m-1}...b_1b_0)_p\ .\end{aligned}$$

Carrying out the addition of these two numbers in the p-adic representation produces carries $\varepsilon_0, \varepsilon_1, \varepsilon_2, ...$ which are either 0 or 1. Formally, they are obtained from

$$\varepsilon_0=\left\lfloor\frac{a_0+b_0}{p}\right\rfloor \quad \text{and} \quad \varepsilon_i=\left\lfloor\frac{a_i+b_i+\varepsilon_{i-1}}{p}\right\rfloor,\quad i=1,2,...$$

The sum of carries in Kummer's theorem is

$$c_p(n,k)=\sum_{i=0}^{\infty}\varepsilon_i\ .$$

Now we consider the sum $n+k$ in p-adic representation, i.e.,

$$n+k=\sum_{i=0}^{\infty}c_ip^i$$

where $c_i \in \{0, 1, ..., p-1\}$. If we define for convenience of notation $\varepsilon_{-1} = 0$ then we can express the p-adic digits c_i of $n + k$ in terms of those of n and k and the carries ε_i as follows.

$$c_i = a_i + b_i + \varepsilon_{i-1} - \varepsilon_i p \text{ for } i = 0, 1, 2, ...$$

Finally, we use Legendre's identity to show (9.11):

$$\begin{aligned}
\nu &= \mu(n+k) - \mu(n) - \mu(k) \\
&= \frac{n+k-\sum_{i=0}^{\infty} c_i}{p-1} - \frac{n-\sum_{i=0}^{\infty} a_i}{p-1} - \frac{k-\sum_{i=0}^{\infty} b_i}{p-1} \\
&= \frac{1}{p-1}\left(\sum_{i=0}^{\infty} a_i + \sum_{i=0}^{\infty} b_i - \sum_{i=0}^{\infty}(a_i + b_i + \varepsilon_{i-1} - \varepsilon_i p)\right) \\
&= \frac{1}{p-1}\sum_{i=0}^{\infty} \varepsilon_i p - \varepsilon_{i-1} \\
&= \frac{1}{p-1}\left(\sum_{i=0}^{\infty} \varepsilon_i p - \sum_{i=0}^{\infty} \varepsilon_{i-1}\right) \\
&= \frac{1}{p-1}\left(\sum_{i=0}^{\infty} \varepsilon_i p - \sum_{i=0}^{\infty} \varepsilon_i\right) \\
&= \frac{1}{p-1}\sum_{i=0}^{\infty} \varepsilon_i (p-1) \\
&= \sum_{i=0}^{\infty} \varepsilon_i \\
&= c_p(n,k) \ .
\end{aligned}$$

With this computation, Kummer's result is established.

Coloring Pascal's Triangle

So far we have only discussed whether a binomial coefficient is divisible by a prime number p or not, which we used for black and white coloring. However, if a coefficient is not divisible by p we could color the respective entry in Pascal's triangle with one of $p-1$ colors (depending on the modulus). Let us close this section with a short remark on the computation of the color. The crucial result which determines the color without having to run the associated linear cellular automaton is attributable to Lucas (1877). Let

$$\begin{aligned}
n+k &= a_0 + a_1 p + a_2 p^2 + \cdots + a_m p^m, \\
k &= b_0 + b_1 p + b_2 p^2 + \cdots + b_m p^m,
\end{aligned}$$

where $a_i, b_i \in \{0, ..., p-1\}$ are the p-adic digits. Then

$$\binom{n+k}{k} \equiv \binom{a_0}{b_0} \cdot \binom{a_1}{b_1} \cdots \binom{a_h}{b_h} \pmod{p} \tag{9.12}$$

Let us look at an example. According to the above criterion

$$\binom{7}{4} \equiv \binom{1}{1} \cdot \binom{2}{1} \pmod{3}$$

because $7 = (21)_3$ and $4 = (11)_3$. In fact,

$$\binom{7}{4} = 35 \equiv 2 \pmod{3}$$

and also

$$\binom{1}{1} \cdot \binom{2}{1} = 1 \cdot 2 \equiv 2 \pmod{3} .$$

Again, the criterion follows from Legendre's identity, as did Kummer's result.[23] We will, however, skip these details and turn now to the description of the global pattern formation in Pascal's triangle.

In summary, we use the mod-p condition to test a binomial coefficient for the divisibility by a prime number, and we resort to Lucas' factorization eqn. (9.12), when we want to know in addition what the value of the coefficient is in the modulo p sense.

[23]For a proof not using Legendre's identity, see also N. J. Fine: *Binomial coefficients modulo a prime number*, Amer. Math. Monthly 54 (1947) 589. Lucas' identity can also be used to analyze the global structure of the colored Pascal triangle, i.e., the color patterns which are obtained if one uses p colors, one for each modulus ($\equiv 0 \pmod{p}$, $\equiv 1 \pmod{p}$, ..., $\equiv p-1 \pmod{p}$). In fact Sved derived Lucas' result from the geometrical patterns of Pascal's triangle mod p. In other words, the fractal patterns in Pascal's triangle are equivalent to number theoretical properties of binomial coefficients, and understanding more about the fractal properties will lead to a wider understanding of these number theoretical properties.

9.3 IFS: From Local Divisibility to Global Geometry

You will recall the surprisingly short 'Program of the Chapter: Sierpinski Gasket by Binary Addresses' from *Fractals for the Classroom, Part One*. Its secret was hidden in just one BASIC statement:

```
IF (X AND Y) = 0 THEN PSET (30+X,Y+30)
```

The logical expression of this statement determined whether a point was drawn or not. The expression 'X AND Y' in the if clause stands for the bitwise logical AND operation. For example, 101 AND 010 $=$ 0 while 101 AND 110 $=$ 1. In other words, the expression is equal to 0 (false) only if no two matching binary digits of X and Y are both 1. Thus, this expression allows us to test for the occurrence of a carry in the binary addition of the coordinates X and Y. In other words this program uses the Kummer criterion for $p = 2$ and $c_2(n,k) = 0$ setting $n =$ X and $k =$ Y. Figure 9.20 gives an impression of the resulting pattern showing more and more details of the Sierpinski gasket. Let us now try to explain why this criterion really is able to generate the Sierpinski gasket.

Mod-2 Pattern and Binary Addresses

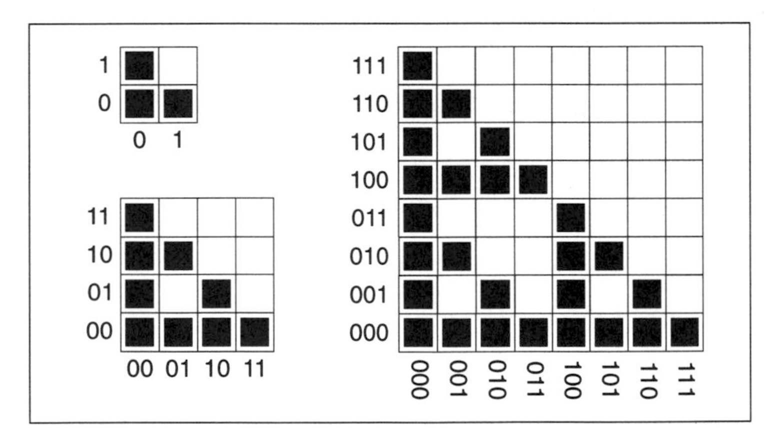

Figure 9.20 : Address testing on a 2 by 2, 4 by 4 and 8 by 8 grid. More and more details of the familiar Sierpinski gasket are revealed.

Sierpinski Gasket and Base 2

Let us consider the unit square in the plane,

$$Q = \{(x,y) \mid 0 \le x \le 1, 0 \le y \le 1\}.$$

Now we expand x and y in base 2, i.e.,

$$x = \sum_{i=1}^{\infty} a_i 2^{-i}, \quad a_i \in \{0,1\}$$

$$y = \sum_{i=1}^{\infty} b_i 2^{-i}, \quad b_i \in \{0,1\}\ .$$

With this notation we can provide a number theoretical description of the Sierpinski gasket:

$$S=\{(x,y)\in Q \mid \text{there are binary expansions of } x \text{ and } y \text{ with } a_i+b_i\le 1, i=1,2,...\} \quad (9.13)$$

Let us look at some examples:

(x,y)	(0,0)	(1,0)	(1/2,1/2)	(3/4,3/4)
x	0.000...	0.111...	0.1000...	0.11000...
y	0.000...	0.000...	0.0111...	0.10111...
	$\in S$	$\in S$	$\in S$	$\notin S$

The first three points, (0,0), (1,0), and (1/2,1/2) are in S. In the last example we also see why we have to say 'there is an expansion...' in the characterization of S. Otherwise we would not have that $(1/2,1/2)$ is in S. On the other hand $(3/4,3/4)$ is not in S, no matter how we expand, because a_1 and b_1 must both be 1, i.e., $a_1+b_1=2$.

Note that there is a direct relation to Kummer's carry-condition above. Indeed, saying that there is a base 2 expansion for x and y with $a_i+b_i\le 1$, is the same as saying that in adding x and y in the binary number system there is no carry.

In section 5.3 of *Fractals for the Classroom, Part One* we used iterated function systems (IFS) to convince ourselves that (9.13) indeed characterizes the Sierpinski gasket. We introduced four contractions w_{00}, w_{01}, w_{10}, and w_{11}, which contract the unit square Q as in figure 9.21 by a factor of 2:

$$\begin{aligned} w_{00}(x,y) &= (x/2, y/2) \\ w_{01}(x,y) &= (x/2, y/2+1/2) \\ w_{10}(x,y) &= (x/2+1/2, y/2) \\ w_{11}(x,y) &= (x/2+1/2, y/2+1/2) \ . \end{aligned}$$

Three of these transformations w_{00}, w_{01}, w_{10} provide a Hutchinson equation for the Sierpinski gasket S:

$$S = w_{00}(S) \cup w_{01}(S) \cup w_{10}(S) \ . \quad (9.14)$$

The Number Theoretical Description

We base the proof of the number theoretical description in eqn. (9.13) on the definition that the Sierpinski gasket is given by the contractions w_{00}, w_{01}, w_{10} and the corresponding Hutchinson equation (9.14). Any object (compact, nonempty set) which satisfies this equation must be the Sierpinski gasket because there is only one solution. Thus, to verify that S from (9.13) is the Sierpinski gasket, we must prove that S as in (9.13) satisfies (9.14). We proceed by showing the two relations

$$w_{00}(S) \cup w_{01}(S) \cup w_{10}(S) \subset S$$

Four Transformations

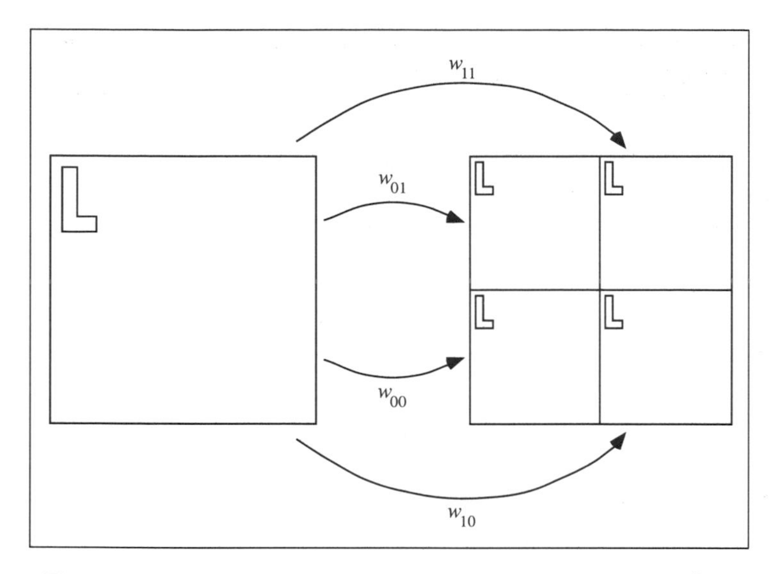

Figure 9.21 : Four similarity transformations of the square Q.

and

$$w_{00}(S) \cup w_{01}(S) \cup w_{10}(S) \supset S\ .$$

For the first, take any point $(x, y) \in S$ and its binary expansion

$$(x, y) = (0.a_1a_2..., 0.b_1b_2...)$$

Following eqn. (9.13), $a_i + b_i \leq 1$ holds for all indices $i = 1, 2, ...$ Now we apply the three transformations, w_{00}, w_{01}, and w_{10},

$$\begin{aligned} w_{00}(0.a_1a_2..., 0.b_1b_2...) &= (0.0a_1a_2..., 0.0b_1b_2...) \\ w_{01}(0.a_1a_2..., 0.b_1b_2...) &= (0.0a_1a_2..., 0.1b_1b_2...) \\ w_{10}(0.a_1a_2..., 0.b_1b_2...) &= (0.1a_1a_2..., 0.0b_1b_2...) \end{aligned}$$

Clearly, all three resulting point are also in S, because the first digits of the x- and y-components of the results are never both equal to 1, and for the remaining pairs of digits the same holds because $a_i + b_i \leq 1$ for $i = 1, 2, ...$

To show the second relation, we again take any point $(x, y) \in S$ with binary expansion as above and have to provide another point in $(x', y') \in S$ such that one of the images $w_{00}(x', y'), w_{01}(x', y')$, or $w_{10}(x', y')$ is equal to the given point (x, y). We may choose

$$(x', y') = (0.a_2a_3..., 0.b_2b_3...)$$

Note that a_1 and b_1 cannot both be equal to 1. Therefore, we immediately obtain

$$(x, y) = \begin{cases} w_{00}(x', y') \text{ if } a_1 = 0 \text{ and } b_1 = 0 \text{ or} \\ w_{01}(x', y') \text{ if } a_1 = 0 \text{ and } b_1 = 1 \text{ or} \\ w_{10}(x', y') \text{ if } a_1 = 1 \text{ and } b_1 = 0 \end{cases}$$

and there are no other cases. This concludes our proof, and, thus, (9.13) characterizes the Sierpinski gasket.

The binary representation also allows us to see how the iteration of the Hutchinson operator, applied to an arbitrary point in the square Q, yields a sequence of points that get closer and closer to the Sierpinski gasket. Observe that if $(x, y) = (0.a_1a_2..., 0.b_1b_2...)$, with arbitrary a_i, and b_i, then applying the maps w_{00}, w_{01}, and w_{10} again and again as in an IFS, yields points with coordinates for which more and more of the leading binary decimals satisfy $a_i + b_i \leq 1$. In other words, starting with

$$A_0 = Q$$

and then running the IFS, generates the sequence

$$A_n = w_{00}(A_{n-1}) \cup w_{01}(A_{n-1}) \cup w_{10}(A_{n-1}), n = 1, 2, ...,$$

where the coordinates of the points of A_n satisfy $a_i + b_i \leq 1$ in the leading n binary decimals. Furthermore, the sequence will lead towards the Sierpinski gasket as an attractor, i.e.,

$$A_\infty = S.$$

The first steps are shown in figure 9.22. Now observe that this would be exactly the result of figure 9.20 if the coordinates used in that figure had been preceded by a decimal point. In this case the patterns found on the 2 by 2, 4 by 4 or 8 by 8 grid would exactly match the steps A_n of our iterated function system. But introducing a decimal point in figure 9.20 simply means that we look at rescaled versions of Pascal's triangle (i.e., scaled by $1/2$, $1/4$ or $1/2^n$ in general). In other words, the mod-2 pattern which we see in Pascal's triangle is exactly the pattern which we obtain when iterating the IFS which encodes the Sierpinski gasket.

Divisibility by Primes

Now we are prepared to look at the patterns obtained from the divisibility of binominal coefficients by primes. Or more formally, we want to describe the global pattern formations in

$$P(p) = \left\{ (n, k) \;\middle|\; \binom{n+k}{k} \text{ not divisible by } p \right\},$$

First we construct an appropriate iterated function system. We consider the unit square Q and subdivide it into p^2 congruent squares $Q_{a,b}$ with $a, b \in \{0, ..., p-1\}$. Then we introduce corresponding contraction mappings

$$w_{a,b}(x, y) = \left(\frac{x+a}{p}, \frac{y+b}{p} \right),$$

Patterns of A_n

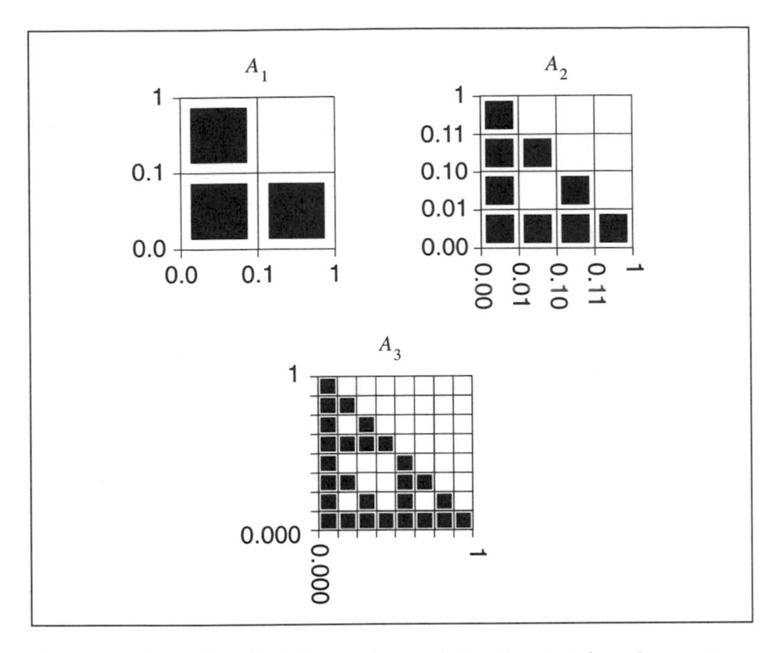

Figure 9.22 : The first three steps of the iterated function system coding the Sierpinski gasket.

where

$$w_{a,b}(Q) = Q_{a,b},\ a, b \in \{0, ..., p-1\} .$$

This is the generalization of what we have already done for the case $p = 2$ in figure 9.21. Now we define a set of admissible transformations by imposing the restriction

$$a + b \leq p - 1 .$$

This yields a total number of $N = p(p+1)/2$ contractions, each with contraction factor $1/p$. We now introduce the Hutchinson operator W_p corresponding to these N contractions,

$$W_p(A) = \bigcup_{a+b \leq p-1} w_{a,b}(A) ,$$

where A is any subset of the plane. With the initial set $A_0 = Q$ we can start the iteration

$$A_m = W_p(A_{m-1}),\ m = 1, 2, ...$$

and figure 9.23 shows the first two steps for the choice $p = 3$.

In order to keep track of the iteration, we subsequently subdivide each of the p^2 subsquares of Q into p^2 even smaller ones, and so on repeatedly. Having indexed the first subdivision of Q by

Mod 3 Machine

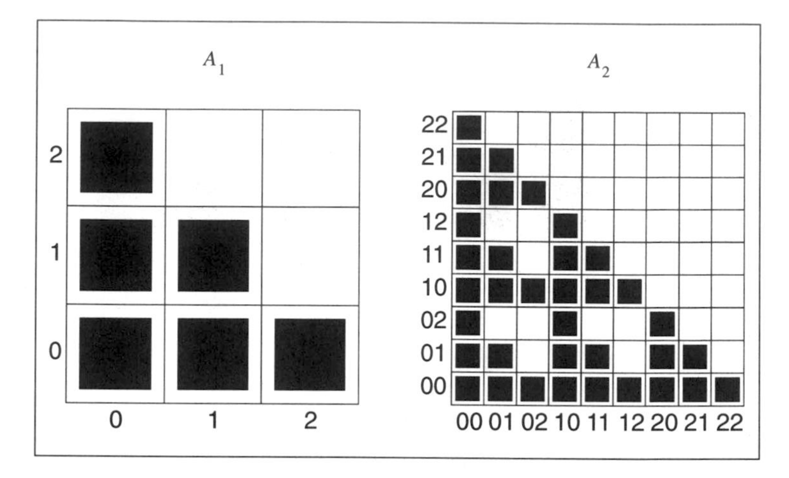

Figure 9.23 : First two steps of the iteration of the function system W_p for $p = 3$.

$Q_{a,b}$, we continue to label the subsquares of the second subdivision by $Q_{ac,bd}$ and so on. For the example $p = 3$ shown in figure 9.23, the square $Q_{10,12}$ are identified in the following way: the pair (1,1), made from the leading digits in the index of $Q_{10,12}$, determines the center square in the first subdivision, and the pair (0,2) determines the upper left corner square therein. In other words, the square

$$Q_{a_{m-1}\ldots a_0, b_{m-1}\ldots b_0}$$

is a square of the mth generation. We find it by reading the double p-adic addresses given by the pair $(a_{m-1}\ldots a_0, b_{m-1}\ldots b_0)$. This natural addressing system helps us to keep track of all the iterations of W_p, the Hutchinson operator. For example, we obtain

$$Q_{a_{m-1}\ldots a_0, b_{m-1}\ldots b_0} = w_{a_{m-1},b_{m-1}}(\cdots w_{a_0,b_0}(Q)) ,$$

where $a_i + b_i \leq p - 1$.

In other words, we now can say that A_m is the collection of all those squares of the m-th subdivision of Q into p^{2m} little squares, whose addresses $(a_{m-1}\ldots a_0, b_{m-1}\ldots b_0)$ satisfy the condition $a_i + b_i \leq p - 1$, i.e.,

$$A_m = \bigcup_{a_i + b_i \leq p-1} Q_{a_{m-1}\ldots a_0, b_{m-1}\ldots b_0} .$$

Rescaling the Pascal Triangle

Let us now relate the subsquares $Q_{a_{m-1}\ldots a_0, b_{m-1}\ldots b_0}$ to the entries of the Pascal triangle. First we generate a geometric model of the divisibility pattern in the Pascal triangle. To this end we equip the first quadrant of the plane with a square lattice so that

The $\mathcal{P}_m(3)$ Subsquares

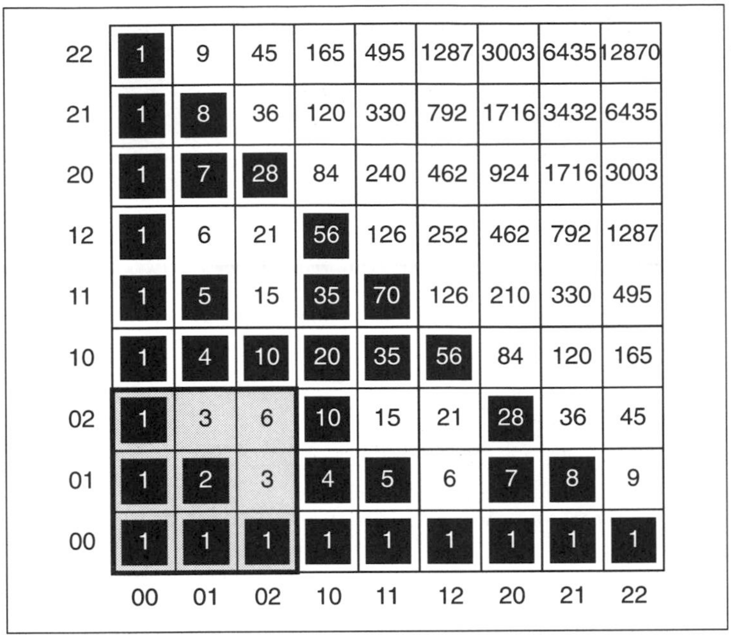

22	1	9	45	165	495	1287	3003	6435	12870
21	1	8	36	120	330	792	1716	3432	6435
20	1	7	28	84	240	462	924	1716	3003
12	1	6	21	56	126	252	462	792	1287
11	1	5	15	35	70	126	210	330	495
10	1	4	10	20	35	56	84	120	165
02	1	3	6	10	15	21	28	36	45
01	1	2	3	4	5	6	7	8	9
00	1	1	1	1	1	1	1	1	1
	00	01	02	10	11	12	20	21	22

Figure 9.24 : The squares $\mathcal{P}_1(3)$ (the six black squares with grey underlay in the lower left hand group) and $\mathcal{P}_2(3)$ (all black squares). Compare with figure 9.23.

each square has side length 1. Thus, each square is indexed by an integer pair (n, k) and we call it $R_{n,k}$.

$$R_{n,k} = \{(x, y) \mid n \leq x \leq n+1, k \leq y \leq k+1\} \ .$$

The geometrical model of $P(p)$ will now be obtained by selecting all squares $R_{n,k}$ for which p does not divide $\binom{n+k}{k}$:

$$\mathcal{P}(p) = \left\{ R_{n,k} \;\middle|\; \binom{n+k}{k} \text{ is not divisible by } p \right\} \ .$$

We will now relate this infinite pattern to the evolution of the Hutchinson operator, i.e., to the sequence of patterns A_m. Note that all A_m are within Q and that A_m is a union of a finite number of squares of side length $1/p^m$. To see the relation between A_m and $\mathcal{P}(p)$ we will look at $\mathcal{P}(p)$ through a sequence of square 'windows' $[0, p^m] \times [0, p^m]$ of side length p^m. Now for $m = 1, 2, ...$ we pick that part from the geometrical model $\mathcal{P}(p)$ which falls in the corresponding window:

$$\mathcal{P}_m(p) = \mathcal{P}(p) \cap [0, p^m] \times [0, p^m] \ .$$

Figure 9.24 displays $\mathcal{P}_1(p)$ and $\mathcal{P}_2(p)$ for $p = 3$. Comparing $\mathcal{P}_1(p)$ and $\mathcal{P}_2(p)$ with the pattern of A_1 and A_2 in figure 9.23 we observe that they are identical, though A_1 and A_2 are in the unit square and $\mathcal{P}_1(p)$ (resp. $\mathcal{P}_2(p)$) fit into a square of side length p (resp. p^2). In other words, if we rescale the patterns $\mathcal{P}_m(p)$ by a factor of $1/p^m$ we obtain an object which we want to show is identical with A_m. To this end we introduce

$$\mathcal{S}_m(p) = \frac{1}{p^m} \cdot \mathcal{P}_m(p)$$

or more explicitly

$$\mathcal{S}_m(p) = \left\{ \frac{z}{p^m} \;\middle|\; z \in \mathcal{P}_m(p) \right\} .$$

Indeed, each subsquare in $\mathcal{S}_m(p)$ is indexed by an integer pair (n, k) such that p does not divide $\binom{n+k}{k}$. In other words each such subsquare is identical with a $Q_{a_{m-1}...a_0,b_{m-1}...b_0}$, where $n = (a_{m-1}...a_0)_p$ and $k = (b_{m-1}...b_0)_p$ according to Kummer's mod-p condition. Summarizing we have that

$$A_m = \mathcal{S}_m(p), \quad m = 1, 2, ...$$

As we let m go to infinity, we know that A_m will converge towards the attractor of the IFS, and consequently the rescaled geometric models $\mathcal{S}_m(p)$ will also converge to the attractor of the IFS.[24] We denote the limit set by $\mathcal{S}(p)$. In this manner we have just seen that the rescaled geometric models have a limit set. It represents a rescaled geometric model of the (infinite) Pascal triangle modulo p which we denoted by $P(p)$.[25]

Fractal Dimension

Figure 9.25 shows the resulting geometric models $\mathcal{S}(p)$ when running the IFSs corresponding to $P(p)$ for $p = 2, 3, 5$. The approach by iterated function systems allows us to compute the fractal dimensions of these sets:

	Dimension
$\mathcal{S}(2)$	log 3 / log 2≈1.585
$\mathcal{S}(3)$	log 6 / log 3≈1.631
$\mathcal{S}(5)$	log 15 / log 5≈1.683
$\mathcal{S}(7)$	log 28 / log 7≈1.712

If p is prime then the formula for the self-similarity dimension of $\mathcal{S}(p)$ is

$$D_s = \frac{\log \frac{p(p+1)}{2}}{\log p} .$$

[24] Convergence is with respect to the Hausdorff metric.

[25] In this regard we also refer to S. J. Willson, *Cellular automata can generate fractals,* Discrete Applied Math. 8 (1984) 91–99. who studied limit sets of linear cellular automata via rescaling techniques.

The First Primes

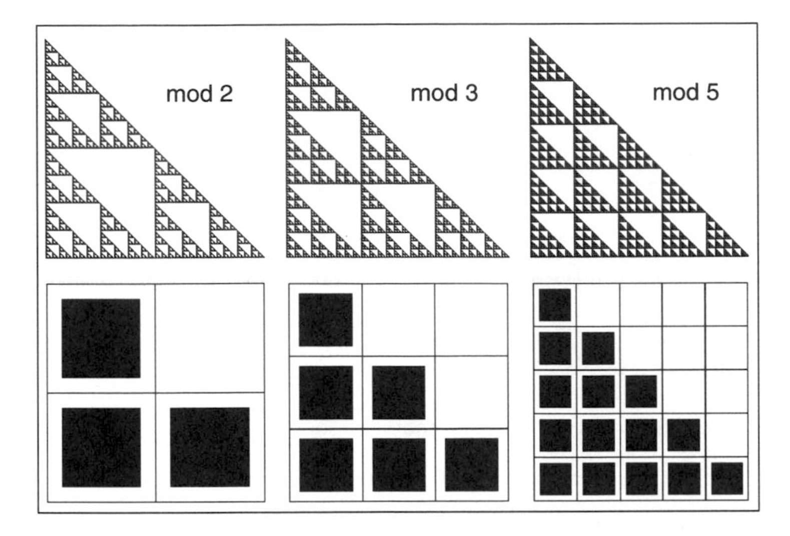

Figure 9.25 : The limit sets of the rescaled geometric models of $P(2)$, $P(3)$, $P(5)$, and their associated IFSs (i.e., a graphical representation of the transformations w_{ab}.

Can you explain why? Just note, that these sets are strictly self-similar and recall the definition of the self-similarity dimension (*Fractals for the Classroom, Part One*, chapter 4).

Divisibility by Prime Powers

We have seen that iterated function systems are in some sense the natural framework in which to decipher the global pattern formation obtained by the divisibility properties in Pascal's triangle (or of pattern formations in linear cellular automata). We have, however, only taken the first step, namely with respect to the divisibility by a prime p. Our next step, considering divisibility by prime powers p^τ, is rather long compared with the first one. In fact, the global patterns arising in Pascal's triangle[26] can be completely understood through hierarchical IFSs.[27] The fractal patterns in

$$P(p^\tau) = \left\{ (n,k) \;\middle|\; \binom{n+k}{k} \text{ is not divisible by } p^\tau \right\} .$$

can be deciphered by a hierarchical IFS whose design and properties are in very close correspondence with Kummer's criterion for finding the largest prime power which divides $\binom{n+k}{k}$. Figure 9.26 shows two examples where we observe that the straight forward self-similarity properties of the sets $P(p)$ have been replaced by hierarchies of self-similarity features.

[26]Sketching some recent work from F. v. Haeseler, H.-O. Peitgen, G. Skordev, *Pascal's triangle, dynamical systems and attractors,* to appear in Ergodic Theory and Dynamical Systems.

[27]See section 5.8 in *Fractals for the Classroom, Part One*.

First Prime Powers

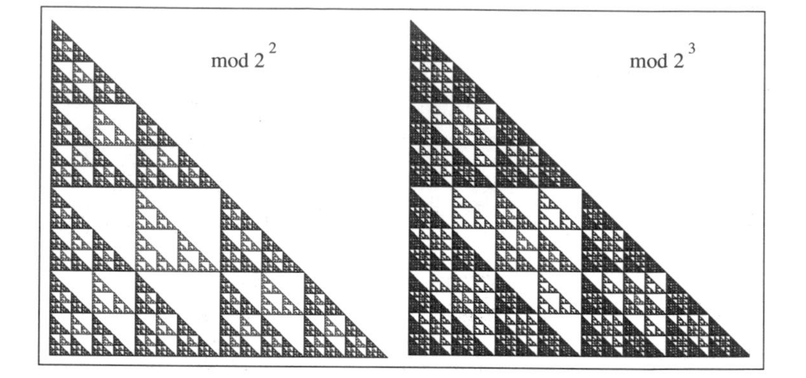

Figure 9.26 : Rescaled geometric models of $P(4)$ and $P(9)$.

The construction of an hierarchical IFS for general prime powers p^μ requires capturing the mechanism of carry production when adding p-adic numbers. In general this results in a rather technical description. Therefore, we restrict ourself to the case $p^\mu = 2^2 = 4$ which already gives quite a good impression. In this case Kummer's criterion implies:

$$P(4) = \{(n,k) \mid c_2(n,k) = 0 \text{ or } c_2(n,k) = 1\} \ .$$

If $c_2(n,k) = 0$, then $\binom{n+k}{k}$ is not divisible by 2. If $c_2(n,k) = 1$, then $\binom{n+k}{k}$ is divisible by 2, but not by 4. Thus, if either one of the conditions is satisfied, then $\binom{n+k}{k}$ is not divisible by 4. Therefore, we also have to take into account those coordinates (n,k) whose p-adic addition have exactly one carry (i.e., where $\binom{n+k}{k}$ is divisible by 2 but not by 4). This can be captured by the rather simple hierarchical IFS which is shown in figure 9.27. Figures 9.29 and 9.28 show the evolving patterns at the three nodes of this hierarchical IFS. Observe how the original Sierpinski gasket is merged into the full pattern of $P(4)$. Our result once again shows very strongly that hierarchical IFSs are not just there to make pretty pictures. They are deeply rooted in pure mathematics. They appear here to be entirely natural for the explanation of the geometrical patterns discussed in the Pascal triangle.

The case $P(4)$

Again we look at the unit square Q and the contractions $w_{00}, \ldots, w_{11}$ of figure 9.21. We have seen that m iterations of these transformations, first applying $w_{a_0 b_0}$ then $w_{a_1 b_1}$, etc. lead to the subsquare

$$w_{a_{m-1} b_{m-1}}(\cdots w_{a_0 b_0}(Q)) = Q_{a_{m-1} \ldots a_0, b_{m-1} \ldots b_0} .$$

So far we have only considered the case where the binary addition of $(a_{m-1} \ldots a_0)_2$ and $(b_{m-1} \ldots b_0)_2$ had no carry (i.e., $a_i + b_i \leq 1$). Now,

HIFS for mod 4

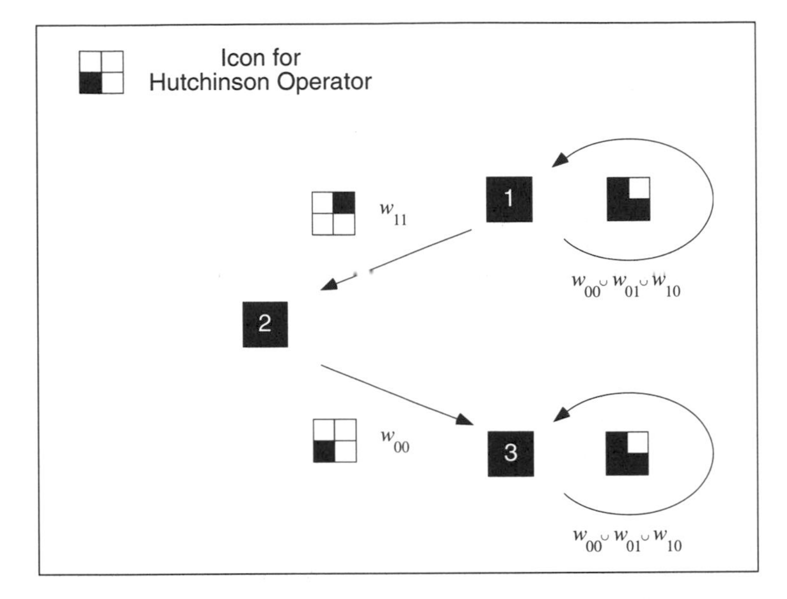

Figure 9.27 : The graph of a hierarchical iterated function system for the mod 4 example. Iterating this system generates 3 images, one for each node (the nodes representing three networked MR-CMs). The final image of node 1 is a Sierpinski gasket. The final image of node 3 is the desired mod 4 pattern.

Powers of Two

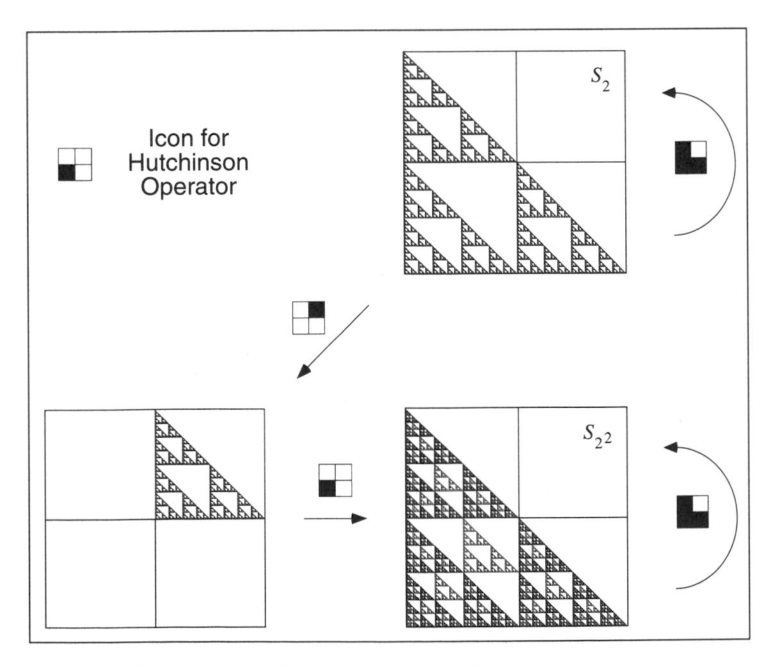

Figure 9.28 : The IFS network for powers of 2.

how can we achieve exactly one carry? At first glance it appears that all we have to use is the transformation w_{11}. For example, applying the sequence $w_{01}, w_{10}, w_{11}, w_{01}, w_{00}$ would lead to

$$Q_{00110,01101} = w_{00}(w_{01}(w_{11}(w_{10}(w_{01}(Q))))).$$

And indeed in this example the addition would provide a carry in the 3rd binary decimal. But wait; there is also a carry in the 4th binary decimal counted from the right! What is wrong? Obviously we have to be a bit more careful. The transformation w_{11} provides a carry — so far okay — but the transformation which follows has to be w_{00}. Otherwise we would obtain another carry. Therefore, having w_{11}, followed by w_{01}, as in our example, is not allowed.

Figure 9.27 shows the graph of a hierarchical iterated function system which reflects our observations. The nodes 1, 2 and 3 represent three networked MRCMs. The first one operates in a feedback loop applying the Hutchinson operator $w_{00} \cup w_{01} \cup w_{10}$. The second one transforms the output of the first one using w_{11}. The third machine again operates in a feedback loop with $w_{00} \cup w_{01} \cup w_{10}$, but additionally it merges the output of the second machine, transformed by w_{00}.

How do we iterate this network? We start the iteration with 3 copies of the unit square Q, one for each node, $A_0(1) = Q$, $A_0(2) = Q$, and $A_0(3) = Q$. The first step provides for these nodes:

$$\begin{aligned}
A_1(1) &= w_{00}(Q) \cup w_{01}(Q) \cup w_{10}(Q) \\
&= Q_{0,0} \cup Q_{0,1} \cup Q_{1,0} \\
A_1(2) &= w_{11}(Q) \\
&= Q_{1,1} \\
A_1(3) &= w_{00}(Q) \cup w_{00}(Q) \cup w_{01}(Q) \cup w_{10}(Q) \\
&= Q_{0,0} \cup Q_{0,1} \cup Q_{1,0}
\end{aligned}$$

and for the second step we obtain

$$\begin{aligned}
A_2(1) &= \bigcup_{a+b\leq 1} w_{a,b}(Q_{0,0} \cup Q_{0,1} \cup Q_{1,0}) \\
&= Q_{00,00} \cup Q_{00,01} \cup Q_{01,00} \cup \\
&\quad\; Q_{00,10} \cup Q_{00,11} \cup Q_{01,10} \cup \\
&\quad\; Q_{10,00} \cup Q_{10,01} \cup Q_{11,00} \\
A_2(2) &= w_{1,1}(Q_{0,0} \cup Q_{0,1} \cup Q_{1,0}) \\
&= Q_{10,10} \cup Q_{10,11} \cup Q_{11,10} \\
A_2(3) &= \bigcup_{a+b\leq 1} w_{a,b}(Q_{0,0} \cup Q_{0,1} \cup Q_{1,0}) \cup w_{0,0}(Q_{1,1}) \\
&= Q_{00,00} \cup Q_{00,01} \cup Q_{01,00} \cup \\
&\quad\; Q_{00,10} \cup Q_{00,11} \cup Q_{01,10} \cup \\
&\quad\; Q_{10,00} \cup Q_{10,01} \cup Q_{11,00} \cup Q_{01,01}
\end{aligned}$$

These steps and the next are visualized in figure 9.29. In step m we obtain in node 1 all subsquares $Q_{a_{m-1}\ldots a_0, b_{m-1}\ldots b_0}$ whose

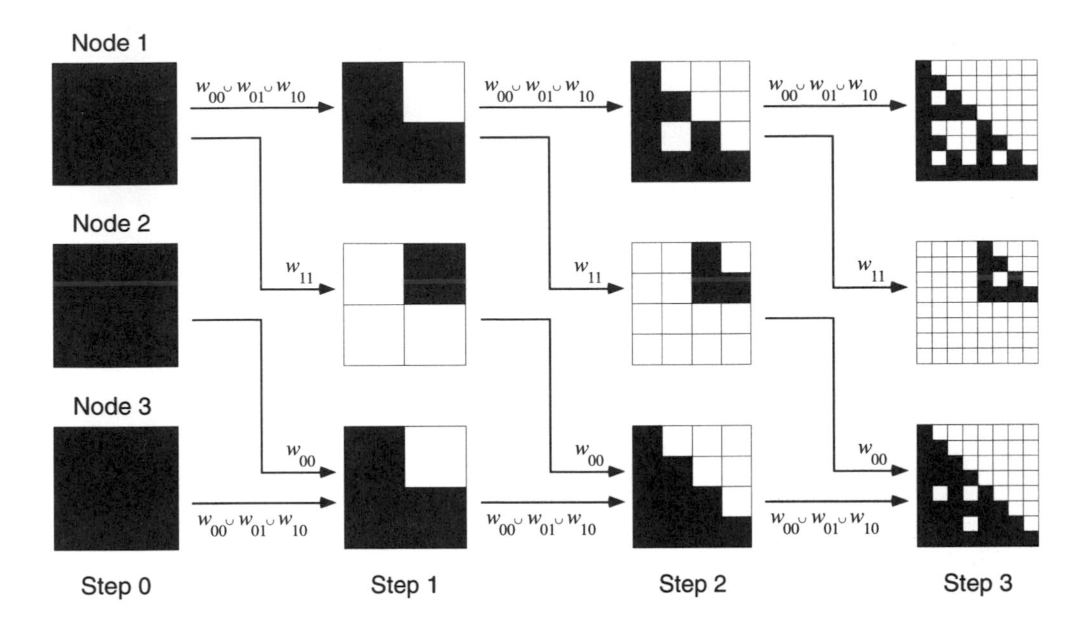

Figure 9.29 : The first three steps of the networked MRCM for $P(4)$ shown in figure 9.27.

indices produce no carry, which yields the rescaled geometric model $\mathcal{S}_m(2)$. For node 2 we obtain subsquares with the carry produced by the leading binary decimal of its indices, and for node 3 we obtain subsquares whose indices provide no carry or just one. In other words, $A_m(3)$ provides the desired geometric model of $P(4)$ after rescaling by $1/2^m$.

9.4 Catalytic Converters or how many Cells are Black?

Pascal's triangle has been in existence for many centuries and has inspired beautiful investigations. We have seen the first step of how it has laid the foundation for the understanding of pattern formation for linear cellular automata in one dimension.[28] But it has also recently sparked the investigation of a problem which at first glance seems to have no relation to the triangle at all.

Assume we were to play darts with a large Pascal triangle as a target. What are the probabilities that we hit a black cell or a white one, or more precisely, an odd or an even number, or a number which is divisible by 3 or one which is not, or a number which is divisible by p^τ or one which is not? Our discussion of the global patterns in the Pascal triangle makes it possible to answer such questions. We just have to evaluate the corresponding areas in the structures corresponding to the rescaled geometric models of $P(p^\tau)$. Depending on the parameters, this may turn out to be a rather technical computation.

Pascal Dart

Figure 9.30 : What is the probability of hitting a point of the modulo 2 pattern in the Pascal triangle?

Number of Black Cells in a Row

Let us look at a related question which allows a more immediate answer. We again use the original coordinate system for the Pascal triangle (see the left option of figure 9.17). How many black cells are there in the rth row? In other words, how many of the numbers which appear in the rth row are not divisible by 2, or 3, or 5, or any other prime number p? There is a remarkably direct procedure to arrive at the result. First, take r and expand it with respect to

[28]For a discussion of higher dimensions see F. v. Haeseler, H.-O. Peitgen, G. Skordev, *On the hierarchical and global structure of cellular automata and attractors of dynamical systems,* to appear.

base p,

$$r = c_0 + c_1 p + c_2 p^2 + \cdots + c_m p^m, \quad c_i \in \{0, ..., p-1\} .$$

Now let $h_p(r)$ be defined by

$$h_p(r) = \prod_{i=0}^{m} (c_i + 1) = (c_0 + 1) \cdot (c_1 + 1) \cdots (c_m + 1) .$$

Then $h_p(r)$ is the number of entries in the rth row of the Pascal triangle which are not divisible by p.

Determining the Count h_p

Let us give an argument using again the modified coordinate system as in figure 9.17 (right). Here the rth row is characterized by $n+k = r$. Thus, we are asking for the cardinality of

$$\left\{ (n,k) \;\middle|\; n + k = r \text{ and } \binom{n+k}{k} \text{ is not divisible by } p \right\} .$$

Consider the p-adic representations of n, k, and r,

$$\begin{aligned} n &= a_m p^m + \cdots + a_1 p + a_0, \quad a_i \in \{0, ..., p-1\} \\ k &= b_m p^m + \cdots + b_1 p + b_0, \quad b_i \in \{0, ..., p-1\} \\ r &= c_m p^m + \cdots + c_1 p + c_0, \quad c_i \in \{0, ..., p-1\} . \end{aligned}$$

According to Kummer's mod-p criterion, $\binom{n+k}{k}$ is not divisible by p if and only if $a_i + b_i \leq p - 1$ for all $i = 0, ..., m$. In this case there is no carry in the p-adic addition of n and k. Thus, the coefficients of the sum $r = n + k$ must satisfy

$$c_i = a_i + b_i, \text{ where } i = 0, ..., m .$$

How many choices of n are there such that this condition is satisfied? For the ith p-adic digit there are $c_i + 1$ such choices, namely $a_1 = 0, ..., c_i$. Thus, the total number of possible choices is the product

$$h_p(n) = \prod_{i=0}^{m} (c_i + 1) .$$

Exactly that many entries (n, k) in row r have the property that $\binom{n+k}{k}$ is not divisible by p.

Towards Catalytic Converters

Figure 9.31 shows $h_p(r)$ for $p = 2$ as a function of r. Now imagine this graph flipped over and compare with figure 9.32, which shows the measurement of the chemical reaction rate as a function of time in a catalytic oxidation process. The remarkable resemblance of the flipped over graph and this kind of measurements provided the motivation to model a catalytic converter by

Number of Entries

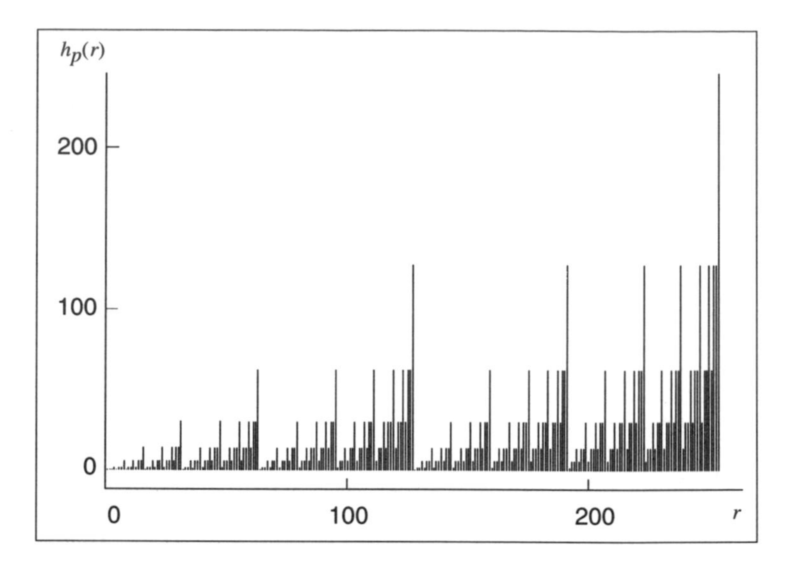

Figure 9.31 : The number of entries in the rth row of the Pascal triangle.

Reaction Rate Measurement

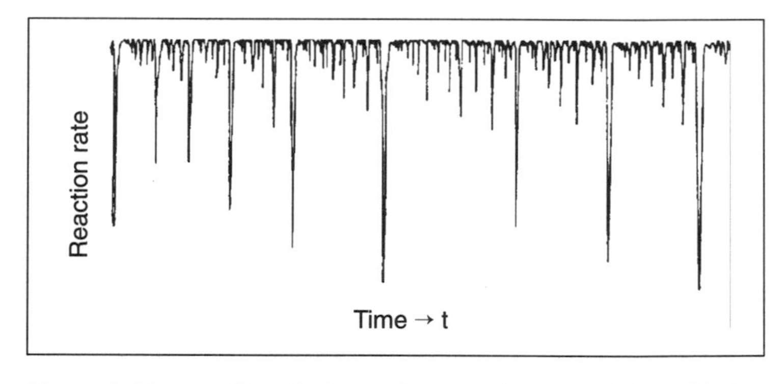

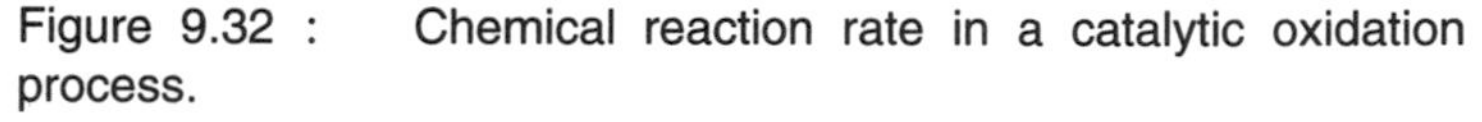

Figure 9.32 : Chemical reaction rate in a catalytic oxidation process.

one-dimensional cellular automata.[29] Thus, we are back to the relation of cellular automata and polynomials which we discussed at the beginning of this chapter. This relation allows us to interpret $h_p(r)$ as a count of 'oxidized' cells in an appropriate cellular automaton. In this sense our discussion has provided a first glimpse of an idea of why modeling a catalytic converter by cellular automata could be a successful approach and exhibits the qualitative behavior found in real chemical experiments.

[29] A. W. M. Dress, M. Gerhardt, N. I. Jaeger, P. J. Plath, H. Schuster, *Some proposals concerning the mathematical modelling of oscillating heterogeneous catalytic reactions on metal surfaces.* In L. Rensing and N. I. Jaeger (eds.), Temporal Order, Springer-Verlag, Berlin, 1984.

9.5 Program of the Chapter: Cellular Automata

One-dimensional linear cellular automata have been the focus of this chapter. We now present a program which allows you to study such automata. For example, it can generate the divisibility patterns of Pascal's triangle. Two approaches have been included which either use look-up tables to determine the change of cell states or directly operate with mod-n arithmetic. When running the program you are asked to enter a number n. If the number is equal to 1, the program uses the look-up tables when generating cell layer by cell layer. Otherwise it computes divisibility patterns for the specified number n.

Screen Image of Program

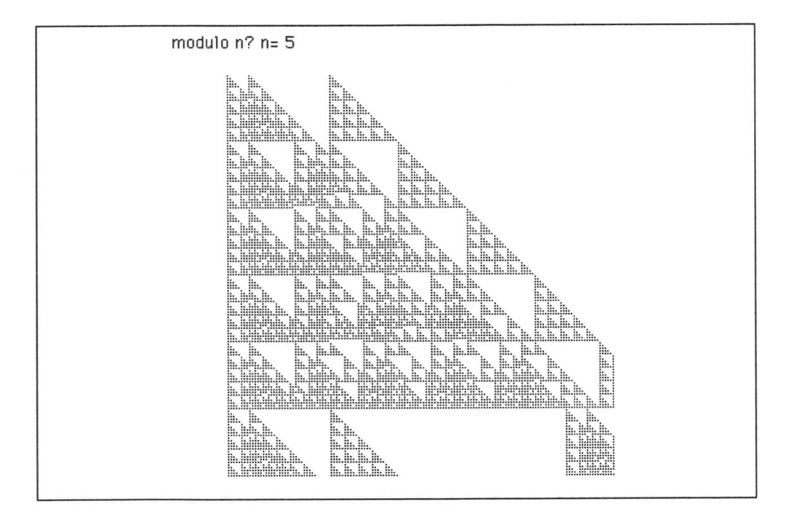

Figure 9.33 : Output of the program 'Cellular Automata'.

At the beginning of the program, the initial state of the one-dimensional string of cells is defined. It is stored in the array `c`. You can change this definition at your convenience. However, if you want to check the mod-n patterns of Pascal's triangle as demonstrated in this chapter you should set only one cell to the initial state 1, for example, `c(10) = 1`.

In the following part of the program, the state of the cells is changed according to the mod-n addition rule. Observe that we do *not* simply use the assignment

```
c(x) = (c(x-1)+c(x)) MOD n.
```

Why would this be a mistake? We have to determine the new cell state at position x from the old cell states at x and the left neighbor at $x - 1$. But since we use the array `c` for storing the old and the new state of cells, `c(x-1)` would already refer to the new state of the left neighboring cell. We therefore introduce a variable `c0`

BASIC Programm **Cellular Automata**
Title Iteration of linear cellular automata

```
DIM c(150), lut(3)
left = 30
w = 300

REM Initialize and plot first cell layer
FOR x = 1 TO w/2
    c(x) = 0
NEXT x
c(7) = 1 : c(15) =1 : c(45) = 1
FOR x = 1 TO w/2
    IF c(x) <> 0 THEN PSET (2*x+left, 2+left)
NEXT x

INPUT "modulo n? n=", n
IF n < 2 GOTO 100

    REM Compute with modulo arithmetic and draw layer by layer
    FOR y = 2 TO w/2
        c0 = 0
        FOR x = 1 TO w/2
            c1 = c(x)
            c(x) = (c0 + c1) MOD n
            c0 = c1
            IF c(x) <> 0 THEN PSET (2*x+left, 2*y+left)
        NEXT x
    NEXT y
    END

    REM Compute by using look-up table and draw layer by layer
100 lut(0) = 1: lut(1) = 0
    lut(2) = 0: lut(3) = 1
    FOR y = 2 TO w/2
        c0 = 0
        FOR x = 1 TO w/2
            c1 = c(x)
            c(x) = lut(2*c0 + c1)
            c0 = c1
            IF c(x) <> 0 THEN PSET (2*x+left, 2*y+left)
        NEXT x
    NEXT y
    END
```

which holds the old state of the left neighbor. In addition, another variable, `c1`, stores the value of the cell under consideration.

In the last part of the program we use look-up tables. First we define a table `lut` with 4 entries for the 4 combinations of states `c0` and `c1` of two cells.

c0	c1	2*c0+c1	Result
0	0	0	lut(0)
0	1	1	lut(1)
1	0	2	lut(2)
1	1	3	lut(3)

For the 4 possible configurations of the cell and its left neighbor the look-up table determines the appropriate new state of the cell. For example, `lut(3) = 1` means that the new state is 1 if the old state `c1` of the cell and the old state of the left neighbor `c0` are both 1. The table defined in the program is one choice of 16 possible ones. If you try some of the others, you will notice that only very few choices generate interesting patterns. We therefore suggest that you try to extend the program such that the look-up table approach allows more than just two cell states.

For example, four different states would require the definition of a 4 by 4 look-up table. Change the dimension statement to

```
DIM c(150), lut(15)
```

and assign values from 0 to 3 corresponding to 4 possible cell states in `c0` and `c1` to all 16 elements `lut(0)` to `lut(15)`. For states `c0` and `c1` the new state for `c1` would be given by `lut(4*c0+c1)` (change the factor 2 to 4 in the corresponding statement `c(x) = lut(2*c0+ c1)`). After these changes you can choose from $4^{16} \approx 5 \times 10^9$ different ways of defining the table.

Chapter 10

Deterministic Chaos: Sensitivity, Mixing, and Periodic Points

A dictionary definition of chaos is a 'disordered state of collection; a confused mixture'. This is an accurate description of dynamical systems theory today — or of any other lively field of research.
Morris W. Hirsch[1]

Mathematical research in chaos can be traced back at least to 1890, when Henri Poincaré studied the stability of the solar system. He asked if the planets would continue on indefinitely in roughly their present orbits, or might one of them wander off into eternal darkness or crash into the sun. He did not find an answer to his question, but he did create a new analytical method, the geometry of dynamics. Today his ideas have grown into the subject called topology, which is the geometry of continuous deformation. Poincaré made the first discovery of chaos in the orbital motion of three bodies which mutually exert gravitational forces on each other.

Others followed Poincaré's pioneering trail. In the former Soviet Union, for example, the mathematician Andrey Kolmogorow made basic advances in the irregular features of dynamics. By the 1960's, the American mathematician Stephen Smale had formulated a plan to classify all the typical kinds of dynamic behavior. Within Smale's world view, chaos found a place as a natural phenomenon completely on a par with such regular behavior as periodic cycles.

Practical applications of the concept of chaos as a natural

[1] In: *Chaos, Fractals, and Dynamics,* P. Fischer, W. R. Smith (eds.), Marcel Dekker, Inc., New York, 1985.

phenomenon followed. For example, sometimes a fluid flows smoothly, but sometimes it becomes turbulent and irregular for no apparent reason. In attempting to explain why, two European mathematicians, David Ruelle and Floris Takens, suggested in 1970 that turbulent flow might be an example of dynamic chaos. At about the same time, chaos began to get attention in the sciences. Experimental scientists, notably the American physicists Harry Swinney and Jerry Golub and the French physicist Albert Libchaber, showed that Ruelle and Takens were partly right. Chaos does occur in turbulent flow but not in precisely the way they suggested.

This work raised important questions. How can chaotic models be tested experimentally, and how can different types of chaos be distinguished? The usual technique for testing a theory is to make a long series of observations and compare the results with the theoretical predictions. With chaos, however, the butterfly effect invalidates the results; they will vary widely because of even the slightest errors in the observations.

Following up on these questions, here are some which we would like to settle in this chapter. How can we make the notion of chaos more precise? How can we be sure that what looks like chaos is really chaotic and not just very complicated but perfectly predictable? For example, when we see a seemingly chaotic time series, how can we be sure that it is not periodic, just with an extremely long period? In other words, what are the signs of chaos? How can they be measured? What is the value of numerical calculations in the presence of chaos? How can we build examples of chaos which can be intuitively understood?

There are many dynamical systems that can produce chaos. However, in this chapter and the next the focus of our presentation is the iteration of only one particular transformation. It is the quadratic transformation which comes in different forms, for example, $x \to ax(1-x)$. This may seem like a rather artificial choice which may not bear much significance for the many other chaotic systems studied in mathematics and observed in physical experiments. However, on the contrary, it has turned out that the qualitative phenomena of the quadratic transformation are in fact the paradigm of chaos in dynamical systems. Moreover, for the quadratic transformation the properties of chaos can be observed and completely analyzed mathematically. Thus, their study is time well spent and later, in chapter 12, we will see how the quadratic transformation reappears as the basic chaos generator in other systems.

10.1 The Signs of Chaos: Sensitivity

Sensitivity Versus Stability — The Main Issue

In Chapter 1 we experienced a big surprise. The computer, paradigm of reliability and precision was knocked out by a simple feedback process, the quadratic iterator. The greatest problem that computers are confronted with when dealing with chaos is the extreme sensitivity of an iterator. It is reflected in a host of unexpected numerical phenomena which lead us to the conclusion that we must be very careful with the interpretation of the computer output. We will now reveal some of the difficulties that the machine cannot cope with in the presence of chaos.

First Time Series

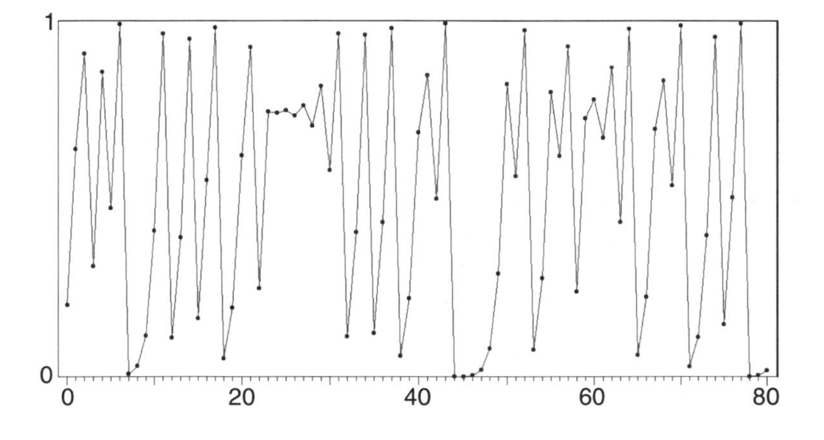

Figure 10.1 : Time series for the quadratic iterator starting with $x_0 = 0.2027$ and parameter $a = 4$.

We have seen three different versions of the quadratic iterator: $x \to x^2 + c$ (for $c = -2$), $p \to p + rp(1-p)$ (for $r = 3$) and $x \to ax(1-x)$ (for $a = 4$). In this chapter we will base our discussion mainly on the quadratic iterator of the form $x \to ax(1-x)$. We start our tour with a couple of very simple time series experiments. Figure 10.1 shows the computed time series of x-values starting at $x_0 = 0.2027$ with the parameter set at $a = 4$. This is called the *orbit* of x_0. On the horizontal axis the number of iterations ('time') is marked, while on the vertical axis the amplitudes (ranging from 0 to 1) are given for each iteration. The points are connected by line segments. The picture shows an irregular pattern much like stock market indices, which are difficult to predict. The only thing that one can be sure of seems to be the fact that the graph cannot escape the bounds 0 and 1. This is clear from the formula $ax(1-x)$: when x is between 0 and 1, then so is $ax(1-x)$ when $0 \leq a \leq 4$.

Second Time Series

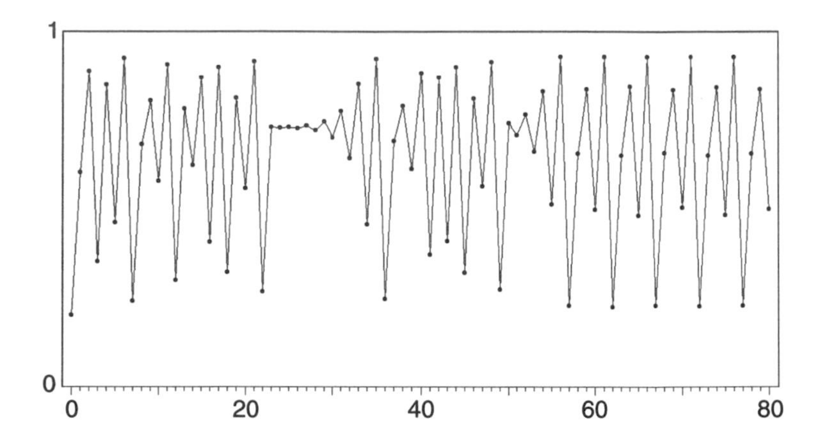

Figure 10.2 : Time series for the quadratic iterator starting with $x_0 = 0.2027$ and parameter $a = 3.742718$.

Now let us look at the second time series in figure 10.2, which is based on the same formula and the same initial value. The only difference lies in the choice of the parameter a. At first glance the graph looks just as chaotic as the previous one. But this is categorically wrong; the two could not be any more different from a qualitative point of view. If you consider the last quarter of the second graph you will notice a periodicity. In fact, the graphics suggest that the orbit settles in on a cycle with an apparent period 5 or 10. Every 5th or 10th iteration seems to give the same amplitude, and the cycle repeats.[2] Although this sort of behavior is not totally trivial, computers will not be seriously troubled by such phenomena. We will discuss the scenario of periodic cycles much more in chapter 11, but at this point let us already make the difference clear.

Figure 10.3 shows a very similar, but simpler, case. Here the two orbits shown converge to a single value x_∞ rather than some periodic orbit. It does not matter where we start the iteration, we will always end up at this same *final state*. The underlying mechanism is well observable by means of *graphical iteration*, which was introduced in chapter 1.[3] We briefly summarize: the left part of figure 10.4 shows the graphical iteration starting at the initial point $x_0 = 0.22$. The parabola is the graph of the iteration function $ax(1-x)$ and is the locus of points (x_n, x_{n+1}), because by definition of the iteration one must have

$$x_{n+1} = ax_n(1 - x_n) \ .$$

[2]Looking more closely at the numerical values of the iterates we see that the period is really 20.

[3]See page 66, *Fractals for the Classroom, Part One.*

Stability

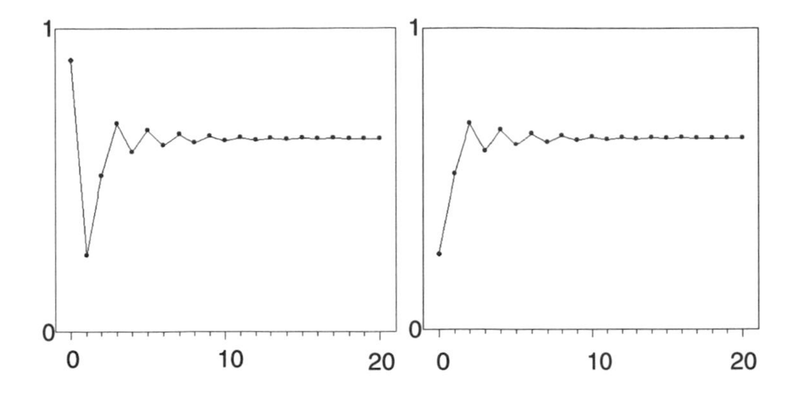

Figure 10.3 : Non-sensitive, stable behavior: all initial values between 0 and 1 lead to the same final state. $a = 2.75$.

Attraction to a Point

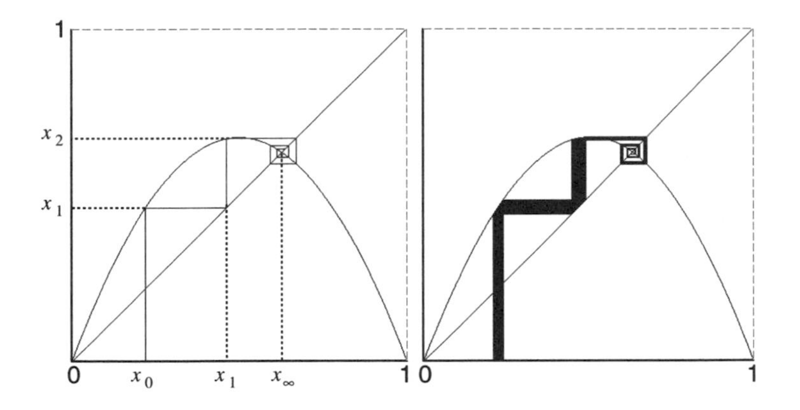

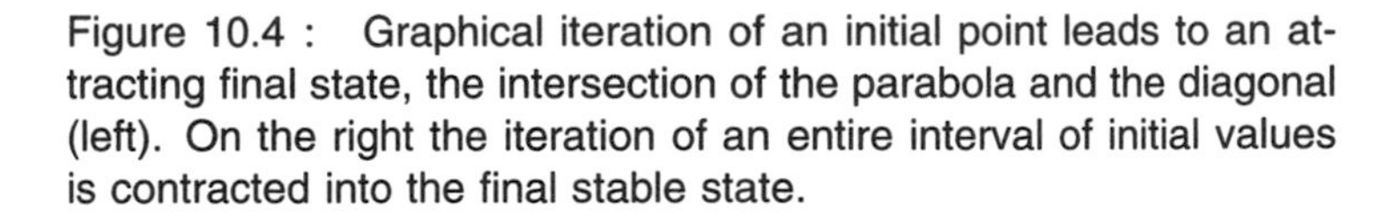

Figure 10.4 : Graphical iteration of an initial point leads to an attracting final state, the intersection of the parabola and the diagonal (left). On the right the iteration of an entire interval of initial values is contracted into the final stable state.

Thus, above x_0 the parabola has height x_1. On the diagonal at the same height we find the point with coordinates (x_1, x_1). Above that point we find the parabola with height x_2, and so on. Continuing the graphical iteration leads us to the intersection of the parabola and the diagonal, this is the point (x_∞, x_∞).

In the right half of figure 10.4 we repeat the experiment with a variation. Not only do we iterate the same initial point $x_0 = 0.22$, but also all values between $x_0 - 0.015$ and $x_0 + 0.015$, in effect iterating an entire interval. We observe that all values in the interval are attracted by the same final state, and this is an example of stability, which definitely cannot survive in the presence of chaos.

Sensitivity

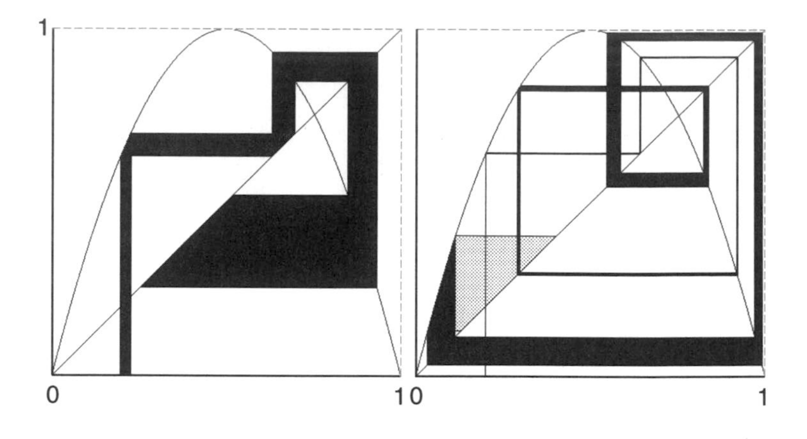

Figure 10.5 : Sensitivity demonstrated by graphical iteration: in the course of the iteration even a small deviation increases substantially (left). The experiment is repeated with an even smaller interval of initial points (right).

Sensitivity Amplifies Even the Smallest Error

The phenomenon of sensitivity, however, magnifies even the smallest error. This is demonstrated dramatically by setting the parameter a to the value 4. Repeating the experiment from figure 10.4 we obtain the plot shown in figure 10.5. The initial small interval has already grown considerably after just a few iterations. Allowing some more iterations would show us that every number from the whole interval [0,1] will be covered. To convince ourselves that this is not an artifact resulting from the width of the initial interval being too large, we repeat the same experiment with an even smaller initial interval. In figure 10.5 (right) the width is only 0.0005, less than in our imitation of the Lorenz experiment, see page 56 in chapter 1. What was shown there in a table as plain numbers can be seen here in graphical iteration: even the smallest deviation will escalate in the course of the iteration.

Let us summarize. Sensitivity implies that any arbitrarily small interval of initial values will be enlarged significantly by iteration. More precisely, this behavior is called *sensitive dependence on initial conditions*. For the quadratic iterator discussed here it is even true that any interval will expand to the full interval from 0 to 1.

Example of a Linear Transformation

The property of sensitivity is central to chaos. Sensitivity, however, does not *automatically* lead to chaos. Indeed, there are sensitive systems which certainly do not behave chaotically. This is demonstrated with the simple example $x \to cx$, where the parameter c is greater than 1. This is a linear transformation, and it exhibits sensitive dependence on initial conditions: any deviation is mag-

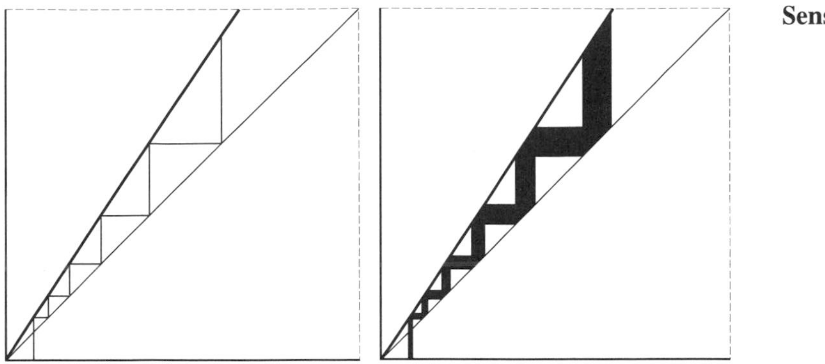

Sensitivity in Linear Systems

Figure 10.6 : Iteration of the linear transformation $x \to 1.5x$. Left: single initial value; right: interval of initial values.

nified during the course of the iteration, see figure 10.6. Given an initial point x_0, then after n iterations we have $x_n = c^n x_0$. The deviations or errors of an orbit started nearby behave in the same fashion. Given an initial error $E_0 = \varepsilon$, i.e., we start with $u_0 = x_0 + E_0$, then after n iterations we have $u_n = c^n(x_0 + E_0)$ and the error has developed to

$$E_n = u_n - x_n = c^n(x_0 + \varepsilon) - c^n x_0 = c^n \varepsilon \ .$$

Therefore, any deviation ε is magnified by the factor $c > 1$ in each iteration. The system is sensitive, but it is certainly not chaotic.

The Characterization by Lorenz

The meteorologist Lorenz described the sensitivity of chaotic systems in a way that differentiates between the quadratic iterator and the tame linear transformation. Given an initial deviation, then in a chaotic system, it will become as large as the true 'signal' itself. In other words, after some iterations the error will be in the same order of magnitude as the correct values. Consider the quadratic iterator ($a = 4$) and the initial value $x_0 = 0.202$ again. Introducing a very small deviation of $\varepsilon = 10^{-6}$, we have another nearby initial value $u_0 = 0.202001$. Figure 10.7 shows the corresponding time series $x_0, x_1, x_2, x_3, \ldots$ (top), $u_0, u_1, u_2, u_3, \ldots$ (center), and the development of the absolute value of deviation, i.e., the error $|u_n - x_n|$ (bottom). In other words, in the top the original signal is shown along with the approximating signal, while in the bottom the difference or error signal is displayed.

In the beginning the error remains very small. For about the first 15 iterations it is not distinguishable from zero in our figure. Although the error is growing during these iterations, it is still too small to be seen. But after it has attained some critical magnitude it seems to explode. For the following iterations the error signal looks just as erratic as the original time series. Moreover, the

Error Development

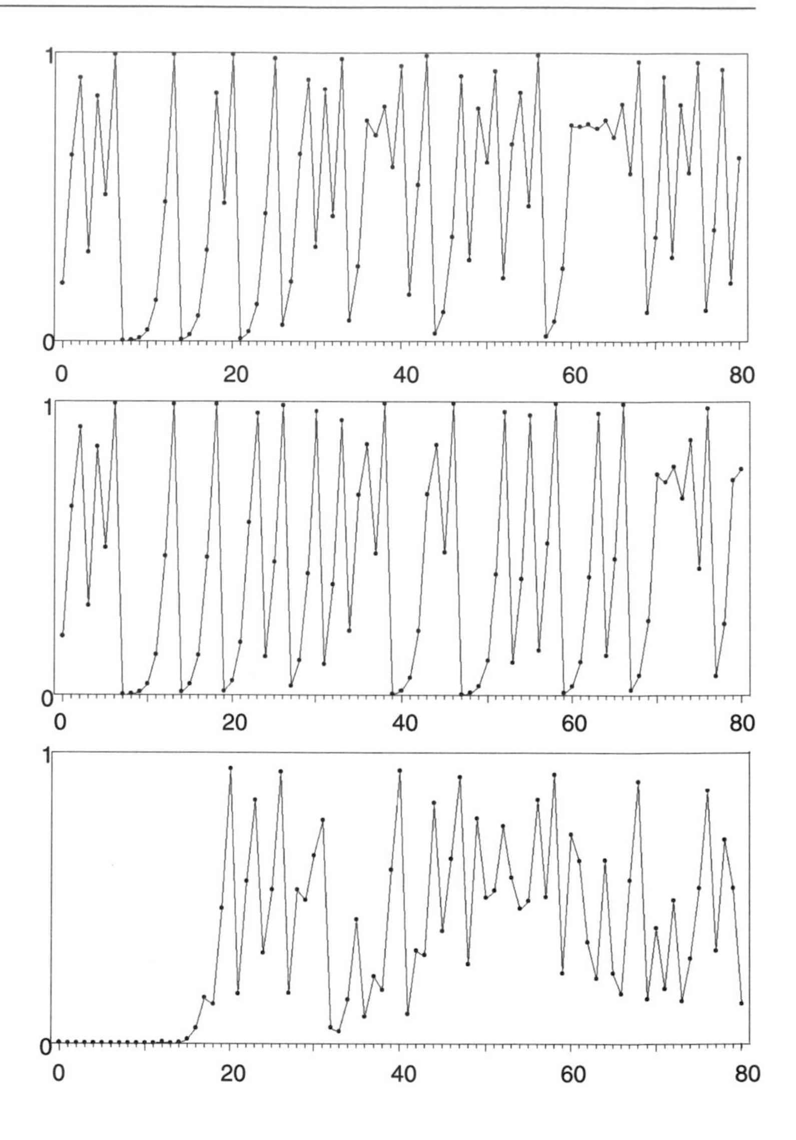

Figure 10.7 : The quadratic iterator $x \longrightarrow 4x(1-x)$ applied to two initial values differing by 10^{-6} (top and center) and the (absolute) difference of the two signals (bottom).

amplitudes of the error are of the same order as the amplitudes of the time series on the left, namely 1. This behavior is typical for chaotic systems.

We can contrast this to the error development in the linear system $x \rightarrow cx$. Here the error and the true values grow in harmony. The relation between the signal, the values of the iteration, and the error remain the same at all times. The *relative* error computes as

$$\frac{E_n}{x_n} = \frac{c^n \varepsilon}{c^n x_0} = \frac{\varepsilon}{x_0} = \text{const} .$$

Thus, the iteration is quite harmless and may be safely carried out on a computer until the limit of the allowable range of numbers of the machine has been reached.

Analysis of Error Propagation

The phenomenon of sensitive dependence on initial conditions as described above is a *quality* that all chaotic systems definitely have. Other systems may or may not exhibit sensitivity. This is already a big step forward in the direction of understanding chaos. However, one of the major driving forces of science has always been to also *quantify* the qualities it has discovered. How would that be possible in the case of error propagation as discussed here?

Let us begin with the error propagation in the simple linear system $x \to cx$. Per iteration the error grows by a factor of c. Therefore

$$\left|\frac{E_n}{E_0}\right| = c^n . \tag{10.1}$$

The experiment in figure 10.7 clearly shows that such a simple law cannot be expected for the error propagation of the quadratic iterator $x \to 4x(1-x)$. On the other hand, during the first 15 iterations or so, the error must have grown more or less uniformly. This motivates another experiment. We count the number of iterations necessary for the error to exceed a certain threshold for the first time, say 0.1. We can do this for various initial values and also for different initial errors. We choose three initial values, 0.202, 0.347, and 0.869, and four possible initial errors, to arrive at a total of 12 cases. Table 10.8 documents the results.

For example, when starting the iteration with the correct initial value 0.202 and comparing the iteration with that of a perturbed initial value 0.203, then the error of 0.001 at the beginning builds up in $n = 9$ steps to $E_n = 0.25622$. To derive a measure of how fast the error grows per iteration we take another look at eqn. (10.1). It relates the ratio E_n/E_0 to the factor c which characterizes the error growth. In our experiment above, we have E_n and E_0 conveniently at our disposal and can now pretend that these data come from a linear system. This allows us to derive the corresponding factor c, which numerically characterizes the growth process of the error. To find the appropriate formula we take natural logarithms on both sides of eqn. (10.1)

$$\ln\left|\frac{E_n}{E_0}\right| = n \ln c .$$

Error Propagation Experiment

x_0	Error E_0	Steps n	Error E_n	Exponent
0.202	0.001000	9	0.25622	0.61623
0.202	0.000100	11	−0.12355	0.64720
0.202	0.000010	15	0.25730	0.67703
0.202	0.000001	17	0.15866	0.70438
0.347	0.001000	7	−0.12331	0.68781
0.347	0.000100	11	−0.18555	0.68417
0.347	0.000010	15	0.31390	0.69028
0.347	0.000001	18	0.19490	0.67668
0.869	0.001000	8	−0.25072	0.69054
0.869	0.000100	10	0.14068	0.72491
0.869	0.000010	13	0.11428	0.71876
0.869	0.000001	18	0.32095	0.70439

Table 10.8 : For three initial values orbits are started nearby using the quadratic iterator $x \to 4x(1-x)$. Four different initial errors are tried. After n steps the error has accumulated to a magnitude exceeding the threshold 0.1 for the first time, see the columns 'Steps n' and 'Error E_n'. The last column lists a measure of how much the error increases per iteration; see the text for an explanation of the exponent.

Dividing by n yields

$$\ln c = \frac{1}{n} \ln \left| \frac{E_n}{E_0} \right| . \tag{10.2}$$

The quantity on the right side thus gives us the logarithm of the error growth constant c. This number is listed in the last column of table 10.8, headed by 'exponent'. It is interesting to note that the results are all about 0.7. Thus, the factor c is

$$c \approx e^{0.7} \approx 2 .$$

The result of the experiment is thus the following:

- *In the quadratic iterator $x \to 4x(1-x)$, small errors will roughly double in each iteration.*

Of course, the errors can double only if they are sufficiently small. Moreover, the error doubling occurs only on the average. There are points x in the unit interval where small errors do not magnify. For example, this is the case near $x = 0.5$, where the graph of the parabola is rather flat, and errors are compressed. On the other hand, near the end points of the unit interval, errors are enlarged by factors up to 4.

Alexander M. Ljapunov

Figure 10.9 : Alexander Michailowitsch Ljapunov, 1857–1918.

The Ljapunov Exponent

The reasoning above directly leads to the concept of *Ljapunov exponents* $\lambda(x_0)$. It quantifies the average growth of infinitesimally small errors in the initial point x_0. Indeed, in eqn. (10.2) and table 10.8 we have derived a method to approximate the exponent. It is interesting that it seems to be independent of the initial value x_0. How can we make the computation more precise? In order to compute the average growth of an error it would be of advantage to consider many more iterations than just a dozen or so as in table 10.8. It seems that this necessarily implies that we start with a *very* small initial error E_0, because it will be roughly doubled in each iteration until it exceeds the threshold 0.1 for the first time. It is clear that this approach does not reach very far. Even if we reduce the initial error to the smallest possible given by the machine precision (which is of the order of 10^{-12} or so) we can expect to be able to perform only a few more iterations, but not hundreds and thousands of them. Thus, a different procedure is asked for. Let us assume that we can work with arbitrarily small initial errors E_0. We rewrite the total error amplification factor $|E_n/E_0|$ as

$$\left|\frac{E_n}{E_0}\right| = \left|\frac{E_n}{E_{n-1}}\right| \cdot \left|\frac{E_{n-1}}{E_{n-2}}\right| \cdots \left|\frac{E_1}{E_0}\right|$$

and attempt to estimate each factor separately! Assume for now that there is a workable method for this task. Now, however, when computing the total error amplification $|E_n/E_0|$ we must multiply all the individual factors which surely will result in a number much too large to be represented by an ordinary computer number (causing an overflow error). Fortunately we can avoid this because what we really are interested in is the geometric mean of the factors or rather the logarithm of the mean (see eqn. (10.2)). In formula,

$$\begin{aligned}\frac{1}{n}\ln\left|\frac{E_n}{E_0}\right| &= \frac{1}{n}\ln\left|\frac{E_n}{E_{n-1}}\cdot\frac{E_{n-1}}{E_{n-2}}\cdots\frac{E_1}{E_0}\right| \\ &= \frac{1}{n}\sum_{k=1}^{n}\ln\left|\frac{E_k}{E_{k-1}}\right| .\end{aligned}$$

Summing up the logarithms of the amplification factors surely avoids the overflow problems.

Now how can we estimate the error amplification factors in each iteration? Let us consider

$$\left|\frac{E_{k+1}}{E_k}\right| .$$

This describes by how much a small error E_k in x_k, the kth iterate, is enlarged (or reduced) in the following iteration. This error amplification factor is in essence independent of the size of the (small) error E_k. For example, if we consider an error in x_k being only half as large, i.e. $E_k/2$, then we can also expect the error in x_{k+1} to be half as large, i.e. $E_{k+1}/2$. Thus, it does not really matter whether we use the precise number E_k or just some arbitrary small error ε, say $\varepsilon = 0.001$.[4] This provides a solution to the problem; we fix an arbitrary small error ε and estimate the error amplification factors $|E_{k+1}/E_k|$ for very small initial errors E_0 by $|\tilde{E}_{k+1}|/\varepsilon$ where

$$\tilde{E}_{k+1} = f(x_k + \varepsilon) - f(x_k)$$

and $f(x) = 4x(1-x)$ is the quadratic iterator. We do not have to worry about errors growing too large and are in the position to compute as many error amplification factors as we desire.

In summary, we have arrived at an improved and feasible method to compute the Ljapunov exponent more reliably by averaging over many iterations. The formula for this procedure is

$$\frac{1}{n}\ln\left|\frac{E_n}{E_0}\right| \approx \frac{1}{n}\sum_{k=1}^{n}\ln\left|\frac{\tilde{E}_k}{\varepsilon}\right| .$$

[4]To justify this argument consider an error $E_k = \varepsilon$ in x_k, i.e. $\tilde{x}_k = x_k + \varepsilon$. Let $f(x) = 4x(1-x)$. Then the error E_{k+1} in x_{k+1} is

$$E_{k+1} = f(\tilde{x}_k) - f(x_k) = \tilde{x}_{k+1} - x_{k+1} = 4(x_k+\varepsilon)(1-x_k-\varepsilon) - 4x_k(1-x_k) = 4\varepsilon(1-2x_k) - 4\varepsilon^2.$$

Dividing the error by E_k yields $E_{k+1}/E_k = 4(1-2x_k) - 4\varepsilon$, a quantity dominated by $4(1-2x_k)$. The term -4ε is comparatively small. Thus, we see that the error amplification factor E_{k+1}/E_k is independent of the error $E_k = \varepsilon$ as long as ε is small. It is given by $4(1-2x_k)$.

The results for the initial values used in table 10.8 are given in table 10.10.

Ljapunov Number Computation

Iterations	$x_0 = 0.202$	$x_0 = 0.347$	$x_0 = 0.869$
10	0.62475	0.68873	0.72803
100	0.69435	0.69360	0.69708
1000	0.69368	0.69199	0.69004
10000	0.69322	0.69290	0.69337
100000	0.69307	0.69306	0.69327

Table 10.10 : For three initial values x_0 (compare table 10.8) Ljapunov exponents are calculated from averaging error amplification factors over varying numbers of iterations. As the number of iterations grows the exponent converges to $\lambda(x_0) = \ln 2 = 0.69314...$ Here an error of 0.001 has been used in each iteration. With smaller values of ε the convergence is even better.

Measuring these exponents in this more careful analysis results in the number $\lambda(x_0) = 0.693$. With that we have succeeded in quantifying the sensitive dependence on initial conditions for the quadratic iterator. Now we are in a position to compare the sensitivity found here with that in other chaotic systems.

The Ljapunov exponent $\lambda(x_0)$ is a powerful experimental device to separate unstable, chaotic behavior from that which is stable and predictable and to measure these properties. Especially where $\lambda(x_0) > 0$ is large, sensitivity with respect to small changes in initial conditions is large. It is important to note that the concept of Ljapunov exponents has been generalized so that it applies to many interesting dynamical systems in mathematics and the sciences. It has become one of the keys to measuring, evaluating and detecting chaotic behavior.[5]

The Ljapunov Exponent for Smooth Transformations

The characterization of the Ljapunov exponent needs a bit of calculus. To this end we assume that we consider the iteration of a smooth transformation f, which certainly is the case when f is a polynomial such as the quadratic $f(x) = ax(1-x)$. In this iteration we have $x_{n+1} = f(x_n)$ for $n = 0, 1, 2, ...$ The equation (10.2) will guide us to an appropriate definition. First we rewrite the relative growth of the error after n steps as a product

$$\frac{E_n}{E_0} = \frac{E_n}{E_{n-1}} \cdot \frac{E_{n-1}}{E_{n-2}} \cdots \frac{E_1}{E_0}$$

[5] See for example H. G. Schuster, *Deterministic Chaos*, Physik-Verlag, Weinheim and VCH Publishers, New York, 1984.

and taking absolute values and natural logarithms we see that

$$\frac{1}{n}\ln\left|\frac{E_n}{E_0}\right| = \frac{1}{n}\sum_{k=1}^{n}\ln\left|\frac{E_k}{E_{k-1}}\right| .$$

By definition of the error terms we have

$$\frac{E_k}{E_{k-1}} = \frac{f(x_{k-1}+E_{k-1}) - f(x_{k-1})}{E_{k-1}}$$

and from calculus we obtain

$$\lim_{E_0\to 0}\frac{E_k}{E_{k-1}} = f'(x_{k-1}) .$$

Thus,

$$\lim_{E_0\to 0}\frac{1}{n}\sum_{k=1}^{n}\ln\left|\frac{E_k}{E_{k-1}}\right| = \frac{1}{n}\sum_{k=1}^{n}\ln|f'(x_{k-1})| .$$

Now letting $n \to \infty$ we obtain the Ljapunov exponent

$$\lambda(x_0) = \lim_{n\to\infty}\frac{1}{n}\sum_{k=1}^{n}\ln|f'(x_{k-1})| . \tag{10.3}$$

Let us present an explicit formula for the Ljapunov exponent $\lambda(x_0)$ for the special case that the orbit of x_0 is periodic with a period $m > 0$. Thus,

$$x_m = f^m(x_0) = x_0 .$$

Because of this periodicity we have

$$\lambda(x_0) = \frac{1}{m}\sum_{k=1}^{m}\ln|f'(x_{k-1})| .$$

In other words, the average of the logarithmic amplification factors, taken over one periodic cycle, is the same as the average taken over two, three, or more periods, and, thus, equal to the limit in eqn. (10.3). Thus, if x_0 is a fixed point, i.e., $m = 1$, we get

$$\lambda(x_0) = \ln|f'(x_0)| .$$

For a 2-cycle the Ljapunov exponent is

$$\lambda(x_0) = \lambda(x_1) = \frac{1}{2}(\ln|f'(x_0)| + \ln|f'(x_1)|) .$$

In the case of the quadratic iterator, where $f(x) = ax(1-x)$ we get

$$\begin{aligned}\lambda(x_0) &= \lim_{n\to\infty}\frac{1}{n}\sum_{k=1}^{n}\ln|a - 2ax_{k-1}| \\ &= \ln a + \lim_{n\to\infty}\frac{1}{n}\sum_{k=0}^{n-1}\ln|1-2x_k| .\end{aligned}$$

Note that since $x = 0$ is a fixed point of the quadratic iterator, we have $\lambda(0) = \ln a$ (and also $\lambda(1) = \ln a$). For $a = 4$ this is approximately 1.39, which indicates the instability of this fixed point. For most other points[6] x between 0 and 1 we compute $\lambda(x) \approx 0.693$.

[6] To be technical and more precise, we note in passing that for $a = 4$ the result $\lambda(x) = \ln 4$ holds for all points x whose orbits eventually end in the fixed point 0. This is a dense subset of the interval $[0, 1]$. On the other hand, almost all points in the interval will have a Ljapunov exponent equal to $\ln 2 \approx 0.693$. This means that if an initial point is picked at random, then the orbit (almost surely) fills the interval densely and the exponent is $\ln 2$.

10.2 The Signs of Chaos: Mixing and Periodic Points

Let us now turn to what is called the *mixing behavior* of the quadratic iterator. In figure 10.5 we saw how a small error is amplified in the course of the iteration. But we can also interpret this figure from a slightly different point of view. The points of the small interval of initial values finally become spread over the whole unit interval (the numbers from 0 to 1). In fact, we can start with an arbitrary interval of initial values. When iterated they become spread over the whole interval.

Mixing

An intuitive way to interpret mixing is to subdivide the unit interval into subintervals and requiring that by iteration we can get from any starting subinterval to any other target subinterval. If this requirement is fulfilled for all finite subdivisions, we say that the system exhibits mixing.

Let us check an example for $f(x) = 4x(1-x)$ using the subdivision of $I = [0,1]$ into 10 subintervals of length 1/10,

$$I_k = \left[\frac{k-1}{10}, \frac{k}{10}\right], \quad k = 1, ..., 10 .$$

Consider the starting subinterval $I_2 = [0.1, 0.2]$. It is transformed by iteration of the quadratic function giving the sequence of intervals listed in table 10.11.

Intervals Reached by Subinterval I_2

Iteration n	Interval	I_1	I_2	I_3	I_4	I_5	I_6	I_7	I_8	I_9	I_{10}
0	$[0.100, 0.200]$		•								
1	$[0.360, 0.640]$				•	•	•	•			
2	$[0.922, 1.000]$										•
3	$[0.000, 0.289]$	•	•	•							
4	$[0.000, 0.822]$	•	•	•	•	•	•	•	•	•	

Table 10.11 : The subinterval $I_2 = [0.1, 0.2]$ is iterated 4 times. The subintervals which are reached at each iteration are marked with a bullet.

After one iteration subintervals I_4 to I_7 are reached. The next iterate produces only points from I_{10}. However, the third and fourth iteration yield points from the remaining intervals I_1, I_3, I_8, and I_9. We would have obtained a similar result if we had taken a starting subinterval different from I_2, or if we had subdivided the unit interval into 100, 1000, or even one million subintervals. This is the essence of mixing. Expressed informally, 'mixing is if we can get everywhere from anywhere'.

The Definition

A bit more formally we describe this mixing property in the following way:

- *For any two open intervals I and J (which can be arbitrarily small, but must have a non-zero length) one can find initial values in I which, when iterated, will eventually lead to points in J.*

Figure 10.12 demonstrates mixing for the quadratic iterator for the value $a = 4$. We show two very small intervals I and J and you can follow the iteration of one initial value taken from I which leads (after 11 steps) to a point in J. You might compare this with stirring a drop of milk into your coffee or mixing a small pocket of spice into dough by kneading it thoroughly. Indeed, this will be discussed in more detail in section 10.4.

From I to J

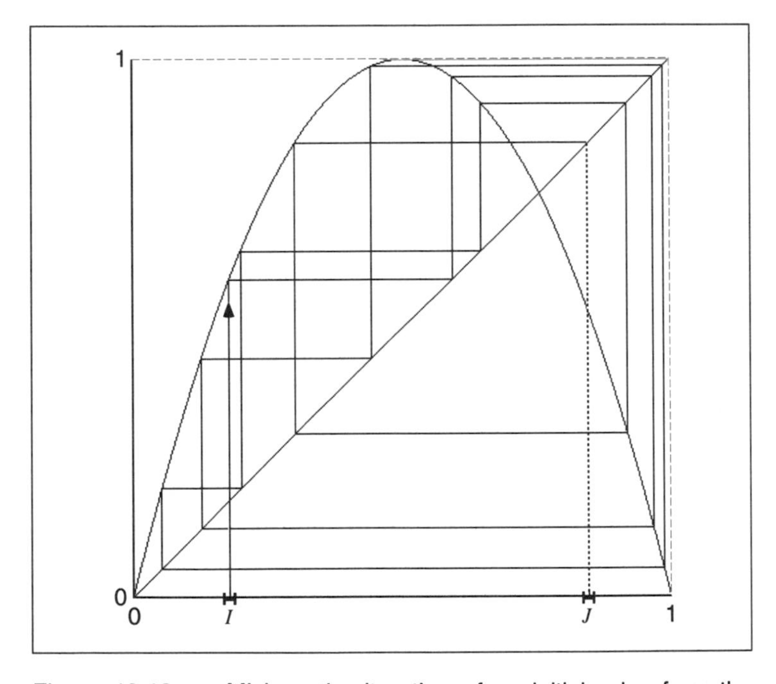

Figure 10.12 : Mixing: the iteration of an initial value from the interval I is spread all over the unit interval. For any interval J there is an initial point in I with an orbit that reaches J. In this example 11 iterations suffice. The parameter is $a = 4$.

Measuring the Mixing Property

Given a small interval I for initial points and a small target interval J we may choose points in I, compute their orbits and check whether they ever enter the target interval. Some orbits will accomplish this sooner than others. Some initial points may have orbits that never reach the target interval. Let us do a straightforward numerical experiment that collects some statistics about the behavior of these

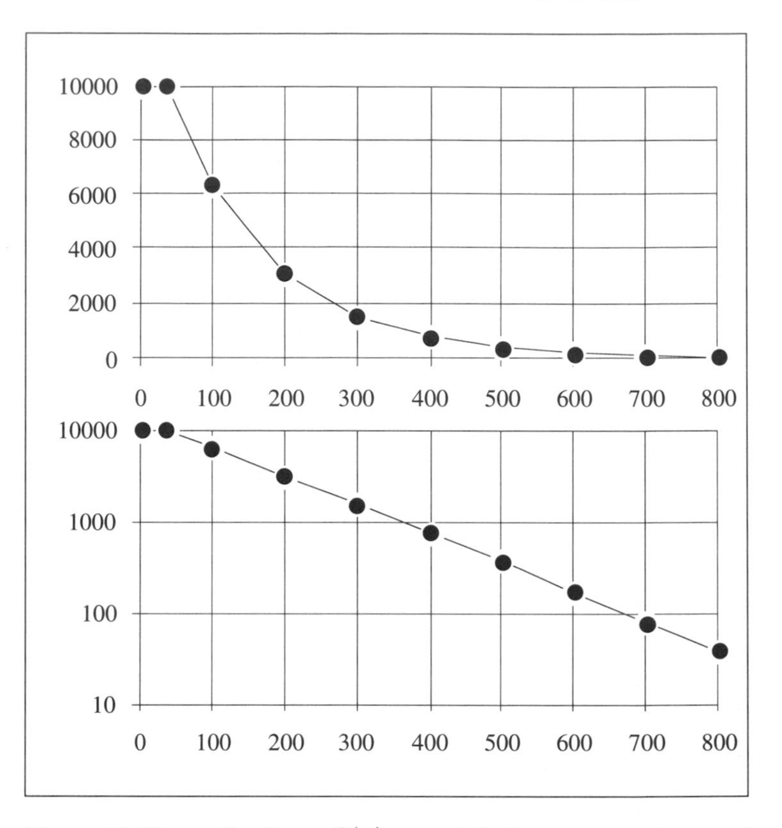

Figure 10.13 : Survivors $S(n)$ are plotted versus number n of iterations. Below the same data are shown logarithmically.

orbits and which leads to an interesting exponential law.[7] We select a family of $10,000$ initial points equally spaced in the interval I and follow their orbits until they have hit the target interval and are discarded. We call the orbits which remain after some number of iterations the *survivors*. How many survivors will there be after, say 100 iterations? After 1000 iterations? Is there some power law by which the number of survivors decays? The answers, of course, depend on the choice of the intervals. For example, we could arrange them so that all orbits fall into the target already in the first iteration. That would be rather uninteresting. Thus, let us make the interval I very small and choose J in such a way so that at least for a few dozen iterations no orbits land in the target interval J. For example, we may choose

$$I = [0.2, 0.2 + 10^{-11}], \quad J = [0.68, 0.69] .$$

During the first 36 iterations none of the initial $10,000$ orbits reach

[7] Such an experiment has been carried out by James Yorke and also Robert Shaw. See R. Shaw, *Strange attractors, chaotic behavior, and information flow*, Z. Naturforsch. 36a (1981) 80–112.

the target. In the following iterations some of the orbits finally succeed; in the 37th iteration 63 of them end up in the target set, then again 63, then 62, 63, and so on. When the number of survivors has decreased to about half the initial number of 10,000, we expect that the decay has dropped to about one half of 63 per iteration. In fact, we obtain a decrease of 31 orbits from 5037 survivors down to 5006 after 133 iterations. This leads us to the hypothesis that the number $S(n)$ of survivors decays exponentially with the number n of iterations according to

$$S(n) \propto e^{-\frac{n}{\tau}} . \tag{10.4}$$

The number τ can be interpreted as an estimate for the number of iterations necessary to reduce the survivors by a factor of $1/e \approx 0.368$ and is also called the *average life time* of the orbits. In order to check this formula we use a semi-logarithmic plot of $\log S(n)$ versus n (see figure 10.13).

After an initial transient period of 36 iterations the logarithm $\log S(n)$ decreases linearly which confirms the conjecture of exponential decay. Based on the results $S(100) = 6346$ and $S(600) = 171$ we estimate that

$$\tau \approx 138 \text{ iterations} .$$

These facts can be summarized as follows. The information contained in the small interval persists for a certain number of iterations depending on the size and location of the interval I. After that it is no longer possible to see that the orbits initially had started very close to each other. Moreover, once this state is achieved the number of the survivors decays exponentially, in this case with the rate of $1/e$ per 138 iterations. The rate depends on the choice of the target interval J. Below in this section we derive histograms and invariant distributions which will reconfirm the numerical result presented here.

Small Intervals Iterated Expand to $[0,1]$

The quadratic iterator exhibits not only mixing as specified in the above definition, but an even stronger type of spreading. Take an arbitrary small subinterval and consider its iteration. In each step the result will be an interval again, however, of different size and location. Small intervals will grow in size on average and after a certain finite number of iterations the whole unit interval is covered. For example, carrying out one more iteration of $I_2 = [0.1, 0.2]$ in table 10.11 would yield the entire unit interval,

$$f^5([0.1, 0.2]) = f([0.000, 0.822]) = [0, 1] .$$

In other words, any given point of the unit interval has preimages in the small initial subinterval. Or, any subinterval will expand to the full unit interval in the course of the iteration. The following question now comes naturally. On the average, how many iterations are necessary for this to happen when considering the collection of subintervals of a given width? Table 10.14 provides the answer. We start with the 2 subintervals of width 1/2. Already the first iteration produces the unit interval. Then we double the number of subintervals over and over again and compute the desired average number of iterations: 2.5, 4.0, 5.12, ...

Subintervals Expand to the Unit Interval

N	k	N	k	N	k	N	k
2	1.00	32	6.19	512	10.27	8192	14.28
4	2.50	64	7.25	1024	11.28	16384	15.28
8	4.00	128	8.25	2048	12.28	32768	16.28
16	5.12	256	9.27	4096	13.28	65536	17.28

Table 10.14 : N denotes the number of subintervals of $[0, 1]$, and k specifies the average number of iterations of $f(x) = 4x(1 - x)$ necessary to expand the intervals to the entire unit interval.

The result is surprisingly clear. When doubling the number of subintervals, i.e., reducing their size by a factor of 1/2, the average number of iterations seems to grow by exactly 1. This is not a coincidence. We recall that the numerical computation of the Ljapunov exponent for this iterator yielded $\lambda = \ln 2$, which states that small errors are amplified by a factor of 2, which is another way expressing the observation above.

Periodic Points are Central to Chaos ...

You already know that there are also points which are of an entirely different nature: *periodic points*. If we start our iteration with a periodic point, this leads only to some few intervals which are visited again and again. Theoretically we can find infinitely many points of this type in each interval. Besides sensitivity and mixing, the existence of periodic points in any given subinterval is regarded as one of the necessary conditions for chaos and will be discussed next. At the end of this chapter we will even learn how the periodic points can be found by a simple formula. This formula applied for $a = 4$ yields that, for example,

$$\sin^2 \frac{\pi}{7} \to \sin^2 \frac{2\pi}{7} \to \sin^2 \frac{4\pi}{7}$$

is a cycle of period three for the quadratic iterator. One of the three points, namely the first, is in the interval I from figure 10.12. Thus, even though the interval I is very small it contains a point from a 3-cycle besides initial points whose orbits reach everywhere. So what kind of behavior will we see if we choose an initial value at random?

... But Cannot be Detected

Well, assume we have chosen $\sin^2(\pi/7) = 0.1882550\ldots$ as an initial point. This is certainly a number which cannot be represented precisely in the computer. The error in representing this

An Unstable 3-Cycle

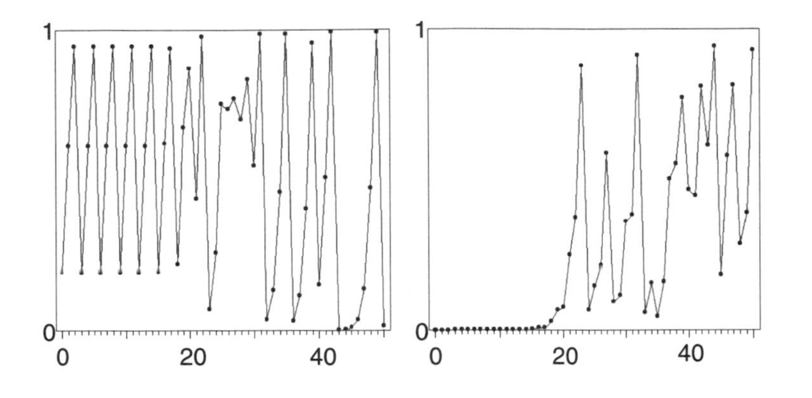

Figure 10.15 : Sensitivity strikes again: the machine computed iteration of a periodic point with period 3. Due to roundoff errors in the computer and the sensitivity the orbit eventually moves away from the exact periodic cycle (at about the middle of the time series on the left). On the right the development of the error is plotted.

number by a single-precision floating point number is approximately 0.00000005.[8] Based on the sensitive dependence on initial conditions, we can predict what will happen. In each step of the iteration the error will become about twice as large. After 20 steps it will have multiplied by a factor of a million: about 1/20. In other words, for at most 20 iterations we will be close to the 3-cycle, and then sensitivity strikes again. This is demonstrated convincingly in figure 10.15. The same problem arises for all the other periodic points which we can compute. And even when a periodic point can be represented exactly in computer arithmetic, after one iteration a minute round-off error will throw the orbit off the periodic one and open the door to letting sensitivity play its destructive part again.[9] Summarizing, we conclude that in a system with sensitivity there is no possibility of detecting a periodic orbit by running time series on a computer.

[8] The number depends on the particular machine.

[9] There are some exceptions to this, namely when the complete periodic orbit is computed exactly by the machine. For the quadratic iterator, this will be the case when starting with $x_0 = 0$: all the following points are also zero, and the machine has no problems with that because there is no round-off error when multiplying by zero.

10.3 Ergodic Orbits and Histograms

Ergodic Orbits

For the discussion of mixing we use small intervals and their iteration, but why don't we simply take single initial points? This question seems to be justified by figure 10.16. Here we have continued the iteration of the initial value which took us from the starting interval I to the target interval J in figure 10.12. It seems that the iteration eventually fills out the whole unit interval similar to the iteration of a small subinterval. Indeed, we will see that in each interval there are infinitely many points of this type.[10] When iterated they are mixed throughout the whole unit interval. The iteration of such a point $x_0, x_1, x_2, \ldots$ gets arbitrarily close to any other point of the unit interval. Such orbits are called *ergodic*. But we must be careful: not all initial points of an interval like I produce ergodic orbits. For example, periodic points and their preimages yield orbits of only finitely many different points which, thus, cannot reach all subintervals of arbitrary small size.

An Ergodic Orbit

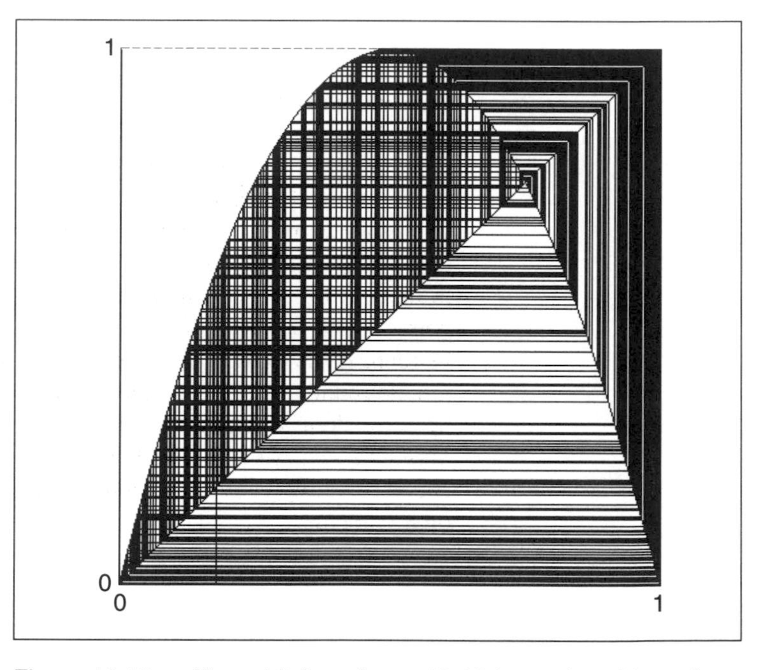

Figure 10.16 : The orbit from figure 10.12 is continued here for a few hundred iterations. It densely covers the unit interval.

Computing a Histogram

So let us look for numerical detection of ergodicity. Here is a simple experiment. We pick an initial point x_0 and iterate our quadratic iterator, say $m = 10^6$ times. Now we would like to see

[10]This is only a theoretical result. On a computer there is only a finite collection of numbers representable. Thus, it is impossible to get arbitrarily close to all of the numbers in the unit interval.

Distribution of an Orbit

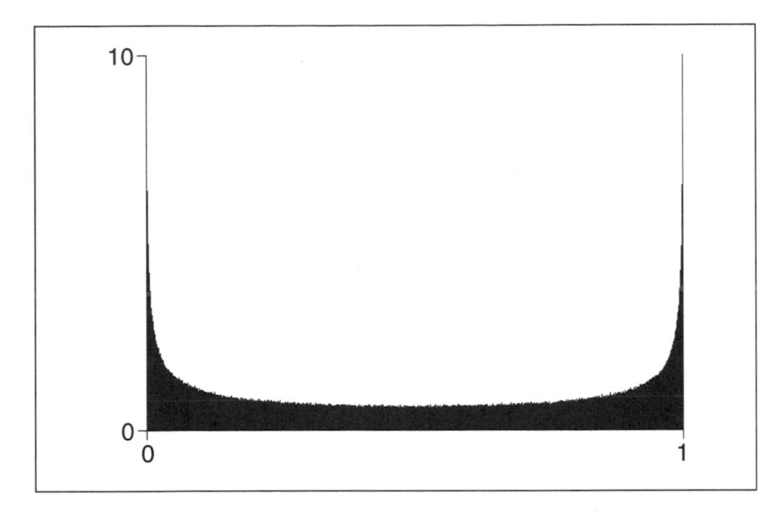

Figure 10.17 : Distribution of the orbit $x_0, ..., x_m$. Hits are counted in 600 different subintervals of $[0, 1]$. The number of hits in an interval is proportional to the area of the columns drawn on this interval. The total area is 1.

which parts of the unit interval are visited by the orbit $x_0, ..., x_m$ and how often. To this end we divide the unit interval again into a large number N of small subintervals given by

$$I_k = \left[\frac{k-1}{N}, \frac{k}{N}\right), \quad k = 1, ..., N .$$

Then we note the number of events in I_k: we count how many of the iterates $x_0, ..., x_m$ fall into each interval I_k. Let this number be n_k. Naturally, however, we want to produce a count which is somewhat independent from the length of the orbit m. Thus, noting that the orbit has $m + 1$ elements, we define

$$\mu_k = \frac{n_k N}{m+1} .$$

These numbers μ_k vary between 0 and N by definition, and sum up to N,

$$\mu_1 + \mu_2 + \cdots \mu_N = N .$$

In other words, μ_k/N can be interpreted as a probability: it is the probability that we guess correctly (without calculations) the interval I_k into which a point falls, randomly chosen from the first $m+1$ points of the orbit of x_0. Now we can plot columns of height μ_k and width $1/N$ into a histogram as in figure 10.17, where we have chosen $m = 10^6$, and $N = 600$. The columns cover an area which is equal to 1.

A Histogram for Mixing

We see a distribution for the μ_k which is symmetric with respect to 1/2, and which is rather flat in the center while having steep boundary layers at 0 and 1. This means that during the course of the quadratic iteration $x \to 4x(1-x)$ the probability that we see a point of the orbit near 0 or 1 is comparatively much higher than that of seeing it in the center of the unit interval. Running the same experiment again for different initial values x_0 results in histograms which are indistinguishable from the one above. As we would increase the number N of subintervals and the length m of the underlying orbit, the effect would be a smoothing of the shape seen in figure 10.17. In the limit we would approximate a well known curve:

$$\nu(x) = \frac{1}{\pi\sqrt{x(1-x)}} . \tag{10.5}$$

The Exponential Decay Rate

Let us apply the invariant density function $\nu(x)$ to estimate the persistence parameter τ in the exponential law in eqn. (10.4) for the decay of the survivors. Consider the collection of initial points and their orbits disregarding the target interval for the moment. After some small number of iterations (here about 36) the points of the orbits are scattered over the entire interval with a distribution which is approximately given by eqn. (10.5). Now we ask the question how many of these points fall into the target interval $J = [0.68, 0.69]$. This can be estimated by the value of the invariant distribution at the center of the interval, multiplied by the width, i.e.,

$$\frac{\nu(0.685)}{100} = \frac{1}{100\pi\sqrt{0.685(1-0.685)}} \approx 0.0068525$$

times the total number of points.[11]

Thus, based on a total of $10,000$ points, approximately 68.525 points are removed per iteration which roughly agrees with our numerical findings of 63 points. To estimate τ we need to compute the number of iterations necessary to reduce the number of remaining points to a fraction of $1/e$ based on the removal rate of a fraction of 0.0068525 of all points per iteration. Thus, we solve

$$(1-0.0068525)^\tau = \frac{1}{e}$$

for τ. The result is $\tau \approx 145$ iterations which again is in accordance with our numerical findings.

---

[11] Precisely the expectation of this number is

$$\int_{0.68}^{0.69} \frac{1}{\pi\sqrt{x(1-x)}}\,dx = \left[\frac{2}{\pi}\arctan\sqrt{\frac{x}{1-x}}\right]_{0.68}^{0.69} = \frac{2}{\pi}\left(\arctan\sqrt{\frac{0.69}{0.31}} - \arctan\sqrt{\frac{0.68}{0.32}}\right) \approx 0.0068527 .$$

times the total number of points.

Testing for Ergodicity

If we want to be sure, at least from an experimental point of view, that a given orbit is in fact ergodic, then we have to look at a whole sequence of experiments refining that of figure 10.17. One way to do this would be to take an arbitrary subinterval, say $[0.4, 0.5]$, and subdivide this interval into N, say again $N = 600$, subintervals and repeat the counting experiment. In other words, we increase the resolution by a factor of 10. Of course, the length of the orbit must be increased correspondingly to provide enough data to support the statistical evaluation we have in mind. If we still see that this subinterval $[0.4, 0.5]$ has positive values μ_k everywhere, then we have gone one step further in supporting the ergodicity hypothesis. Figure 10.18 shows the result.

Let us summarize what we have seen in our experiments so far. Starting the iteration with any number different from 0, 1, and 1/2 the behavior will always be ergodic. In particular, we do not see the periodic points which are theoretically present. But now with the following experiments let us turn everything upside down. And this really should show you how delicate it is to reason about chaos based on computer simulations.

A Bad Histogram

When we did the first quick tests to compute histograms, as in figure 10.17, we were not careful enough and stumbled into a big surprise. Our computations were done in single-precision arithmetic BASIC, and we picked an initial value which should reflect some arbitrariness, namely $x_0 = \sin(2)$. When we looked at the resulting histogram reproduced in figure 10.19 it did not at all show the distribution which we expected.

Very distinct gaps in the histogram are visible. They correspond to small subintervals which have not been visited by the orbit. We repeated the experiment with another initial value, this time $x_0 = 0.55$, and obtained something more satisfactory (see figure 10.20). But shouldn't the histogram be independent from the initial value?

Maybe we did not iterate enough times? We increased m from $m = 10^5$ to $m = 10^6$ and obtained exactly the same images. Now it was clear that something was really wrong. But what? The bad histogram and its stability versus an increase in the number m of iterations suggested that for $x_0 = \sin(2)$ the iteration had run into a periodic cycle. But didn't that contradict what we have discussed so far? Shouldn't small unavoidable errors in the computation guarantee that we never run into a periodic cycle? Unfortunately, not at all!

Our next experiment confirmed that. We simply asked the following questions: Given initial sampling values spread over the entire unit interval, $x_0 = i/600, i = 1, ..., 599$, is there a periodic cycle into which the iteration will run if started with x_0? And if

Histogram Detail

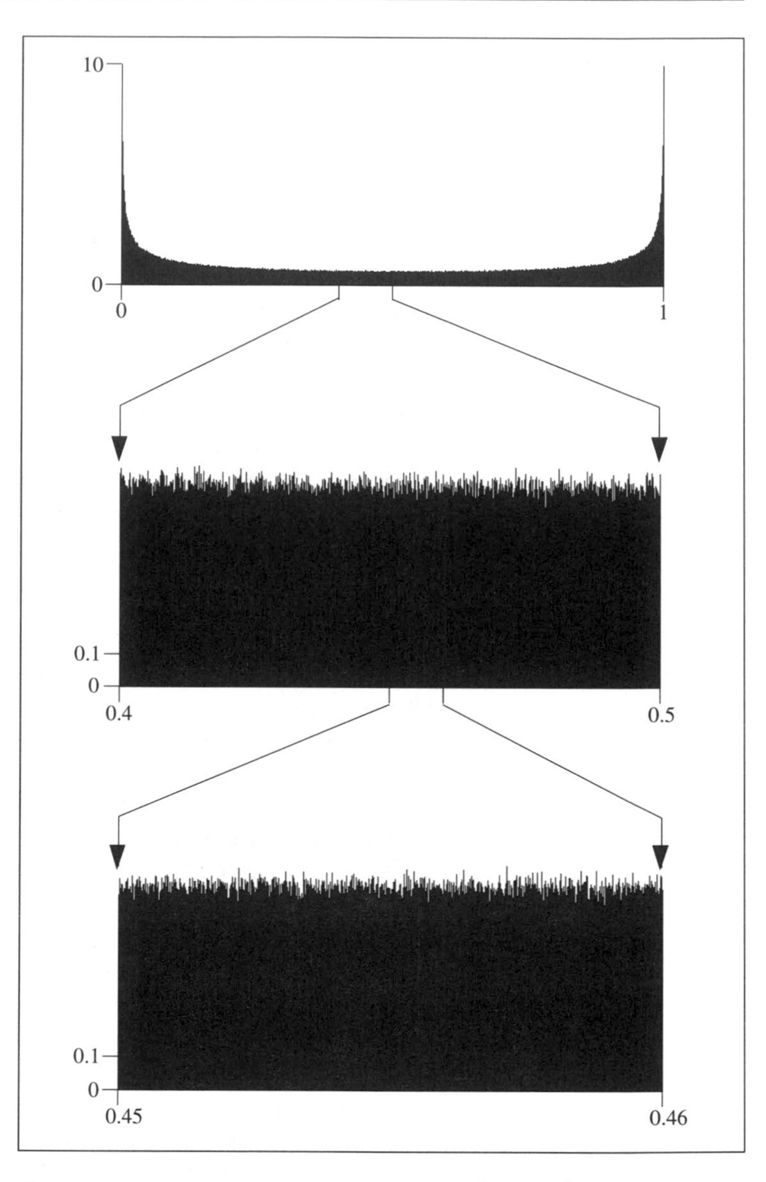

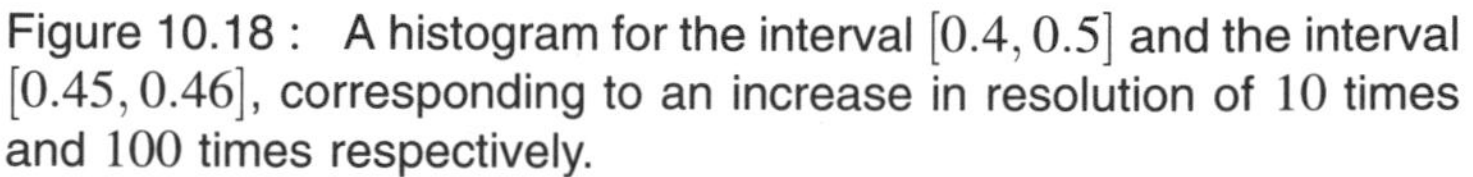

Figure 10.18 : A histogram for the interval $[0.4, 0.5]$ and the interval $[0.45, 0.46]$, corresponding to an increase in resolution of 10 times and 100 times respectively.

so, what is the corresponding period? The setup for the experiment was very simple. Starting at x_i, 5000 iterations were carried out, and we assumed that the last point x_{5000} was already a periodic point. To compute the corresponding period, we performed another 10000 iterations and checked at each iteration whether x_n was

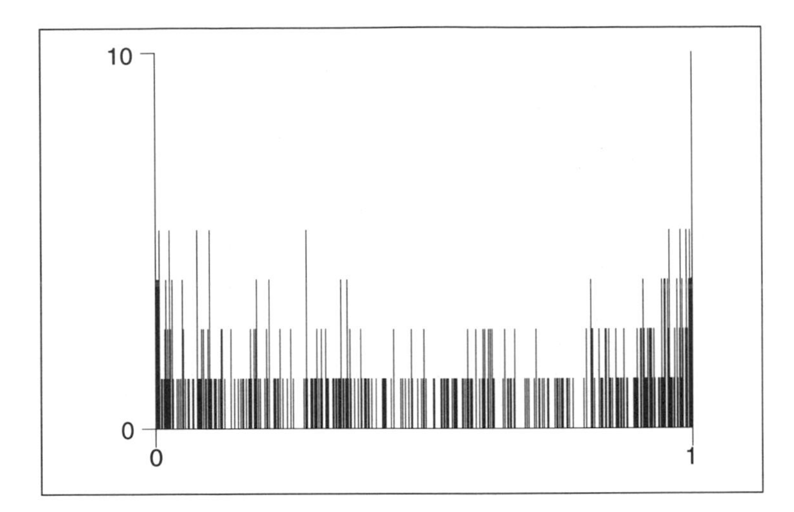

Single-Precision Histogram

Figure 10.19 : Histogram of 5000 iterations for $a = 4$ with intial point $x_0 = \sin(2)$ using single-precision arithmetic.

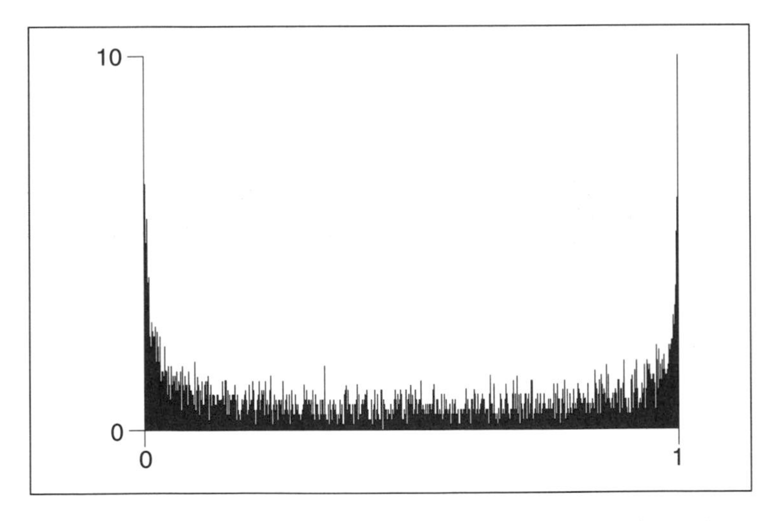

Another Single-Precision Histogram

Figure 10.20 : Histogram of 5000 iterations for $a = 4$ with intial point $x_0 = 0.55$ using single-precision arithmetic.

equal to x_{5000}. The first time that such an x_n was found, the period was determined as $n - 5000$.

Periodic Cycles

Indeed, it turned out that any of these sampling initial values eventually led to a periodic cycle of period 1, 436, 836, or 4344. Figure 10.21 (top) shows which periodic cycle was eventually reached for the different initial values. The majority of initial values led to the cycle of length 4344. By the way, $x_0 = \sin(2)$ led to a cycle of length 436, which was short enough to remain visible

Periodicity After All

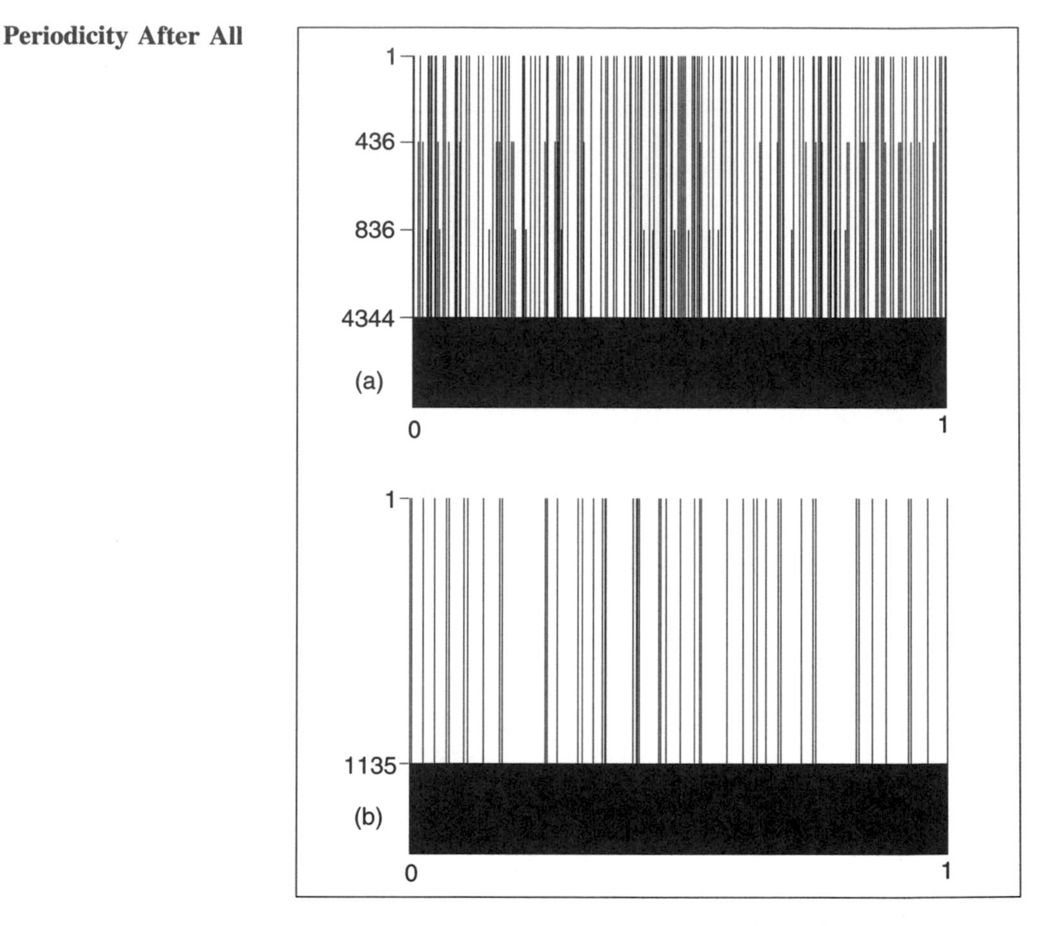

Figure 10.21 : (a) Implementation of $4x(1-x)$: Periods 1, 436, 836, 4344; (b) Implementation of $4x-4x^2$: Periods 1, 1135.

in the histogram reproduced in figure 10.19. Once this became clear we changed the implementation of the quadratic function in the code. Instead of $4x(1-x)$ we then used $4x-4x^2$, which is the same mathematically, of course, but makes a big difference in the experiment, as can be seen in the bottom part of the figure. Then almost all initial values led to a periodic cycle of period 1135, and a few to the fixed point 0, which corresponds to the number 1 in the figure.

So what is the conclusion? Once again we have experienced extreme sensitivity in the iteration which renders the computer results worthless and even misleading. These periodic orbits are artifacts of the computation, the result of systematic rounding-off or truncation errors. They seem to vanish as soon as the precision of the arithmetic is increased, as for example in figure 10.17, where

Grid Arithmetic

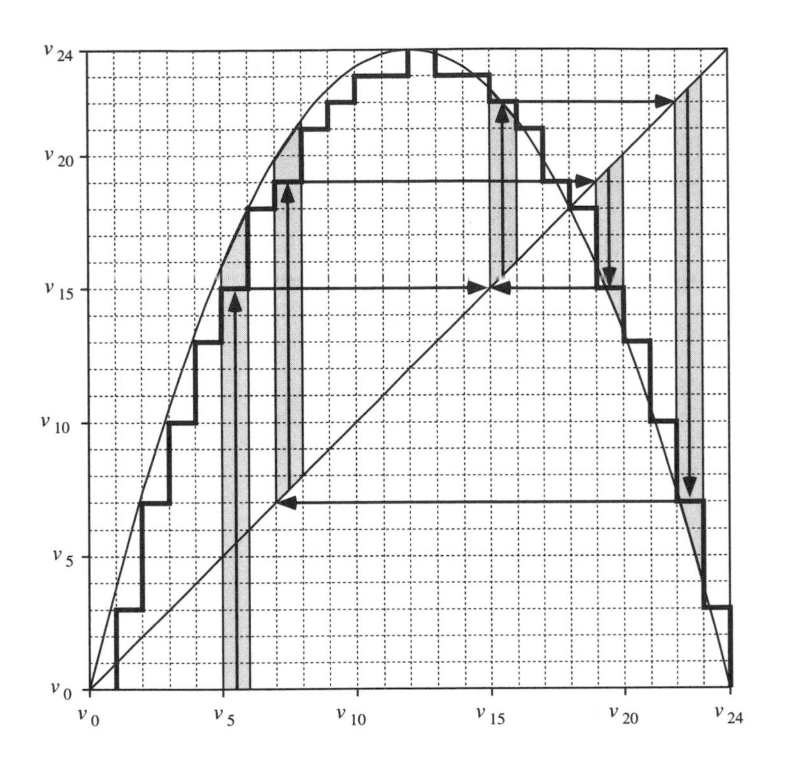

Figure 10.22 : Graphical iteration on a coarse grid. The parabola $4x(1-x)$ becomes a staircase function.

double-precision arithmetic is used.

Very Low Precision Arithmetic

Let us investigate the mechanism, which produces the periodic orbits, a bit further. To make things clearer we decrease the precision dramatically. In figure 10.22 we show a situation where we distinguish only 25 different values $v_0, \ldots, v_{24}$ in the range from $v_0 = 0$ to $v_{24} = 1$. Our purpose is to demonstrate the effect of truncation.[12] In this example all real numbers between v_n and v_{n+1} are truncated to the value v_n. Since the corresponding truncation is carried out for the evaluation of $4x(1-x)$, we obtain a staircase function as an approximation of the parabola as shown in our figure 10.22 and table 10.23.

Let us examine the iteration of v_5 which represents all real numbers from $5/24 \approx 0.2083$ to $6/24 = 0.25$ (not including the last number).[13] The quadratic iterator transforms this range of

[12] Truncating a number is the process of keeping the first significant digits, discarding or chopping off all others. Most computers nowadays use rounding instead which may modify the last possible significant digit depending on the following digits. For example, in a machine working with a precision of 5 decimal digits, the number 3.14159265 would be stored as 3.1415 when truncation is used, while the result would be 3.1416 with rounding.

[13] In general, v_k represents all numbers $k/24 + r$, where $r \in [0, 1/24)$.

Quadratic Transformation in Low Precision Arithmetic

0	0	5	15	10	23	15	22	20	13
1	3	6	18	11	23	16	21	21	10
2	7	7	19	12	24	17	19	22	7
3	10	8	21	13	23	18	18	23	3
4	13	9	22	14	23	19	15	24	0

Table 10.23 : This table captures the essence of the staircase function approximating the parabola in figure 10.22. For the 25 points $v_0, \ldots, v_{24}$ the indices of the transformed points are listed. Thus, $v_0 \to v_0$, $v_1 \to v_3$, $v_2 \to v_7$, and so on.

numbers to the interval from $95/144 \approx 0.6697$ to $3/4 = 0.75$. Thus, the size of the interval is expanded by the factor $13/6 \approx 2.17$, which agrees with our expectation from the analysis of the sensitivity to initial conditions. But what we observe using our low precision arithmetic is drastically different: v_5 is simply mapped onto v_{15}. We can interpret this as the introduction of an additional contraction in the iteration process: the range of real numbers represented by v_5 is contracted to v_{15}. The next step of the iteration leads us to v_{22}, then to v_7, v_{19} and again to v_{15}. Thus we have run into a periodic 4-cycle. We can also find a 3-cycle (v_3, v_{10}, v_{23}) and two fixed points (v_0 and v_{18}). There are no other periodic cycles, especially no periodic 2-cycles. All initial values of the iteration lead to one of these periodic orbits or fixed points. These facts are summarized and illustrated in table 10.24 and figure 10.25.

Long Term Behavior in Low Precision Arithmetic

Final state		Corresponding initial values
fixed point	0	0, 12, 24
fixed point	18	6, 18
3-cycle	3–10–23	1, 3, 4, 8, 10, 11, 13, 14, 16, 20, 21, 23
4-cycle	7–19–15–22	2, 5, 7, 9, 15, 17, 19, 22

Table 10.24 : The very low precision arithmetic allows only four different long term behaviors of the quadratic iteration, two fixed points and two periodic cycles of period three and four. For the cycles and the corresponding initial values only the corresponding indices are given. For example, 15 corresponds to v_{15}.

Periodic Orbits are Artifacts

If we increase the precision of our arithmetic (perhaps to 100 or 1000 different values) everything changes: the periodic orbits vanish and new ones arise. In other words, the periodic orbits are artifacts of the computation with limited precision. Indeed, essentially the same mechanisms can be found if we use floating point arithmetic. When iterating $4x(1-x)$ more and more orbits fall onto each other and only a few periodic cycles survive. This

The Complete Picture

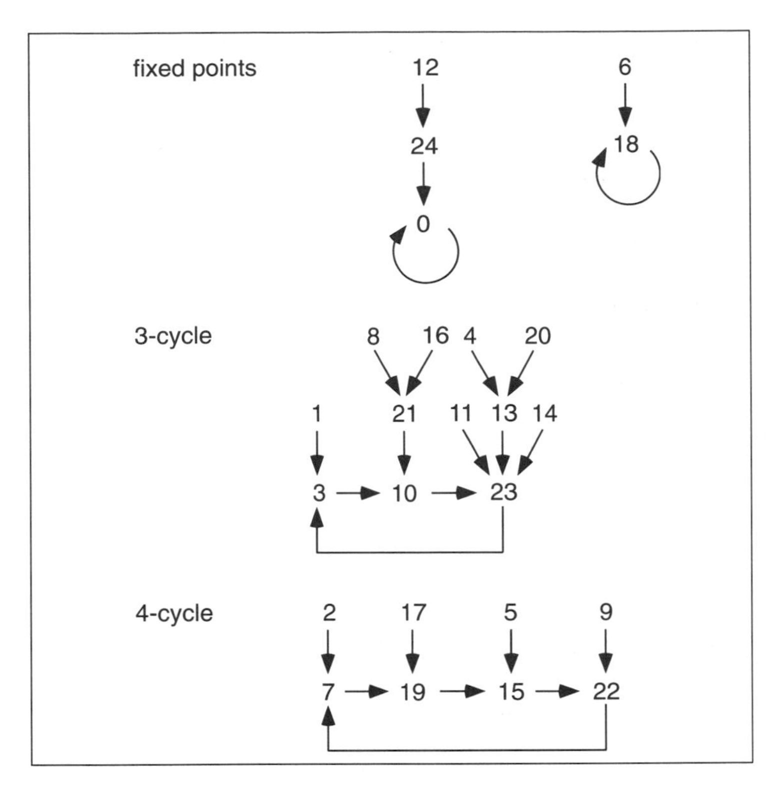

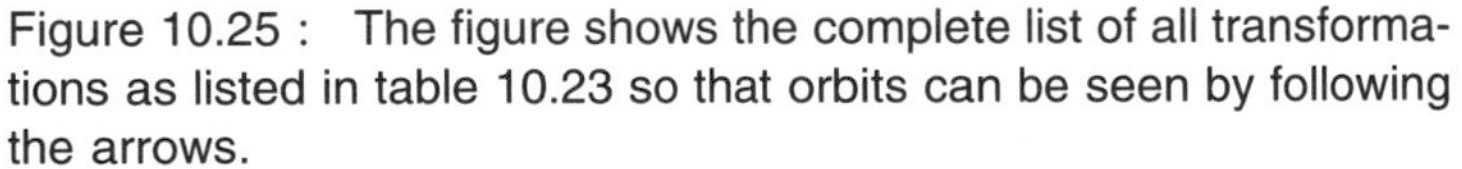
Figure 10.25 : The figure shows the complete list of all transformations as listed in table 10.23 so that orbits can be seen by following the arrows.

result explains figure 10.21.

Do you recall how we obtained the 'good' histogram in figure 10.17? We used double-precision arithmetic rather than single precision. Indeed, we still have that any iteration will eventually run into a periodic cycle — there is no way out simply because there are only finitely many, though very many, machine numbers — but these cycles are different from those found for single-precision arithmetic; and the length of the cycles will be so large that it doesn't destroy the experiment immediately. In fact, we have tested for periodic cycles for some samples of initial values and have not found that they become periodic for the first billion iterations.

10.4 Paradigm of Chaos: The Kneading of Dough

Generally speaking, the analysis of chaos is extremely difficult. However, in the specific model case of the quadratic iterator there is a beautiful and illuminating way to really understand chaotic behavior through and through. First we prepare the ground for one of the most important metaphors in chaos theory: the kneading[14] of dough. The kneading process guarantees that a pocket of spices inserted into the dough will be mixed thoroughly throughout the mass.

A Definition of Chaos

While a general definition for chaos applicable to most cases of interest is still lacking, mathematicians agree that for the special case of iteration of transformations there are three common characteristics of chaos.[15] These are:

1. sensitive dependence on initial conditions,
2. mixing, and
3. dense periodic points.

The kneading of dough provides an intuitive access to all of these mathematical properties of chaos. Moreover, we will see that the kneading process is closely related to the quadratic iterator! In this way the scenario of chaos that we have been discussing so far can be completely understood without having to resort to some higher mathematics.

Kneading with a Rolling Pin

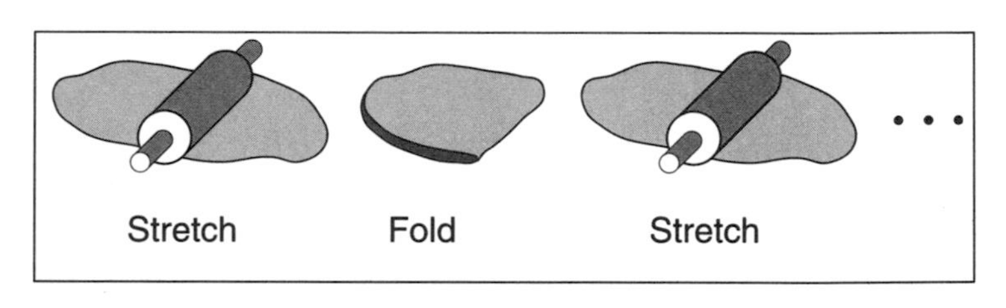

Figure 10.26 : Kneading as a feedback process: stretch, fold, stretch, and so on.

Kneading, a Deterministic Process

There is nothing random about the kneading process itself. Rather, a baker applies a certain action over and over again. We imagine kneading as the process of stretching the dough and folding it over, repeated many times. But in spite of this deterministic definition, the results have many features in common with randomness. Let us see why.

[14]Kneading = to mix and work a substance into a uniform mass, especially to fold, press and stretch dough with the hands (from *The American Heritage Dictionary of the English Language*, Houghton Mifflin, 1980).

[15]See R. L. Devaney, *An Introduction to Chaotic Dynamical Systems,* Addison-Wesley, Redwood City, 1989. Devaney uses the notion of *transitivity* for mixing.

Stretch-and-Fold

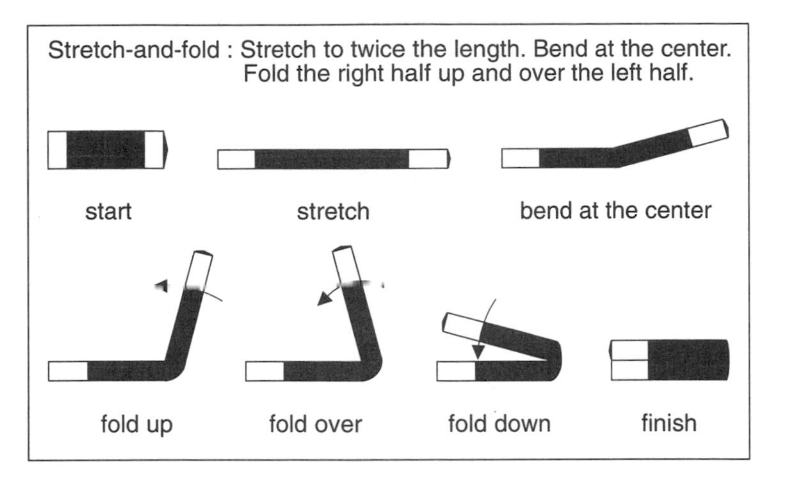

Figure 10.27 : Uniform kneading by stretch-and-fold.

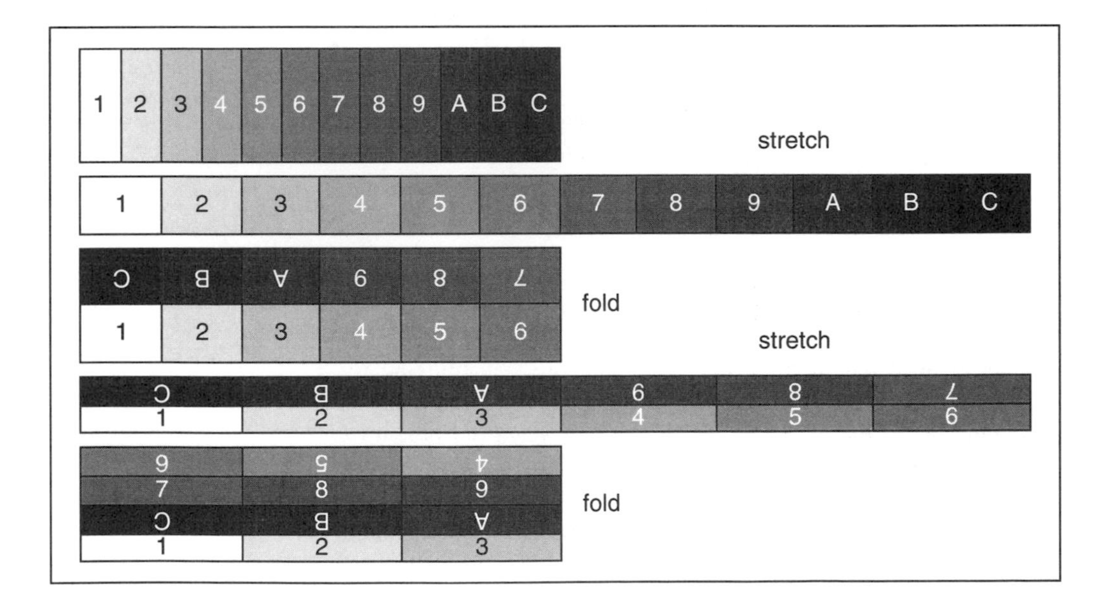

Figure 10.28 : Two operations of stretch-and-fold kneading applied to 12 textured blocks of dough.

Stretch-and-Fold Kneading

We look at an idealized situation, which certainly only a highly trained baker can achieve. The dough is homogeneously stretched to twice the length. Then it is bent at the center and folded over. Figure 10.27 shows a side view of this operation. Let us see how this kneading works on different parts of the dough by dividing the dough into 12 blocks which then are put through two stretch-

Kneading

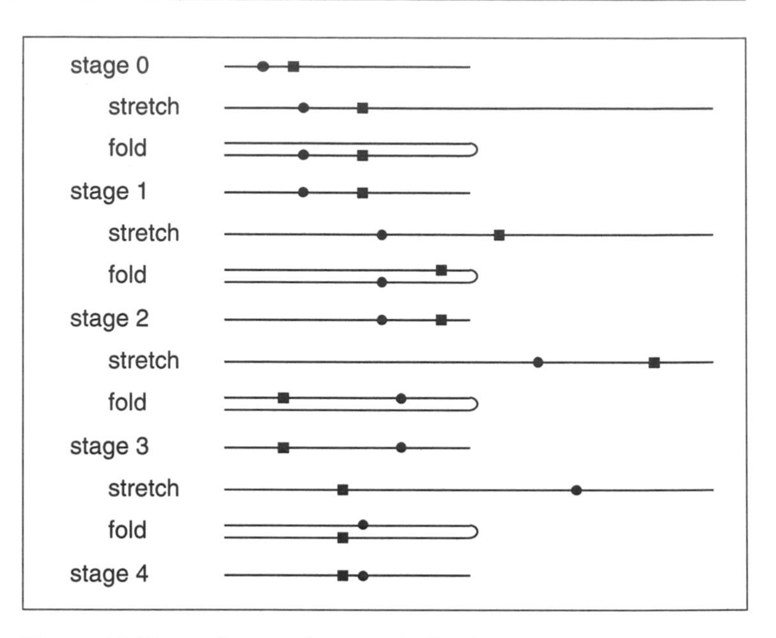

Figure 10.29 : Two grains, symbolized by a dot and a square, are subjected to four stages of the stretch-and-fold. They are mixed throughout the dough.

and-fold operations (see figure 10.28). The resulting layers of transformed blocks of dough already look a bit mixed.

We idealize the situation even one step further. Imagine we work with infinitely thin layers of dough. Folding these layers does not change the thickness and we can represent the dough by a line segment. Figure 10.29 shows some stages of the corresponding representation of the kneading process. We mark two grains of spice and follow their paths. The two grains are rather close together initially. But where will they be after a dozen kneadings? It is very likely that we will find them in very different places in the dough. In fact, that would be a consequence of the mixing properties of kneading. In other words, kneading destroys *neighborhood relations*. Grains which are very close initially will likely not be close neighbors after a while. This is the effect of *sensitive dependence on initial conditions*. Small deviations in initial positions lead to large deviations in the course of the process.

The stretch-and-fold operation leads to an easy and practical demonstration which allows us to see the effects of kneading. In figure 10.30, we propose a simple setup which simulates the stretch-and-fold kneading operations in eqn. (10.6).

The Chaos Machine

The machine has three parts, two needles, a rubber band of length d and an unstretchable thread of length $2d$. Both the rubber band and the thread are tied together at their end points, and one end is fixed with a needle to a piece of cardboard at point A. The other needle is fixed at distance d from the first needle at point B. Now the machine works in the following four steps:

Step 1: Mark an arbitrary point x_0 between A and B.
Step 2: Extend the rubber band from A to B and copy the point (and in subsequent steps, the last point $x_k, k = 0, 1, 2, ...$) marked between A and B to the rubber band.
Step 3: Extend the thread to its full length, thereby stretching the rubber band to double its original length and lead both from A to B and then back to A.
Step 4: Transfer the point from the stretched rubber band onto the cardboard, thus creating a new point (x_{k+1}) between A and B.

Now we can perform several interesting experiments by returning to step 2 after step 4 has been completed and run some number of such cycles. One experiment would be to just run the process, say, 50 times to see how the points marked between A and B distribute. You will find that they seem to spread around everywhere.

Another experiment would be best conducted by two different

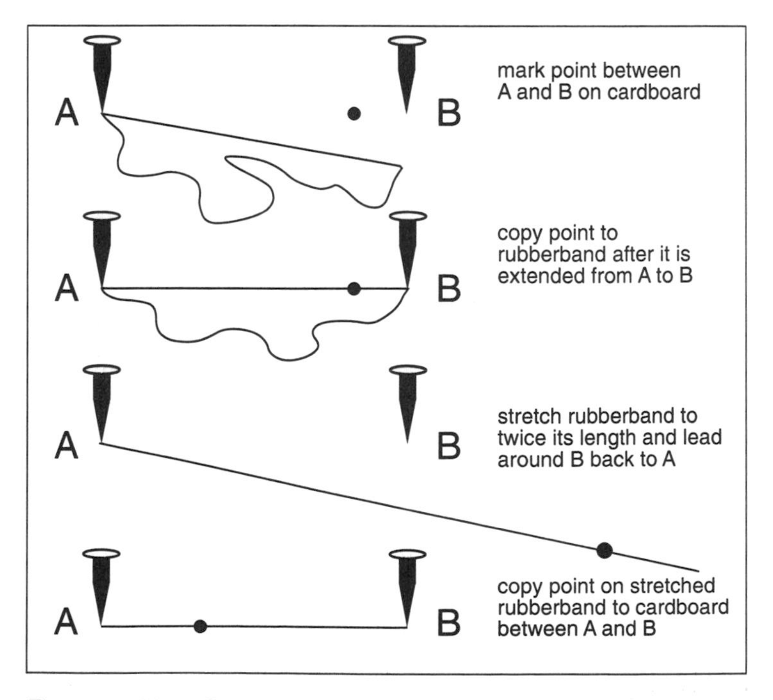

Figure 10.30 : Setup and operation of a simple chaos machine.

players. Both players use the same machine and start at the same initial point. Then they iterate, say 20 times, and thereafter compare their results. The surprise will be that the two results (e.g, the points which were copied last) will most likely be very different. After this last experiment, it will be very interesting to watch step by step how the two players start at the same position but then soon produce ever more deviating results. The machine displays chaos! In terms of the kneading of the dough, these experiments demonstrate that a tiny little pocket of spice will get mixed throughout the dough.

Stretch-Cut-and-Paste

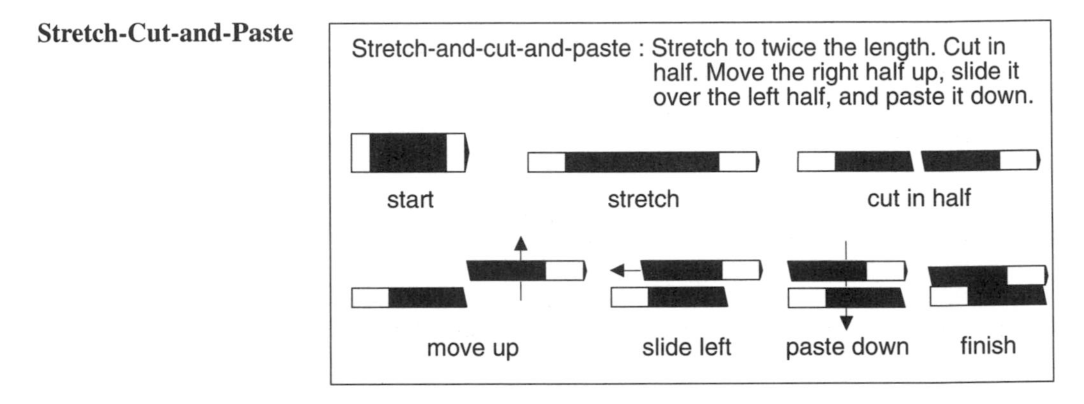

Figure 10.31 : Uniform kneading by stretch-cut-and-paste.

Stretch-Cut-and-Paste Kneading

Let us discuss a second kneading operation. Here we again stretch the dough uniformly to twice its length. But then we cut the dough at the center into two parts and paste them on top of each other (see figure 10.31).

When comparing the stretch-and-fold operation with the stretch-cut-and-paste operation, our intuition would be that both kneading operations apparently mix particles around, but in a very different manner generating quite distinct iteration behaviors. The surprise is, however, that both kneadings are essentially the same! A first idea of this fact can be obtained from figure 10.32. Again we divide the dough into 12 blocks. Then we apply the stretch-cut-and-fold operation followed by one stretch-and-fold operation. The result is compared with the one obtained for two succeeding stretch-and-fold operations in the bottom part of the figure. We observe that they are identical when ignoring the vertical order of the pieces!

This again suggests to neglect any thickness of the dough. Thus, from now on we think of the dough being represented by a line segment. And after a kneading operation — for example stretch-cut-and-paste — the result will be represented again by an

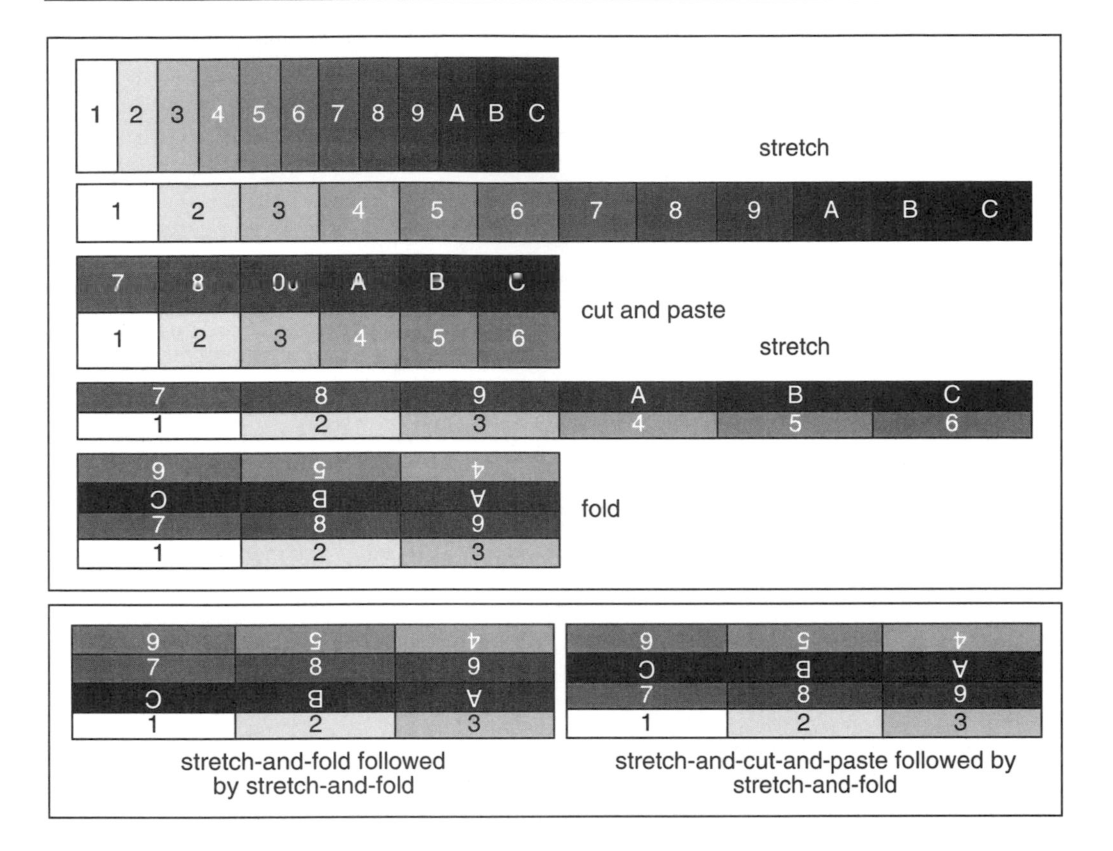

Figure 10.32 : Stretch-cut-and-paste followed by stretch-and-fold applied to 12 textured blocks of dough. The resulting horizontal order of the blocks is identical to the one obtained from the application of the two succeeding stretch-and-fold operations in figure 10.28.

immaterial straight line. This is the first step towards a mathematical model of the kneading operations. Taking the interval $[0, 1]$ as the original line segment modeling the dough, we can now check how the two different kneading operations act. Let us use the symbol T for the stretch-and-fold operation and the symbol S for stretch-cut-and-paste.

As an example, let us follow a particle through several kneading steps once applying the stretch-and-fold operation T three times (see figure 10.33, left) and another time using two stretch-cut-and-paste transformations S followed by one stretch-and-fold operation T (see figure 10.33, right). We observe in both experiments that the particle arrives exactly at the same position, though the route in between is different. Thus, if the particle is initially at position

Tracing a Particle

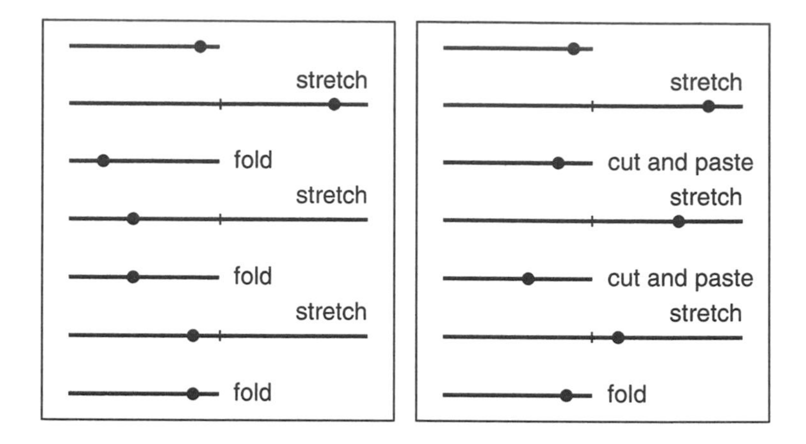

Figure 10.33 : The interval $[0,1]$ at the top is the model for the dough. We follow the path of a particle at $x = 9/10$ when the following kneading operations are applied: left T, T, T; right S, S, T.

x, we have seen that

$$T(T(T(x))) = T(S(S(x))) .$$

This experience along with the result in figure 10.32 motivates us to conjecture an substitution property of the two kneading operations. In fact, we will prove this conjecture below starting on page 156.

Substitution Property

Fact. *N kneading steps using the stretch-and-fold operation T, i.e.,*[16]

$$T^N = \underbrace{TT \cdots T}_{N \text{ times}}$$

yield the same material in each vertical column as $N-1$ kneading steps by the stretch-cut-and-paste operation S followed by one kneading step of T, i.e.,

$$TS^{N-1} = T \underbrace{SS \cdots S}_{N-1 \text{ times}} .$$

Formula for Stretch-and-Fold Kneading

The mathematical model for kneading of the one-dimensional ideal of dough is a function. The stretch-and-fold kneading operation is represented by the following transformation, for which we use the symbol T again:

$$T(x) = \begin{cases} 2x & \text{if } x \le 0.5 \\ -2x + 2 & \text{if } x > 0.5 . \end{cases} \tag{10.6}$$

[16]It is common mathematical notation to interpret a composition of operators such as $TSSS$ from right to left. In this case S is applied three times followed by one application of T.

Tent Transformation

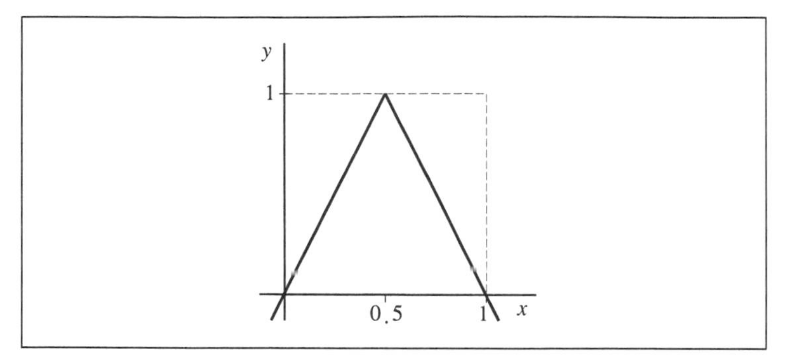

Figure 10.34 : Graph of piecewise-linear tent transformation eqn. (10.6) corresponding to stretch-and-fold transformation.

Saw-Tooth Transformation

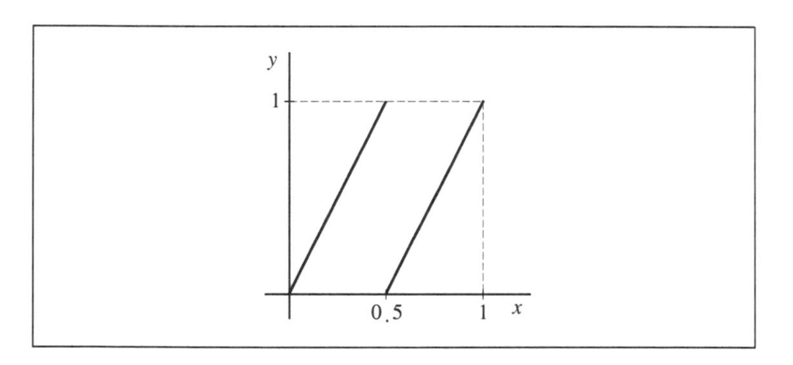

Figure 10.35 : Graph of saw-tooth transformation eqn. (10.7) corresponding to the stretch-cut-and-paste transformation.

Figure 10.34 shows the graph of this transformation, which is called the *tent transformation* because of its shape. This graph looks like a simplification of the parabola.

The justification of this model is almost self-evident. The dough is modeled by the unit interval $[0, 1]$. The stretching operation is taken care of by the factor 2 in front of x. The first half interval of $[0, 1]$ is only stretched and not folded. Thus, the first part of the definition of T is in order, $T(x) = 2x$ for $x \leq 1/2$. The second half interval becomes $[1, 2]$ after the stretching, and must be folded over its left end point. This is equivalent to folding at $x = 0$, i.e., multiplying by -1 and shifting to the right by two units. Thus, $T(x) = -2x + 2$ for $x \geq 1/2$.

Formula for Stretch-Cut-and-Paste Kneading

The model for the second procedure, the stretch-cut-and-paste kneading operation, is another elementary mathematical transformation, the *saw-tooth transformation* S, defined for numbers x

from the unit interval $[0, 1]$:

$$S(x) = \begin{cases} 2x & \text{if } \ x < 0.5 \\ 2x - 1 & \text{if } \ x \geq 0.5 \end{cases} \tag{10.7}$$

Its graph (see figure 10.35) justifies the name. Again the verification of the model is evident, and we omit the details.[17] It is this transformation and its close relatives which is one of the most powerful keys to understanding chaos.

Substitution Property of the Kneading Operations

As announced we can now show the substitution property of the kneading operations using their appropriate mathematical descriptions. Let T denote the piecewise-linear tent transformation (10.6) and S the saw-tooth transformation (10.7). Then we can verify by straightforward calculation that for any x in $[0, 1]$

$$TT(x) = TS(x) \; . \tag{10.8}$$

This is visualized in figure 10.36.
Let us look at an example: $x = 0.9$. Then

$$\begin{aligned} T(0.9) &= -2 \cdot 0.9 + 2 = 0.2 \\ T(0.2) &= 2 \cdot 0.2 = 0.4 \; . \end{aligned}$$

On the other hand,

$$\begin{aligned} S(0.9) &= 2 \cdot 0.9 - 1 = 0.8 \\ T(0.8) &= -2 \cdot 0.8 + 2 = 0.4 \; . \end{aligned}$$

Now let us check eqn. (10.8) formally. On the left side we compute the result in four cases as follows.

$$\begin{array}{llll} T(T(x)) = & T(2x) & = \;\; 4x & \text{for } \ 0 \leq x \leq 0.25 \\ T(T(x)) = & T(2x) & = -4x + 2 & \text{for } \ 0.25 < x \leq 0.5 \\ T(T(x)) = & T(-2x+2) & = \;\; 4x - 2 & \text{for } \ 0.5 < x \leq 0.75 \\ T(T(x)) = & T(-2x+2) & = -4x + 4 & \text{for } \ 0.75 < x \leq 1 \end{array}$$

and the corresponding right hand sides yield

$$\begin{array}{llll} T(S(x)) = & T(2x) & = \;\; 4x & \text{for } \ 0 \leq x \leq 0.25 \\ T(S(x)) = & T(2x) & = -4x + 2 & \text{for } \ 0.25 < x < 0.5 \\ T(S(x)) = & T(2x-1) & = \;\; 4x - 2 & \text{for } \ 0.5 \leq x \leq 0.75 \\ T(S(x)) = & T(2x-1) & = -4x + 4 & \text{for } \ 0.75 < x \leq 1 \end{array}$$

which completes the proof. (Note that $-4x+2 = 4x-2$ for $x = 0.5$.)

[17] Only for the point $x = 1/2$ it is not self-evident what the corresponding value, $S(1/2)$, of the saw-tooth function should be. We cut the dough precisely at $x = 1/2$. Now the right end point of the left portion of the dough, which is 1/2, moves to the point 1. However, the left end point of the right dough segment also corresponds to 1/2, but it is moved to 0. Thus, we may choose either $S(1/2) = 0$ or $S(1/2) = 1$ in the definition of S. However, all of the following arguments can be carried through with either definition using only very minor modifications. For the presentation in this book we have chosen the first option which leads to a slightly simpler discussion.

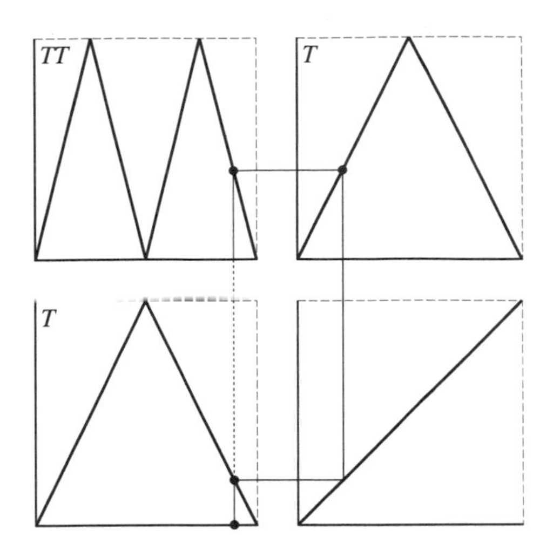

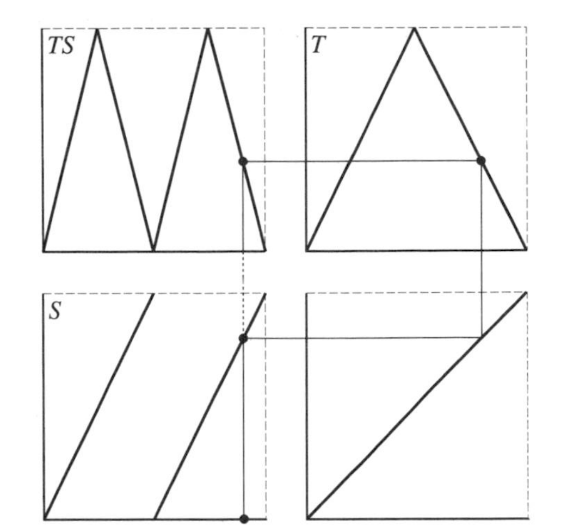

Figure 10.36 : The composition of two tent functions (left) and the saw-tooth function followed by the tent function (right). Both approaches lead to the same result, a double tent. The lower left graph in each part provides for a given x the value $T(x)$ (left) or $S(x)$ (right) which then is fed to the upper right graph leading to the value $T(T(x))$ (left) and $T(S(x))$ (right). The upper left graph is the graph of the composition TT (left part) and TS (right part). Both are the same demonstrating the substitution property.

The identity eqn. (10.8) is the key to obtaining identities for higher iterations very elegantly: for example

$$TTT(x) = TSS(x) ,$$

or

$$TTTT(x) = TSSS(x) ,$$

and so on. Indeed, when we apply T to both sides of

$$TT(x) = TS(x) ,$$

we obtain

$$TTT(x) = TTS(x) .$$

Now we can substitute TS for TT on the right side and obtain

$$TTT(x) = TSS(x) ,$$

and so on. This actually means that N iterations of T — stretch-and-fold — lead to the same result as $N-1$ iterations of S — stretch-cut-and-paste — followed by one application of T. This is the substitution property.

The Generic Parabola

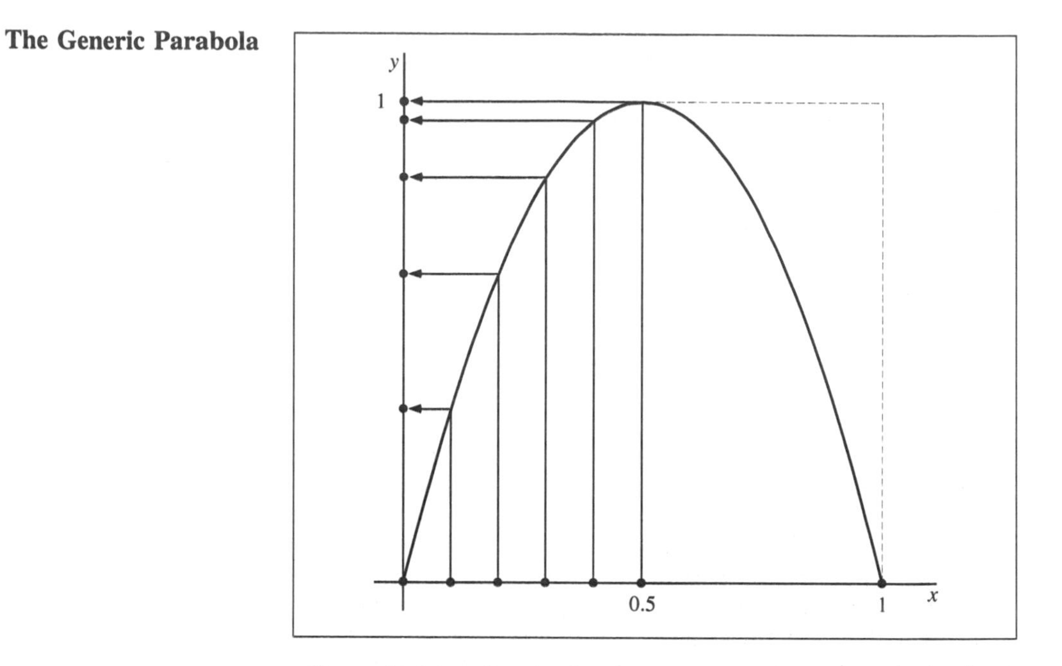

Figure 10.37 : Graph of the generic parabola $y = 4x(1 - x)$.

Kneading and the Quadratic Iterator

Let us introduce an argument which makes a connection between the feedback system $x \to 4x(1 - x)$ and the kneading of dough. When we graph the transformation $y = 4x(1 - x)$ in an xy-coordinate system, we obtain the generic parabola shown in figure 10.37.

Here we are interested only in x-values ranging from 0 to 1. Note, that corresponding y-values also range from 0 to 1. More precisely, we have that y monotonically increases for $x < 1/2$ and monotonically decreases for $x > 1/2$. We observe that the interval $[0, 1/2]$ on the x-axis is stretched out to the interval $[0, 1]$ on the y-axis, and the same is true for the interval $[1/2, 1]$. In other words, the transformation $4x(1 - x)$ stretches both intervals to twice their length.

Non-Uniform Stretching

The stretching, however, is non-uniform. In fact, small intervals which are close to 0 or 1 are stretched a great deal, while intervals, which are close to the midpoint $1/2$ are compressed. To show this we have put markers on the x-axis which are equally spaced and observe how their corresponding y-values are not uniformly spaced in figure 10.37.

Now we can get to the point of making a connection with kneading. What happens if we apply the transformation $4x(1 - x)$ to the interval $[0, 1]$? We already know that each half of the interval is stretched to twice its length. Moreover, checking the

Non-Uniform Stretching With Parabola

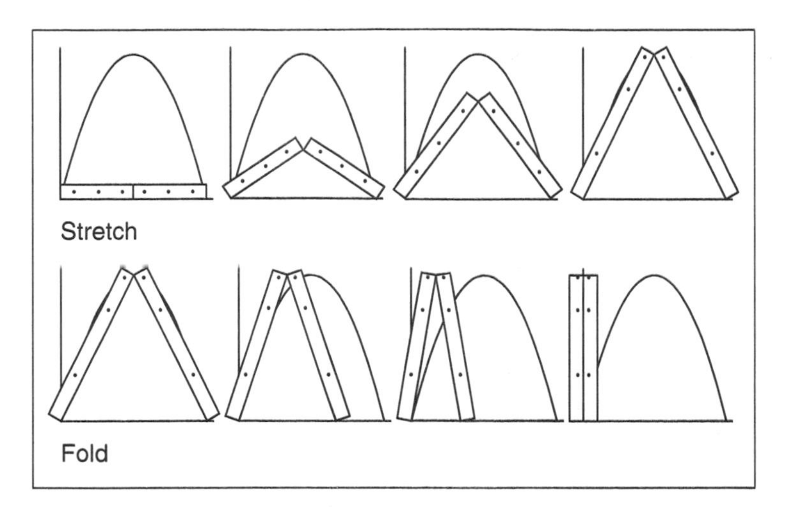

Figure 10.38 : Interpretation of the quadratic transformation as a stretch-and-fold operation using elastic bars.

end points of the intervals, we find: $0 \rightarrow 0$, $1/2 \rightarrow 1$, $1 \rightarrow 0$. This means that the result of one application of the transformation $4x(1-x)$ to the interval $[0,1]$ can be interpreted as a combination of a stretching and folding operation (see figure 10.38). In other words, the iteration of $x \rightarrow 4x(1-x)$ is a relative of the uniform stretch-and-fold kneading operation. Actually we will see that it is a very close relative.

To understand the properties of chaos for the quadratic iterator we will study those of the iteration of the tent function and show that both are equivalent. The goal of this section is to derive the central tool for this purpose, a formula that allows the direct computation of any iterate for the tent transformation without having to carry out the iteration process over and over again. In other words, we will obtain an explicit expression for the result $T^k(x_0)$ for all initial points x_0 and all iteration stages k. The first piece of the solution of this problem is the substitution property of the tent and the saw-tooth transformations explained above. This yields $T^k(x_0) = TS^{k-1}(x_0)$. The remaining piece is an explicit formula for the iteration of the saw-tooth function needed for the evaluation of $S^{k-1}(x_0)$ and is demonstrated next.

A New Notation for the Saw-Tooth Function

There is an elegant formula for the saw-tooth function which is different from the original definition in eqn. (10.7). It uses a function which computes the fractional part[18] Frac (x) of a number x.

$$\text{Frac}\,(x) = x - k \quad \text{if} \quad k \leq x < k+1, \quad k \text{ integer}\,. \tag{10.9}$$

[18] In the mathematical literature the fractional part Frac (x) of a number x is usually expressed using the modulo function, written as $x \bmod 1$.

Some examples for the evaluation of this function are

$$\begin{aligned} \text{Frac}\,(0.4) &= 0.4 \\ \text{Frac}\,(5.123) &= 0.123 \\ \text{Frac}\,(18) &= 0 \\ \text{Frac}\left(24/7\right) &= 3/7\;. \end{aligned}$$

With this notation the saw-tooth transformation can be written as

$$S(x) = \text{Frac}\,(2x) \quad \text{for} \quad 0 \le x < 1\;. \tag{10.10}$$

This is straightforward to check. If $0 \le x < 1/2$, then $0 \le S(x) = 2x < 1$ and $\text{Frac}\,(2x) = 2x$, yielding the same result. On the other hand, if $1/2 \le x < 1$, then $0 \le S(x) = 2x - 1 < 1$, thus, $1 \le 2x < 2$ and $\text{Frac}\,(2x) = 2x - 1$, giving the same result again. Only for the point $x = 1$ the formula in eqn. (10.10) does not work. But this is not significant, because $x = 1$ is a fixed point of the operator S and, moreover, there are no other points in the unit interval which are transformed to this fixed point. In other words, the dynamics of the iteration of the saw-tooth transformation S in the unit interval can be split into two independent parts, the interval $[0, 1)$, which does not include the point $x = 1$, and the singleton $\{1\}$. Of course, all of the interesting dynamics happens in $[0, 1)$. Thus, it is no loss to neglect the fixed point $x = 1$ in the following discussion, where we will use mostly the representation of eqn. (10.10) for the saw-tooth function.

The first advantage of the Frac-version of the saw-tooth transformation is the fact that iterating this transformation is possible in a closed form. What does that mean? Assume we start with $0 \le x_0 < 1$ and compute

$$x_1 = \text{Frac}\,(2x_0)\;,$$

and then

$$x_2 = \text{Frac}\,(2x_1)\;,$$

and so on. In other words

$$x_{k+1} = \text{Frac}\,(2x_k)\,, \quad k = 0, 1, 2, 3, \ldots$$

Assume we want to know what x_k will be for some very large value of k. Do we have to carry out the iteration process k times? Not at all. We can express x_k explicitly in terms of x_0 alone. The result of the following straightforward derivation is the simple closed form

$$x_k = \text{Frac}\left(2^k x_0\right)\;. \tag{10.11}$$

Let us look at an example. We iterate $x_0 = 8/25$ ten times using the saw-tooth function and check the closed form.

k	x_k	$2^k x_0$	Verify $x_k = \text{Frac}\left(2^k x_0\right)$
0	8/25	8/25	$8 = 0 \cdot 25 + 8$
1	16/25	16/25	$16 = 0 \cdot 25 + 16$
2	7/25	32/25	$32 = 1 \cdot 25 + 7$
3	14/25	64/25	$64 = 2 \cdot 25 + 14$
4	3/25	128/25	$128 = 5 \cdot 25 + 3$
5	6/25	256/25	$256 = 10 \cdot 25 + 6$
6	12/25	512/25	$512 = 20 \cdot 25 + 12$
7	24/25	1024/25	$1024 = 40 \cdot 25 + 24$
8	23/25	2048/25	$2048 = 81 \cdot 25 + 23$
9	21/25	4096/25	$4096 = 163 \cdot 25 + 21$
10	17/25	8192/25	$8192 = 327 \cdot 25 + 17$

Derivation of the Closed Form Iterate

The closed form eqn. (10.11) of the iteration of the saw-tooth function eqn. (10.10) follows from two basic properties of the operator Frac. Let m be any integer. Then

$$\begin{aligned} \text{Frac}\,(x+m) &= \text{Frac}\,(x) \\ \text{Frac}\,(mx) &= \text{Frac}\,(m\text{Frac}\,(x)) \ . \end{aligned}$$

The first formula follows directly from the definition in eqn. (10.9). To verify the other we let k be an integer such that $k \le x < k+1$. Thus, $\text{Frac}\,(x) = x - k$. Then

$$\begin{aligned} \text{Frac}\,(m\text{Frac}\,(x)) &= \text{Frac}\,\big(m(x-k)\big) = \text{Frac}\,(mx - mk) \\ &= \text{Frac}\,(mx) \end{aligned}$$

where the last equality follows from the first property. With these results we show the validity of the closed form iterate eqn. (10.11) by induction. The first iterate

$$x_1 = \text{Frac}\,(2x_0) \ ,$$

is already in the closed form as claimed. For the induction step from k to $k+1$ let us assume the hypothesis

$$x_k = \text{Frac}\,\left(2^k x_0\right) \ .$$

Then we compute

$$\begin{aligned} x_{k+1} &= \text{Frac}\,(2x_k) = \text{Frac}\,\left(2\text{Frac}\,\left(2^k x_0\right)\right) \\ &= \text{Frac}\,\left(2^{k+1} x_0\right) \end{aligned}$$

using the second property of the Frac-operator in the last equality. This concludes the proof by induction.

We do not recommend the closed form of the saw-tooth transformation for the numerical computation of the iterations because the required powers of 2 rapidly grow large and become untractable. However, the closed form is of great value for the theoretical discussion of chaos in the iteration as we will see in the next section.

Direct Computation of Iterates of the Tent Transformation

We now have collected the two necessary tools for the final result of this section, namely the substitution property of the kneading operations and the closed form of the iteration of the saw-tooth transformation. Assume we have x_0 and would like to know the result x_k after k stretch-and-fold operations T. Without the substitution property there would be no other way to do this than to compute iteration after iteration. But with the aid of the substitution property, we would first compute $k-1$ iterations of the stretch-cut-and-paste operation S, based on the shortcut according to the explicit formula eqn. (10.11), which gives

$$y = S^{k-1}(x_0) = \text{Frac}\left(2^{k-1}x_0\right)$$

and then apply the stretch-and-fold operation T once,

$$x_k = T(y) = \begin{cases} 2y & \text{if } y \leq 0.5 \\ -2y+2 & \text{if } y > 0.5 \ . \end{cases}$$

In other words, rather then iterating k times, we just have to compute a power of 2, and carry out two multiplications and one addition! Extracting the fractional part, i.e., evaluation the function Frac costs almost nothing because it just means neglecting the integer part of the result.

Let us present an example. The initial point $x_0 = 8/25$ is a periodic point of the tent transformation with period 10. We can now check this fact without iterating ten times! Instead we compute x_{10} directly, using the above method for $k = 10$. We obtain

$$\begin{aligned} y = S^9(x_0) &= \text{Frac}\left(2^9 \cdot 8/25\right) = \text{Frac}\left(4096/25\right) \\ &= \text{Frac}\left(163 + 21/25\right) = 21/25 \ . \end{aligned}$$

The result $y = 21/25$ is greater than 1/2. Thus,

$$x_{10} = T(y) = -2y + 2 = -42/25 + 2 = 8/25 \ .$$

Voilà! $x_{10} = x_0$.

Summary of Results

Let us summarize the mathematical results of this section. They will be used in the following. The saw-tooth function

$$S(x) = \begin{cases} 2x & \text{if } x < 0.5 \\ 2x - 1 & \text{if } x \geq 0.5 \end{cases}$$

is the mathematical model for the stretch-cut-and-paste kneading. Using the notation of the fractional part

$$\text{Frac}\,(x) = x - k \;\text{ if }\; k \le x < k+1, \;\; k \text{ integer}$$

we can put the saw-tooth function into the simplified form

$$S(x) = \text{Frac}\,(2x) \quad \text{for} \quad 0 \le x < 1 \; .$$

There is an explicit formula for the kth iterate,

$$S^k(x_0) = \text{Frac}\left(2^k x_0\right) \; .$$

The tent transformation

$$T(x) = \begin{cases} 2x & \text{if } \; x \le 0.5 \\ -2x+2 & \text{if } \; x > 0.5 \end{cases}$$

is the mathematical model for the stretch-and-fold kneading. The substitution property relates the two kneading transformations in the sense that $k-1$ applications of S followed by one transformation T yield the same result as the tent transformation T, applied k times,

$$T^k(x) = TS^{k-1}(x) \; .$$

This relation allows the direct computation of iterates also for the tent transformation. Let

$$y = S^{k-1}(x_0) = \text{Frac}\left(2^{k-1} x_0\right)$$

and compute

$$x_k = T^k(x_0) = T(y) = \begin{cases} 2y & \text{if } \; y \le 0.5 \\ -2y+2 & \text{if } \; y > 0.5 \; . \end{cases}$$

10.5 Analysis of Chaos: Sensitivity, Mixing, and Periodic Points

We are now prepared to carry out the next step in the plan of attack for unraveling the chaos for the quadratic iterator. We start with the chaotic properties for the iteration of the saw-tooth transformation using the closed form description to derive the central properties of chaos: sensitivity, mixing, and dense periodic points. The substitution property allows us to carry over these features to the iteration of the tent transformation. In the following section we conclude the analysis of chaos by exploiting another equivalence, namely between the tent transformation and the quadratic iterator.

We begin with the saw-tooth transformation

$$S(x) = \text{Frac}\,(2x) \quad \text{for} \quad 0 \le x < 1$$

and reveal a new interpretation by passing to binary representations of the real number x between 0 and 1. Recall that any real number x from the unit interval can be written as $x = 0.a_1a_2a_3...$, where the a_k are *binary digits*, i.e., each a_k is either 0 or 1 and

$$x = a_1 2^{-1} + a_2 2^{-2} + a_3 2^{-3} + ...$$

For example $1/2 = 0.100...$, $3/4 = 0.1100...$, $1/3 = 0.\overline{01}$ (overlining means periodic repetition), $1/7 = 0.\overline{001}$. One useful observation related to binary expansions is the following. Let x and y be numbers in the unit interval having binary expansions $x = 0.a_1a_2a_3...$ and $y = 0.b_1b_2b_3...$ Then

$$|x - y| \le 2^{-k} \tag{10.12}$$

provided that x and y agree in the first k binary digits, i.e., $a_i = b_i$ for $i = 1, ..., k$.[19]

The Shift Operator

Now what does the iteration of the saw-tooth function mean in terms of binary expansions? Very simply, multiplication by 2 means passing from $0.a_1a_2a_3...$ to $a_1.a_2a_3...$ Therefore one application of the transformation is accomplished by first shifting all binary digits one place to the left and then erasing the digit that is moved in front of the point,

$$x = 0.a_1a_2a_3... \rightarrow S(x) = 0.a_2a_3a_4... \tag{10.13}$$

Because of the type of this almost mechanical procedure the transformation is also called the *shift operator* when interpreted in the context of binary representations. For the examples above, we list the results in the following table 10.39.

[19]To derive this property, we assume $x > y$ and compute

$$|x - y| = |0.a_1a_2a_3... - 0.b_1b_2b_3...| \ge |0.a_1...a_k\overline{1} - 0.b_1...b_k\overline{0}| = |0.0...0\overline{1}| = 2^{-k}\ .$$

x_0	Binary	Transformed Binary	x_1
1/2	0.1	0.0	0
3/4	0.11	0.1	1/2
1/3	$0.\overline{01}$	$0.\overline{10}$	2/3
1/7	$0.\overline{001}$	$0.\overline{010}$	2/7

Table 10.39 : Four examples for the transformation of the shift operator.

Resolving the Ambiguity

There is a technicality which we must address here, namely the ambiguity of the binary representations. For example, the decimal number 0.5 has two possible binary versions, 0.1 and $0.0111... = 0.0\overline{1}$. The application of the shift operator will have different results for the two numbers, 0 and $0.\overline{1} = 1$. To arrive at the correct result, we require that binary numbers must not end with repeating digits '1'. Thus, we represent $1/2 = 0.1$ and $1/4 = 0.01$, but not as $0.0\overline{1}$ or $0.00\overline{1}$.[20]

Binary Encoding of a Number

How can one obtain the binary representation of a decimal or rational number? There are several small algorithms for this purpose. It may come as a surprise that the shift operator provides a direct method to do the transformation. Interpreted this way it is a binary encoder. Here are the details.

Assume the binary representation of a number x with $0 \leq x < 1$ is desired. Set $x_0 = x$ and start the iteration of the shift operator. Thus, we compute

$$x_{n+1} = \text{Frac}\,(2x_n)\,, \quad n = 0, 1, ...$$

Then the binary representation of x is given by

$$0.a_0 a_1 a_2 ...$$

where the binary digits a_k are related to the kth iterate x_k:

$$a_k = \begin{cases} 0 & \text{if } x_k < 1/2 \\ 1 & \text{otherwise}\,. \end{cases}$$

Let us take two examples, $3/4$ and $1/7$. In the first case the iteration gives $3/4$, $1/2$, 0, 0, 0, ... Thus, the binary encoding of $3/4$ is 0.11000... For $1/7$ we obtain a periodic cycle $1/7$, $2/7$, $4/7$, $1/7$, ... Thus, the binary encoding is also periodic, namely $0.001001... = 0.\overline{001}$.

[20] Two remarks are in order here. Firstly, our convention to outlaw all binaries ending with repeating digits '1' implies that we cannot represent the number 1.0 in the form 0.111... But we have already argued that the point $x = 1$ is uninteresting and irrelevant for the dynamics of the iteration of the saw-tooth transformation. Secondly, referring to the footnote on page 156, note that in order to comply with the alternative definition of the saw-tooth function, where $S(1/2) = 1$, we would have to forbid all finite binaries, i.e., numbers $0.a_1 a_2 a_3...$ that end with repeating digits '0'.

The motion of a spice particle in the dough (when kneading by the stretch-cut-and-paste operation) can now be studied by investigating eqn. (10.13), which is also called the *shift on two symbols*. As simple as it may look, the dynamics which it generates are extremely complicated. We now turn to the three characteristics of chaos in iterated transformations: sensitivity, mixing, and periodic points merged in everywhere. All three of them can be verified for the stretch-cut-and-paste kneading.

Sensitivity to Initial Conditions

Imagine that we pick an initial number $x_0 = 0.a_1a_2a_3...$ but only specify it up to N digits, say $N = 100$. Then the true number will differ from the specified one by at most 2^{-100}, a difference so small that we would say it should not matter at all. In any event, we can consider this difference to be something like an error of measurement. Since we don't know the digits $a_{101}a_{102}a_{103}...$, we can assume that in each step of our calculation somebody flips a coin and thereby determines those digits a_{101} and so on (head = 0, tail = 1). Thus, we could say that our initial number is known to us only up to some degree of uncertainty — sometimes called noise — in the data which only affects the digits at position 101 and higher. Now let us run the iteration in eqn. (10.13).

At the beginning everything is tame. But as we continue iterating the noise creeps closer and closer to the decimal point, and after precisely 100 iterations, the results will become perfectly random. That is the phenomenon of sensitive dependence on initial conditions, but it is simultaneously an accurate and solid argument for the properties of the corresponding kneading operation on the dough.

Moreover, we can now provide an argument for the uniform distribution of spices in the dough after kneading. If the spice originally comes in a clump, the coordinates of its particles should be given as $0.a_1a_2...a_ka_{k+1}...$, where the first k digits are the same for all particles, because they are clustered. The remaining digits are uniformly distributed, modeling the random mixing of the spice in the cluster. After k applications of the shift, the common coordinates are gone, and only the random digits are left, which yields a uniform distribution of spice throughout the entire dough.

Sensitivity — A Closer Look

A more precise definition of sensitivity is the following: given any point x_0 between 0 and 1, there exists a point y_0 arbitrarily close to x_0 such that the outcome of the iteration started at points x_0 and y_0 will eventually differ by a certain threshold. This threshold must be the same for all points x_0 in the interval and is called the *sensitivity constant.* Note, that it is not required that all orbits started close to x_0 will develop this deviation exceeding the threshold.

Let us argue that this definition of sensitivity holds for the iteration of the shift operator. We claim that the threshold in this case can be as large as $1/2$. Let us pick an arbitrary sample starting point in binary representation by rolling a die writing 0 for an even roll and 1 for an odd one. The result might be

$$x_0 = 0.01011010111000110001...$$

Now we try to find a starting point y_0 close by which should develop a difference to the orbit of x_0 which reaches the threshold $1/2$ at some point. For y_0 we may pick the same number as x_0 except for one of the binary digits, which we change. For example, if we require that y_0 has a distance to x_0 of at most 2^{-5}, then it suffices to flip the sixth digit of x_0,

$$y_0 = 0.01011111011100011001...$$

After five iterations we have

$$x_5 = 0.01011100011001... \leq \frac{1}{2}$$

and

$$y_5 = 0.11011100011001... = x_5 + \frac{1}{2}$$

Thus,

$$|y_5 - x_5| = \frac{1}{2}$$

as required. Clearly, we can find starting points y_0 arbitrarily close to x_0 with the same property. All we need to do is to just flip one of the binary digits which must be of sufficiently high order. At one point in the iteration this digit will be the most significant one and the difference with the orbit of x_0 will be $1/2$ again. Note that all further iterates in both orbits are identical. A difference also in those iterations is not required by the definition. Of course one may devise other strategies for the choice of y_0 that produce an orbit which is different from that of x_0 in all iterations.

Periodic Points ...

Let us proceed to an understanding of another phenomenon which goes with chaos. What happens if we specify

$$x_0 = 0.\overline{a_1 a_2 a_3 ... a_k} \ .$$

In other words, we have an infinite string of binary digits which repeats after k digits. Now running the iteration means that after k steps we will see x_0 again, and then after another k steps again, and again, and so forth. In other words, we see a cycle of length k. We call x_0 periodic with respect to the binary shift. Clearly, we can produce cycles of any length. But more importantly, for any given number x_0, we can find a number w_0 arbitrarily close to x_0, which is periodic. Let us see how this works.

A Cycle of Period 4

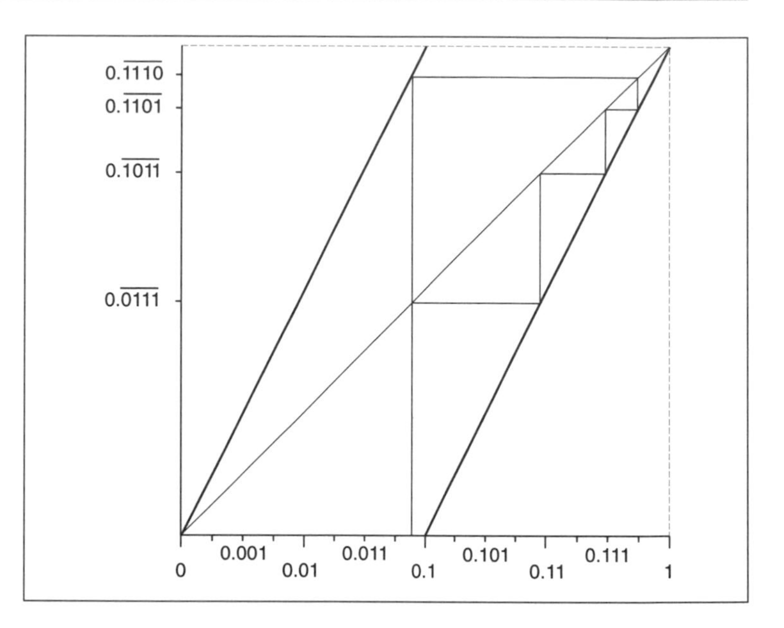

Figure 10.40 : The point $0.\overline{0111}$ is a periodic point. The binary expansion allows to immediately read off the iterative behavior (here visualized as graphical iteration).

... Are Dense

Well, if $x_0 = 0.a_1a_2a_3...$ then choose

$$w_0 = 0.\overline{a_1a_2a_3...a_k} \ .$$

for some k. Then x_0 and w_0 differ only by (at most) 2^{-k}, and w_0 is periodic. This means that *periodic points are dense.* An illustrative example for a periodic point is given in figure 10.40. Let us discuss another example in more detail. We want to approximate the irrational number

$$\frac{1}{\pi} = 0.3183098861...$$

better and better by periodic points. In binary representation $\frac{1}{\pi}$ is

$$\frac{1}{\pi} = 0.01010001011111001100000110 1101...$$

Allowing longer and longer periods we can approach the initial point as closely as we like. In the table below we use the first 5, 10, 15, 20, and 25 binary digits and repeat them periodically to generate the approximations, which we list also as decimal fractions.

Initial Point		Difference	Period
Binary	Fraction	$\lvert x_0 - 1/\pi \rvert$	
0.0101000101111100110000011...	$1/\pi$	$0.000 \cdot 2^{0}$	aperiodic
$0.\overline{01010}$	$10/31$	$0.137 \cdot 2^{-5}$	5
$0.\overline{0101000101}$	$325/1023$	$0.632 \cdot 2^{-10}$	10
$0.\overline{010100010111110}$	$10430/32767$	$0.060 \cdot 2^{-15}$	15
$0.\overline{01010001011111001100}$	$333772/1048575$	$0.211 \cdot 2^{-20}$	20
$0.\overline{0101000101111100110000011}$	$10680707/33554431$	$0.113 \cdot 2^{-25}$	25

Mixing

The final property is *mixing*, see page 133. Choose any two arbitrarily small intervals I and J. For mixing, one requires that one can find a starting point x_0 in I, whose orbit will enter the other interval at some iteration (see figure 10.41). It is straightforward to check this property for the shift operator.

Reaching an Interval J

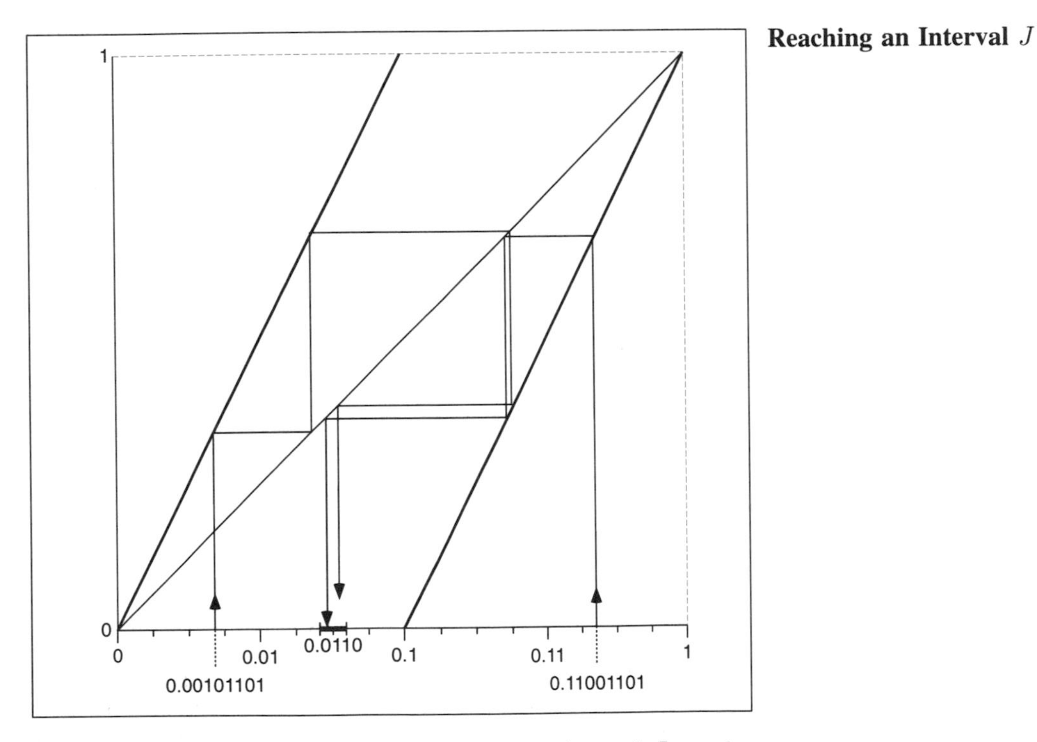

Figure 10.41 : Mixing requires that any given interval J can be reached from any other interval. Here two examples are shown how we can reach a small interval at 0.0110.

Derivation of Mixing

Choose any two intervals I and J and let $n > 0$ such that interval I has a length greater than $1/2^{n-1}$. Further, let

$$0.a_1a_2a_3...$$

be the binary representation of the midpoint of the first interval I. Moreover, let

$$y = 0.b_1b_2b_3...$$

be the binary representation of a point y of the second interval J. Now we construct an initial point x_0 in I which, after exactly n iterations of the shift operator, will be equal to y, thus, providing the required point in the target interval J. To define x_0 we copy the first n digits of the center of I and then append all digits of the target point y:

$$x_0 = 0.a_1...a_nb_1b_2b_3...$$

Now we check: x_0 differs from the center of I by at most 2^{-n} which is at most half the width of interval I. Thus, it must be contained in the interval I. Secondly, after n iterations we have

$$x_n = 0.b_1b_2b_3... = y\ .$$

So we have even over-fulfilled the requirement. In the case of the shift operator (saw-tooth transformation) we can hit *any* target point in the interval J.

Ergodic Behavior

So far we have verified the three properties of chaos for the saw-tooth transformation: sensitivity, dense periodic points, and mixing. Closely related to mixing is *ergodic* behavior. Ergodicity means that if we pick a number x_0 in the unit interval at random, then almost surely[21] the results of the shift operation will produce numbers which will get arbitrarily close to any number in the unit interval. Numbers x_0 with a periodic pattern in their binary expansion do not show such behavior and in some way they are extremely scarcely populated in the unit interval.

Example for Ergodicity

Typical numbers are more like this number:

$$x_0 = 0.0100011011000001010011100101110111 0000...$$

Can you identify the pattern? Here is the rule: first write all numbers which need 1 binary digit, i.e., 0 and 1, then all strings of two binary digits, i.e., 00, 01, 10, and 11. Continue in this fashion with strings of three digits (000, 001, 010, 011, 100, 101, 110, 111) and so forth. Clearly by construction the resulting number will get close, and in fact arbitrarily close, to any given number under iteration of the shift operator. Indeed, let us take any number y in the interval and expand it in binary form

$$y = 0.a_1a_2a_3...$$

[21] The technical term 'almost surely' means that the probability for the following assertion is equal to one. For example, a number picked at random from the unit interval almost surely is irrational.

Cutting off the expansion after k digits results in a number

$$z = 0.a_1a_2...a_k$$

which is very close to y:

$$|z - y| \leq 2^{-k}.$$

Now we observe that the string of digits $a_1a_2...a_k$ must appear in x_0 by construction at some place, and, therefore, sufficiently many shifts will bring this string to the leading digits. This provides a number that agrees with y and z in the first k leading digits, thus, it is as close to y as z.

It may seem that our example of the initial point x_0 is rather artificial. However, selecting the binary digits for x_0 at random has the same effect. The resulting orbit almost surely is ergodic. To see this just note that any block of binary digits must appear at some point in the binary representation of x_0 and the same reasoning as above applies.

The Next Step: Chaos for the Tent Transformation

We have seen that the iteration of the saw-tooth transformation S (or the shift operator, or the stretch-cut-and-paste kneading) exhibits the three properties of chaos. Now we proceed to the next stage and unfold the chaos for the tent transformation T (or the stretch-and-fold kneading). We recall that by means of the substitution property the iteration of T can be reduced to the iteration given by the saw-tooth transformation S. The kth iterate x_k is obtained by $k-1$ binary shifts followed by a single stretch-and-fold operation T. Since the first part is just a shift by $k-1$ binary digits, we now can easily carry all the complicated dynamic behavior — sensitive dependence, denseness of periodic points, and mixing — of the shift transformation over to the stretch-and-fold transformation. As a first example, let us see how a periodic point for the shift transformation generates a periodic point for the stretch-and-fold transformation. Or to phrase this differently, let us see how a seemingly impossible question turns into a very simple one.

Periodic Points for Stretch-and-Fold

Assume that we ask: what are periodic points for the iteration of the tent transformation T? Or, more precisely, find x_0, so that $x_n = x_0$ for a given integer n where $x_i = T(x_{i-1})$ for $i = 1, ..., n$. We claim that all we have to do is to take a point w_0 which is periodic for the shift transformation with period n, and to apply the tent transformation to obtain a periodic point $x_0 = T(w_0)$ of T. Indeed, let $w_0 = S^n(w_0)$ be a periodic point of S. Then we check whether $x_n = x_0$ using our definition of x_0 and the substitution property of the two kneading transformations.

$$\begin{aligned} x_n &= T^n(x_0) = T^n(T(w_0)) = T^{n+1}(w_0) \\ &= T(S^n(w_0)) = T(w_0) = x_0 \ . \end{aligned}$$

Hence, it is true: if w_0 is a periodic point for the binary shift, then $x_0 = T(w_0)$ is a periodic point for the stretch-and-fold transformation T with the same period.[22]

Using this result it is not hard to reason that periodic points of T are dense in the unit interval. Likewise it is a bit technical but not difficult to derive sensitivity and mixing for the tent transformation. The details are given in the following technical section.

Chaos for the Tent Transformation

The goal of this technical section is to derive the three central properties of chaos for the iteration of the tent transformation T: dense periodic points, sensitivity, and mixing.

Binary Representation of the Tent Transformation

As for the analysis of chaos for the saw-tooth function, the representation of the transformation in terms of binary expansions is essential for the discussion of the three properties of chaos. We recall that the tent transformation is given by

$$T(x) = \begin{cases} 2x & \text{if } x \leq 0.5 \\ -2x+2 & \text{if } x > 0.5 \ . \end{cases}$$

Let

$$x = 0.a_1a_2a_3...$$

be a binary expansion of $x \in [0,1]$. Clearly, if $x < 1/2$ the tent transformation is identical to the saw-tooth transformation, thus

$$T(x) = 0.a_2a_3a_4... \ \text{if} \ x < 1/2 \ .$$

If $x \geq 1/2$, then $S(x) = 2x - 1$, and

$$T(x) = -2x + 2 = 1 - (2x - 1) = 1 - S(x) = 1 - 0.a_2a_3a_4...$$

Introducing the dual binary digit

$$a^* = \begin{cases} 1 & \text{if } a = 0 \\ 0 & \text{if } a = 1 \end{cases}$$

we have in that case, $x \geq 1/2$, $T(x) = 0.a_2^*a_3^*a_4^*...$ because

$$0.a_2a_3a_4... + 0.a_2^*a_3^*a_4^*... = 0.111... = 1 \ .$$

The *binary representation of the tent transformation* therefore is

$$T(0.a_1a_2a_3...) = \begin{cases} 0.a_2a_3a_4... & \text{if } a_1 = 0 \\ 0.a_2^*a_3^*a_4^*... & \text{if } a_1 = 1 \ . \end{cases}$$

To deal with the ambiguous binary representations of rational numbers we only have to require to use 0.1 for $1/2$. The above binary version of T works also for $x = 1$ when representing 1 as $0.111...$

[22] To ensure that n is also the *minimal* period, i.e., $x_m \neq x_0$ for $m = 1, ..., n-1$, certain restrictions on the choice of the binary digits in $w_0 = 0.\overline{a_1...a_n}$ must be applied.

Dense Periodic Points

Let us start the discussion of chaos with periodic points. We have already shown that a periodic point $w \in [0, 1)$ of the saw-tooth function S with $S^n(w) = w$ induces a periodic point $x = T(w)$ of T with $T^n(x) = x$. To show the denseness of these periodic points of T we demonstrate that we can find periodic points whose binary expansion start with an arbitrary sequence $a_1...a_n$. The point $w = 0.\overline{0a_1...a_n} < 1/2$ is periodic under S with period $n + 1$. Thus, recalling that T is the shift transformation when the first digit of the argument is 0, we obtain $x = T(w) = 0.\overline{a_1...a_n 0}$, and x is periodic under T with period $n + 1$.

For an example we take the interval $I = [27/32, 28/32]$ consisting of all binary numbers whose expansion starts with 0.11011 and look for a periodic point in I. The midpoint of the interval is $55/64 = 0.110111$. Then, according to the above we may choose

$$x = T(0.\overline{0110111}) = 0.\overline{1101110} = \frac{110}{127} \in I \, .$$

Indeed, x is in I and periodic under T with period 7.

Sensitive Dependence on Initial Conditions

We continue with the *sensitive dependence on initial conditions.* Thus, let $x_0 = 0.a_1a_2a_3...$ be an arbitrary initial point from the unit interval and its binary expansion. For $n > 0$ given we search for a point z_0 near x_0, such that $|z_0 - x_0| < 2^{-n}$ and such that the orbits of x_0 and z_0 differ at some iterate by $\varepsilon = 1/2$. We choose

$$z_0 = 0.a_1...a_n a_{n+1}^* a_{n+1} a_{n+2} a_{n+3} \cdots$$

i.e., we flip bit number $n + 1$. Using the estimate (10.12), we obtain $|z_0 - x_0| < 2^{-n}$ and claim that $|z_n - x_n| = 1/2$. For the proof we consider the two cases $a_n = 0$ and $a_n = 1$ separately.
Case $a_n = 0$. Then

$$\begin{aligned} x_n &= T^n(x_0) = T(S^{n-1}(x_0)) = T(0.a_n a_{n+1} a_{n+2} \cdots) \\ &= T(0.0 a_{n+1} a_{n+2} a_{n+3} \cdots) = 0.a_{n+1} a_{n+2} a_{n+3} \cdots \\ z_n &= T^n(z_0) = T(S^{n-1}(z_0)) = T(0.a_n a_{n+1}^* a_{n+2} a_{n+3} \cdots) \\ &= T(0.0 a_{n+1}^* a_{n+2} a_{n+3} \cdots) = 0.a_{n+1}^* a_{n+2} a_{n+3} a_{n+4} \cdots \end{aligned}$$

and the claim follows.
Case $a_n = 1$. Then

$$\begin{aligned} x_n &= T(0.1 a_{n+1} a_{n+2} a_{n+3} \cdots) = 0.a_{n+1}^* a_{n+2}^* a_{n+3}^* \cdots \\ z_n &= T(0.1 a_{n+1}^* a_{n+2} a_{n+3} \cdots) = 0.a_{n+1} a_{n+2}^* a_{n+3}^* \cdots \end{aligned}$$

and again, the claim follows also for this case.

For an example let us reconsider the interval $I = [27/32, 28/32]$ and its midpoint $x_0 = 0.110111\overline{0}$. We choose $n = 6$ so that we are looking for an initial point z_0 in I which will drift away from the orbit of x_0 during the course of iteration of the tent transformation. We flip the 7th bit of x_0 yielding $z_0 = 0.1101111$ and compute

$$\begin{aligned} x_6 &= T^6(x_0) = T(S^5(0.110111\overline{0})) = T(0.1\overline{0}) = 0.\overline{1} = 1 \\ z_6 &= T^6(z_0) = T(S^5(0.1101111\overline{0})) = T(0.11\overline{0}) \\ &= 0.0\overline{1} = 0.1\overline{0} = 1/2 \end{aligned}$$

Thus, $|z_6 - x_6| = 1/2$, as required.

Mixing

We conclude with the derivation of the *mixing* property. Given are two arbitrary open intervals I and J in the unit interval. It is always possible to choose n large enough and bits $a_1 ... a_n$ and $b_1 ... b_n$ so that all binary numbers in $[0, 1]$ whose binary expansion begins with $a_1 ... a_n$ are in the interval I and all binaries starting with $b_1 ... b_n$ are in J. Now we specify an initial point $x_0 \in I$ such that the nth iterate is in J, $x_n = T^n(x_0) \in J$. Again we treat the two cases $a_n = 0$ and $a_n = 1$ separately.

Case $a_n = 0$. Then we choose $x_0 = 0.a_1 ... a_n b_1 ... b_n$ and verify

$$\begin{aligned} x_n &= T^n(x_0) = T(S^{n-1}(x_0)) = T(0.a_n b_1 ... b_n) \\ &= T(0.0 b_1 ... b_n) = 0.b_1 ... b_n \in J \; . \end{aligned}$$

Case $a_n = 1$. Then we choose $x_0 = 0.a_1 ... a_n b_1^* ... b_n^*$ and verify

$$\begin{aligned} x_n &= T^n(x_0) = T(S^{n-1}(x_0)) = T(0.a_n b_1^* ... b_n^*) \\ &= T(0.1 b_1^* ... b_n^*) = 0.b_1 ... b_n 111 ... \in J \; . \end{aligned}$$

For an example we use $I = [27/32, 28/32]$, as before, and $J = [14/32, 15/32]$. Here $n = 5$ and all binaries of the form $0.11011...$ are in I and all binaries $0.01110...$ are in J. Since $a_5 = 1$, we choose

$$x_0 = 0.110110^*1^*1^*1^*0^* = 0.1101110001 \; .$$

Then

$$\begin{aligned} x_5 &= T^5(x_0) = T(S^4(x_0)) = T(0.110001) \\ &= T(0.01110\overline{1}) \in J \; . \end{aligned}$$

Thus, the knowledge of the shift operator and its relation to the tent transformation indeed provides us with the key to derive the three properties of chaos for the tent transformation.

This concludes the theoretical discussion of chaos for the iteration of the kneading operations and the binary shift operator.

10.6 Chaos for the Quadratic Iterator

The Last Step

What we have learned in the previous sections may seem to be a rather special case, but in fact the contrary is true. This section is devoted to the *equivalence* of the iteration of the uniform kneading operator given by the tent transformation and the quadratic iteration $x \to 4x(1-x)$ (the non-uniform kneading). In other words, all the complex behavior which we were able to show first for the shift operator and then for the tent transformation can also be found in the quadratic iterator. And in some respects this gives a theoretical background for what we have learned in our experiments in the first part of this chapter: there is sensitivity on initial conditions; there is mixing; we can determine ergodic and also periodic points. Thus, we have made a full circle in our story about chaos and kneading.

Coordinate Transformation

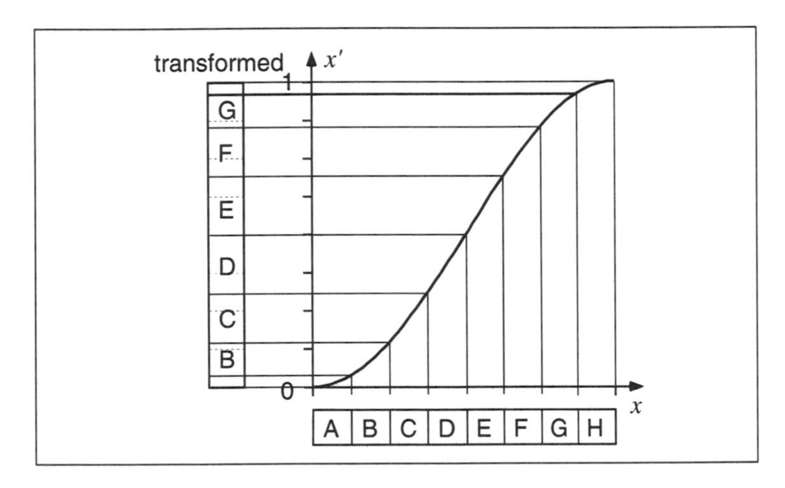

Figure 10.42 : The function $x' = h(x) = \sin^2(\pi x/2)$ is a coordinate transformation. For each x there is a corresponding value x' and vice versa. As indicated along the axes, intervals do not retain their lengths when subjected to the transformation.

Tent Function Versus Parabola

The equivalence of the iteration of the tent transformation $T(x)$ and the quadratic parabola $4x(1-x)$ is established by a nonlinear change of coordinates given by

$$x' = h(x) = \sin^2\left(\frac{\pi x}{2}\right) .$$

Before we show why this is true let us first explain what it means. The S-shaped graph of the function h is shown in figure 10.42. Note that h transforms the unit interval $[0,1]$ to itself in a one-to-one fashion, i.e., for all $x' \in [0,1]$ there is exactly one $x \in [0,1]$ with $x' = h(x)$. How does this function translate the dynamics of the tent transformation to that of the quadratic iterator? Assume we

Coordinate Transformation of Graphical Iteration

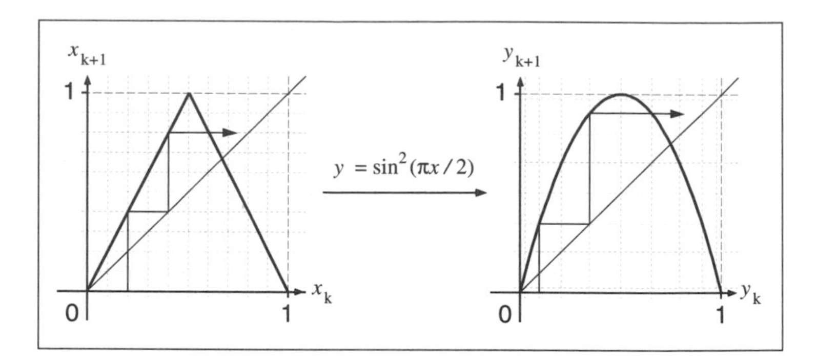

Figure 10.43 : Changing the coordinates according to the function $h(x)$ transforms the graph of the tent transformation to that of the quadratic function $f(x) = 4x(1-x)$. Furthermore, graphical iteration for T (left) is transformed into graphical iteration for f (right) also using h. The two orbits are equivalent.

are looking at an initial point x_0 and its orbit $x_0, x_1, x_2, \ldots$ under the tent transformation. Thus,

$$x_1 = T(x_0), x_2 = T^2(x_0), \ldots, x_k = T^k(x_0), \ldots \qquad (10.14)$$

The transformed initial point is

$$x_0' = h(x_0) \ .$$

This is — so to speak — the initial point x_0 in new coordinates, namely those belonging to the iteration of the parabola. Now we compute the iteration of $y_0 = x_0'$, using the quadratic transformation $f(x) = 4x(1-x)$, obtaining

$$y_1 = f(y_0), y_2 = f^2(y_0), \ldots, y_k = f^k(y_0), \ldots$$

The claim is that this is really the same orbit as the above in eqn. (10.14), however, given in the modified coordinates (see figure 10.43). In other words, not only is $y_0 = h(x_0)$, but also

$$y_1 = h(x_1), y_2 = h(x_2), \ldots, y_k = h(x_k), \ldots$$

Thus, iterating x_0 under T produces an orbit which is — after change of coordinates — the same as that of $y_0 = x_0' = h(x_0)$ under the quadratic f. In terms of the functions f and T this equivalence can be put in the form of the *functional equation*

$$f^k(h(x)) = h(T^k(x)), \quad k = 1, 2, \ldots \qquad (10.15)$$

for all $x \in [0, 1]$.

k	$x_k = T^k(x_0)$	$x'_k = \sin^2(\pi x_k/2)$
0	8/25	0.232
1	16/25	0.713
2	18/25	0.819
3	14/25	0.594
4	22/25	0.965
5	6/25	0.136
6	12/25	0.469
7	24/25	0.996
8	2/25	0.016
9	4/25	0.062

$y_k = f^k(y_0)$
0.232
0.713
0.819
0.594
0.965
0.136
0.469
0.996
0.016
0.062

Table 10.44 : The table lists the first 9 iterates of $x_0 = 8/25$ under T, the corresponding iterates of x'_0 under f, and makes the check, computing the transformed numbers $x'_k = h(x_k)$.

Example

Let us present an example, the iteration of $x_0 = 8/25$. The data for the first 9 iterations is collected in table 10.44 and the equivalence is visualized in figure 10.45. Indeed we find that the transformed coordinates of the orbit for the tent transformation perfectly matches the orbit of the quadratic iterator started at x'_0.

Sensitivity May Destroy the Equivalence

However, this result must be interpreted with caution! Although the mathematics seem to provide the definite and assuring claim that the two rightmost columns in table 10.44 are identical for as many iterations as we wish, we must not conclude that this equivalence holds in practice when computing more and more entries with a calculator or computer program. The reason, of course, lies in the sensitive dependence on initial conditions. Although $x_0 = 8/25$ is exact, x'_0 cannot be exactly represented in the machine, and, moreover, there will be additional small errors introduced at every step of the iteration. Thus, after a finite number of iterations, depending on the precision of the arithmetic, we end up producing a numerical orbit that has no resemblance to the true orbit belonging to x'_0, and, thus, the numerical observation of the equivalence is in fact destroyed by chaos.

The Equivalence of Tent Transformation and Quadratic Iterator

In this section we present the mathematics behind the equivalence of the iteration of the tent transformation and the quadratic iterator. To see that they are really the same, all the tools we need are two familiar trigonometric identities,

$$\cos^2 \alpha = 1 - \sin^2 \alpha \ ,$$
$$\sin 2\alpha = 2 \ \sin \alpha \cos \alpha \ .$$

Iterating an initial point x_0 under the tent function and iterating the transformed point $x'_0 = \sin^2(x_0\pi/2)$ under the parabola $f(x) =$

$4x(1-x)$ produces iterations that correspond to each other by means of the transformation $x' = h(x) = \sin^2(x\pi/2)$.

To establish this algebraically, we start with x_0 for the tent function and use $y_0 = x'_0$ for the parabola. Thus, $x_0, x_1, \ldots$ is the iteration under the tent function and $y_0, y_1, \ldots$ is the corresponding iteration under the parabola. We can show by induction that, in fact, $y_k = x'_k = h(x_k)$ for *all* numbers $k = 0, 1, 2, \ldots$, proving the equivalence.

We start with the transformation $y_0 = \sin^2(x_0\pi/2)$, where $0 \leq x_0 \leq 1$. We substitute for y_0 in the formula for the quadratic iteration.

$$y_1 = 4y_0(1-y_0) = 4\sin^2\left(\frac{\pi x_0}{2}\right)\left(1-\sin^2\left(\frac{\pi x_0}{2}\right)\right) .$$

We substitute using the trigonometric identity $\cos^2\alpha = 1 - \sin^2\alpha$.

$$y_1 = 4\sin^2\left(\frac{\pi x_0}{2}\right)\cos^2\left(\frac{\pi x_0}{2}\right) .$$

Simplify using the double-angle identity $\sin 2\alpha = 2\sin\alpha\cos\alpha$.

$$y_1 = \sin^2(x_0\pi) .$$

The first iterate of x_0 under the tent function is $x_1 = T(x_0)$. We now show that y_1 above is in fact identical to x_1 after change of coordinates, i.e.,

$$x'_1 = h(x_1) = y_1 .$$

We begin with the case $0 \leq x_0 \leq 1/2$. Thus, $x_1 = T(x_0) = 2x_0$ and

$$x'_1 = \sin^2\left(\frac{\pi x_1}{2}\right) = \sin^2(x_0\pi) = y_1 .$$

Now we do the other case, $1/2 < x_0 \leq 1$. First we substitute $x_1 = T(x_0) = 2 - 2x_0$.

$$x'_1 = \sin^2\left(\frac{\pi x_1}{2}\right) = \sin^2(\pi - x_0\pi) .$$

Then we simplify, first using $\sin^2(\alpha + \pi) = \sin^2(\alpha)$, and then $\sin^2(-\alpha) = \sin^2\alpha$.

$$x'_1 = \sin^2(-x_0\pi) = \sin^2(x_0\pi) = y_1 .$$

The result shows $x'_1 = y_1$ and the conclusion $x'_k = y_k$ for all k now follows by induction. Thus, since $x'_k = h(T^k(x_0))$ and $y_k = f^k(h(x_0))$, we have shown the functional equation (10.15).

By the way it is also possible to define a transformation that relates the iteration of the saw-tooth transformation S to that of the quadratic iterator. For this purpose use $h(x) = \sin^2(x_0\pi)$. The corresponding functional equation again is $f^k(h(x)) = h(S^k(x))$. Note, however, that h does not induce an equivalence relation, because h is not one-to-one. This means that we only can conclude from properties of S to properties of f, but not vice versa. We omit the details.

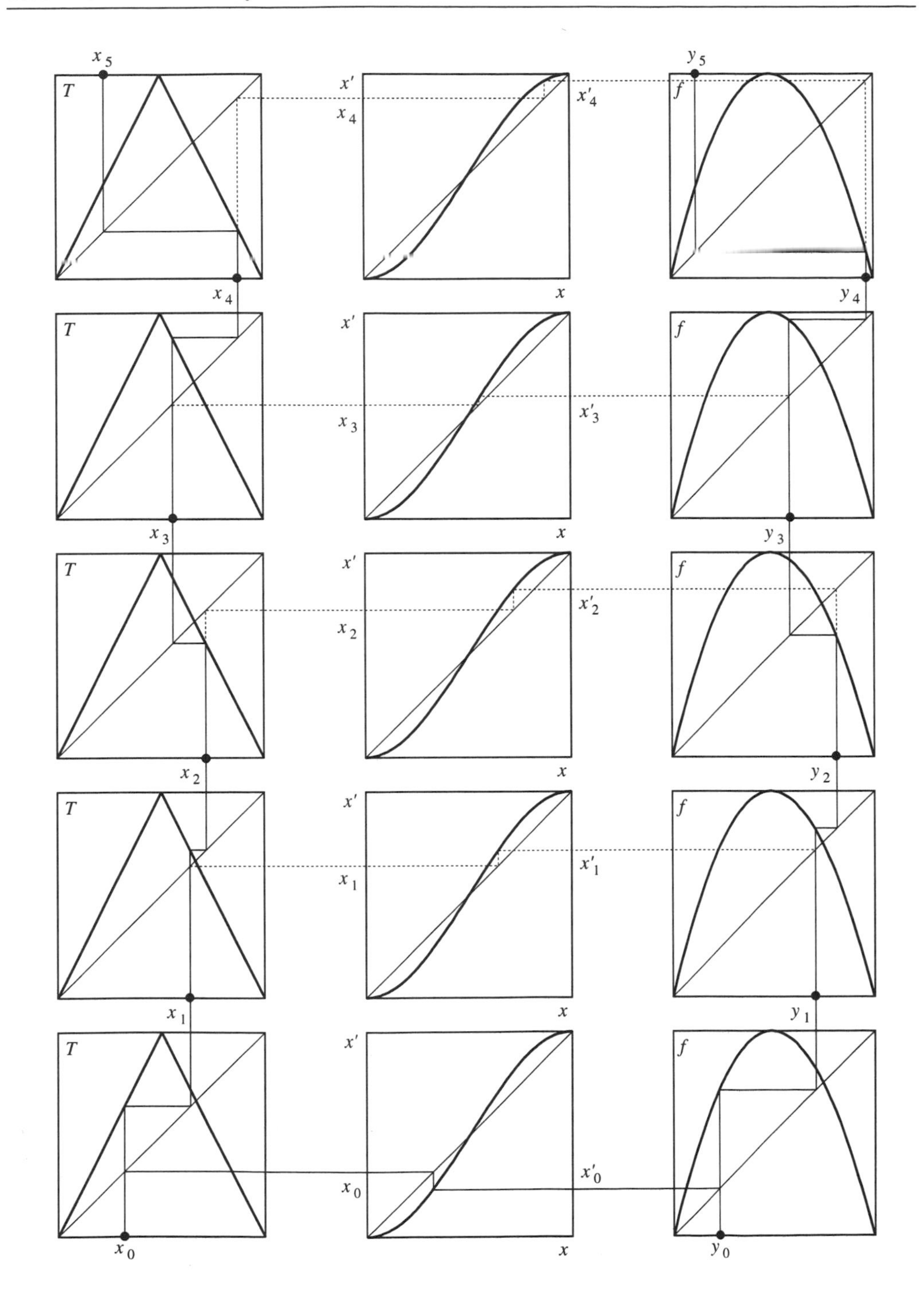

Figure 10.45 : The first few iterations of the tent transformation in table 10.44 and their counterpart for the parabola are visualized here together with the coordinate transformation establishing the equivalence.

Exploiting the Equivalence: Explicit Periodic Points

As an application of the above formulas for the change of coordinates we can easily name periodic points for the quadratic iterator. All we need is a periodic point for the tent transformation, say x_0. Then we apply the equivalence transformation to obtain $\sin^2(\pi x_0/2)$ which is guaranteed to be periodic in the quadratic iterator. For example, $x_0 = 2/7$ is periodic with period three for the tent transformation,

$$2/7 \to 4/7 \to 6/7 \to 2/7 \ .$$

Therefore, the initial value $\sin^2(\pi/7) = 0.188255099...$ is also a point from a periodic orbit of period three in the quadratic iterator. This is what we have used in the first section of this chapter on page 136.

From Chaos for T to Chaos for f

Thus, the iteration of the tent transformation and the parabola are totally equivalent. All the signs of chaos are found when iterating $f(x) = 4x(1-x)$.

- Points that are periodic for the tent transformation correspond to points that are periodic for the parabola.
- Points that show mixing by leading from one given interval to another for the tent transformation correspond to points that have the same behavior for the parabola.
- Points that exhibit sensitivity for the tent transformation correspond to points that show sensitivity for the parabola.

However, we remark that these conclusions are not self-evident. In the following technical section we present the proof for the first two properties.

Dense Periodic Points and Mixing

Let T be the tent transformation, $f(x) = 4x(1-x)$ the quadratic transformation, and $h(x) = \sin^2(\pi x/2)$ the transformation for the change of coordinates. The functional equation $f^k(h(x)) = h(T^k(x))$ for $k = 1, 2, ...$ and $x \in [0,1]$ has been shown above. Furthermore, we know that periodic points of T are dense in $[0,1]$ and T is mixing.

(a) We claim that periodic points of f are dense in $[0,1]$. Let $y \in [0,1]$. We will show that there is a sequence of periodic points of f with limit y. We may choose x as a preimage of y under h, i.e., $h(x) = y$, because h is onto. Since periodic points of T are dense in $[0,1]$ we find a sequence of points $x_1, x_2, ...$ with limit x and such that each point x_k is a periodic point of T of some period, say p_k.[23] Thus, $T^{p_k}(x_k) = x_k$ for $k = 1, 2, ...$ We claim that the sequence $y_1, y_2, ...$ with $y_k = h(x_k)$ has limit y and is a sequence of periodic points of f. The first claim is true because h is continuous.[24]

[23] Note that — in contrast to our standard use of notation — the sequence $x_1, x_2, ...$ is *not* an orbit of T.

[24] One way of defining what continuity for a function $f : X \to X$ (where X is, for example, a subset of the real line) means is the following. The function f is said to be continuous provided that for any $x \in X$ and any sequence

The second claim follows from the functional equation $f^k h = hT^k$. Indeed,

$$f^{p_k}(y_k) = f^{p_k}(h(x_k)) = h(T^{p_k}(x_k)) = h(x_k) = y_k \ .$$

(b) We claim that the transformation f is mixing. Let U and V be two open intervals in $[0,1]$. We have to find a point $y \in U$ and a natural number k so that $f^k(y) \in V$. We start with taking the preimages $A = h^{-1}(U) = \{x \in [0,1] \mid h(x) \in U\}$ and $B = h^{-1}(V)$. Note that A and B are open, because h is continuous. Thus there exists a natural number k and $x \in A$ such that $T^k(x) \in B$, since T is mixing. Set $y = h(x)$. Now, using the functional equation we obtain $f^k(y) = f^k(h(x)) = h(f^k(x))$. And since $f^k(x) \in B$ we conclude that $h(f^k(x)) \in h(B) = V$. Thus, f is mixing.

These proofs are rather straightforward and require only to properly use the appropriate definitions together with the functional equation $f^k h(x) = hT^k(x)$. Dense periodic points for the tent transformation and the equivalence of T and f yield that also f has dense periodic points. Mixing for T and the equivalence yield that also f is mixing. Now, this approach does *not* work for the third property of chaos, sensitivity. This can be demonstrated by the following rather simple counterexample.

Sensitivity: Counterexample

The iterations of the functions

$$f(x) = 2x, \quad x \in [1, \infty)$$

and

$$g(y) = y + \log 2, \quad y \in [0, \infty)$$

are equivalent by means of the coordinate transformation

$$h(x) = \log x \ .$$

Indeed, $h : [1,\infty) \to [0,\infty)$ is a continuous, one-to-one and onto transformation and the inverse of h, $h^{-1}(y) = e^y$, is continuous as well. Moreover,

$$h(f(x)) = g(h(x)) \ \text{ for all } \ x \in [1,\infty) \ ,$$

since

$$\log 2x = \log x + \log 2 \ .$$

$x_1, x_2, \ldots$ with limit x we have that the sequence $f(x_1), f(x_2), \ldots$ has also a limit which is $f(x)$. An alternative and equivalent definition is the following. The transformation f is continuous provided for any open set U in X the preimage $f^{-1}(U) = \{x \in X \mid f(x) \in U\}$ is open in X. A subset U of the real line **R** is said to be open provided for any $x \in U$ there is an open interval I containing x which is entirely in U.

From this we get the functional equation $h(f^k(x)) = g^k(h(x))$ for $k = 1, 2...$ Note that f has sensitive dependence on initial conditions but g has not, because g is just a translation. Initial errors are magnified as powers of 2 in the course of the iteration of f, while initial errors remain constant using g. As a consequence of this observation, sensitive dependence on initial conditions is not generally inherited from one dynamical system to another which has iterations that are equivalent by change of coordinates. In contrast, the properties of mixing and dense periodic points are passed over to the equivalent system.

Therefore, the derivation of sensitivity for the quadratic transformation requires more than just the sensitivity of T and the equivalence of f and T. In this case we must exploit the fact that the underlying space is just a (compact) interval. This is the crucial difference to the counterexample presented above.

Derivation of Sensitivity

We claim that f has sensitive dependence on initial conditions. Let y be an arbitrary point in $[0, 1]$. We will show that there is a sequence of initial points $y_1, y_2, ...$ with limit y such that the corresponding orbits will drift away from that of y by at least a distance of some certain $\delta_f > 0$. We may choose x as the preimage of y under h, i.e., $h(x) = y$. Now T is sensitive, and, thus, there is a constant $\delta_T > 0$ and a sequence of initial points $x_1, x_2, ...$ for T with limit x, such that the corresponding orbits under T will drift away from that of x by at least a distance of δ_T. Precisely, this means that for each initial point x_k there is an iteration count n_k such that the n_k-th iterate of x_k differs from the n_k-th iterate of x by at least δ_T,

$$|T^{n_k}(x_k) - T^{n_k}(x)| \geq \delta_T \ . \tag{10.16}$$

We define

$$\delta_f = \inf\{|h(x) - h(y)| \mid |x - y| \geq \delta_T, \ x, y \in [0, 1]\} \ .$$

Since $h(x) = \sin^2 (\pi x/2)$ is a strictly monotonically increasing function we conclude that δ_f is the minimum of the continuous function $h(x) - h(x - \delta_T)$ defined for $\delta_T \leq x \leq 1$. In our case this minimum is $\delta_f = h(\delta_T) > 0$.[25]

Now we consider the sequence $y_1, y_2, ...$ with $y_k = h(x_k)$, i.e., we apply the change of coordinates to the sequence of initial points for T. Firstly, since the transformation h is continuous, we have that this sequence has a limit, namely,

$$\lim_{k \to \infty} y_k = \lim_{k \to \infty} h(x_k) = h(x) = y \ .$$

[25] Note that this definition of δ does not work in the counterexample presented further above where the change of coordinates is given by $h(x) = \log x$. In that case $\inf\{|h(x) - h(y)| \mid |x - y| \geq \delta_T, \ x, y > 1\} = 0$, which is the point where the proof would collapse if applied to the counterexample.

Secondly, we apply the functional equation $f^{n_k}h = hT^{n_k}$ to obtain

$$\begin{aligned}|f^{n_k}(y_k) - f^{n_k}(y)| &= |f^{n_k}(h(x_k)) - f^{n_k}(h(x))| \\ &= |h(T^{n_k}(x_k)) - h(T^{n_k}(x))| \ .\end{aligned}$$

Because of the inequality (10.16) and the definition of δ_f we get

$$|f^{n_k}(y_k) - f^{n_k}(y)| \geq \delta_f \ .$$

Thus, the orbit with initial point y_k achieves a distance greater than or equal to δ_f after n_k iterations. Since $y_k \to y$, we have found initial points arbitrarily close to y that have this property. Thus, f has sensitivity at the point y.

Mixing and Dense Periodic Points Imply Sensitivity

There is an alternative and elegant solution of the problem of deducing sensitivity which works not only for the quadratic transformation but for all similar cases. The result has recently been worked out by a group of five Australian mathematicians.[26] They showed in a theorem that the properties of mixing and dense periodic points already suffice to show the third property of chaos, sensitivity. In other words, if f is chaotic and f and g are equivalent via a change of coordinates h, i.e., $f(h(x)) = h(g(x))$, then also g is chaotic. Thus, we do not need to undertake the task of deriving sensitivity of f from that of T. The proof of this theorem is not difficult but technical and we refer the interested reader to the reference.

[26] J. Banks, J. Brooks, G. Cairns, G. Davis, P. Stacey, *On Devaney's definition of chaos,* American Math. Monthly 99.4 (1992) 332–334.

10.7 Numerics of Chaos: Worth the Trouble or Not?

In chapter 1 (and also here) we demonstrated that the theoretical and computational behavior of the iteration of chaotic systems will almost always differ significantly. This is a consequence of two central aspects of these systems. The first one is the extreme sensitivity which is inherent to chaotic systems. The other reason is due to the limited precision when doing floating point arithmetic on a computer. It is only natural to question whether it is worthwhile at all to compute any orbits for chaotic systems. Strictly speaking, all computed values clearly will be wrong. The first numbers will have only small errors, but soon the errors become as large in magnitude as the points in the true orbit. So what is the use of the computed orbit? The answer is stunning and unbelievable at first sight. Although the computed orbit is fundamentally wrong, it is also equally correct in that it approximates a true orbit of the same chaotic system very well. Moreover, this approximation holds not only for the first few iterations but for all of them! This means that the computed orbit always stays close to a true one, like the hiker's shadow on the ground stays close to the hiker wherever he chooses to go. How can that be true in the presence of sensitivity? There seems to be some contradiction. But if the statement of the shadowing of an orbit is really true — and we will in fact show that further below — then the question raised in the title of this section can be answered affirmatively.

Before we go into details of this *shadowing lemma*, as it is called in the initiated circles of chaos researchers, we would like to raise your attention to another problem illustrating again the fact that computed orbits must be carefully interpreted. In the last section we clearly made the point that the shift operator is the central chaos generating mechanism. Perhaps some readers have already felt inspired to try a quick and straightforward implementation of the shift operator on a computer or calculator. But to their dismay, the orbits will certainly have been anything but chaotic. What is your guess as to what we will see if we iterate the shift operator on a computer?

Shift Operator on the Calculator

Most people who know a little bit (or more) about the internal works of computers regard this as a rather boring question. They reckon that independent of the initial value, after only a few iterations we end in the fixed point 0. The reasoning is simple. Computers encode real numbers in the binary system, but of course they can only use a finite number of digits. Therefore they can only represent binary numbers like:

$$x_0 = 0.a_1 a_2 a_3 ... a_m 000...$$

where the digits a_k are 0 or 1 and m is a constant which depends on the chosen precision and the type of the computer. Thus, after m iterations of the shift operator we reach $x_m = 0$. In other words, on the computer we do not expect to see ergodic or periodic behavior. But let us be safe and try to confirm our argument by an experiment on the computer.

Most people take it for granted that a calculator is a computer, just smaller. We enter (for example) 0.6 into a pocket calculator. With this initial value x_0 we start some iterations of the shift operator. First, we multiply by 2, subtract 1 and obtain $x_1 = 0.2$. For the next step we obtain $x_2 = 0.4$, then $x_3 = 0.8$, $x_4 = 0.6$ and again $x_5 = 0.2$, etc. Yes, we see a perfect periodic behavior. If you want to check this, go ahead. We tried several pocket calculators and always found the same result. If you should not have a pocket calculator at hand but rather a personal computer on your desk, you might want to check it on that machine. Perhaps you can write a small BASIC program which can be done in a couple of minutes. The outcome of the experiment will depend on the BASIC interpreter on the machine. Some interpreters really compute in binary arithmetic and after 22 iterations you have reached 0 but some others work more like a pocket calculator and show the periodic behavior. So, what is the trick?

The Trick of Pocket Calculators ...

Most pocket calculators use binary coded decimal arithmetic. This means that the individual digits of decimal numbers are binary coded in four bits per digit. In that way up to a certain number of digits it is possible to encode decimals exactly. For example the number 0.6 is encoded by

$$0.0110\ .$$

Note that 110 is simply the binary encoding of 6. Let us compare this with the ordinary binary encoding of 0.6 which is the periodic extension

$$0.100110011001\ldots$$

The exact binary encoding on a computer would require infinitely many binary digits which of course are not available. Therefore 0.6 cannot be encoded exactly in binary form whereas it can be encoded exactly as a binary coded decimal. And this carries over to the iteration of the shift operator. Using (plain) binary encoding we can represent only those decimals exactly which have a finite encoding (e.g., 0.375 is binary encoded 0.011), thus the iteration will always lead to 0. Using binary coded decimal arithmetic we can also represent decimals exactly which have an infinite binary encoding (like 0.6), and therefore we can see some periodic behavior.

... Knocked Out by Chaos

But be warned. This trick does not really change the problems. Sensitivity is present, and its problems cannot be circumvented by any machine with finite precision. Just recall the experiments from chapter 1. Somehow pocket calculators only fake precision; they do not really provide it. And even a simple choice like 1/3 taken as initial value will knock out the tricky pocket calculator.

The Shadowing Lemma

Let us now discuss why any computed orbit of a chaotic system, although exposed to sensitivity, manages to be a close approximation of a truthful orbit of the same system. Since the quadratic iterator is equivalent to the shift transformation, we concentrate on the latter and only remark on the implications in other systems towards the end.

Given an initial point x_0 and the corresponding exact orbit under the shift operator

$$x_{k+1} = \mathrm{Frac}\,(2x_k)\ , \quad k = 0, 1, \ldots$$

we allow errors to be made in each step of the computation.[27] First of all, the initial point x_0 may not be represented exactly. A machine can only use a number y_0 close to x_0. Let us call the error made in this approximation ε_0,

$$y_0 = x_0 + \varepsilon_0\ .$$

Based on this value y_0 an orbit is calculated, for which we use the notation $y_0, y_1, y_2, \ldots$ However, this is not the exact orbit for y_0 because in each iteration step there will be an error. Let us call the error in the kth step ε_k. More precisely, we define

$$y_k = \mathrm{Frac}\left(2y_{k-1} + \varepsilon_k\right)\ .$$

From our discussion of sensitivity it is clear that any error introduced anywhere doubles in each iteration. After only a few steps, there is absolutely no correlation between what is computed and either of the true orbits started at x_0 or y_0. Still we can show that there is some exact orbit started at some initial point near x_0 and y_0, say z_0, which is approximated closely for all of the computed iterations! The situation is sketched in figure 10.46. The only assumption we must require to prove that fact and to derive the initial point z_0 is that the errors are bounded by some constant $\varepsilon > 0$,

$$|\varepsilon_k| \leq \varepsilon, \quad k = 0, 1, 2, \ldots$$

The conclusion is that for any iteration the exact orbit generated out of z_0 will fall within ε-distance of the computed orbit. In the kth iteration we have

$$|z_k - y_k| \leq \varepsilon\ . \tag{10.17}$$

[27] These errors are not to be confused with the accumulated (amplified) errors from the previous iterations.

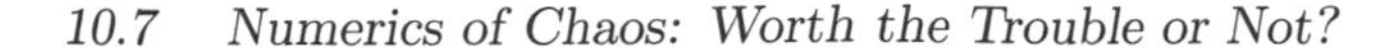

The Shadow of the Computed Orbit

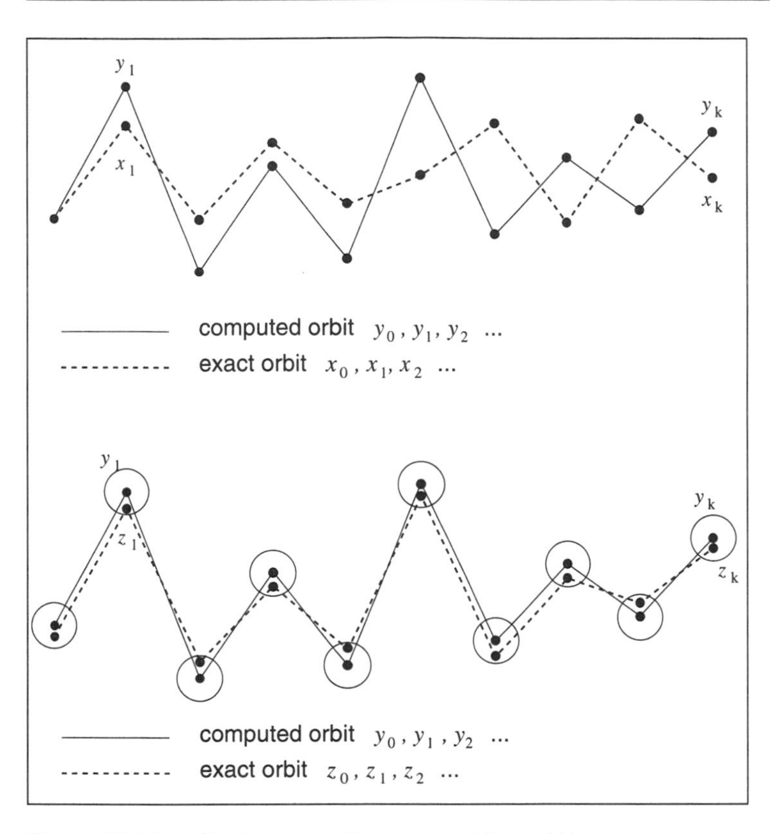

Figure 10.46 : Starting at x_0 the exact orbit would be $x_0, x_1, ..., x_n$. The computed orbit starts within ε-distance at y_0 and soon will deviate from the exact orbit. However, in the ε-shadow of the computed orbit $y_0, y_1, ..., y_n$ there will be an exact orbit which starts at z_0.

Derivation of the Shadowing Lemma

The proof for the shadowing lemma as stated in the text requires a fair amount of formula writing. This should not distract the reader, because everything is elementary, and the underlying idea is simple. To begin with the derivation let us consider a finite sequence of n computed points for the shift operator as already stated above. For a given initial point x_0 we get the computed orbit which is listed explicitly here for completeness:

$$\begin{aligned} y_0 &= x_0 + \varepsilon_0 \\ y_1 &= \text{Frac}\,(2y_0 + \varepsilon_1) \\ &\vdots \\ y_n &= \text{Frac}\,\big(2y_{n-1} + \varepsilon_n\big) \ . \end{aligned}$$

The errors ε_k are bounded by $\varepsilon > 0$

$$|\varepsilon_k| \le \varepsilon, \quad k = 0, 1, ..., n \ .$$

We can express the points y_k in terms of x_0 and errors alone using the properties of Frac on page 161:

$$\begin{aligned} y_0 &= x_0 + \varepsilon_0 \\ y_1 &= \text{Frac}\left(2(x_0 + \varepsilon_0) + \varepsilon_1\right) \\ &= \text{Frac}\,(2x_0 + 2\varepsilon_0 + \varepsilon_1) \\ y_2 &= \text{Frac}\left(2(2x_0 + 2\varepsilon_0 + \varepsilon_1) + \varepsilon_2\right) \\ &= \text{Frac}\,(4x_0 + 4\varepsilon_0 + 2\varepsilon_1 + \varepsilon_2)\ . \end{aligned}$$

For each iteration we must multiply the previous result by two and add the corresponding error term. Thus, in the kth step we obtain

$$y_k = \text{Frac}\left(2^k x_0 + 2^k \varepsilon_0 + 2^{k-1}\varepsilon_1 + \cdots + 2\varepsilon_{k-1} + \varepsilon_k\right)\ .$$

More conveniently this is expressed in mathematical shorthand as

$$y_k = \text{Frac}\left(2^k x_0 + \sum_{i=0}^{k} 2^{k-i}\varepsilon_i\right)\ . \qquad (10.18)$$

In particular, in the last iteration ($k = n$) we obtain

$$y_n = \text{Frac}\left(2^n x_0 + \sum_{i=0}^{n} 2^{n-i}\varepsilon_i\right)\ .$$

Now we can specify z_0, the initial point of the exact orbit whose shadow we have computed as the orbit $y_0, ..., y_n$. The idea is simply to define z_0 as that point, which when multiplied with 2^n will yield exactly the argument of the Frac-function in the last formula. This is

$$\begin{aligned} z_0 &= \text{Frac}\left(x_0 + \sum_{i=0}^{n} 2^{-i}\varepsilon_i\right) \\ &= \text{Frac}\left(x_0 + \varepsilon_0 + \frac{\varepsilon_1}{2} + \frac{\varepsilon_2}{4} + \cdots + \frac{\varepsilon_n}{2^n}\right)\ . \end{aligned} \qquad (10.19)$$

With that choice we have made sure that $z_n = y_n$. Thus, the error between the computed and the exact orbit $z_0, ..., z_n$ is zero in the last iterate. To analyse the deviation between the orbit of z_0 and the computed orbit $y_0, ..., y_n$ in the other iterates we first derive an explicit formula for the exact k^{th} iterate z_k.

$$\begin{aligned} z_k &= \text{Frac}\left(2^k z_0\right) = \text{Frac}\left(2^k\left(x_0 + \sum_{i=0}^{n} 2^{-i}\varepsilon_i\right)\right) \\ &= \text{Frac}\left(2^k x_0 + \sum_{i=0}^{n} 2^{k-i}\varepsilon_i\right)\ . \end{aligned} \qquad (10.20)$$

To discuss the difference in the kth iterate between the exact orbit of z_0 and the computed one of y_0 we consider first the difference

Δ_k in the arguments of the Frac-function in eqn. (10.18) and eqn. (10.20),

$$\begin{aligned}\Delta_k &= \left(2^k x_0 + \sum_{i=0}^{n} 2^{k-i}\varepsilon_i\right) - \left(2^k x_0 + \sum_{i=0}^{k} 2^{k-i}\varepsilon_i\right) \\ &= \sum_{i=k+1}^{n} 2^{k-i}\varepsilon_i \ .\end{aligned}$$

To obtain an error bound for this expression we use the bound ε on the errors ε_i and compute

$$\begin{aligned}|\Delta_k| &\le \sum_{i=k+1}^{n} 2^{k-i}|\varepsilon_i| \le \sum_{i=k+1}^{n} 2^{k-i}\varepsilon \\ &= \varepsilon\left(\frac{1}{2} + \frac{1}{4} + \cdots + \frac{1}{2^{n-k}}\right) < \varepsilon \ .\end{aligned}$$

Here we must include a word of caution because from these inequalities we cannot conclude that $|z_k - y_k| \le \varepsilon$ holds for $k = 0, ..., n$, although in most cases this will be true. The reason lies in the discontinuities of the Frac-function at integer points. Thus, for example, in the case that the argument in eqn. (10.18) is slightly below an integer, and that in eqn. (10.20) is slightly above an integer we arrive at $|z_k - y_k| \approx 1 > \varepsilon$. Formally, in order to resolve this problem correctly, we would have to define a new metric for the unit interval which identifies the points 0 and 1. In other words, we think of the unit interval as a circle with circumference equal to 1. Then the closeness of the arguments in equations (10.18) and (10.20) carries over to that of z_k and y_k, as claimed. Speaking rigorously, the statement in eqn. (10.17) must be modified in this sense to be true.

Moreover, a straightforward modification is in order to show that the claim holds for all integers $k = 0, 1, 2, ...$ We just have to consider the limit as $n \to \infty$ in equations (10.19). Then

$$\begin{aligned}|\Delta_k| &= \left|\sum_{i=k+1}^{\infty} 2^{k-i}\varepsilon_i\right| \le \sum_{i=k+1}^{\infty} 2^{k-i}|\varepsilon_i| \\ &\le \varepsilon \sum_{i=k+1}^{\infty} 2^{k-i}|\varepsilon_i| = \varepsilon\end{aligned}$$

and the rest follows as above.

Let us summarize and interpret this result. We have learned that when we compare computed orbits with exact orbits then the deviation due to accumulated error propagation will soon amplify so rapidly in the course of the computation that typically any correlation between exact orbits and computed orbits will vaporize. Nevertheless, the iteration behavior of our system is so enormously rich that within the shadow of the computed orbit there will be some exact orbit traveling along. This is a truly amazing fact, in particular if we remind ourselves that the errors in the computed orbit may be chosen randomly, as long as they remain bounded overall, i.e., the individual errors ε_k in each step of the computation could be thought of as being even chosen by a random number generator, as long as $|\varepsilon_k| < \varepsilon$ for some appropriate choice of ε.

On the other hand, the shadowing lemma should not mislead us to think that it provides us with a way to escape the consequences of sensitivity to initial condition and some value for the long-term *prediction* of orbits in our chaotic system. However, the shadowing lemma does ensure us that statistical properties measured by computer experiments are in fact significant.

The assertions of the shadowing lemma hold true in many chaotic systems. They can be interpreted in a rather dramatic way: Under the circumstances of shadowing a deterministic model supports almost any prediction.

10.8 Program of the Chapter: Time Series and Error Development

We have discussed Lorenz' comparison of the iterative behavior of signals and errors in chaos.[28] A chaotic system shows an error development which essentially looks like the development of the signal itself. Our program of the chapter highlights this point of view for the quadratic iterator

$$x_{n+1} = f_a(x_n) \ ,$$

where

$$f_a(x) = ax(1-x) \ , 1 \leq a \leq 4 \ .$$

When running the program the user is prompted for three inputs:

- the mode `Signal` (0) or `Error` (1),
- the parameter a of the function $f_a(x) = ax(1-x)$,
- the initial value x_0.

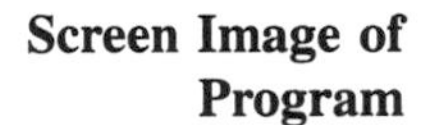

Screen Image of Program

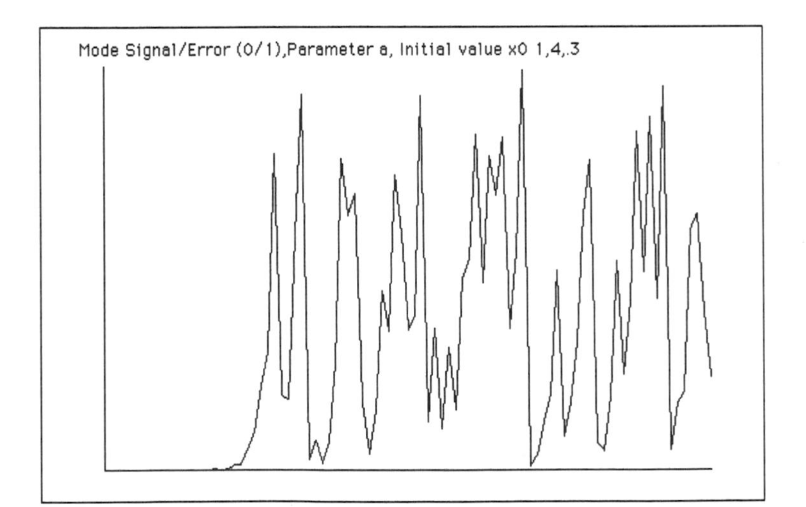

Figure 10.47 : Output of the program 'Time Series'. Here the program is in the `Error` mode, computing the iterations for the parameter $a = 4$ and initial point $x_0 = 0.3$. The difference between results of single and double-precision calculations is shown.

If you set the mode to 0, i.e., choose the mode `Signal`, the program will compute the iteration of the specified initial value using single-precision arithmetic. As output, a time series will be

[28] See page 123.

BASIC Programm **Time Series**
Title Time series and error development for the quadratic iterator

```
REM double-precision variables dxn#, da#
left = 20
h = 300
w = 1.5*h
dw = 5
m = 1
INPUT "Mode (0/1), Parameter a, Initial value x0", mode, a, x0
xn = x0
dxn# = x0
da# = a

REM Draw coordinates and start value
LINE (left,left) - STEP (0,h)
LINE - STEP (w,0)
IF mode = 0 THEN PSET (left,left+(1-xn)*h)
IF mode = 1 THEN PSET (left,left+h)

FOR i = 1 TO w/dw
    REM Evaluate iteration
    FOR k = 1 TO m
        xn= a * xn * (1 - xn)
    NEXT k
    IF mode = 1 GOTO 100
        REM Draw iteration
        LINE - (left+i*dw,left+(1-xn)*h)
        GOTO 200
100 REM Evaluate second iteration
        FOR k = 1 TO m
            dxn#= da# * dxn# * (1 - dxn#)
        NEXT k
        REM Draw difference
        LINE - (left+i*dw,left+(1-ABS(xn-dxn#))*h)
200 NEXT i
END
```

produced, plotting the values x_n versus the step counts n. If you set the mode to 1, i.e., choose the mode `Error`, the program will compute the iteration of x_0 in two different versions, the first one using single and the other using double-precision arithmetic. The output will consist of a graph of the difference of the two results. In other words, if $x_0, x_1, x_2, ...$ denotes the computed orbit for single precision, and $y_0, y_1, y_2, \ldots$ is the corresponding orbit with $y_0 = x_0$, computed with double precision, the absolute differences

$$|y_0 - x_0|, |y_1 - x_1|, |y_2 - x_2|, \ldots, |y_n - x_n|, \ldots$$

will be plotted against the index n on the horizontal axis. For parameter values well below the Feigenbaum point $a = s_\infty = 3.5699456\ldots$ these difference should be indistinguishable from 0 because there is no sensitivity in the iteration. The long term behavior is the same regardless of the errors involved in the computation in single and double-precision arithmetic. But for parameters like $a = 3.7$ or $a = 4.0$, there is sensitivity and the outcome will be dramatically different from those below the Feigenbaum point (see figure 10.47).

First the program draws a coordinate system and the start value (which is at the origin if you have set the mode to 1). Note that this program uses a rectangular screen area. The width `w` is 1.5 times the height `h`. Next, we step through the iteration (this is the `FOR` loop for `i`). Observe that the evaluation of the iteration step is done in a `FOR` loop, too, allowing us to use the m times iterated function f_a^m in place of f_a. For this to happen the variable `m` should be set to the appropriate value at the beginning of the program.

This program is very similar to the *graphical iteration* program of chapter 1.[29] It is interesting to do comparative experiments with both of them using the same parameters and initial values. The program of chapter 1 gives you the view of graphical iteration while this program shows you the corresponding time series.

In the `Error` mode, the program also performs the iteration in double precision (see the code starting at label 100). Note that the suffix `#` in the names `dxn#` and `da#` defines these variables as double-precision type variables. Finally, the absolute value of the difference between the single-precision iteration value `xn` and the double-precision iteration value `dxn#` is drawn.

We propose changing the program to do another experiment which we also discussed in this chapter. Try to compare the effect of the different implementations $f_a(x) = a \cdot x \cdot (1-x)$ and $f_a(x) = a \cdot x - a \cdot x \cdot x$. Both implementations should use single-precision arithmetic (i.e., delete the suffix `#` in the names `dxn#` and `da#`) or both should use double precision. Can you predict what the qualitative difference will be?

[29] See section 1.6 in *Fractals for the Classroom, Part One*.

Chapter 11

Order and Chaos: Period-Doubling and its Chaotic Mirror

... there is a God precisely because Nature itself, even in chaos, cannot proceed except in an orderly and regular manner.
Immanuel Kant

Routes to Chaos

Chaos theory began at the end of last century with some great initial ideas, concepts and results of the monumental French mathematician Henri Poincaré. Also the more recent path of the theory has many fascinating success stories. Probably the most beautiful and important one is the theme of this chapter. It is known as the *route from order into chaos*, or *Feigenbaum's universality*.

Chaos and order have long been viewed as antagonistic in the sciences. Special methods of investigation and theory have been designed for both. Natural laws like Newton's law or Kepler's law represent the domain of order. Chaos was understood to belong to a different face of nature where simple — or even complicated — laws would not be valid. In other words, chaos was seen not just as a higher degree of complexity or as a more complex form of order, but as a condition in which nature fails to obey laws. Even more challenging was the observation that natural systems seem to have no difficulty switching from one state into the other, from laminar flow into turbulent flow, from a regular heart beat into a fibrillating heart beat, from predictability into unpredictability.

Time Series and Final State

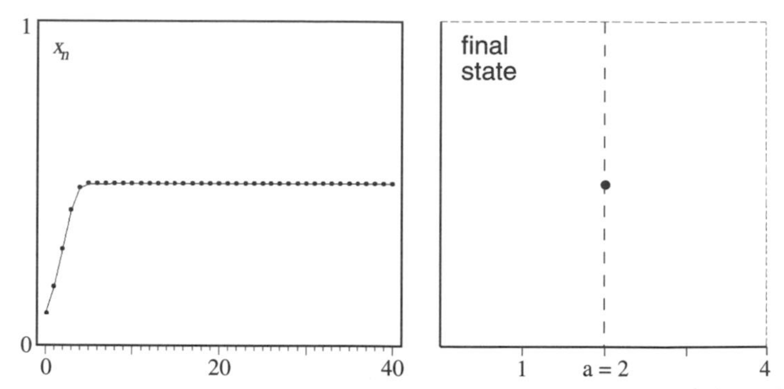

Figure 11.1 : The long term behavior of the quadratic iterator for $a = 2$: plot of time series (left) and final state marked in a final-state diagram (right).

One of the great surprises revealed through the studies of the quadratic iterator

$$x_{n+1} = ax_n(1 - x_n), n = 0, 1, 2, ... \tag{11.1}$$

is that both antagonistic states can be ruled by a *single* law. An even bigger surprise was the discovery that there is a very well defined 'route' which leads from one state — order — into the other state — chaos. Furthermore, it was recognized that this route is *universal*. 'Route' means that there are abrupt qualitative changes — called bifurcations — which mark the transition from order into chaos like a schedule, and 'universal' means that these bifurcations can be found in many natural systems both qualitatively and quantitatively.

Long Term Behavior

The following crucial computer experiment turns out to be loaded with marvellous scientific discoveries. In fact, we will soon see that it raises more questions than we can answer. Some of them are still open today and present tough research problems. Here is the experiment. We want to explore the behavior of the quadratic iterator (11.1) for all values of the parameter a between 1 and 4. To be more precise, we are only interested in the long term behavior, i.e., we would like to know what happens to the iterates x_n when the dependence on the initial choice x_0 is diluted to almost zero.

Clearly, the iteration produces values x_n which remain in the interval $[0, 1]$ as long as the initial value x_0 is from that interval. This is true for all parameters a between 1 and 4. Let us look at an example. What do we obtain for $a = 2$? Figure 11.1 (left) shows the time series for this parameter and a randomly chosen initial value x_0. After a transient phase of a few iterations the orbit

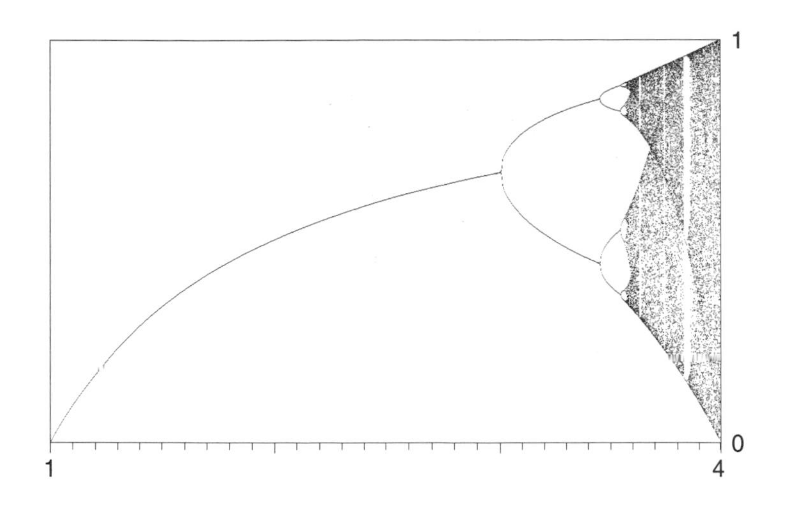

Figure 11.2 : Final-state diagram for the quadratic iterator (11.1) and parameter a between 1 and 4.

Final-State Diagram

settles down to a fixed point, which we call the *final state*. If we repeat this experiment for different initial values between 0 and 1 we always reach the same final state, the value 0.5. Let us enter this point into a *final-state diagram:* we draw the final state versus the value of the chosen parameter a. This is done in figure 11.1 (right) where we have marked the point ($a = 2$, final state $= 0.5$).

The Feigenbaum Diagram

Now let's complete the diagram. But how can we compute the final state of the iteration for a particular parameter a? Here is a procedure which works fine to obtain a first draft:

1. We choose an initial value x_0 at random from the interval $[0, 1]$ and carry out, say 200, iterations computing $x_1, x_2, ..., x_{200}$.
2. We drop the first 100 iterations $x_1, x_2, ..., x_{100}$.
3. We plot the remaining iterations $x_{101}, ..., x_{200}$ in the diagram.

Applying this procedure only for the parameter $a = 2$ we obtain the diagram shown in figure 11.1 (right). Thus, all plotted points x_{101} to x_{200} fall onto just one dot. Figure 11.2 shows the result for all parameters between 1 and 4. We note that for parameters $a > 3$ the final state is not a mere point but a collection of 2, 4, or more points. For $a = 4$, of course we have the chaos discussed in the previous chapter, and the points of the final state densely fill up the complete interval. Sometimes this image is also called the *Feigenbaum diagram* because it is intimately connected with the ground breaking work of the physicist Mitchell Feigenbaum. Indeed this diagram is a remarkable fractal, and later we will even see that it is closely related to the famous Mandelbrot set.

The Feigenbaum diagram has become the most important icon of chaos theory. It will most likely be an image which will remain as a landmark of the scientific progress of this century. The image is a computer generated image and is necessarily so. That is to say that the details of it could have never been obtained without the aid of a computer. Consequently, the beautiful mathematical properties attached to the structure would definitely be still in the dark and would probably have remained there if the computer had not been developed. The success of modern chaos theory would be unimaginable without the computer.

One essential structure seen in the Feigenbaum diagram 11.2 is that of a branching tree which portrays the qualitative changes in the dynamical behavior of the iterator $x \to ax(1-x)$. Out of a major stem we see two branches bifurcating, and out of these branches we see two branches bifurcating again, and then two branches bifurcating out of each of these again, and so on. This is the *period-doubling regime* of the scenario.

The Period-Doubling

Let us explain very crudely what period-doubling means. Where we see just one branch the long term behavior of the system tends towards a fixed final state, which, however, depends on the parameter a. This final state will be reached no matter where — at which initial state x_0 — we start. When we see two branches this just means that the long term behavior of the system is now alternating between two different states, a lower one and an upper one. This is called *periodic* behavior. Since there are two states now, we say that the *period* is two. Now, when we see four branches all that has happened is that the period of the final state behavior has increased from two to four. That is period-doubling: $1 \to 2 \to 4 \to 8 \to 16 \to \cdots$ Beyond this period-doubling cascade at the right end of the figure we see a structure with a lot of detailed and remarkable designs. Chaos has set in, and eventually, at $a = 4$, chaos governs the whole interval from 0 to 1.

The Feigenbaum diagram has features that are both of a qualitative nature and a quantitative one. The qualitative features are best analyzed through the methodology of fractal geometry. The structure in figure 11.2 has self-similarity properties, which, we will now show, means that the route from order to chaos is one with infinite detail and complexity.

Figure 11.3 shows a sequence of close-ups. We start this sequence with a reproduction of figure 11.2 and magnify the rectangular window indicated in the initial diagram, but showing it upside-down. This is our first close-up image, which indeed looks like the whole diagram. Again we make a magnification of the rectangle indicated and show the result upside-down obtaining the

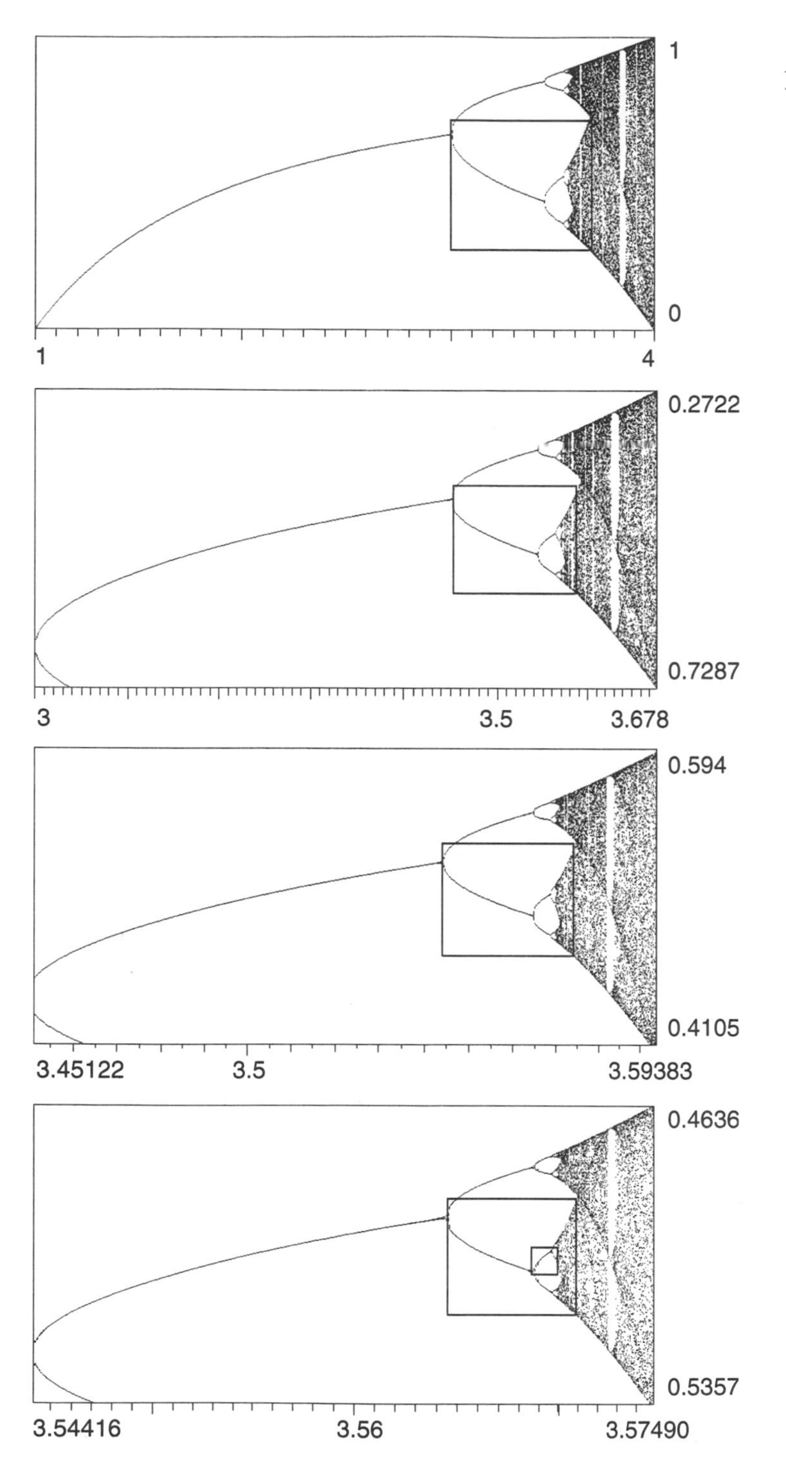

Figure 11.3 : A close-up sequence of the final-state diagram of the quadratic iterator reveals its self-similarity. Note that the vertical values in the first and third magnifications have been reversed to reflect the fact that the previous diagram has been inverted. The second magnification is, of course, also a vertical inversion of the first; the values, however, are in their 'normal' relationship.

second close-up. The third close-up is the last one in our demonstration. Theoretically, we could go on infinitely often, which we have symbolized by drawing the next two succeeding close-up windows into the bottom image. In other words, the final-state diagram is a self-similar structure.

To get the flavor of some quantitative features let us note that the branches in the period-doubling regime become shorter and shorter as we look from left to right. It is therefore a tempting thought to imagine that the lengths of the branches (in direction of the a-axis) relative to each other might decrease according to some law, perhaps a geometric law. This idea leads to several consequences.

The Feigenbaum Point

First of all, if true, it would constitute a threshold, i.e., a parameter a beyond which the branches of the tree could never grow. This would mark the end of the period-doubling regime. Indeed, there is such a threshold, which has become known as the *Feigenbaum point* $a = s_\infty = 3.5699456...$ It is precisely the a-value at which the sequence of rectangles shown in figure 11.3 converge. The Feigenbaum point splits the final-state diagram into two very distinct parts, the period-doubling tree on the left and the area governed by chaos on the right. However, the right part is not simply a region of utter chaos; it hides a variety of beautiful structures which we will reveal in this chapter.

The Feigenbaum Constant and Universality

Secondly, if there is a rule that quantifies the way the period-doubling tree approaches the Feigenbaum point, one could try to compare it with the laws which one might observe in related iterators. In fact, experiments regarding these fascinating ideas were carried out by Feigenbaum around 1975; and to the great surprise of the scientific community, he found that the law could be isolated from the branching behavior and that this law, in fact, was exactly the same for many different iterators. Naturally, his discoveries stimulated a whole new mathematical research direction. Actually, in a very precise sense the law can be captured in just one number which Feigenbaum measured at first by numerical experiments to be $\delta = 4.6692...$ and then found that this number was the same for related iterators. This number δ became known as the *Feigenbaum constant* and its appearance in many different systems was called *universality*.

Roughly the meaning of the constant δ is this: if we measure the length of two succeeding branches (in the a-axis direction) then their ratio turns out to be approximately δ (see figure 11.4). In fact, this number is also reflected in the sequence of magnifications in figure 11.3; δ is the magnification factor from one enlargement to the next.

Feigenbaum's δ

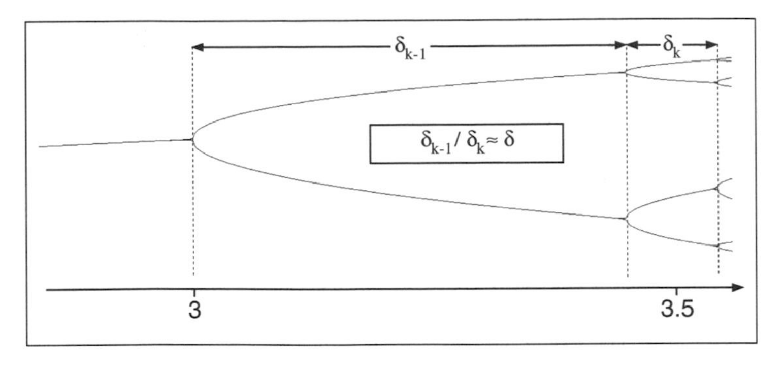

Figure 11.4 : Interpretation of the Feigenbaum constant δ.

The number $\delta = 4.6692...$ is a constant of chaos comparable only to the fundamental importance of numbers like π. Feigenbaum's discovery was the first of many footprints by which the tracks of chaos are now recognized. The number δ has been observed in systems as varied as dripping faucets, the oscillation of liquid helium, and the fluctuation of gypsy moth populations. It is a predictable constant in the world of chaos.

One possible and very useful interpretation of the universality of δ would be by using it for predictions. For example, by just measuring two successive bifurcations we would be able to predict the bifurcations thereafter and also predict where the threshold would be. This interpretation became, in fact, an achievement of incredible consequences some years after Feigenbaum's work, when experimental physicists discovered that the scenario of period-doubling and the value of δ manifest themselves in real physical experiments like the dynamic behavior in certain fluid flows. In other words, the meaning of universality was suddenly covering not just very primitive mathematical models but also real physical phenomena. Soon a gold rush in the experimental sciences set in and the region of validity for the universality became larger and larger. Simultaneously, it was understood that there was not just one universal aspect in the structure of figure 11.2 but many other quantitative and qualitative ones. In essence that means whenever a system (even a very complex real system) behaves in a period-doubling fashion then it is very likely that one will see the full structure of the Feigenbaum diagram in it. In other words, although the quadratic iterator in some sense is certainly much too simple to carry any information about real systems, in a very striking and general sense it, in fact, does carry the essential information about how systems may develop chaotic behavior.

This discovery is one of first rate importance and gives testimony to the beauty and adventure of mathematical thinking.

11.1 The First Step From Order to Chaos: Stable Fixed Points

The portion of the final-state diagram to the left of the Feigenbaum point s_∞ is a self-similar fractal tree (see figure 11.5). It describes the *period-doubling scenario* of the quadratic iterator, which leads from a very simple and orderly behavior of the dynamics right to the beginning of the chaotic region. Let us try to understand the mechanism lying at the base of its generation and leading to the self-similarity of the tree.

The Period-Doubling Tree

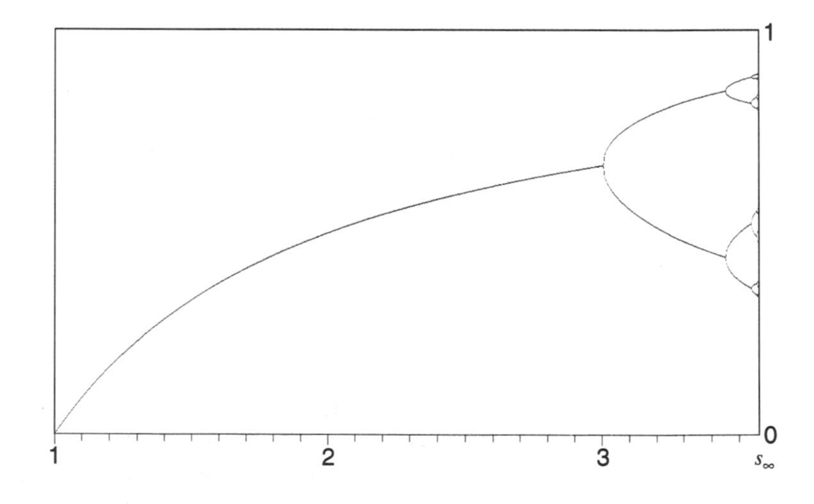

Figure 11.5 : The first portion of the final-state diagram — the period-doubling tree.

At the Stem ...

We start our discussion at the 'stem' of the tree, the part which lies between $a = 1$ and $a = 3$. This part represents rather simple dynamics, a stable situation where the iteration is always led to a rest point. In figure 11.6 we show two typical examples. The left plot shows the time series of the initial point 0.1 for the parameter $a = 1.75$. The iteration settles down on a final state which we denote by $A(1.75)$. In this case the final state is simply the value $3/7$. Indeed, starting with any initial value between 0 and 1, we approach this limiting value. The right graph shows the situation for $a = 2.75$. In this case the final state $A(2.75)$ is $7/11$, but the course of the iteration does not approach this value directly. Rather, it oscillates around the final state while settling down.

In both cases we have a situation which, from the point of view of making predictions, could not be better. No matter where we choose an initial value x_0, we can predict that in the long run we

Stability of Final State

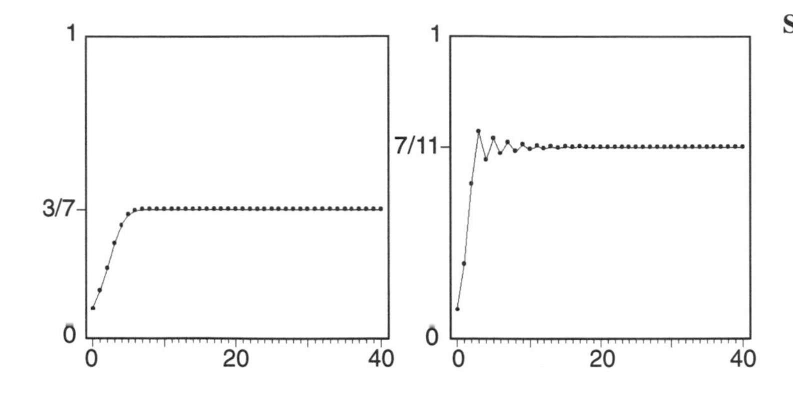

Figure 11.6 : Stable behavior of time series for initial value 0.1 and $a = 1.75$ (left) and $a = 2.75$ (right).

will be near or at the attractor $A(a)$. It seems as if that point is equipped with a magic force which attracts iterations independent of x_0.

For ease of notation in this chapter let us introduce the symbol f_a for the quadratic function

$$f_a(x) = ax(1-x) \ .$$

... a Stable Fixed Point

What would we observe when starting the iteration of the quadratic iterator f_a exactly with the value of the attractor, i.e., $x_0 = A(a)$? Then we would have for all iterates $x_0 = x_1 = x_2 = \ldots$ In other words, x_0 would be a fixed point of f_a (sometimes also called a rest point). Thus x_0 would solve the fixed point equation $f_a(x) = x$,

$$ax(1-x) = x \ . \tag{11.2}$$

There are two solutions, which we call p_0 and p_a,

$$p_0 = 0 \text{ and } p_a = \frac{a-1}{a} \ .$$

Moreover, we note that if $x_0 = 1$, then $x_1 = 0$ and $x_1 = x_2 = \ldots = 0$. We say that 1 is a *preimage* of the fixed point $p_0 = 0$. But $x_0 = 0$ and $x_0 = 1$ are the only initial values which lead to 0; all other values are attracted by $p_a = (a-1)/a$. Indeed, if we take any x_0 between 0 and $(a-1)/a$ then $x_1 > x_0$ for all parameters $a > 1$.[1] In other words, the iteration is pushed away from the rest point $p_0 = 0$. We say that p_0 is a *repeller* or an *unstable fixed point.* On the other hand $p_a = (a-1)/a$ is a *stable fixed point* (or

[1] Assume $0 < x_0 < p_a = (a-1)/a$. We compute $x_1 - x_0 = ax_0(1-x_0) - x_0 = x_0(a - ax_0 - 1)$. From the assumption it follows that $a - 1 > ax_0$, thus, $a - ax_0 - 1 > 0$ and $x_1 - x_0 > 0$. In other words, $x_1 > x_0$.

Graphical Iteration Near a Stable Fixed Point

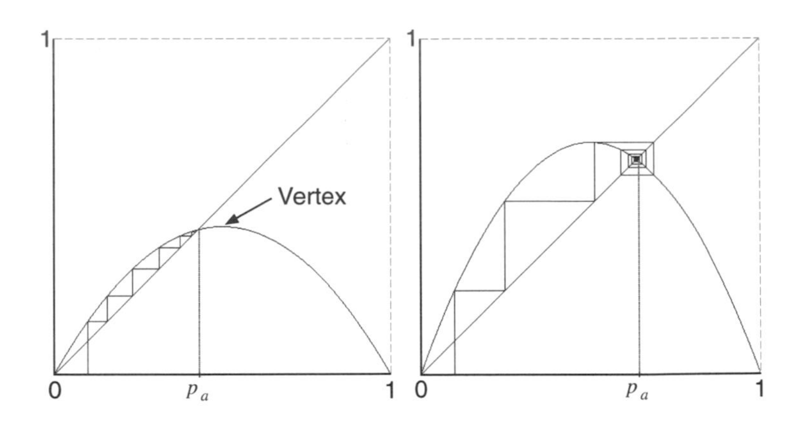

Figure 11.7 : Graphical iteration for $a = 1.75$ (left) and $a = 2.75$ (right). In both cases we start with $x_0 = 0.1$. The iteration settles down at $p_a = (a-1)/a$ which is $3/7$ (left) and $7/11$ (right).

attractive fixed point) for all parameters a between 1 and 3.[2] Let us verify these facts using graphical iteration as shown in figure 11.7. You will recall that here the iteration is represented by a path with horizontal and vertical steps, which for convenience, we will now call a poly-line (see chapter 1, page 66). The left image shows the situation for $a = 1.75$. The parabola $ax(1-x)$ intersects the bisector at the fixed points $p_0 = 0$ and $p_a = (a-1)/a$. Between these values the parabola lies above the bisector, but its vertex lies beyond the intersection. Thus the iteration must be repelled away from 0. On the other hand the poly-line representing the iteration is trapped between the parabola and the bisector and thus is led directly to the second intersection point at $(a-1)/a$.

The right graph in figure 11.7 shows the situation for $a = 2.75$. In this case the bisector intersects the parabola beyond its vertex. Thus the poly-line, which represents the iteration, begins to spiral around the point of intersection. This spiraling is directed inwards; the process again settles down at $(a-1)/a$. In other words, the fixed point is still attractive although the local behavior (i.e., the way orbits are attracted) has changed. The spiraling of the poly-line explains the oscillations which we observed in figure 11.6.

Four Types of Local Dynamics

There are four basic types of iteration behavior near a fixed point. As an example we consider the linear feedback process

$$x_{n+1} = sx_n$$

i.e., the iteration function is sx, and its graph is a straight line in-

[2]However for $a > 3$ this point also looses its stability, and this will be discussed later.

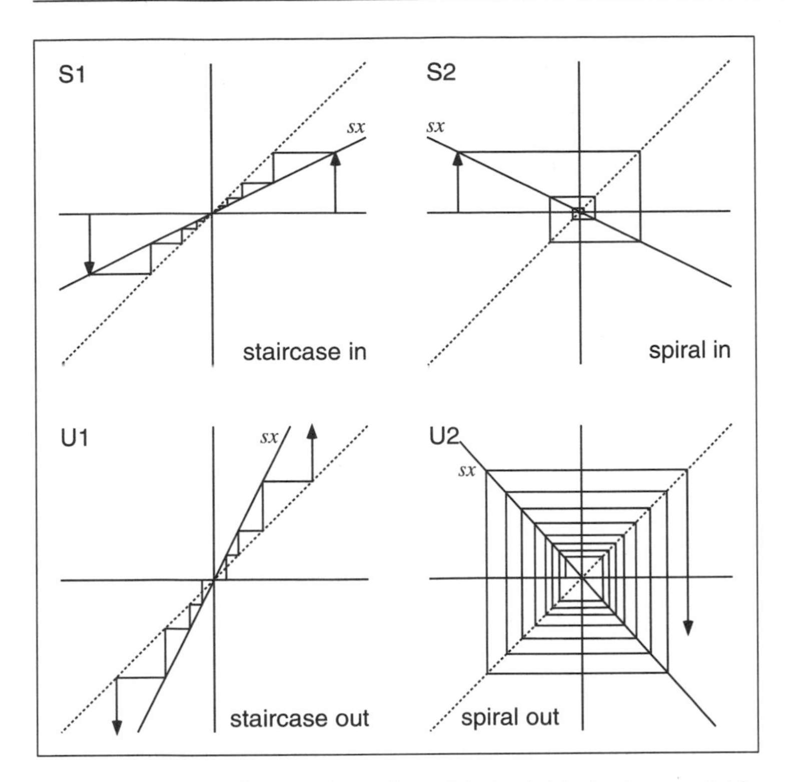

Figure 11.8 : The fixed point at the origin is stable (top) or unstable (bottom) depending on the slope of the function at the fixed point.

tersecting the bisector at the origin, which is the only fixed point if $s \neq 1$. The slope of the straight line determines the way it intersects the bisector (see figure 11.8). The slope is given by the parameter s. Accordingly we classify the different iteration behavior into four cases:

S1: $0 < s < 1$ iteration as staircase inwards to stable fixed point 0

S2: $-1 < s < 1$ iteration as spiral inwards to stable fixed point 0

U1: $s > 1$ iteration as staircase outwards from unstable fixed point 0

U2: $s < -1$ iteration as spiral outwards from unstable fixed point 0.

The cases $s = \pm 1$ are special. If $s = 1$, every point is a fixed point, while for $s = -1$, all orbits have the form $x, -x, x, -x, \ldots$

We proceed by considering *nonlinear* systems. The type of the fixed point x^* of a nonlinear feedback process $x_{n+1} = f(x_n)$ can be characterized by the derivative $f'(x^*)$, which can be interpreted as the slope of the graph of f at x^*. In the case of a linear transformation $x \rightarrow sx$, the type of dynamics near the fixed point are characterized by the parameter s. For a nonlinear transformation the same characterization is possible by the derivative $f'(x^*)$. In particular, the fixed point is attractive if

$$|f'(x^*)| < 1$$

and repelling if

$$|f'(x^*)| > 1 \, .$$

Let us apply this fact to the two fixed points $p_0 = 0$ and $p_a = (a-1)/a$ of the quadratic iterator $x \to f_a(x) = ax(1-x)$. The derivative of the transformation f_a is given by

$$f'_a(x) = a(1-2x) \, .$$

Consider the first fixed point $p_0 = 0$. The derivative is $f'_a(0) = a$. Thus,

$$\begin{aligned} |f'(p_0)| &< 1 \quad \text{for} \quad 0 \le a < 1 \\ |f'(p_0)| &= 1 \quad \text{for} \quad a = 1 \\ |f'(p_0)| &> 1 \quad \text{for} \quad a > 1 \, . \end{aligned}$$

Thus, as soon as the parameter a passes from $a < 1$ to $a > 1$ the fixed point p_0 loses its stability and becomes a repelling fixed point (type U1, staircase out).

To discuss the other fixed point $p_a = (a-1)/a$ we compute the derivative

$$f'_a(p_a) = 2 - a \, .$$

Thus, for this fixed point we obtain:

Parameter	Derivative	Type of fixed point
$0 \le a < 1$	$1 < f'(p_0) \le 2$	repelling
$1 < a < 2$	$0 < f'(p_0) < 1$	stable (staircase in)
$2 < a < 3$	$-1 < f'(p_0) < 0$	stable (spiral in)
$3 < a \le 4$	$-2 \le f'(p_0) < -1$	repelling (spiral out)

Now we can ask: is there a parameter a for which the behavior at the rest point changes to spiraling (or oscillation)? Yes, the spiraling sets in as soon as the vertex of the parabola surpasses the right intersection point of the parabola and the bisector. In other words, we are looking for the case where the intersection point and the vertex of the parabola fall together. Since the parabola has its maximum at $x_{\max} = 0.5$, we must solve the equation $0.5 = 0.5a(1-0.5)$.

The Super Attractive Case

The solution is $a = 2$. Figure 11.9 shows this situation. The left graph is the plot of the time series for the initial value 0.1.

Iteration to Super Attractive Fixed Point

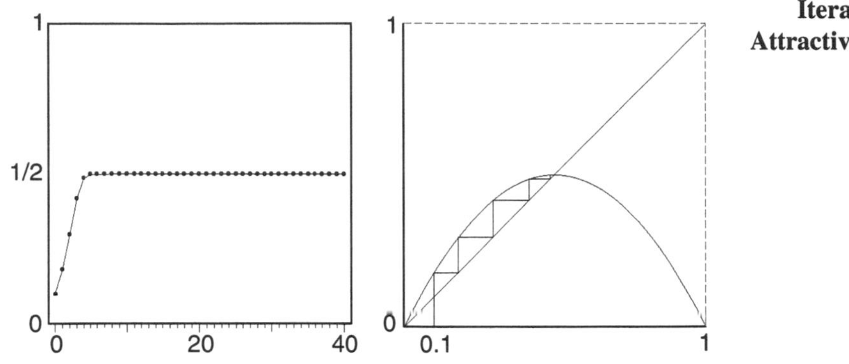

Figure 11.9 : For $a = 2$ we observe a super attractive situation. The graphical iteration demonstrates how the orbit rushes into the fixed point.

A Direct Approach to the Fixed Point

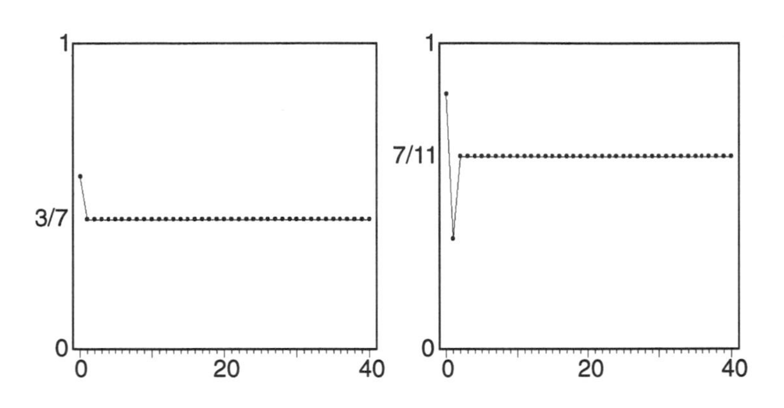

Figure 11.10 : Examples of initial values which directly lead to the fixed points: $x_0 = 0.5714...$ and $a = 1.75$ (left), $x_0 = 0.8432...$ and $a = 2.75$ (right).

Comparing it with the time series for $a = 1.75$ and $a = 2.75$ (figure 11.6), we find that for a = 2 the attractive fixed point is approached much faster. This impression is also supported by the graphical iteration; when the poly-line representing the iteration approaches the vertex of the parabola it rushes to the fixed point very quickly. Indeed, this point is called *super attractive*. This case is very special; in the interval of parameters from 1 to 3 it occurs only at one point, $a = s_1 = 2$.

It seems reasonable to conjecture that for a given starting point x_0 the iteration reaches the attractive fixed point fastest when the parameter a is of the super attractive type. In fact, this is true for most initial values x_0, but not for all; there are a number of exceptions. Figure 11.10 shows just two of them (compare figure 11.6). On the left (for $a = 1.75$) we start with $x_0 = 4/7$. This leads within one step directly to the fixed point.[3] On the right ($a = 2.75$) we set $x_0 = 0.8432...$, which leads within two steps exactly to the fixed point. Somehow it seems to be a paradox, but in the super attractive case $a = 2$ this behavior is not possible; there are no starting points x_0 which end up right in the fixed point after a finite number of iterations.

Super Attractive Fixed Points

We have characterized a fixed point x^* of a nonlinear feedback process $x_{n+1} = f(x_n)$ by its derivative $f'(x^*)$. It is stable provided $|f'(x^*)| < 1$. It is said to be *super attractive* if $f'(x^*) = 0$. Let us see what happens if the iteration approaches such a rest point for the example of the quadratic iterator $f_a(x) = ax(1-x)$. The fixed point in question is $x^* = p_a = (a-1)/a$. Let us start the iteration close to p_a; we set

$$x_0 = \frac{a-1}{a} + \varepsilon$$

equal to the fixed point plus a small perturbation ε. Then we obtain $x_1 = ax_0(1 - x_0)$ and calculate

$$x_1 = \frac{a-1}{a} + 2\varepsilon - a\varepsilon - a\varepsilon^2 .$$

Now the fixed point is attractive provided

$$|2\varepsilon - a\varepsilon - a\varepsilon^2| < |\varepsilon| .$$

Then x_1 will be closer to x_0, and — using the same line of arguments — x_2 will be even closer, and so on. This is always the case when $1 < a < 3$ and ε is sufficiently small. In the special case when $a = 2$ the term on the left becomes $2\varepsilon^2$, which means that the initial perturbation ε is reduced quadratically. This implies that the number of digits which agree when we compare p_a with x_k will double in each step. However, for $a \neq 2$ the linear term $2\varepsilon - a\varepsilon$ remains and therefore the reduction is less powerful. Thus, it becomes apparent, why we call the fixed point in the case $a = 2$ super attractive.

Time Profiles

These observations raise the question how long it will take the iteration of an arbitrary initial value to settle down at the fixed point. Let us set up an experiment to investigate this question. From the unit interval, we choose equally spaced initial values

$$x_0 = \frac{k}{600}, \quad k = 1, 2, ..., 599 .$$

[3] We check that: for $a = 7/4$ the fixed point is $p_a = (a-1)/a = 3/7$. With the starting point $x_0 = 4/7$ we get $x_1 = 7/4 \cdot 4/7 \cdot (1 - 4/7) = 3/7 = p_a$.

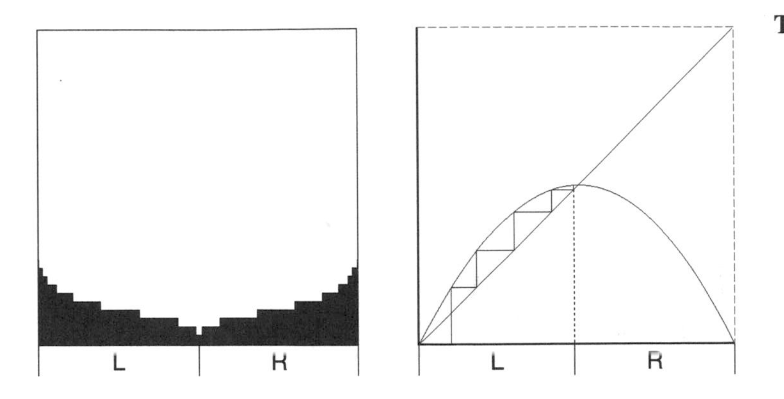

Time Profile for Super Attractive Case

Figure 11.11 : Time profile for the super attractive case $a = s_1 = 2$ and the graphical iteration of a typical orbit ($x_0 = 0.1$).

For each of these starting values we begin the iteration and count how many steps it takes to get to the fixed point $(a-1)/a$ within a short distance, say $1/6000$. The result of each sequence of iterations is a count $N_a(x_0)$, which depends on the initial value x_0 and the parameter a. We present the result of the experiment in a diagram which we call the *time profile*. On the horizontal axis the initial values are marked. For each starting value x_0, we plot a column whose height represents the computed iteration count $N_a(x_0)$. Figure 11.11 shows the time profile for the super attractive case $a = 2$.

The diagram shows a valley shape which agrees with a first intuitive guess. The attractor $A(2) = 0.5$ lies in the center of the interval and exactly where the valley has its deepest point. In other words, the number $N_a(x_0)$ increases as x_0 moves from the center towards either one of the endpoints of the unit interval. Note that the parabola of the quadratic iterator and its time profile exhibit the same symmetry. We can verify this property if we compare the iteration of an initial value from the left part L of the unit interval $x_0 = c$ and its symmetric counter part $x_0 = 1 - c$ from the right side R. In both cases the first step of the iteration gives $x_1 = ac(1-c)$ (which lies on the left side). Thus $N_a(c)$ and $N_a(1-c)$ are the same; the valley must be symmetric.

Let us now examine the time profile for a parameter a below the super attractive case, say $a = 1.75$, see figure 11.12. Surprisingly, we observe two small sub-valleys within the large valley and one intermediate flat-topped mountain. Further experiments for other values of a between 1 and 2 provide the same result: two sharp sub-valleys within one large valley. Again the reason for this behavior can be studied with graphical iteration, also shown in figure 11.12.

Time Profile $a = 1.75$

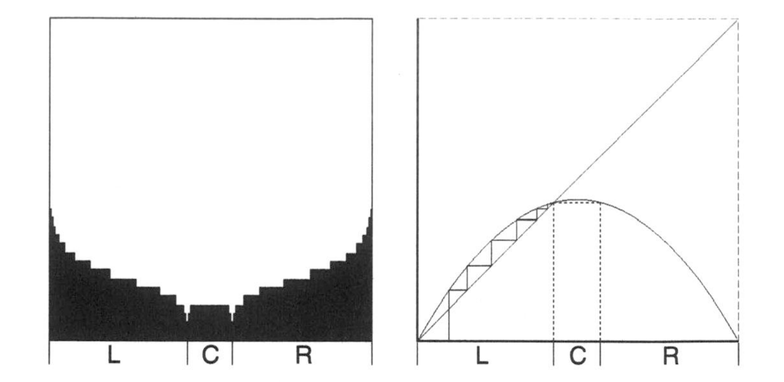

Figure 11.12 : Time profile for $a = 1.75$ and graphical iteration of a typical orbit ($x_0 = 0.1$).

Every x_0 chosen in the left interval L generates a poly-line which creeps up towards the fixed point. In other words, the iteration monotonically approaches the fixed point $p_a = (a-1)/a = 3/7$. The number of steps of the staircase which it takes to get close to the attractor increases as we move x_0 closer to 0. This explains the steep angle of the slope on the extreme left of the time profile. What, however, about initial points x_0 in the remaining part of the unit interval? The graphical iteration makes it clear that there is exactly one new choice $x_0 > 1/2$ such that the first step of the iteration

$$x_1 = \frac{7}{4}x_0(1 - x_0)$$

hits the fixed point $x_1 = 3/7$. In other words, x_0 is a preimage of the fixed point. Solving the quadratic equation

$$\frac{3}{7} = \frac{7}{4}x_0(1 - x_0)$$

we obtain the two solutions, $x_0 = 3/7$, which is the fixed point itself, and $x_0 = 4/7$. In terms of graphical iteration, this means that the parabola has the same height at both points. The two sub-valleys of the time profile are located precisely at these two points.

The Center Peak in the Time Profile

More generally, let us ask, when does the first iterate x_1 coincide with the fixed point $(a-1)/a$? The answer is obtained from the equation

$$\frac{a-1}{a} = ax_0(1 - x_0)$$

or

$$x_0^2 - x_0 + \frac{a-1}{a^2} = 0$$

Time Profile $a = 2.75$

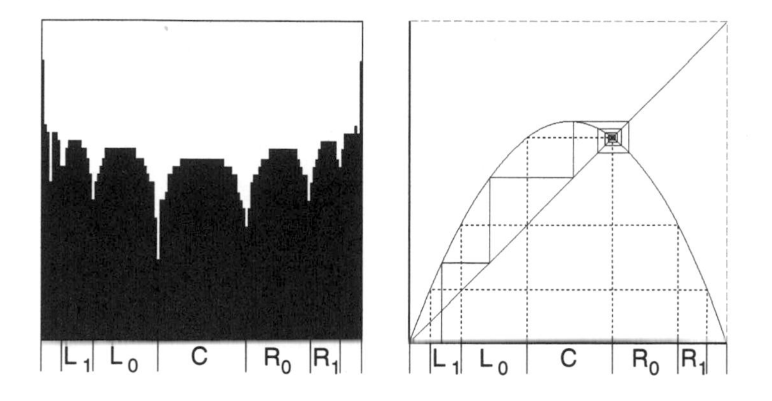

Figure 11.13 : Time profile for $a = 2.75$ and graphical iteration of a typical orbit ($x_0 = 0.1$).

which is solved by

$$x_0 = \frac{a-1}{a} \text{ and } x_0 = \frac{1}{a} \ .$$

Based on these results let us rearrange the left and right intervals L and R and introduce a third interval C:

$$L = \left(0, \frac{a-1}{a}\right) \quad C = \left(\frac{a-1}{a}, \frac{1}{a}\right) \quad R = \left(\frac{1}{a}, 1\right) \ .$$

The graphical iteration for initial values x_0 in the center interval C produces poly-lines that creep down to the intersection point of parabola and bisector, which represents the attractor. The number of iterations needed to get close to p_a will go up as x_0 is chosen closer to 1/2, and it will decrease as x_0 gets closer to the endpoints of the interval C. This explains the central mountain around $x = 1/2$ in the time profile. Finally, we note that initial points x_0 chosen from the interval R will be carried in one step of the iteration into the interval L and then follow the pattern which we have discussed already. This explains the extreme right slope in the time profile.

Beyond the Super Attractive Case

In the super attractive case ($a = 2$) the two end points $(a-1)/a$ and $1/a$ of the center interval are identical. Thus the interval C vanishes and the two sub-valleys form one big valley. Let us now see what happens when we increase the parameter a further. Figure 11.13 shows the result for $a = 2.75$. Here we see many sub-valleys (theoretically there are infinitely many) and intermediate mountain tops. Further experiments would show that for all parameters a between 2 and 3 there are similar results. Again let us see what the graphical iteration can teach us in this situation.

While for parameters a below 2 the fixed point $p_a = (a-1)/a$ is always to the left of 1/2 (i.e., on the left side of the vertex of the parabola in the graphical iteration), it is now to the right of 1/2 (i.e., beyond the apex of the parabola); and that has remarkable consequences. Also the preimage of the fixed point has reversed sides. Now $1/a$ is left of 1/2, and this implies that the bisector lies below the parabola at this point. This suggests that now there are preimages of $1/a$ (i.e., solutions to the equation $ax(1-x) = 1/a$). If x_0 is such a solution, then $x_1 = 1/a$, and $x_2 = (a-1)/a$. In other words, starting the iteration with x_0 would lead exactly to the fixed point in just two steps. Indeed, for $a > 2$ the quadratic equation $ax(1-x) = 1/a$ yields the solutions

$$\alpha_1 = \frac{1}{2} - \frac{1}{2a}\sqrt{a^2-4}$$

$$\beta_1 = \frac{1}{2} + \frac{1}{2a}\sqrt{a^2-4}$$

both symmetrically located around 1/2 for $a > 2$. Note that these solutions are defined only for parameters $a \geq 2$. This establishes three interesting intervals:

$$L_0 = \left(\alpha_1, \frac{1}{a}\right) \quad C = \left(\frac{1}{a}, \frac{a-1}{a}\right) \quad R_0 = \left(\frac{a-1}{a}, \beta_1\right)$$

which belong to the three central mountains of the time profile. Now we can look for the preimages of α_1: α_2 and β_2, then for the preimages of α_2, and so on. The valleys of the time profile are precisely at theses values α_k and β_k. Given an initial point $x_k = \alpha_1$ or $x_k = \beta_1$, the first $k+1$ iterations are

$$\alpha_{k-1}, \alpha_{k-2}, ..., \alpha_1, \frac{1}{a}, \frac{a-1}{a} \ .$$

Thus, after $k+1$ iterations, the fixed point is reached.

Preimages of p_a and Iteration Patterns

For a between 2 and 3 the iteration behavior is as follows: if we are in C then the following step will be in R_0 and the next step back in C, and so on. This explains the spiral part of the poly-line in figure 11.13. If we start in L_0 then the next step will also be in C. We can extend this observation using the iterated preimages of p_a to obtain a complete decomposition of the unit interval. With α_0 and β_0 as above, we define sequences $\alpha_1, \alpha_2, ...$ and $\beta_1, \beta_2, ...$ implicitly by

$$f_a(\alpha_k) = f_a(\beta_k) = \alpha_{k-1}, \quad \alpha_k < \beta_k, \quad k = 1, 2, ...$$

These equations can also be solved explicitly,

$$\alpha_k = \tfrac{1}{2} - \sqrt{\tfrac{1}{4} - \tfrac{\alpha_{k-1}}{a}}$$
$$\beta_k = \tfrac{1}{2} + \sqrt{\tfrac{1}{4} - \tfrac{\alpha_{k-1}}{a}}$$

for k = 1,2,... There is an ordering

$$\cdots < \alpha_3 < \alpha_2 < \alpha_1 < \alpha_0 < \beta_0 < \beta_1 < \beta_2 < \beta_3 < \cdots$$

These preimages are the locations of the small valleys in the time profile. These values define intervals $L_k = (\alpha_{k+1}, \alpha_k)$ and $R_k = (\beta_k, \beta_{k+1})$. The intervals L_k cover the left part of the unit interval from 0 to $1/a$, while the intervals R_k cover the right part from $(a-1)/a$ to 1.

Together with these intervals the iteration behavior is easily described:

$$\begin{aligned} &L_k \to L_{k-1} \to L_{k-2} \to \cdots \to L_0 \to C \\ &C \to R_0 \to C \\ &R_k \to L_{k-1} \; . \end{aligned}$$

This yields an interpretation of the infinite number of mountains in our time profile. Each one corresponds to one of the intervals L_k, R_k, or C. Let us contrast this behavior to the dynamics of the quadratic iterator for parameters below 2. For $1 < a < 2$ we have:

$$\begin{aligned} &L \to L \\ &C \to C \\ &R \to L \; . \end{aligned}$$

For the super attractive case ($a = 2$) we have only

$$\begin{aligned} &L \to L \\ &R \to L \; . \end{aligned}$$

The iteration behavior switches to much more complicated patterns when the parameter a crosses the super attractive case $a = 2$.

11.2 The Next Step From Order to Chaos: The Period Doubling Scenario

Having studied the dynamics of the quadratic iterator f_a in detail for parameters between 1 and 3, we continue to increase a beyond 3. For such large parameters the fixed point $p_a = (a-1)/a$ is not stable anymore; it is a repeller. Is there a different attractor that takes over the role of p_a?.

Stable Oscillation

Let us see what happens, for example, if $a = 3.1$, see figure 11.14. We obtain a time series which exhibits an entirely new behavior. There is oscillation as in the case of a between 2 and 3, but it does not finally settle down to one single point. Rather, it stabilizes in oscillating between two values, a low number $x_l(a)$, and a high value $x_h(a)$. Thus, in the final-state diagram we obtain just these two points at parameter $a = 3.1$.

Time Series $a = 3.1$

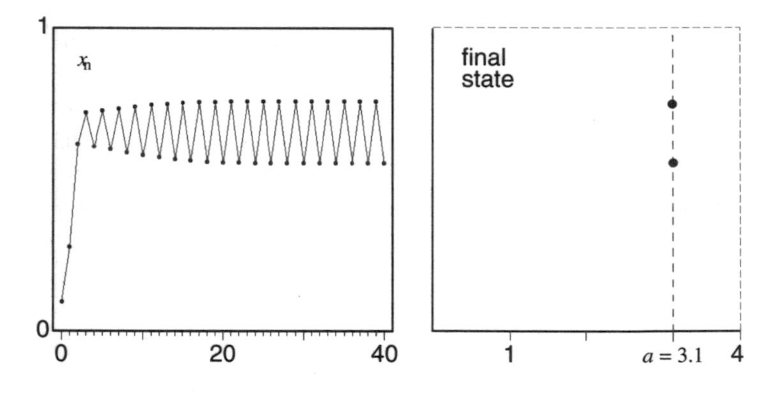

Figure 11.14 : Time series for $a = 3.1$ and initial value 0.1. The iteration leads to a final state which consists of two points $x_l(a)$ and $x_h(a)$.

Again we turn to graphical iteration to obtain some further insight. In figure 11.15 we consider initial values $x_0 = 0.075$ (left) and $x_0 = 0.65$ (right). Indeed, we observe a new phenomenon. In the left graph we first notice a familiar staircase. But then the polyline turns into an inward spiral which slowly runs into a repeating loop. In the right plot the initial value x_0 is close to the unstable fixed point p_a. Its orbit spirals outward towards the same loop as seen on the left. In other words, while for $a < 3$ the fixed point p_a attracts all iterations, it turns into a repeller when a grows larger than 3. Close to p_a, iterations will be pushed away. The fixed point p_a loses its stability as a crosses the border $b_1 = 3$. This particular parameter value is called a *bifurcation point*.

Let us summarize our findings for the final-state diagram (see

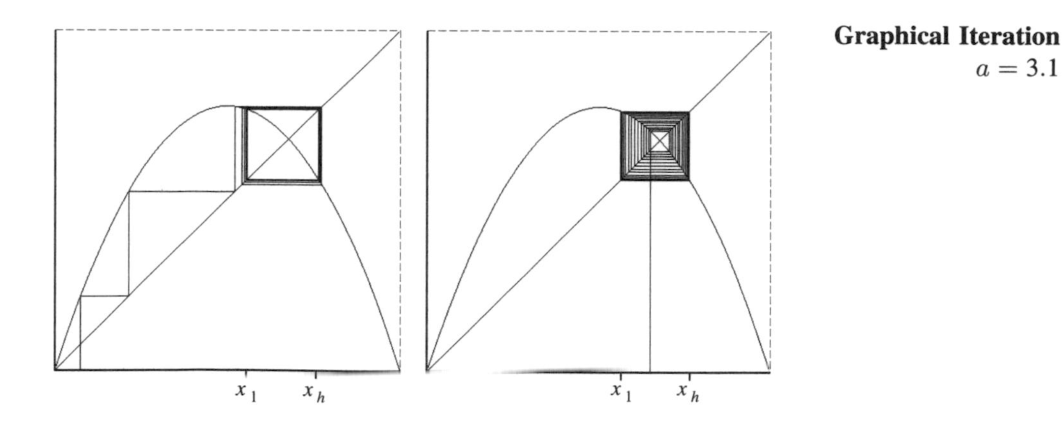

Graphical Iteration
$a = 3.1$

Figure 11.15 : The periodic cycle $\{x_l, x_h\}$ is the attractor for the quadratic iterator at parameter $a = 3.1$. On the left the initial point for the iteration is $x_0 = 0.075$, while on the right $x_0 = 0.65$.

figure 11.16); $p_a = (a-1)/a$ is an attractor for all iterations starting in the interval (0,1) for parameters $1 < a < 3$. Formally, the attractor is a single point,

$$A(a) = \{p_a\} \text{ for } 1 < a < 3 .$$

For $a > 3$, p_a still exists. Thus, an iteration started precisely at this point remains there forever,[4] $p_a = x_0 = x_1 = \cdots$. However, p_a is a repeller, and therefore is *not* part of the final state $\boldsymbol{A(a)}$. The attractiveness has been taken over by the loop which oscillates between the two values $x_l(a)$ and $x_h(a)$. Thus, the final state is the attractor made of two points,

$$A(a) = \{x_l(a), x_h(a)\} \text{ for } a > 3 .$$

We call the pair $\{x_l(a), x_h(a)\}$ a *2-cycle* or an *orbit of period two*. It is characterized by the fact that x_l is transformed into x_h and vice versa. This can be used for an explicit calculation of these numbers, see below. This periodic orbit exists for all parameters $3 < a \leq 4$. However, as the fixed point p_a loses stability at $b_1 = 3$, also the 2-cycle loses stability at a certain parameter value $b_2 > 3$, and this will be discussed below.

Explicit Calculation of Bifurcations

We have already performed explicit calculations for the fixed points of the quadratic iterator. Also for the 2-cycle we can carry out a direct calculation. Writing again $f_a(x) = ax(1-x)$ we need to find

[4]The same holds for the other fixed point of the iterator, $p_0 = 0$, which is unstable for parameters $a > 1$.

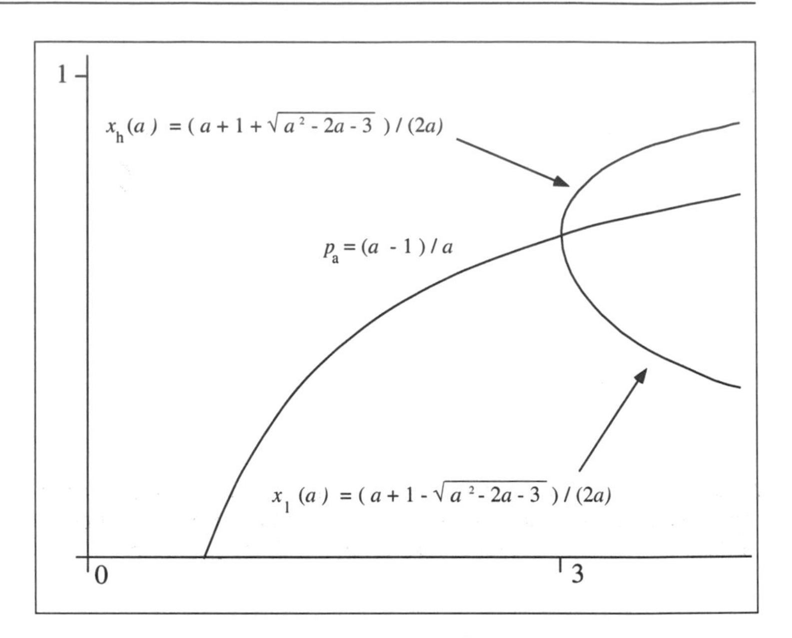

Figure 11.16 : In this bifurcation diagram the fixed points p_a and the 2-cycles $\{x_l(a), x_h(a)\}$ are shown.

the solutions of the equation $f_a(f_a(x)) = x$. This is a 4th order equation:

$$-a^3x^4 + 2a^3x^3 - (a^2 + a^3)x^2 + (a^2 - 1)x = 0 \ .$$

This looks complicated. But luckily we already know two solutions namely the fixed points of the iteration which solve $f_a(x) = x$ (i.e., 0 and $(a-1)/a$). The solution $x = 0$ allows us to simplify the equation to a third order equation

$$-a^3x^3 + 2a^3x^2 - (a^2 + a^3)x + (a^2 - 1) = 0$$

Knowing the second solution, $(a-1)/a$, we now divide this equation by $x - (a-1)/a$, which gives the second order equation

$$-a^3x^2 + (a^2 + a^3)x - (a^2 + a) = 0$$

or, dividing by $-a^3$,

$$x^2 - \frac{a+1}{a}x + \frac{a+1}{a^2} = 0 \ .$$

The roots of this quadratic equation are

$$x_h(a) = \frac{a + 1 + \sqrt{a^2 - 2a - 3}}{2a}$$

and

$$x_l(a) = \frac{a + 1 - \sqrt{a^2 - 2a - 3}}{2a}$$

Considering only parameters between 0 and 4, we note that these solutions are defined only for $a \geq 3$. Moreover, at $a = 3$ we get $x_l(a) = x_h(a) = (a-1)/a$, i.e., the two solutions bifurcate from the fixed point p_a. Figure 11.16 shows a bifurcation diagram of the explicitly calculated solutions.

Note that in the final-state diagram you do not see the solution $p_a = (a-1)/a$ for $a > 3$. This corresponds to the fact that although the fixed point continues to exist, it has become unstable (i.e., the iteration is pushed away from this point).

More Time Profiles

Figure 11.17 shows an experiment which is related to the time profiles we have computed so far. Here we take again sample starting points from the unit interval and run the iteration, counting the number of iterations until the orbit comes close to the 2-cycle. In practice, the iteration can be stopped as soon as x_k and x_{k+2} have become very close,

$$|x_{k+2} - x_k| < \varepsilon$$

where ε is some small number, for example $\varepsilon = 1/6000$. In the figure we choose three parameters,

$$a = 3.1, \quad a = 1 + \sqrt{5} \approx 3.236, \quad a = 3.4\ .$$

The results are much more complex than the time profiles for parameters between 1 and 3. In particular, there are two striking phenomena.

- The sub-valleys found for parameters just below $a = 3$ (see figure 11.13) have turned into spikes.
- The central areas between the two inner spikes at $1/a$ and $(a-1)/a$ look like compressed versions of the complete diagrams found in figures 11.11, 11.12, and 11.13, corresponding to the present cases $a = 3.1$, $a \approx 3.236$, and $a = 3.4$.

The Spikes at α_k and β_k

There must be a reason for this striking similarity. First, we recall that 0 is a repeller for the iteration if $a > 1$, and 1 is its preimage. For $a > 3$ we now find that $p_a = (a-1)/a$ is also a repeller, and $1/a$ is its preimage. This already explains to some degree that the 'spikes' at 0 and 1 in figures 11.11 to 11.13, can now be found at $1/a$ and $(a-1)/a$. The other 'spikes' in the new diagrams are at the points α_k and β_k, which we already have introduced as iterated preimages of the fixed point p_a. Spikes correspond to high iteration counts, which, however, seem to raise a contradiction. Starting the iteration exactly at one of the points $x_0 = \alpha_k$ or $x_0 = \beta_k$, we arrive exactly at the fixed point p_a after only $k+1$ iterations, $x_{k+1} = p_a$. Then the next two iterations are $x_{k+2} = x_{k+3} = p_a$, and, thus, the test

$$|x_{k+3} - x_{k+1}| < \varepsilon$$

Time Profile $a > 3$

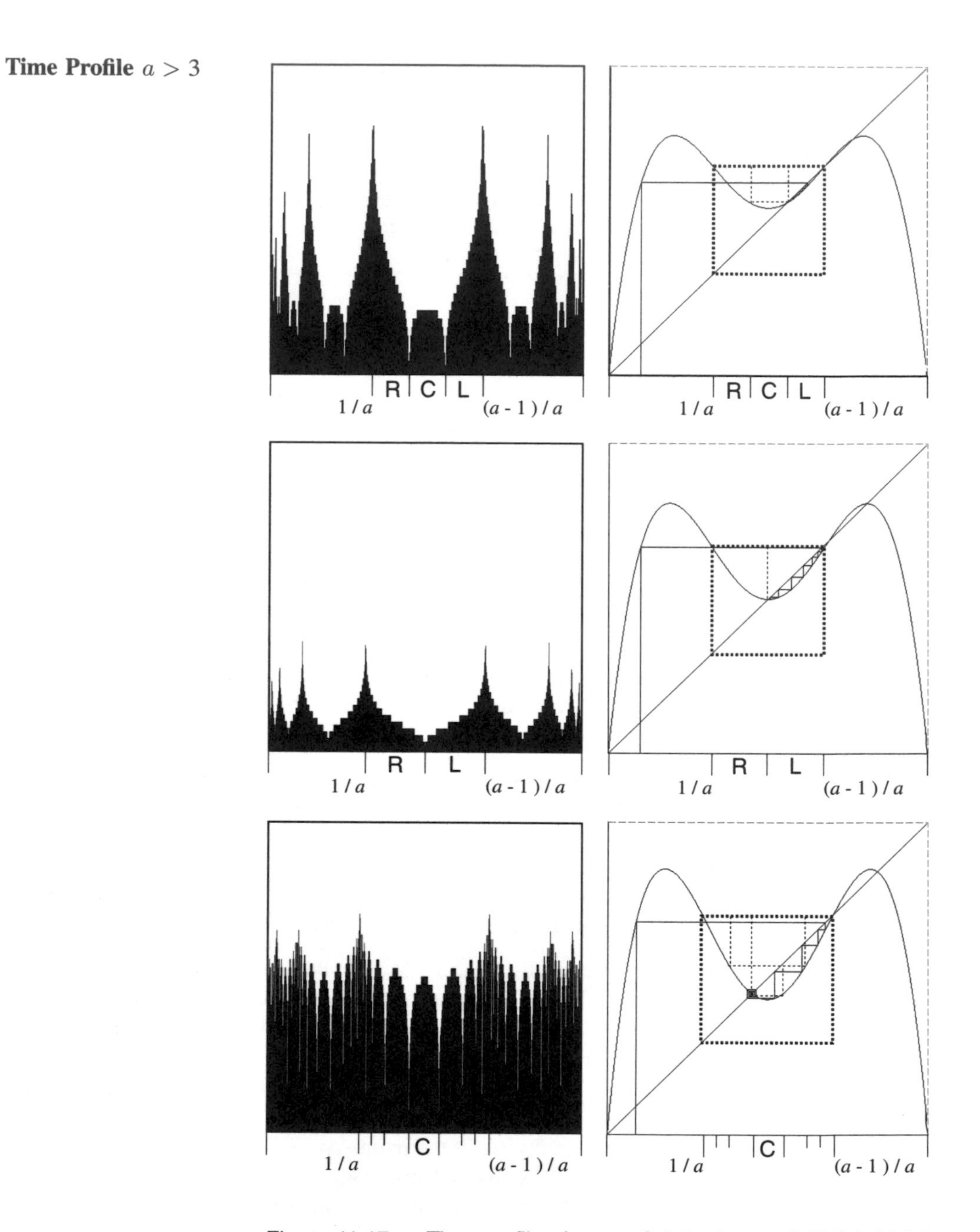

Figure 11.17 : Time profiles for $a = 3.1$ (top), $a = 3.236$ (middle) and $a = 3.4$ (bottom). Note the similarities of the portion between $1/a$ and $(a-1)/a$ to the preceding time profiles 11.11, 11.12, and 11.13 which implies the reversal of the labels L and R.

f_a and f_a^2

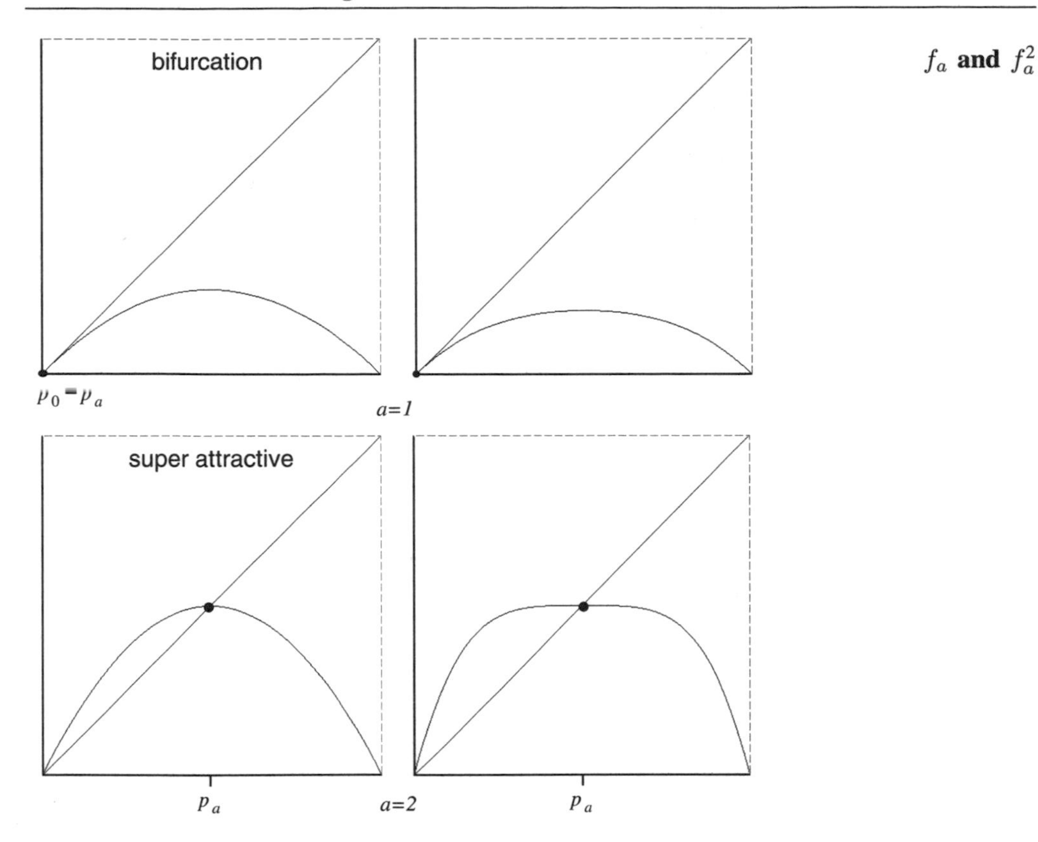

Figure 11.18 : Comparing $f_a(x)$ (left) and $f_a^2(x)$ (right) for $a = 1$ and $a = 2$.

is satisfied. This yields an iteration count

$$N_a(\alpha_k) = N_a(\beta_k) = k + 3,$$

a rather low number compared to the spikes observed in the numerical studies. But since the fixed point is unstable, this behavior is not observable. To be visible, infinitely precise computation would be required, but also the specification of the starting points with infinite precision. This is not possible, and therefore it is not visible. On the contrary, starting the iteration close to one of the points α_k or β_k produces a $(k+1)^{\text{st}}$ iterate close to the fixed point p_a. The closer it gets, the more further iterations it takes to move the orbit away from the repulsive fixed point and to come close to the attractive periodic cycle. This results in the spikes as seen in the figures.

The Iterate of f_a

But there is more to our observation of similarity, and that is revealed in the graphical iterations to the right of the time profiles.

f_a **and** f_a^2

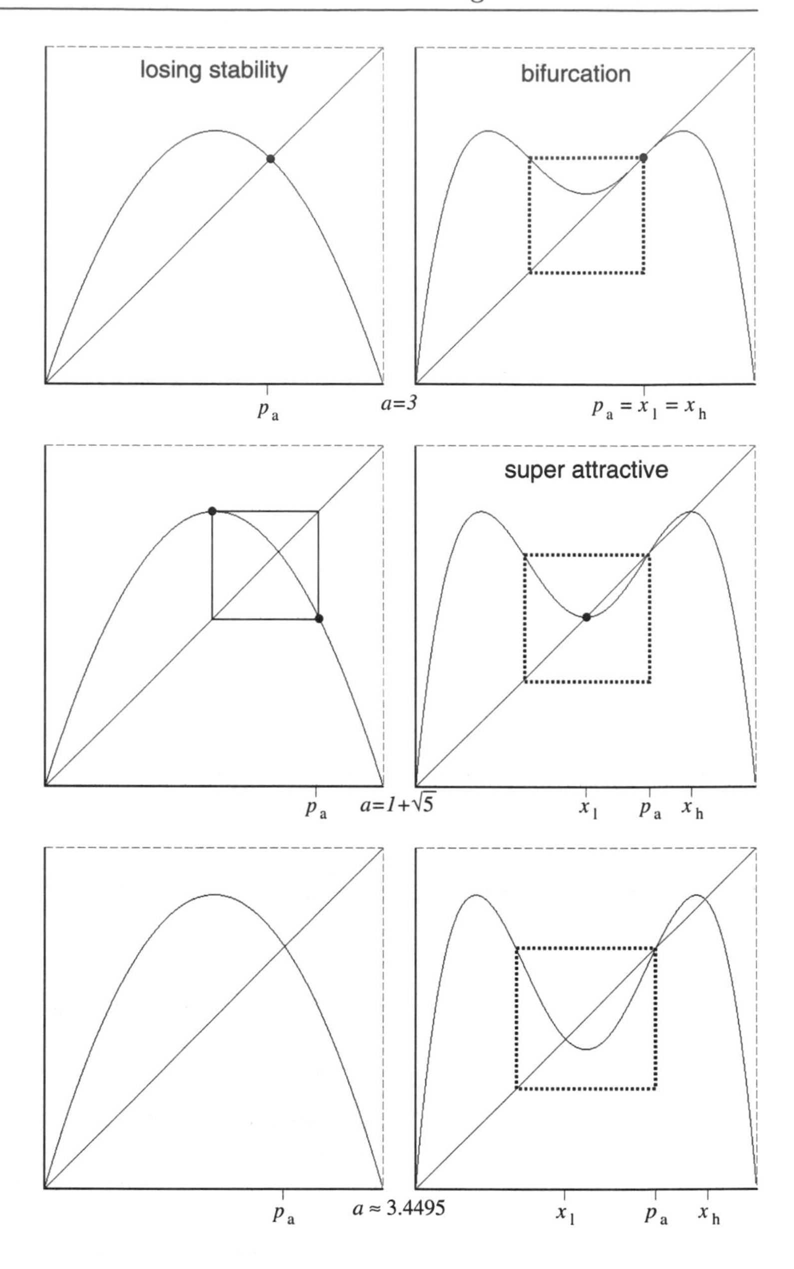

Figure 11.19 : Comparing $f_a(x)$ (left) and $f_a^2(x)$ (right) for $a = 3$ to $a = 3.4496$.

They show a typical orbit, not for the quadratic iterator $x_n = f_a(x_{n-1})$, but for its second order composition $f_a(f_a(x))$ (i.e., we look at the iterator $x_n = f_a(f_a(x_{n-1}))$). For short we call this composition the *second iterate of* f_a and write this as

$$f_a^2(x) = f_a(f_a(x)) \ .$$

You should not, however, confuse this with the square of the values of $f_a(x)$! The graph of $f_a^2(x)$ is given by the 4th degree polynomial

$$\begin{aligned} f_a^2(x) &= a^2x(1-x)(1-ax(1-x)) \\ &= -a^3x^4 + 2a^3x^3 - (a^2+a^3)x^2 + a^2x \end{aligned}$$

shown in the figures. The second iterate of f_a has four fixed points: $\{0, p_a, x_l(a), x_h(a)\}$. These are the fixed points of f_a and the two elements of the 2-cycle. They can be seen in the figures as intersections of the graph with the bisector. Between $1/a$ and $(a-1)/a$ we have outlined a square to draw your attention to this section of the graph. Doesn't it look like the familiar parabola of $f_a(x)$ upside-down? In fact, we also observe that the poly-line representing the iteration behaves in this region similar to what we have seen for the iteration of $f_a(x)$.

Let us compare the graphs of $f_a(x)$ and $f_a^2(x)$ more systematically (see figures 11.18 and 11.19). We start at $a = 1$, which is the parameter where the fixed point $p_0 = 0$ of $f_a(x)$ becomes unstable and a new fixed point $p_a = (a-1)/a$ begins to exist (for $a > 1$). Here the graph of $f_a^2(x)$ looks a bit lower and the fixed points are identical to those of $f_a(x)$. Then at $a = s_1 = 2$ we reach the super attractive case for $f_a(x)$. The new fixed point p_a and the critical point $x_{\text{max}} = 1/2$ coincide. The graph of $f_a^2(x)$ has now reached the same height, but its top is almost flat.

At $a = b_1 = 3$ the fixed point p_a for $f_a(x)$ looses its stability.[5] Also for $f_a^2(x)$ the fixed point p_a looses its stability; but here two new, additional fixed points begin to exist (for $a > 3$): $x_l(a)$ and $x_h(a)$. Note that the portion of the graph which is enclosed by the dashed square looks like the graph of $f_a(x)$ for $a = 1$. The bifurcation at $b_1 = 3$ is called a *period-doubling bifurcation*; a fixed point becomes unstable and gives birth to a 2-cycle.

Then at $a = s_2 = 1 + \sqrt{5}$ we obtain the super attractive case for $f_a^2(x)$: the fixed point $x_l(a)$ and the critical point $x_{\text{crit}} = 0.5$ are identical. For $f_a(x)$ this means that iterating x_{crit} for two steps brings us back to x_{crit}. If we increase the parameter further to $a = b_2 \approx 3.4495$, the fixed point $x_l(a)$ of $f_a^2(x)$ also becomes unstable. In other words, all the changes which we observe for $f_a(x)$ while varying the parameter a between 1 and 3 can also be found for $f_a^2(x)$ in the parameter range $a = 3$ to $a = b_2$. This is recapitulated in the following table:

[5]The derivative is $|f_a'(p_a)| = 1$.

a	$f_a(x)$	$f_a^2(x)$
1	bifurcation of p_a	
2	super attrac., $p_a = x_{\text{crit}}$	
3	p_a becomes unstable	bifurcation of x_l and x_h
3.236	$(f_a(f_a(x_{\text{crit}})) = x_{\text{crit}})$	super attrac., $x_l = x_{\text{crit}}$
3.449		x_l becomes unstable

Now you can guess what happens if we increase a beyond b_2. Yes, we find fixed points of $f_a^2(f_a^2(x))$ which bifurcate off from $x_l(a)$ (and also from $x_h(a)$). This composition of f_a^2 (applied to a point x) is nothing else but the fourth iterate of f_a, i.e., $f_a(f_a(f_a(f_a(x))))$, and will be written for short f_a^4. The new stable fixed points of f_a^4 are equivalent to the birth of a stable cycle of period four for f_a. If we increase the parameter even further, stability is lost again, which marks the birth of a period 8 cycle for f_a and so on ... Again, the bifurcations are called period-doubling bifurcations. The periods of the attractive cycles are 1, 2, 4, 8, 16, 32, ...

Period-Doubling Bifurcation Series

This process establishes two sequences of important parameters.

- the parameters $s_1, s_2, s_3, \ldots$ for which we obtain a super attractive case for $f_a, f_a^2, f_a^4, \ldots$ For these the critical point $x_{\text{crit}} = 0.5$ is a fixed point of $f_{s_1}, f_{s_2}^2, f_{s_3}^4$ etc.
- the sequence $b_1, b_2, b_3, \ldots$ of parameter values for which we have a period-doubling bifurcation .

We have seen $s_1 = 2, s_2 = 3.236\ldots$, $b_1 = 3$ and $b_2 = 3.449490\ldots$ and we already know where these sequences lead to: the Feigenbaum point s_∞ (recall our observation of self-similarity in the final-state diagram). Let us discuss the sequence of period-doubling bifurcations a bit further. It appears that the distance d_k between two successive bifurcation points

$$d_k = b_{k+1} - b_k, \ k = 1, 2, 3, \ldots,$$

decreases rather rapidly. This is also visible in figure 11.20 where we have enlarged the period-doubling tree to show some more of its bifurcation points. A first guess would be that the decrease is geometric, i.e. that

$$\frac{d_k}{d_{k+1}} = \delta \ .$$

In that case the bifurcation values b_k would form a converging sequence

Period-Doubling Tree

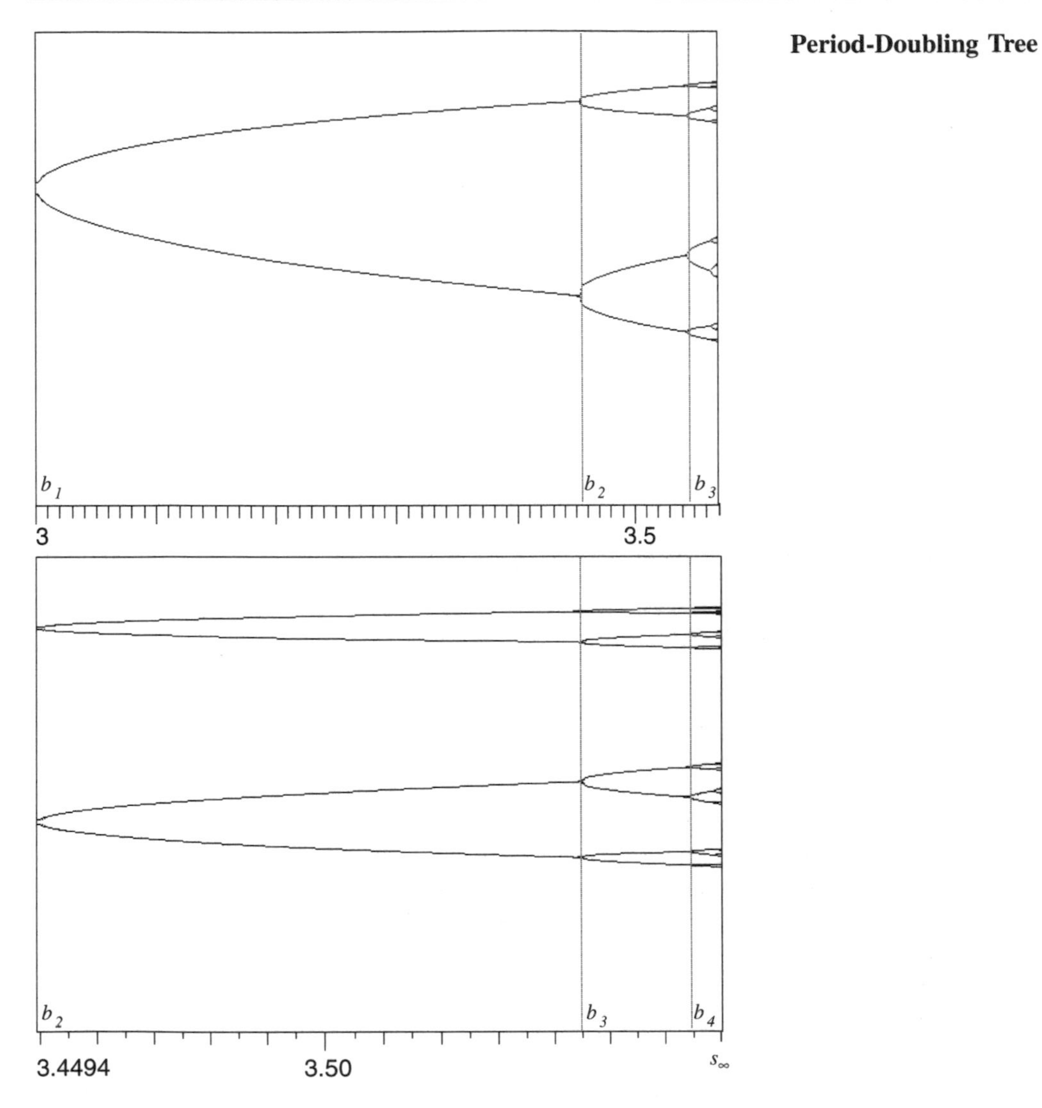

Figure 11.20 : Two close-ups of the period-doubling tree. Note the relative position of the points b_1, b_2, b_3 and b_4, which indicates that $(b_2 - b_1)/(b_3 - b_2) \approx (b_3 - b_2)/(b_4 - b_3)$.

$$b_k = b_1 + d_1\left(1 + \frac{1}{\delta} + \cdots + \frac{1}{\delta^{k-2}}\right), \quad k = 2, 3, \ldots$$

with limit

$$b_\infty = b_1 + d_1 \frac{\delta}{\delta - 1} \, .$$

Unfortunately things are not that easy. For example,

$$\begin{aligned}
d_1 &= b_2 - b_1 = 3.449489... - 3.0 &&\approx 4.4949 \cdot 10^{-1}\\
d_2 &= b_3 - b_2 = 3.544090... - 3.449490... &&\approx 9.4611 \cdot 10^{-2}\\
d_3 &= b_4 - b_3 = 3.564407... - 3.544090... &&\approx 2.0316 \cdot 10^{-2}\\
d_4 &= b_5 - b_4 = 3.568759... - 3.564407... &&\approx 4.3521 \cdot 10^{-3}\\
d_5 &= b_6 - b_5 = 3.569692... - 3.568759... &&\approx 9.3219 \cdot 10^{-4}\\
d_6 &= b_7 - b_6 = 3.569891... - 3.569692... &&\approx 1.9964 \cdot 10^{-4}\ .
\end{aligned}$$

Based on these values we compute

$$\begin{aligned}
d_1/d_2 &= 4.7514...\\
d_2/d_3 &= 4.6562...\\
d_3/d_4 &= 4.6682...\\
d_4/d_5 &= 4.6687...
\end{aligned}$$

Thus, the decrease in d_k from bifurcation to bifurcation is not exactly geometric but only approximately geometric (i.e., the ratio $\delta_k = d_k/d_{k+1}$ converges with increasing k) and Feigenbaum computes[6]

$$\lim_{k\to\infty} \delta_k = \delta = 4.6692016091029...$$

A Universal Constant

At first, it seems that this is just another number which somehow documents the behavior of our particular example, the quadratic iteration. It certainly has to be expected that the number δ depends on this specific model, just as the value of the Feigenbaum point $s_\infty = 3.5699456...$ does. Indeed, if we just changed the scale of a, then naturally the value of s_∞ would turn out different. However, the nature of Feigenbaum's number $\delta = 4.669202...$, discovered in October 1975, is quite different. It is universal, i.e., it is the same for a wide range of different iterators. Due to his achievement, this number is now called the *Feigenbaum constant.*

Computation of the Feigenbaum Constant

At first glance it may seem very difficult to compute the bifurcation parameter sequence $b_1, b_2, b_3, ...$ from which δ can be estimated. However, Feigenbaum did his first experiments merely using a pocket calculator, and it is not black magic at all. In this technical part we present an algorithm to compute δ from the sequence $s_1, s_2, s_3, ...$ of super attractive parameters, which are more accessible than the bifurcation parameters.[7] The computation is based on the formula

$$\delta = \lim_{n\to\infty} \frac{s_n - s_{n-1}}{s_{n-1} - s_{n-2}}$$

[6] See M. J. Feigenbaum, *Quantitative universality for a class of nonlinear transformations,* J. Stat. Phys. 19 (1978) 25–52.

[7] We assume here that the reader is familiar with basic calculus (in particular the product rule) and Newton's method in one dimension.

where $s_1, s_2, s_3, \ldots$ are the super attractive parameters corresponding to periodic cycles of length $1, 2, 4, 8, \ldots$ In other words, we may replace the bifurcation parameters in the usual approach described in the main text above by the super attractive parameters and arrive at the same constant, δ. We already know the first two parameters

$$s_1 = 2, \quad s_2 = 1 + \sqrt{5} \approx 3.236067978 .$$

The following parameters $s_3, s_4, \ldots$ can be computed numerically using Newton's method for solving the appropriate nonlinear equations. We recall that the super attractive parameter s_n is characterized by a corresponding periodic orbit of minimal period 2^{n-1} with the property that the critical point $x_{\text{crit}} = \frac{1}{2}$ is a member of the cycle. Thus, s_n is a solution of the equation

$$f_a^{2^{n-1}}\left(\frac{1}{2}\right) = \frac{1}{2} \tag{11.3}$$

where the unknown variable is the parameter a. However, note that also $a = s_1, s_2, \ldots, s_{n-1}$ solve the same equation, because at these parameter values the point $x_0 = \frac{1}{2}$ is also periodic — however, with a minimal period smaller than 2^{n-1}. Thus, a numerical method for solving eqn. (11.3) for $a = s_n$ must be carefully tuned in order not to produce one of the irrelevant solutions.

By Newton's method we can approximate zeroes of differentiable functions, and in the case at hand the function can be taken as

$$g(a) = f_a^{2^{n-1}}\left(\frac{1}{2}\right) - \frac{1}{2} .$$

The method consists of iterating the transformation

$$\mathcal{N}(a) = a - \frac{g(a)}{g'(a)}$$

where $g'(a)$ denotes the derivative of the function g with respect to the variable a. In other words, for a given initial guess $a^{(0)}$ we compute the sequence $a^{(0)}, a^{(1)}, a^{(2)}, \ldots$ where

$$a^{(k+1)} = \mathcal{N}(a^{(k)}), \quad k = 0, 1, 2, \ldots$$

If this sequence converges to some number, then the limit is a solution to the equation $g(a) = 0$. Under suitable conditions, convergence can be guaranteed provided that the initial guess is sufficiently close to the solution. At this point we need to address the following three questions:

- Given some parameter a, how do we compute $g(a)$ and $g'(a)$?
- What is a good starting point $a^{(0)}$ near the (unknown) solution s_n?
- How do we decide to stop the iteration in Newton's method? In other words, under what conditions do we accept $a^{(k)}$ as a sufficient approximation of s_n?

Let us discuss these questions one by one. We begin by introducing the notation $x_0, x_1, x_2, ...$ for the critical orbit, i.e.,

$$x_0 = \frac{1}{2}, \quad x_{k+1} = f_a(x_k) = ax_k(1 - x_k) \tag{11.4}$$

for $k = 0, 1, 2, ...$ Note that these numbers depend on the parameter a. More precisely, we may think of them as *functions* of a, $x_k = x_k(a)$. Setting

$$N = 2^{n-1}$$

we can now write

$$g(a) = x_N - \frac{1}{2}$$

and, thus, $g(a)$ may be computed using the iteration in eqn. (11.4). To compute the derivative of g we need to differentiate $x_N = x_N(a)$. We use the prime notation to indicate derivatives with respect to the parameter a,

$$g'(a) = x'_N \ .$$

Again we have to use some iteration. First we note that x_0 is the constant $\frac{1}{2}$ and does not depend on a. Thus, $x'_0 = 0$. Then we consider $x_{k+1} = ax_k(1 - x_k)$, use the product rule applied to the three factors a, x_k, and $1 - x_k$, and obtain

$$\begin{aligned} x'_{k+1} &= \big(ax_k(1 - x_k)\big)' \\ &= x_k(1 - x_k) + a(1 - x_k)x'_k - ax_kx'_k \\ &= x_k(1 - x_k) + a(1 - 2x_k)x'_k \end{aligned}$$

for $k = 0, 1, 2, ...$ In summary, we have to simultaneously carry out the iterations

$$\begin{aligned} x_{k+1} &= ax_k(1 - x_k) & x_0 &= \tfrac{1}{2} \\ x'_{k+1} &= x_k(1 - x_k) + a(1 - 2x_k)x'_k & x'_0 &= 0 \end{aligned}$$

for $k = 0, ..., N - 1$.

Good starting points for the method are easy to obtain. If the sequence of super attractive parameters has already been computed up to $s_1, ..., s_n$, then we can estimate the Feigenbaum constant by

$$\delta_n = \frac{s_{n-1} - s_{n-2}}{s_n - s_{n-1}} \ .$$

Using this result, we can produce an estimate $s^{(0)}_{n+1}$ for the next parameter s_{n+1} to be computed,

$$s^{(0)}_{n+1} = s_n + \frac{s_n - s_{n-1}}{\delta_n} \ .$$

This is very close to the true value of s_{n+1} and only a couple of Newton iterations suffice to bring the estimate to the highest possible precision. Initially, before any computations are done, we have only two numbers available, s_1 and s_2, which are not sufficient to compute the estimate δ_2 of δ. Here we simply set $\delta_2 = 4$.

The last point is related to error estimation in Newton's method. The usual approach is to make use of the relation

$$s_n - s_n^{(k)} \approx s_n^{(k+1)} - s_n^{(k)}$$

Thus, we may stop the iteration as soon as the estimate $(s_n^{(k+1)} - s_n^{(k)})/s_n^{(k)}$ for the relative error is of the same order as that of the machine unit.[8] Further iterations would not improve the quality of the result anymore. Thus, we set $s_n = s_n^{(k+1)}$.

The following table records the results using double-precision calculations. The column labeled # lists the number of Newton iterations that were necessary to compute the super attractive parameters.

n	s_n	#	δ_n
1	2.000000000000000000		
2	3.236067977499789696		4.0000000000
3	3.498561699327701520	6	4.7089430135
4	3.554643880189573995	1	4.6805191559
5	3.566667594798299166	1	4.6642974062
6	3.569243531637110338	4	4.6677055227
7	3.569795293749944621	4	4.6685641853
8	3.569913465422348515	3	4.6691571813
9	3.569938774233305491	3	4.6691910025
10	3.569944194608064931	3	4.6691994706
11	3.569945355486468581	3	4.6692011346
12	3.569945604111078447	3	4.6692015094
13	3.569945657358856505	3	4.6692015880
14	3.569945668762899979	3	4.6692016018
15	3.569945671205296863	2	4.6692016148

Convergence is very rapid; only a few Newton iterations suffice. The major computational burden lies in the number of iterations for obtaining x_N and x'_N. Already at $n = 15$ we obtain an approximation of δ correct in the leading 8 digits. The computer used in this experiment has a machine unit of $2^{-63} \approx 1.08 \cdot 10^{-19}$, i.e., 19 significant decimal digits. At this precision it is not sensible to continue the table for larger indices n, because the change in consecutive values of s_n, i.e., $s_n - s_{n-1}$ will have less than half as many significant digits, which signal large errors in the computation of the approximations δ_n due to cancellation of digits.

[8] The machine unit is the smallest number 2^i such that the expression $1 + 2^i$ can still be distinguished from the number 1 using machine arithmetic.

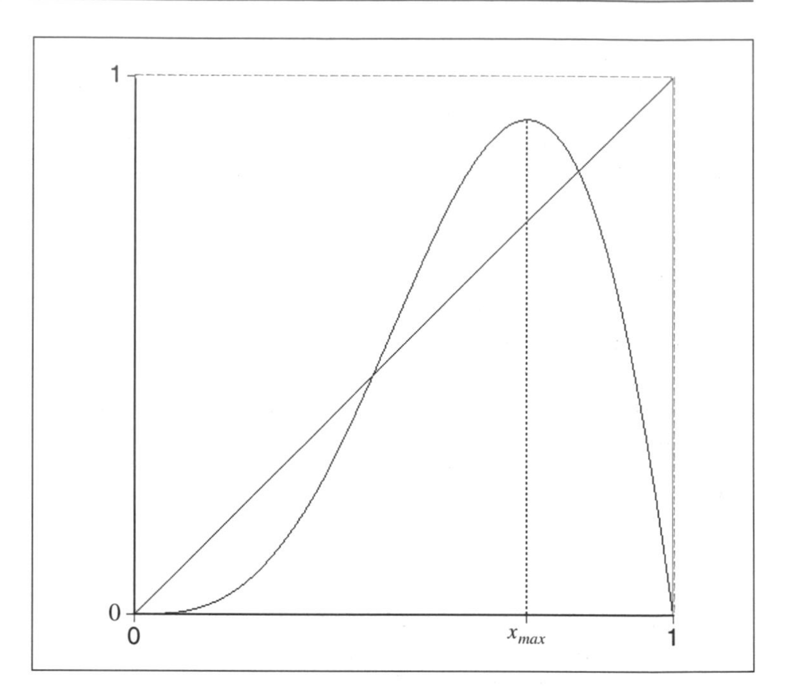

Figure 11.21 : Graph of $g_a(x) = ax^2 \sin(\pi x)$ for $a = 2.3$.

Universality of δ

The Feigenbaum constant $\delta = 4.6692...$ is universal, i.e., it is exactly the same for a whole class of iterations generated by functions similar to a quadratic function, such as $f(x) = \sin(\pi x)$ or $g_a(x) = ax^2 \sin(\pi x)$, to take an example which is not symmetric (figure 11.21 shows this function for $a = 2.3$).

This class is better described by the following properties:

- f is a smooth function from $[0, 1]$ into the real numbers.
- f has a maximum at x_m, which is quadratic, i.e., $f''(x_m) \neq 0$.
- f is monotone in $[0, x_m)$ and in $(x_m, 1]$.
- f has a negative Schwarzian derivative, i.e., $S_f(x) < 0$ for all x in $[0, 1]$, where

$$S_f(x) = \frac{f'''(x)}{f'(x)} - \frac{3}{2}\left(\frac{f''(x)}{f'(x)}\right)^2 .$$

If one computes the bifurcation points of the period-doubling tree for a function from this class the difference d_k of its bifurcation parameters b_k will form a sequence which is asymptotically geometric. Again the ratios d_k/d_{k+1} converge to δ:

$$\lim_{k\to\infty} \frac{d_k}{d_{k+1}} = \delta = 4.6692...$$

Final-State Diagram

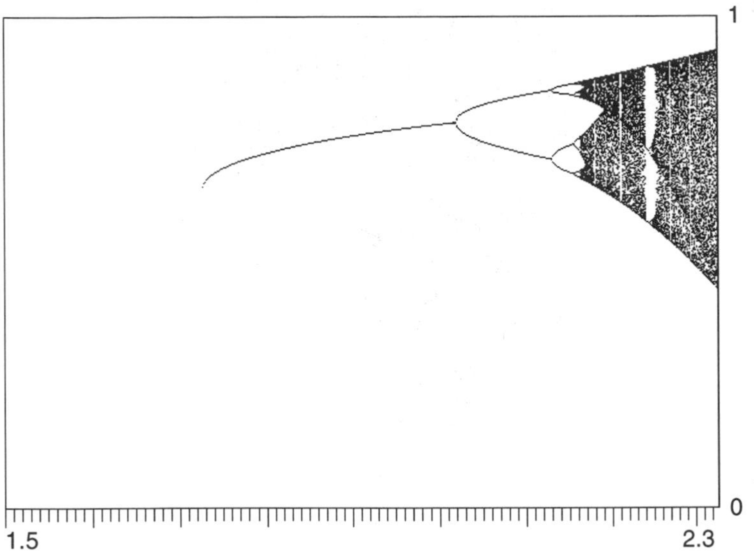

Figure 11.22 : The final-state diagram of $g_a(x) = ax^2 \sin(\pi x)$ for a between 1.5 and 2.3265. Again we observe the scenario familiar from the quadratic iterator, see figure 11.2. There is a break point at about $a = 1.7263$, where the period-doubling tree begins. For parameters less than that the final state is the attractive fixed point 0.

Figure 11.22 shows the final-state diagram for the iteration of $g_a(x)$ from figure 11.21.

For a while this universality appeared as a mathematical mystery, but it turned out that this number was much more important than that. It was conjectured that this number should also be verifiable in real physical experiments. This seemed to be really a little too far fetched because there is no reason whatsoever that a real physical experiment should have anything in common with the simple minded iteration process $x_{n+1} = ax_n(1 - x_n)$. However, the idea was right. In the early 1980's physicists carried out a whole variety of extremely sophisticated experiments in hydrodynamics, electronics, laser physics, and acoustics and found period-doubling bifurcations, with the surprising result that the associated numbers d_k/d_{k+1} did in fact numerically show a remarkable degree of agreement with Feigenbaum's constant $\delta = 4.669...$, see table 11.24.

Mitchell J. Feigenbaum

Figure 11.23 : Mitchell J. Feigenbaum.

Universality of the Feigenbaum Constant

Experimental Measurements of Period-Doublings		
Experiment	Number of period doublings	δ
Hydrodynamic:		
water	4	4.3 ± 0.8
helium	4	3.5 ± 0.15
mercury	4	4.4 ± 0.1
Electronic:		
diode	5	4.3 ± 0.1
transistor	4	4.7 ± 0.3
Josephson	4	4.4 ± 0.3
Laser:		
laser feedback	3	4.3 ± 0.3
Acoustic:		
helium	3	4.8 ± 0.6

Table 11.24 : Results from experiments wherein period-doubling plays a role. The numbers in the third column are to be compared with the Feigenbaum constant $\delta = 4.669...$ Table adapted from P. Cvitanović, *Universality in Chaos,* Adam Hilger, Bristol, 1984.

11.3 The Feigenbaum Point: Entrance to Chaos

We started the discussion of this chapter with the self-similarity features in the final-state diagram of the quadratic iterator, see figure 11.3. This kind of self-similarity is already contained in the first part of the diagram, the period-doubling tree, ranging from $a = 1$ to $a = s_\infty$, the Feigenbaum point. However, the self-similarity in either case is not strict: although the branches of the tree look like small copies of the whole tree there are parts, like the stem of the tree, which clearly do not. Moreover, even the branches of the tree are not exact copies of the entire tree. Here we have to use the term 'self-similarity' in a more intuitive sense without being precise. By contrast, the self-similarity of the address tree of the Sierpinski gasket or the Cantor set was described very precisely in chapters 2, 3 and 5. In those cases the 'small copies' really are exact copies of the whole.

Self-Similarity of the Period-Doubling Tree

For the period-doubling tree everything is more complicated. First, we have noted that the sequence of differences d_k between the parameters of the bifurcation points is not precisely geometric. In other words, when we make close-ups as in figure 11.3, the scaling factor slightly changes from close-up to close-up approaching the factor $\delta = 4.669...$ But this is only true for the scaling in the horizontal direction of the parameter a. With respect to the vertical direction we have to scale (in the limit) with approximately 2.3.

In figure 11.25 we have used these scaling factors 4.669... and 2.3 to obtain a schematic representation of the period-doubling tree which exhibits these limiting scaling properties in all stages. Note that the leaves of this tree form a strictly self-similar Cantor set.[9]

When comparing this tree with the original bifurcation tree, the non-linear distortion becomes apparent. Here branches of the same stage are exactly the same. In the original period-doubling tree, branches have different sizes. Nevertheless, we can identify corresponding branches. Also, the leaves of the original tree form a Cantor set. This happens right at the Feigenbaum point s_∞, where the final-state diagram reaches a new stage which is much more delicate than the situation for parameter values less than s_∞.

Dynamics at the Feigenbaum Point

For all parameters a between 3 and s_∞ we observe stable periodic orbits as final states. Now the natural question arises: what kind of dynamics do we have for $a = s_\infty$? This is a difficult problem. Considering a starting point x_0 from the final state, the Cantor set, the complete orbit of x_0 is also in the Cantor set. This reminds us of the chaos game for the Sierpinski gasket. In chapter 6 we used infinite symbolic addresses to identify points of the

[9]The fractal dimension D of of this special Cantor set has been estimated as $0.5376 < D < 0.5386$ by Peter Grassberger. See P. Grassberger and I. Procaccia, *Measuring the strangeness of strange attractors,* Physica 9D (1983) 189–208.

Period-Doubling Tree

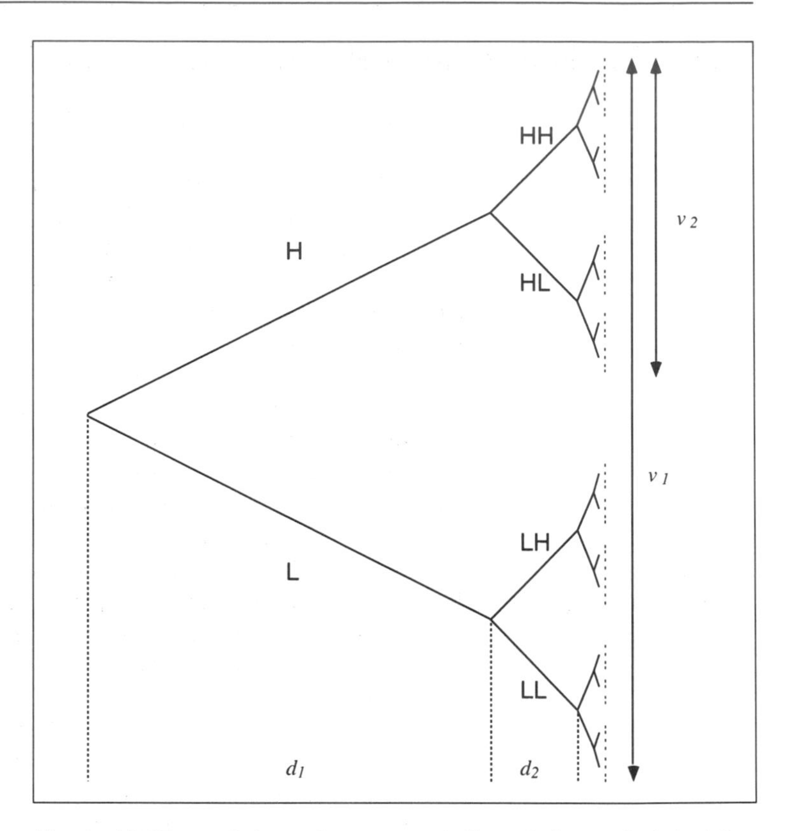

Figure 11.25 : Schematic representation of the period-doubling tree with scaling factors $d_1/d_2 = 4.669$. The vertical range of the complete tree is v_1 and the range of the upper main branch v_2. The ratio is $v_1/v_2 = 2.3$. Note that the leaves of this tree form a strictly self-similar Cantor set.

Sierpinski gasket and to describe the iteration process. It turns out that the same technique can be used to analyse the dynamics on the Cantor set at the Feigenbaum point.

Addresses for the Branches

We introduce addresses for the branches and the leaves of the period-doubling tree as we did previously for other binary trees. First, we label the lower main branch of the tree with L (for low) and the top branch with H (for high). When the two branches split into four we label the upper two parts with HH and HL and the lower two parts with LH and LL. This is the second stage of our addressing hierarchy. In figure 11.25 we have already indicated these labels for the schematic tree. The branches of the third stage would obtain the labels HHH, HHL, HLH, HLL, LHH, LHL, LLH, and LLL. In general we would obtain 2^k sub-branches labeled with k-letter addresses for stage k. The leaves would have infinite address strings. The first k letters of these

Period Two Dynamics
Stage 1

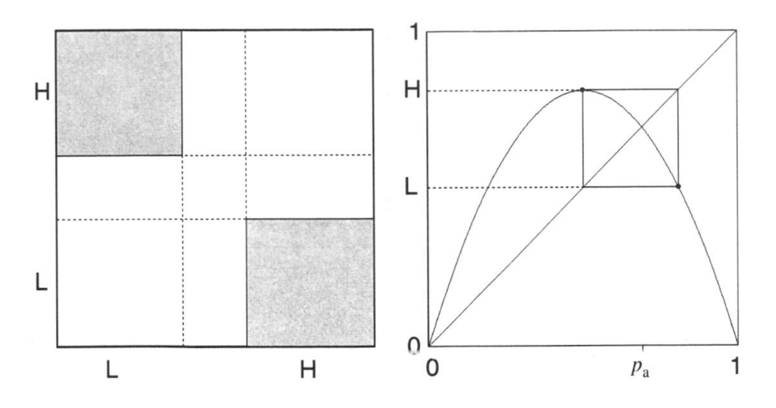

Figure 11.26 : Period two oscillation: L is mapped to H and H is mapped to L.

addresses indicate the sub-branch of stage k to which a particular leaf belongs.

Dynamics in Terms of Addresses

Now we can start to discuss the dynamics of orbits on the Cantor set in terms of addresses. For all parameters a between $b_1 = 3$ and $b_2 = 3.4495$ we have a stable periodic oscillation of period two. This is the range of parameters were one-letter addresses are sufficient: we oscillate between the H-branch and the L-branch.

In the right half of figure 11.26 we show the super attractive case as an example of this oscillation. The diagram on the left shows the corresponding mapping of addresses: L is mapped to H (this relation is indicated by the upper left hand grey box) and H is mapped to L (which is marked by the lower right hand grey box).

Now we turn to the next stage ($b_2 < a < b_3$), which represents the oscillation between four different values. Here we need two-letter addresses. The right half of figure 11.27 shows the super attractive case. If we trace the poly-line, which represents the iteration, we can read off the mapping of addresses

$$HH \to LL \to HL \to LH \to HH \ .$$

This transformation is shown in the left diagram.

Now you might wonder whether there is a reason why we see precisely this sequence of addresses, and not, for example, $HH \to HL \to LH \to LL \to HH$ or $HH \to LH \to HL \to LL \to HH$? Indeed, we can rule out these other cases as follows.

First, we observe that the iteration always has to oscillate up and down between addresses which start with H and addresses which start with L. This carries over from the oscillation between

Period Four Dynamics Stage 2

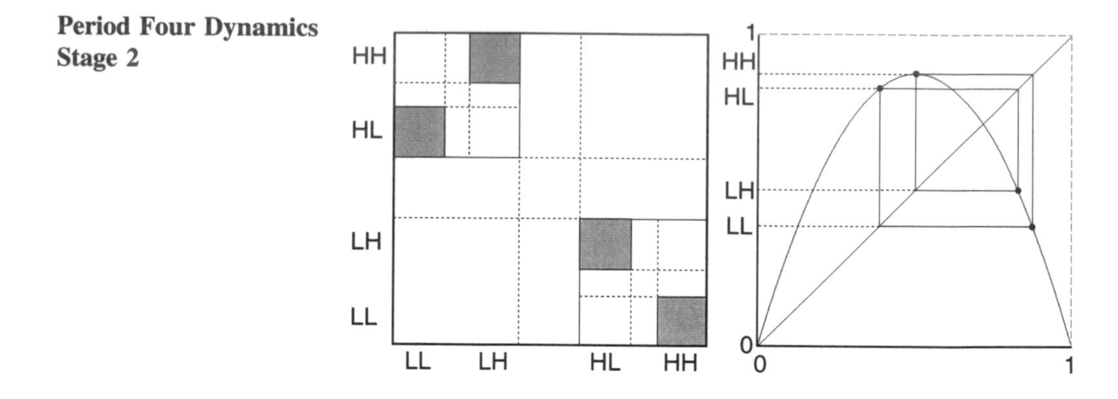

Figure 11.27 : Period four oscillation: $HH \rightarrow LL \rightarrow HL \rightarrow LH \rightarrow HH$.

H and L from which the four-cycle bifurcated off. Next we note that the top most address HH must be followed by the lowest, i.e. LL. These two observations already determine the complete sequence. Note that these facts are clearly visible in the diagram. The grey boxes of the preceding diagram 11.26 now are refined, corresponding to the fact that the orbit oscillates between the main high to the main low branch as before. Furthermore, the box in the lower right hand corner indicates that the highest address is mapped to the lowest.

The same kind of argument allows us to determine the address sequence which describes the orbit of period eight, which bifurcates from the four-cycle. In figure 11.28 we show the related diagram (upper-left) and the diagrams for the following two stages. The last diagram shows the transformation of infinite addresses, a transformation from points in a Cantor set to other points in the Cantor set. It describes the symbolic dynamics of our quadratic iterator at the Feigenbaum point.

Diagrams and Symbolic Dynamics

Let A_k be the set of all k-letter addresses (formed by H and L) and A_∞ the set of all infinite addresses. The dynamics of a periodic orbit with respect to point addresses is described by a transformation f_k: $A_k \rightarrow A_k$. For example, the period-four cycle can be described by

$$\begin{aligned} f_2(HH) &= LL\ , \\ f_2(LL) &= HL\ , \\ f_2(HL) &= LH\ , \\ f_2(LH) &= HH\ . \end{aligned}$$

This transformation is visualized in the stage-two transformation diagram of figure 11.27. For each stage, we divide the axis of the transformation diagrams as in a typical Cantor set construction. Thus in

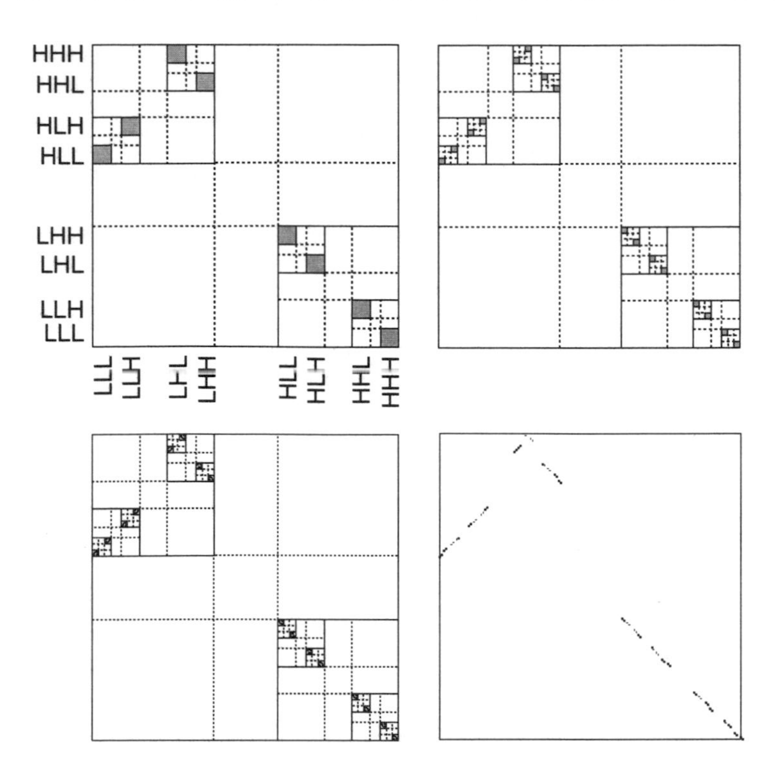

Figure 11.28 : Diagrams for period eight (upper-left), 16 (upper-right), and 32 (lower-left). The lower right hand diagram shows the transformation of infinite addresses, i.e., describes the symbolic dynamics at the Feigenbaum point.

the limit we obtain a diagram which visualizes the transformation $f_\infty : A_\infty \to A_\infty$ as a transformation of points of a Cantor set. Already when comparing the transformation diagrams of the first stages, it becomes apparent that there is a clear structure in these diagrams. First, let us discuss the refinement of the lower right grey box of the stage-one diagram. In stage three it becomes apparent that here a diagonal of boxes begins to form. Thus, for the corresponding addresses starting with H we have the transformation rule:

$$f_\infty(HX_2X_3X_4...) = LX_2^TX_3^TX_4^T... \tag{11.5}$$

where X^T denotes the complement of X (i.e., X^T is H if X is L and X^T is L if X is H). Next, let us examine the refinement of the two upper boxes from the stage-two diagram 11.27. The refinement of the left-most grey box again leads to a diagonal of boxes, see figure 11.28. Again we can write down a transformation rule, now for all addresses which start with LL:

$$f_\infty(LLX_3X_4X_5...) = HLX_3X_4X_5... \tag{11.6}$$

The refinement of the top-most grey box is not as simple, but even more striking since here self-similarity is built in. In fact, the refinement of this box shown in stage three is just a scaled down copy of the stage one diagram, and in general at stage k this is a scaled down copy of the complete stage diagram at stage $k-2$. In the limit this leads to the self-similarity of the transformation diagram for f_∞: the graph of f_∞ for the addresses which start with LH is a scaled down copy of the complete graph. To compute the transformation of an address beginning with $LHX_3X_4X_5...$ we first write down HH, then we drop the first two letters of the original address and apply f_∞ to the remaining letters, thus we compute $f_\infty(X_3X_4X_5...)$. Finally we append the result of this evaluation to the initial letters HH. Thus,

$$f_\infty(LHX_3X_4X_5...) = HHf_\infty(X_3X_4X_5...) \ . \tag{11.7}$$

Let us demonstrate the transformation of an address keeping track only of the first five digits.

Step	Address	Rule	Step	Address	Rule
0	LLLLL	(11.6)	16	$LLLLH$	(11.6)
1	$HLLLL$	(11.5)	17	$HLLLH$	(11.5)
2	$LHHHH$	(11.7)	18	$LHHHL$	(11.7)
3	$HHLLL$	(11.5)	19	$HHLLH$	(11.5)
4	$LLHHH$	(11.6)	20	$LLHHL$	(11.6)
5	$HLHHH$	(11.5)	21	$HLHHL$	(11.5)
6	$LHLLL$	(11.7)	22	$LHLLH$	(11.7)
7	$HHHLL$	(11.5)	23	$HHHLH$	(11.5)
8	$LLLHH$	(11.6)	24	$LLLHL$	(11.6)
9	$HLLHH$	(11.5)	25	$HLLHL$	(11.5)
10	$LHHLL$	(11.7)	26	$LHHLH$	(11.7)
11	$HHLHH$	(11.5)	27	$HHLHL$	(11.5)
12	$LLHLL$	(11.6)	28	$LLHLH$	(11.6)
13	$HLHLL$	(11.5)	29	$HLHLH$	(11.5)
14	$LHLHH$	(11.7)	30	$LHLHL$	(11.7)
15	$HHHHL$	(11.5)	31	$HHHHH$	(11.5)
			32	LLLLL	(11.6)

The recursive rule (11.7) is applied twice in steps 14 and 30. Moreover, in step 30 ($LHLHL$) the rule may possibly be applied even more times depending on the letters following the first five. Note that after 32 iterations the initial 5-letter address repeats. However, this is only true for the first five digits; the others have changed. In fact, it is true, that any number n of leading digits in the address of an orbit must go through a cycle of all possible 2^n combinations. Therefore, the orbit is not periodic. Moreover, the orbit gets arbitrarily close to any point in address space.

Similarity of the Parabola

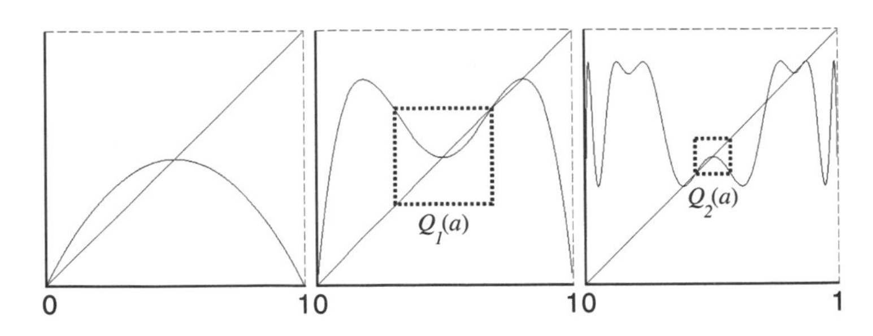

Figure 11.29 : $f_{s_1}(x)$ and parts of the graph of $f^2_{s_2}(x)$ and $f^4_{s_3}(x)$ at the super attractive parameters s_1, s_2 and s_3.

Enlargements

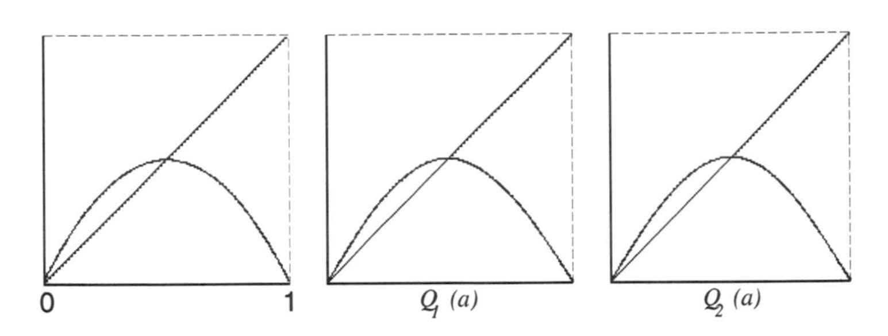

Figure 11.30 : Similarity between the graph of f_{s_1} and 'close-ups' of $f^2_{s_2}$ (inverted) and $f^4_{s_3}$. (See the dotted squares in figure 11.29). Only the left graph, however, is a parabola, the other two represent 4th -order or 8th -order polynomials.

At the Feigenbaum point the final state of the iterator is given by an infinitely long, non-periodic orbit in a Cantor set which gets arbitrarily close to every point of the Cantor set. We may say that this is a first sign of chaos; we are at the entrance to chaos.

Let us now turn to the self-similarity features related to the change of the dynamics as the parameter a increases. You will recall that for a in the interval $(1,3)$ we have just one attractive fixed point at $p_a = (a-1)/a$, and all orbits belonging to initial values between 0 and 1 converge to that attractor. The way initial values are attracted changes at $a = s_1 = 2$ which marks the super attractive case. For parameters below s_1 initial values are attracted directly (monotonically) while for parameters above s_1 the orbit spirals around the fixed point. At $a = b_1 = 3.0$ the fixed point p_a becomes unstable, and an attractive two-cycle is born. The old fixed point p_a continues to exist, but it now repels. Now the two-cycle undergoes all the changes which we have seen for the fixed point. Especially to be noted is the fact that there is again the super attractive case at $a = s_2$. The two-cycle finally becomes unstable at $a = b_2$. Then the story repeats for a four-cycle, and so on.

Similarity Mechanisms in the Change of Dynamics

At each period-doubling bifurcation the dynamics of the iteration becomes dramatically more complex, though the mechanism is always the same. This is related to the 'similarity' of the graph of $f_a(x) = ax(1-x)$ (parabola) to sections of the graphs of the iterated transformations $f_a^2(x)$, $f_a^4(x)$, $f_a^8(x)$, and so on at higher parameters a. Figure 11.29 illustrates this similarity between the graphs of f_a (for $a = s_1 = 2$), $f_a^2(x)$ (for $a = s_2 \approx 3.236$) and $f_a^4(x)$ (for $a = s_3 \approx 3.498$).

A New Universal Function

We can make the similarity of the graphs even more apparent if we make a close-up of the squares outlined in figure 11.29. We enlarge the squares such that they match the unit square which encloses the whole graph. This is demonstrated in figure 11.30, which shows, from left to right, the graph of $f_{s_1}(x)$ and magnifications of $f_{s_2}^2(x)$ and $f_{s_3}^4(x)$. Note that when magnifying $f_{s_2}^2(x)$ we also flipped the graph horizontally and vertically. If we did this for all values s_k and the corresponding compositions of f_{s_k}, we would obtain a sequence of close-ups settling down on the graph of a new function f_∞. When comparing the graphs shown in 11.30, it is already hard to see differences. The fact is that this new function is as universal as the constant $\delta = 4.669...$

A Rescaling Operator

Let us describe the magnification process a bit more carefully and from a slightly different point of view: how do we get from f_{s_1} to the 'close-up-graph' of $f_{s_2}^2(x)$? This is shown in figure 11.31.

We can formalize this process using an operation on functions like f_a. This operator would turn the graph of f_a into the graph of $\Phi(f_a)$. In this respect it is like the Hutchinson operator (see chapter

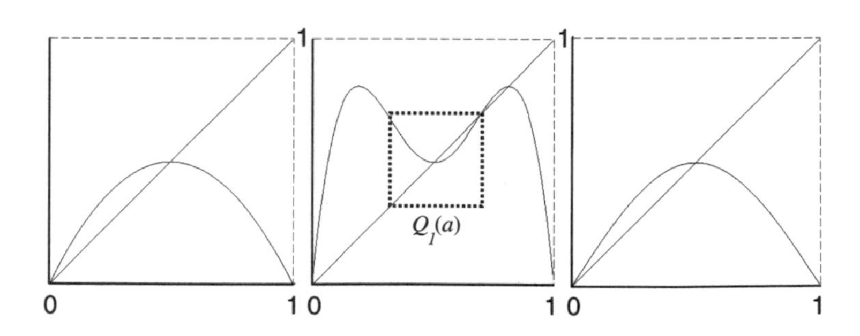

Figure 11.31 : We start with the graph of f_{s_1} (left). In the next step (center) we go to f_{s_2} and consider the composite of the function (i.e., we form $f_{s_2}(f_{s_2}(x))$). The square outlined is enlarged on the right using the magnification factor $s_2/(s_2-2)$ and additionally reflected both horizontally and vertically.

5), which also is not a plain function of numbers but operated on images. But there is another similarity between these operators. When iterated, a particular Hutchinson operator H leads to a final image I which does not depend on the starting image. Rather, it is determined by the operator H alone, namely as its fixed point $I = H(I)$. For Φ there also exists such an invariance property and this is:

$$f_\infty = \Phi(f_\infty) .$$

Furthermore, we could start with graphs of other functions (like $g_a(x) = ax^2 \sin \pi x$), the iteration of Φ would lead to the universal function f_∞ again. And δ is also related to the operator. It is a so called *eigenvalue* and controls the way the iteration approaches the fixed point.

In other words, the secret behind the universality of f_∞ and δ (and therefore also behind the self-similarity of the final-state diagram at the Feigenbaum point) is this operator Φ. It captures the essentials of this process. But we have to skip the mathematical details since the required techniques are definitely beyond the scope of this book.[10]

Again this means that if we had investigated for example the iteration $x_{n+1} = ax_n^2 \sin(\pi x_n)$ instead of the quadratic iteration, then all our observations made so far would have had exact analogies, except that the crucial parameters of a, b_k and s_k would have different values. But the constant δ and the function f_∞ would be the same. To show this kind of universality would require some very deep methods from current mathematics and would need the ingenuity of several of the best living mathematicians and physicists aided by extensive computer studies.

Let us return to our initial observation of self-similarity in the final-state diagram. We have seen that the scaling which is necessary to make the close-ups (as shown in figure 11.3) for the direction of the parameter a is essentially the universal constant $\delta = 4.669...$ Now we can also point out what lies behind the scaling in the vertical direction. In fact, these are exactly the same scaling factors which we used to make the close-ups of $f_{s_2}^2$ or $f_{s_3}^4$ to be able to compare f_{s_1} and $f_{s_2}^2$, $f_{s_2}^2$ and $f_{s_3}^4$, etc.

[10] See P. Collet and J.-P.Eckmann, *Iterated Maps on the Interval as Dynamical Systems,* Birkhäuser, Boston, 1980, and M. Feigenbaum, *Universal behavior in nonlinear systems,* Physica 7D (1983) 16–39, also in: D. Campbell and H. Rose (eds.), *Order in Chaos* North-Holland, Amsterdam, 1983.

11.4 From Chaos to Order: a Mirror Image

Let us now turn to the second part of the final-state diagram, the parameter range between the Feigenbaum point s_∞ and the value $a = 4$, see figure 11.32. We call this part the chaotic mirror image of the period-doubling tree. Indeed, there are features of the period-doubling, though in reversed order, but that is not all. Where chaos reigns, everything becomes infinitely more complicated. While for the first part of the final-state diagram we can predict for each parameter a exactly what the dynamics are, here we have great difficulty even distinguishing stable periodic from chaotic behavior.

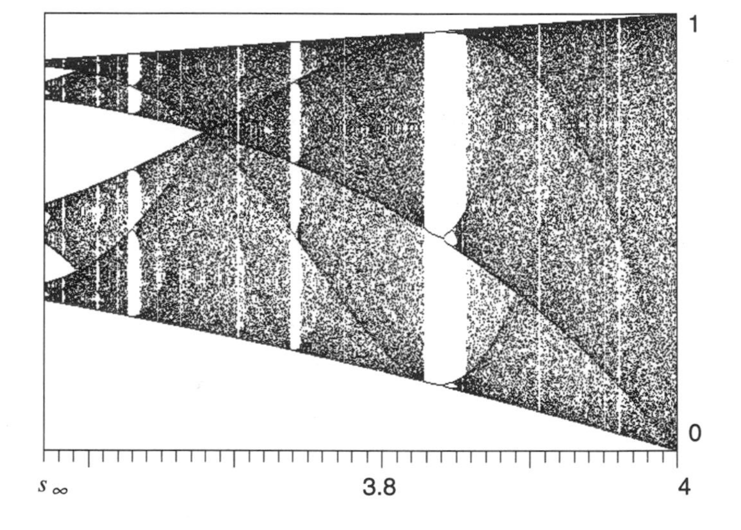

Figure 11.32 : The second part of the final-state diagram of the quadratic iterator.

Band Splitting

We have already investigated the situation for the parameter value $a = 4$ in great detail in chapter 10. This is the parameter where the graph of $f_a(x) = ax(1 - x)$ spans the unit square and we can observe chaos in the whole unit interval. In the final state-diagram this is represented by the random looking distribution of dots which vertically span the range between 0 and 1. This kind of chaotic dynamics is not present for all parameters in the second part of the diagram. The chaos seems to be interrupted by windows of order where the final state again collapses to only a few points, corresponding to attractive periodic orbits. Furthermore, there seems to be an underlying structure of bands resulting from points not being uniformly distributed in each vertical line. Points seem to condense at certain lines which border bands that encapsulate the chaotic dynamics. For $a = 4$ there is only one band spanning the whole unit interval. As a decreases this band slowly

Band Splitting

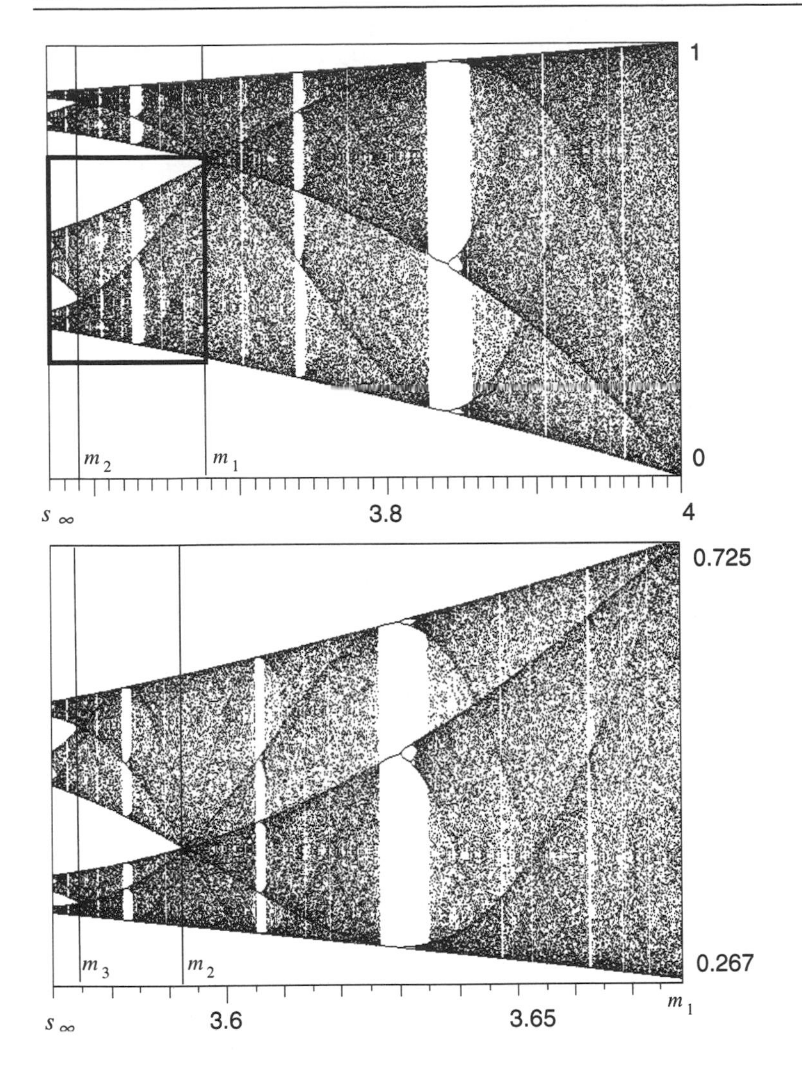

Figure 11.33 : Magnifying the diagram at the window framed in black next to the Feigenbaum point reveals further band splitting parameters.

narrows. Then at the parameter labeled with m_1, it splits into two parts; and at $a = m_2$ the two split into four parts (see figure 11.33).

Merging Points

Now we magnify the diagram between the parameters s_∞ and m_1 at the window shown in figure 11.33. There are more band splitting points. In fact, there is an infinite, decreasing sequence of parameter values $m_1, m_2, m_3, \ldots$ at which one observes the splitting into 2, 4, 8, ... (in general 2^k) bands. This can be interpreted as another consequence of the self-similarity of the final-state diagram at the Feigenbaum point. Thus, this sequence leads exactly to the

Time Series for
$a = 3.67$

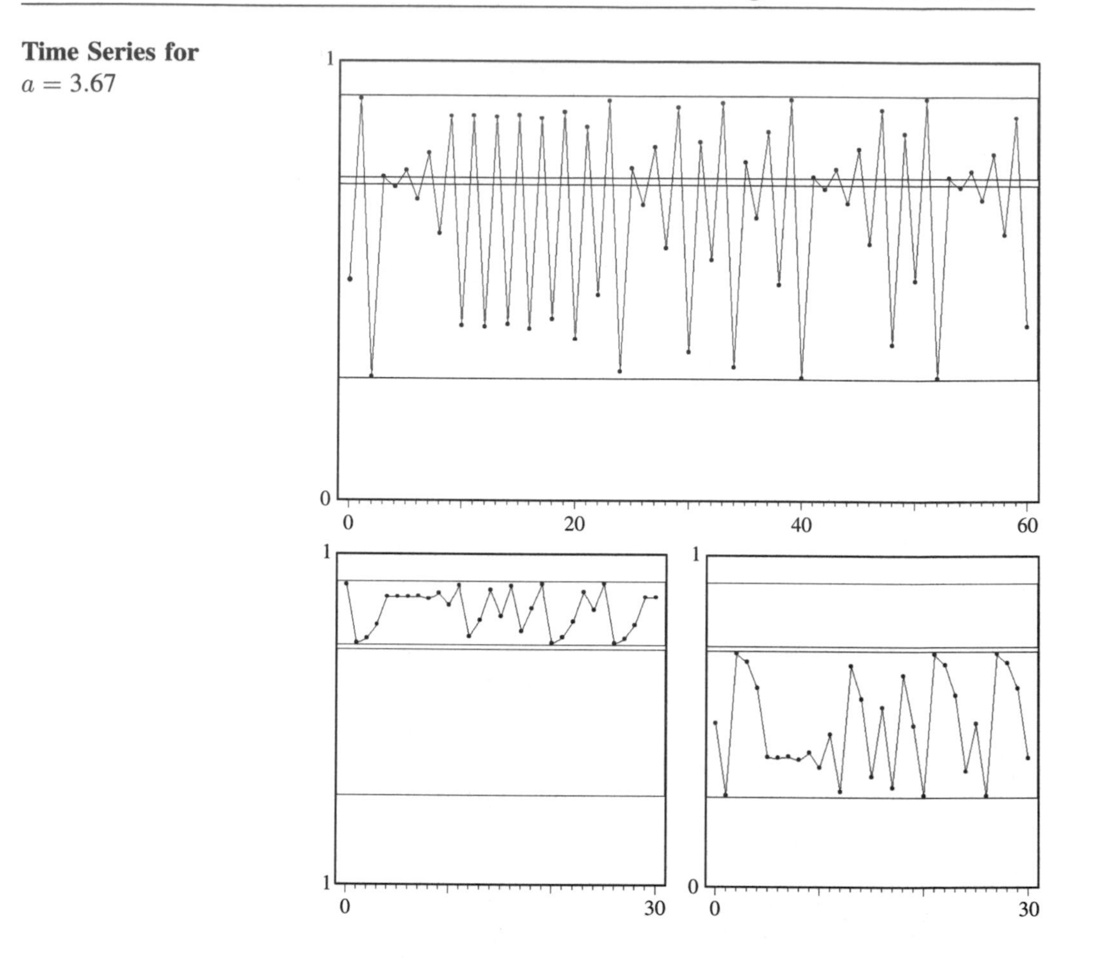

Figure 11.34 : Time series for $a = 3.67$ (just below m_1). The orbit oscillates from step to step between the two marked bands (top). Within each band the dynamics look chaotic. In the lower two images f_a^2 is shown.

limit $m_\infty = s_\infty$ (i.e., to the Feigenbaum point). Moreover, we can guess what the result of the following experiment by Großmann and Thomae should be.[11] They tried to find out whether the distances of the band merging points $d_k = m_{k+1} - m_k$ obey a particular growth law. They guessed that this would be similar to what had been found for the sequences s_k (the parameters of super attractiveness) and b_k (the parameters of the period-doubling bifurcations). And indeed, they were able to confirm that the ratio d_k/d_{k+1} converges to the universal constant $\delta = 4.669...$ as the number k increases.

[11] See S. Großman and S. Thomae, *Invariant distributions and stationary correlation functions of one-dimensional discrete processes* Zeitschrift für Naturforschg. 32 (1977) 1353–1363.

Mixing Histograms

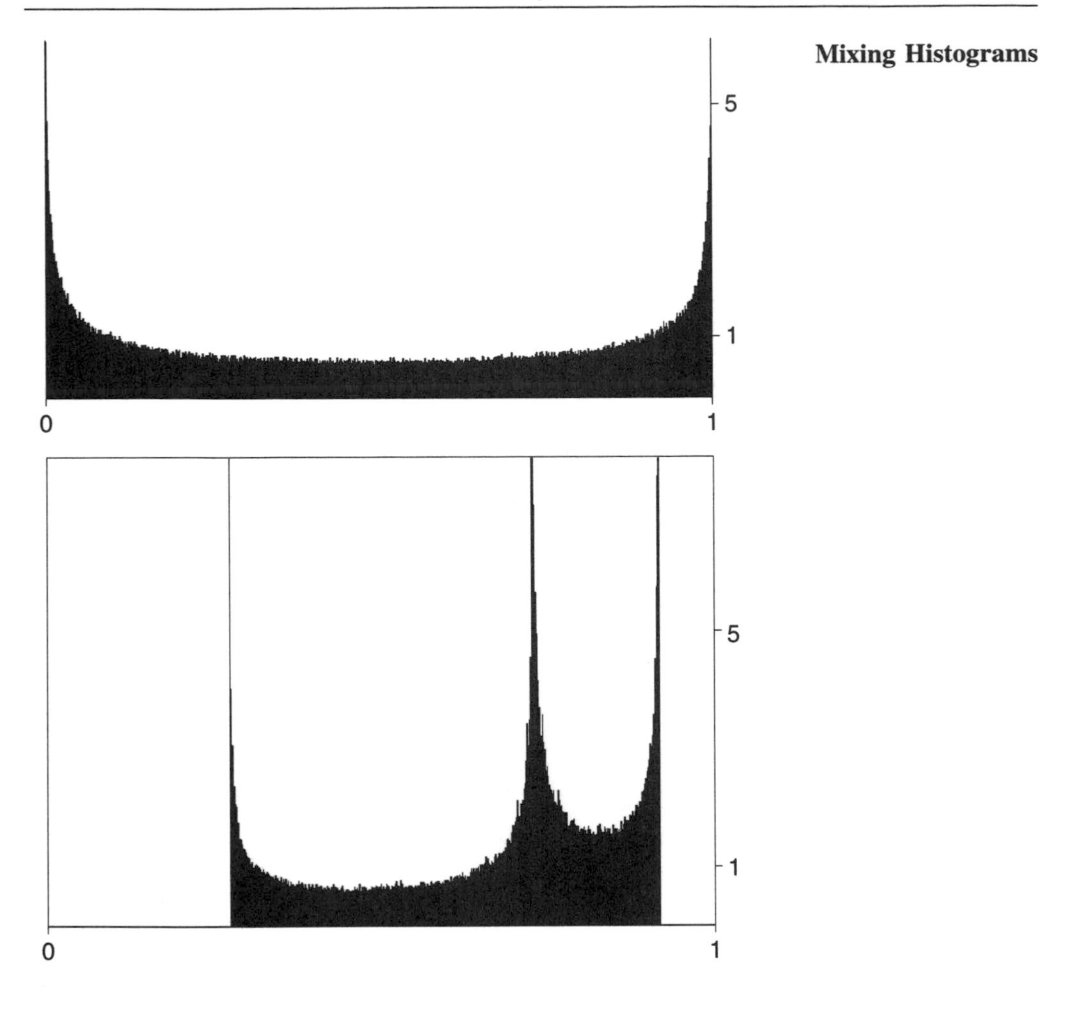

Figure 11.35 : The histogram for $a = m_1 = 3.6785$ (bottom) looks like two scaled down copies of the histogram for $a = 4$ (top) fitted together.

Let us explore these bands a little bit further. What kind of change lies behind the splitting (or merging) of bands? Figure 11.34 (top) shows a typical time series of f_a for the parameter $a = 3.67$, which is slightly below $a = m_1$. It becomes immediately apparent what the two bands mean. Although the dynamics behave chaotically, it oscillates from step to step back and forth between two distinct bands. In other words, when we look at the dynamics of f_a^2 we see points only moving chaotically either in the upper or in the lower band. This is shown in the lower two images of figure 11.34. In summary, the first band splitting is also a kind of

Another Histogram

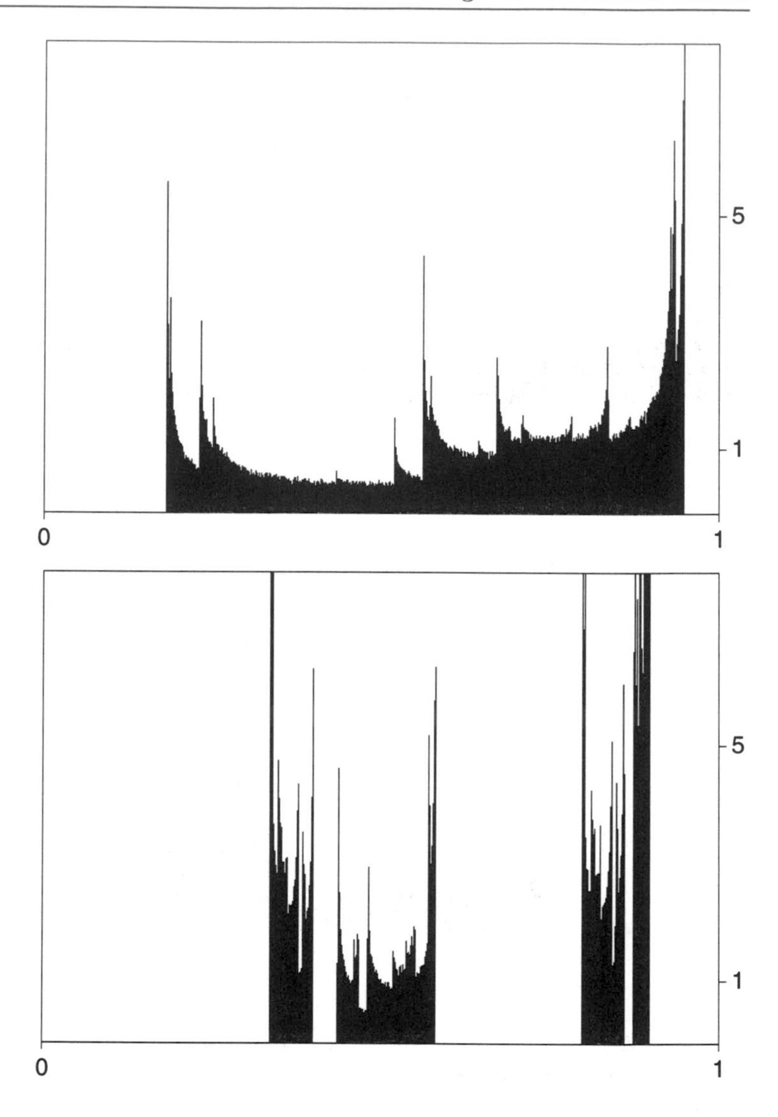

Figure 11.36 : Mixing histograms for a = 3.7 (above) which is a bit above m_1 and for $a = 3.585$ (below) which is smaller than m_2.

period-doubling bifurcation.

You will recall our histogram experiment from chapter 10. We visualized the mixing property of the chaos parabola by measuring how frequently the iteration of an orbit visits the different points of the unit interval (we had broken the unit interval into equally spaced sub-intervals, then we counted how often a given orbit had hit each of these intervals). Figure 11.35 shows the same experiment for $a = m_1$ and compares it with the histogram for

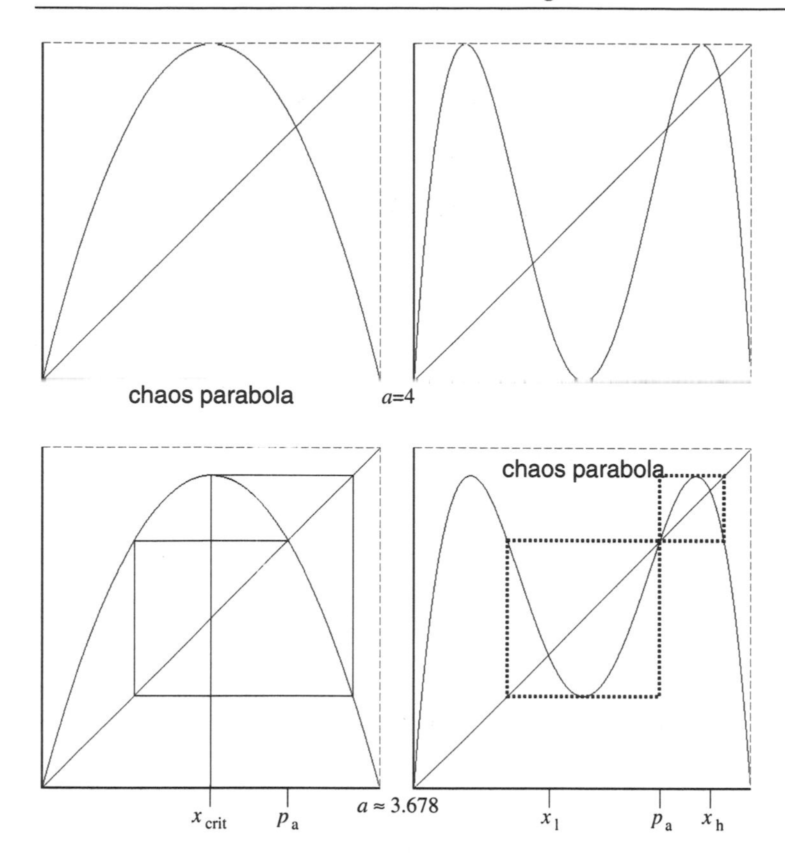

Figure 11.37 : Comparing f_a and f_a^2 at $a = 4$ and at $a = m_1$. f_a^2 forms two small versions of graphs similar to the parabola for $a = 4$ (enclosed by the dashed square).

$a = 4$. The range of the two bands at $a = m_1$ is clearly visible. Indeed, this looks a bit like two differently scaled copies of the diagram for $a = 4$ fitted together.

Comparing f_a^2

Let us again compare f_a and f_a^2 using graphical iteration. Figure 11.37 shows the result for $a = 4$ (top) and for $a = m_1$ (bottom). Let us call the parabola in the upper left graph ($4x(1-x)$) a *generic parabola*. Generic parabolas are characterized by the fact that their graph precisely fits into a square which has one of its diagonals on the bisector of the x-y-coordinate system. Note that for $a = m_1$ we can also find a generic parabola in the graph of f_a^2 (lower right of figure 11.37). However, this is not quite correct, because f_a^2 is not really a parabola, but rather a fourth degree polynomial having a graph that only looks parabolic in the outlined region enclosed by a dashed square. Once the iteration of $f_{m_1}^2(x)$ has led into this region it is trapped, and we should expect to see chaotic behavior

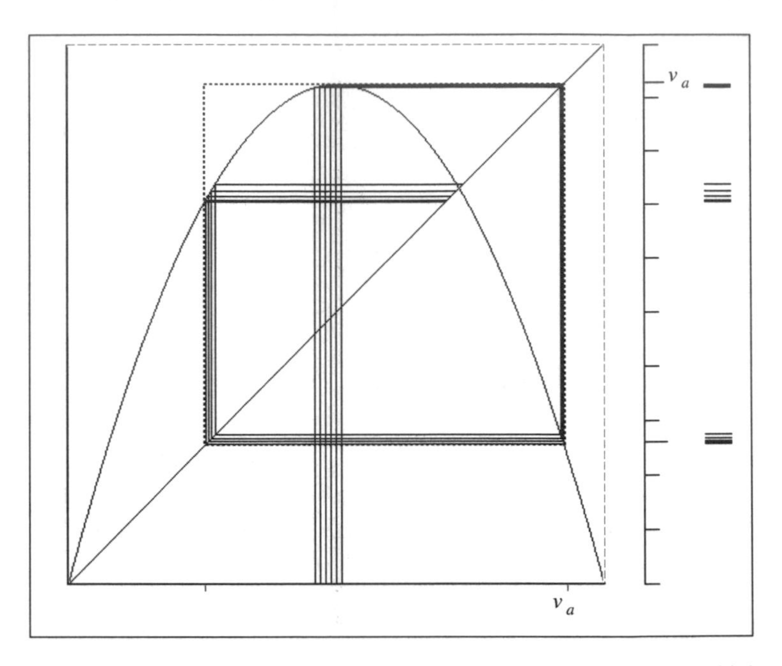

Figure 11.38 : Graphical iteration of some equally spaced initial values near 0.5. The first three iterates condense at $v_a = f_a(0.5)$, $f_a(v_a)$ and $f_a^2(v_a)$

which spans the interval $(1/a, (a-1)/a)$. This corresponds to the lower band visible in the final-state diagram right at $a = m_1$. The upper band corresponds to the part of the graph of f_a^2 enclosed by the small dotted square. Also in this region the iteration is trapped and spans the interval $((a-1)/a, a/4)$.

Now you guess what the situation for all other parameters $a = m_k$ will be. In all these cases we find a generic parabola (i.e., in the graph of $f_{m_2}^4(x)$, in the graph of $f_{m_3}^8(x)$, etc.). Of course this explains only what we see at the special parameter values $a = m_k$. On the other hand it seems possible to trace these bands also in between. Somehow they shine through the whole second part of the final-state diagram; there is a mechanism behind this observation.

Critical Value Lines

In figure 11.38 we show the graphical iteration of a few initial values which we have chosen to be equally spaced near 0.5. For each initial value we have performed three iterations and drawn the corresponding outcome on the right side of the graph. First we note, that the iteration never leaves the outlined square (i.e., the points of the final-state diagram have to lie within the interval between the critical value $v_a = f_a(0.5)$ and $f_a(v_a)$). Furthermore, we observe that the values of the iteration condense a bit at these points. This happens because the parabola has its vertex at 0.5 which squeezes nearby orbits together.

In the histogram, we thus expect a spike at $v_a = f_a(0.5)$. Moreover, there should be another spike at the next iterate, $f_a(v_a) = f_a^2(0.5)$, and also at the following one, $f_a^2(v_a)$, and so on. For $a = 4$, however, $v_a = 1$ and all further iterates are 0. Thus, it is reasonable to expect only the two spikes at 0 and 1. For $a = m_1$, on the other hand, we have $f_a^2(v_a) = p_a = (a-1)/a$, the fixed point of f_a, and all further iterates are the same. Therefore, there should be three spikes, at v_a, $f_a(v_a)$, and $f_a^2(v_a) = p_a$. This, in fact, is just what figure 11.35 shows. In summary, this leads to the conjecture that what we see shining through as lines of condensation in figures 11.32 and 11.33 could be the trace of the iterates of the critical value v_a.

Figure 11.39 shows the experiment which confirms this conjecture. We compute the first eight iterates of 0.5 for the parameter range from s_∞ to 4. The upper plot shows the first four iterates (i.e., v_a to $f_a^3(v_a)$). These lines apparently correspond to the main bands (or stripes) which shine through the final-state diagram in figures 11.32 and 11.33. The lower plot shows all eight iterates exhibiting more of the relation to finer band structures.

Although these critical lines (i.e., the iteration $f_a^k(v_a)$ of the critical value v_a) explain our perception of a band structure in the final-state diagrams, this does not mean that the complete lines as shown are part of the final state. Already for v_a this is not true, as demonstrated in figure 11.40, which shows a close-up of the final-state diagram next to the line of critical values v_a. The final states are bound by this line, but we undoubtedly can see that from a certain parameter value[12] (about $a \approx 3.82843$) the final states consist of a stable attracting periodic cycle, of which only one point is shown in the blow-up.

Periodic Windows

In fact, this close-up shows a small part of one of the white windows which interrupt the chaotic region of the final-state diagram.[13] There are an infinite number of such windows, which all correspond to stable periodic cycles. This one between $a \approx 3.828$ and $a \approx 3.857$ is the most prominent one; it is the so called period-three window. In figure 11.41 we have indicated not only this window but also the windows of period 5, period 7 and the window of period 6. But let us first examine the period-3 window a bit closer. The bottom part of the figure shows two successive close-ups of the part which is marked by the black frame.

[12]More precisely, the parameter is $a = 1 + \sqrt{8} \approx 3.828427125$ (compare section 11.5).

[13]The name 'window' was used first in R. M. May's remarkable paper, *Simple mathematical models with very complicated dynamics,* Nature 261 (1976) 459–467.

Critical Lines

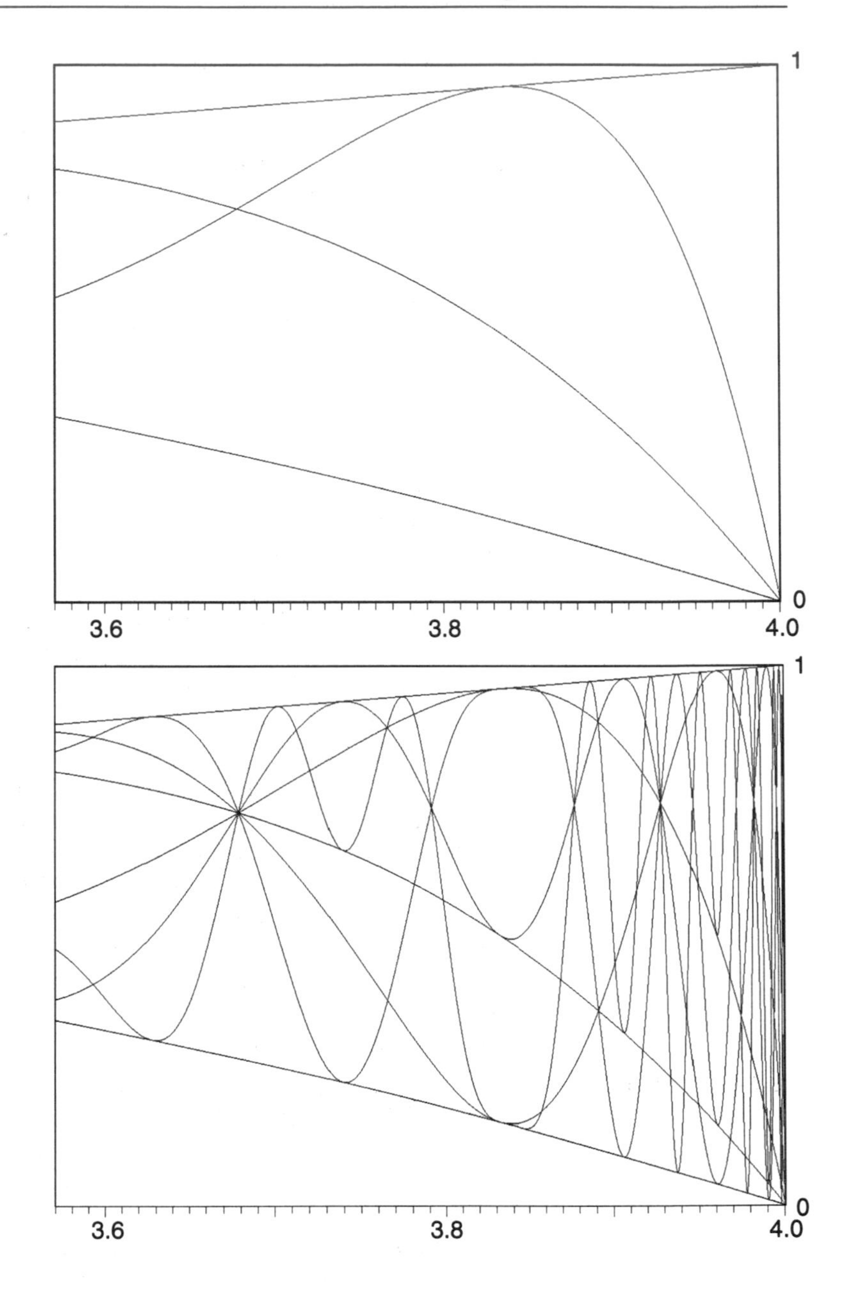

Figure 11.39 : Images of the critical value v_a for the second part of the final-state diagram: (top) v_a, $f_a(v_a)$, $f_a^2(v_a)$ and $f_a^3(v_a)$, (bottom) v_a to $f_a^7(v_a)$.

Self-Similarity Again

Again we discover self-similarity. We see smaller and smaller copies of the whole final-state diagram. And indeed we can find

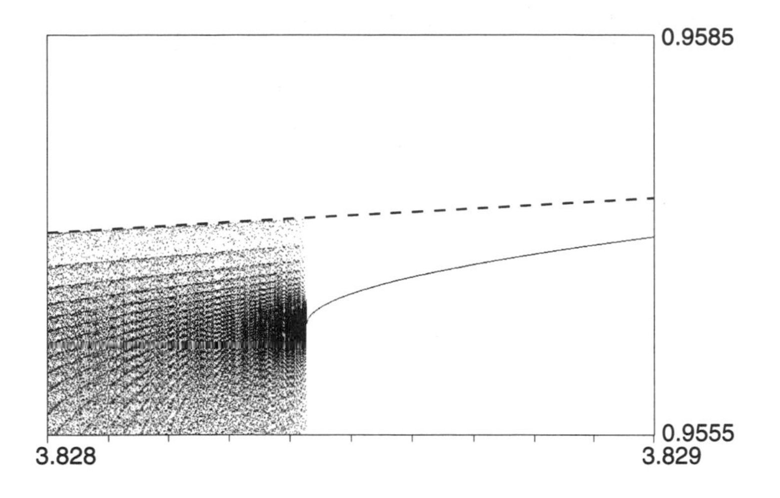

Figure 11.40 : Close-up of the final-state diagram between $a =$ 3.828 and $a = 3.829$ next to the line of critical values v_a which is shown in dashed patterns.

the complete scenario of period-doubling, chaos and band splitting again, however, on a much smaller scale. And again the mechanisms behind this are the same as before. There is only one important difference; instead of $f_a(x)$, here everything is based on $f_a^3(x)$. The period-doubling begins when the three fixed points of $f_a^3(x)$ loose their stability and six new fixed points of $f_a^6(x)$ are born (i.e., for f_a we have a 6-cycle). As a increases further, each of the stable fixed points of f_a^6 will undergo a period-doubling bifurcation (i.e., for f_a we will obtain attracting cycles of length $3 \cdot 2^2$), and so on. The relative length of the intervals for which these stable cycles exist will be governed once more by the universal number $\delta = 4.669...$ At the end of this period-doubling scenario, near $a = 3.8415...$, there will again be a transition to chaotic behavior very much like that at the Feigenbaum point s_∞.

Let us take a look at some graphs of f_a^3. In figure 11.42 (left) we have drawn the super attractive case. At the center again we observe a segment which looks like a small parabola. Indeed, the changes of this small part are responsible for the complete scenario of period-doubling which ends at $a \approx 3.857$ in fully developed chaos as it is shown in the center part of figure 11.41. The corresponding graph of f_a^3 is shown on the right hand side of figure 11.42. And indeed, at the center a generic parabola is visible.

If we magnified any of the other periodic windows, we would indeed make exactly the same finding; but everything would be on an even smaller scale. In fact, between the period-3 window

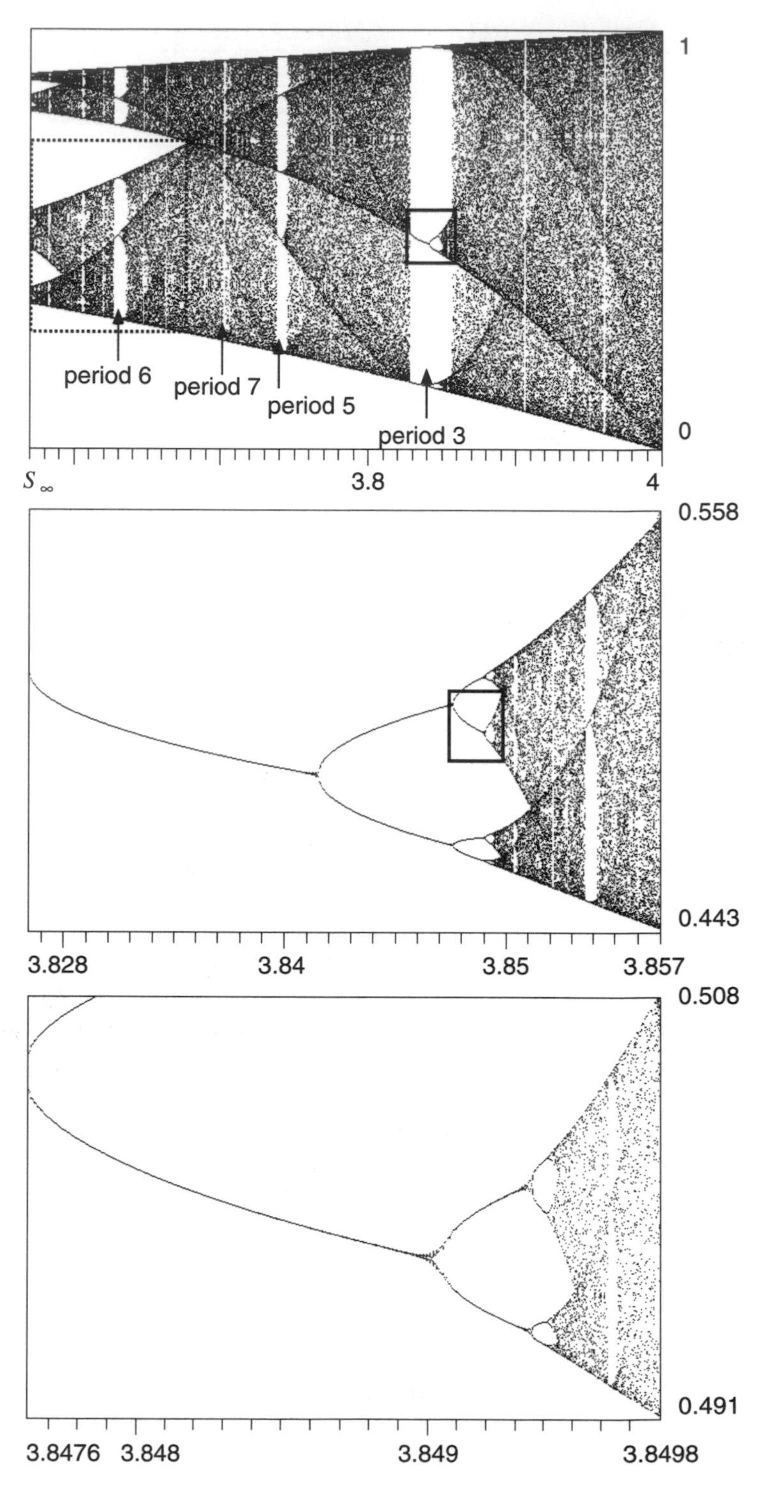

Figure 11.41 : Two successive close-ups of the period-3 window. It starts at $a = 1 + \sqrt{8} \approx 3.8284$ and extends up to $a \approx 3.857$.

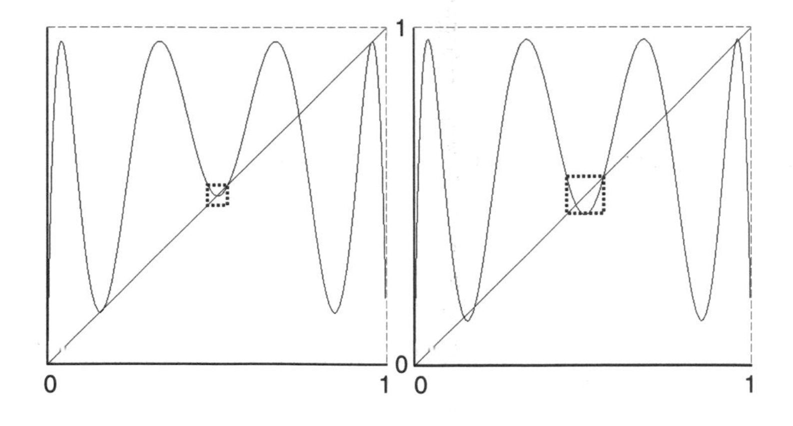

Figure 11.42 : The graphs of the third iterate f_a^3 of f_a. Left: super attractive case, right: fully developed chaos.

and the band merging point m_1 there are an infinite number of windows for all odd integers, i.e., for 3, 5, 7, 9, 11, ..., which can be found in reversed order (i.e., 3 is right of 5, etc.). But as the period increases the size of these windows rapidly decreases and the period-9 window is already hard to find.

The Charkovsky Sequence

Are these all the periodic windows? Certainly not! You have heard of self-similarity — haven't you? Look at the left part of the diagram in figure 11.41 (top) which we have enclosed by a dashed rectangle. Here we find everything once again, but now with a doubled period. In other words, it starts with a period-6 window, then we find a period-10 window, etc. Expressed generally, in this step we find windows of period $2 \cdot k$, for all odd integers $k \geq 3$. In summary, self-similarity reveals a sequence of windows with period

$$\begin{array}{ll}
3, 5, 7, 9, 11, 13, ... & \text{(all odd integers)} \\
2 \cdot 3, 2 \cdot 5, 2 \cdot 7, 2 \cdot 11, ... & (\text{all } 2 \cdot k,\ k \text{ odd}) \\
4 \cdot 3, 4 \cdot 5, 4 \cdot 7, 4 \cdot 11, ... & (\text{all } 2^2 \cdot k,\ k \text{ odd}) \\
... & (\text{all } 2^n \cdot k,\ k \text{ odd}) \\
...2^4, 2^3, 2^2, 2, 1 & \text{(all powers of 2)}
\end{array}$$

exactly in this order from right to left in the final-state diagram. Actually the last row of this sequence is special. It represents the period-doubling at the beginning of the diagram (we could say that this is the last periodic window). This strange sequence of numbers is named after the Russian mathematician Alexander N. Charkovsky and is the heart of some remarkable results on periodic points of feedback systems. But let us return to the question: are these all the periodic windows? You probably already suspect that this is not everything. Indeed, it is known that theoretically in any parameter interval we can find a periodic window (i.e., a stable periodic cycle).

Charkovsky Sequence and Periodic Points

In 1964 Charkovsky introduced his famous sequence

$$\begin{array}{l} 3 \rhd 5 \rhd 7 \rhd 9 \rhd \cdots \rhd \\ 2\cdot 3 \rhd 2\cdot 5 \rhd 2\cdot 7 \rhd 2\cdot 9 \rhd \cdots \rhd \\ 4\cdot 3 \rhd 4\cdot 5 \rhd 4\cdot 7 \rhd 4\cdot 9 \rhd \cdots \rhd \\ \cdots \rhd 2^5 \rhd 2^4 \rhd 2^3 \rhd 2^2 \rhd 2 \rhd 1\,. \end{array}$$

It is ordered by the symbol $\rhd$ (i.e., a comes before b is written as $a \rhd b$). He was able to prove some very remarkable results such as the following.

Assume that f transforms an interval I onto itself and has a point of period k (i.e., there is x in I such that $f^k(x) = x$). Then f has points of period m for every m such that $k \rhd m$.

The consequences of this amazing result are manifold: if f has a point of period 3, then f has periodic points of any period. If f has a point of period $k \neq 2^n$, then f has infinitely many periodic points.

11.5 Intermittency and Crises: The Backdoors to Chaos

Periodic windows interrupt the chaotic region in an extreme way. At the beginning of such a window there is a sudden and dramatic change in the long term behavior of the quadratic iterator. For example, let us look at the neighborhood of the parameter

$$a = w_3 = 1 + \sqrt{8} \approx 3.82843$$

at the start of the period-3 window. For parameters a slightly above w_3, a perfectly stable cycle of period 3 exists. On the other hand, as soon as we decrease the parameter a bit below the value w_3, we immediately stumble into chaos spanning the whole interval $(f_a(v_a), v_a)$.[14] This is quite a different route to chaos compared to the orderly path of period-doubling bifurcations.

The Vexatious Path to Chaos

But the situation is even more troublesome. The chaos for parameters below w_3 reveals itself only in the long term. The short term behavior wrongly suggests stable periodic orbits. This is demonstrated in figure 11.43, which shows the time series of $x_0 = 0.5$ for $a = 3.82812 < w_3$. For more than 50 iterations the orbit looks perfectly stable and predictable, but then it rapidly falls into chaotic oscillation.

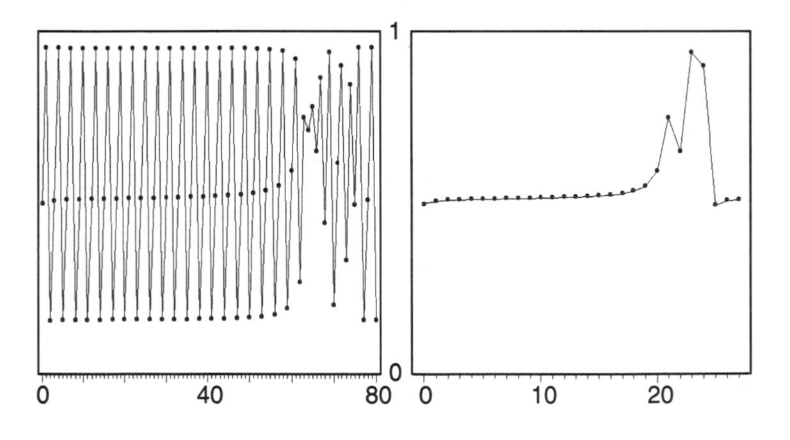

Figure 11.43 : Time series of $x_0 = 0.5$ for $a = 3.82812$: (left) iteration of f_a, (right) iteration of f_a^3. In the right hand graph the time axis is scaled up by 3 to allow a better comparison.

This experiment gives us a first impression of what is called *intermittency* — a new type of iterative behavior. What are the ingredients of intermittency? The first one is a so called *tangent*

[14]Recall that v_a is the critical value $v_a = f_a(0.5)$.

Graphical Iteration for $g_a(x)$

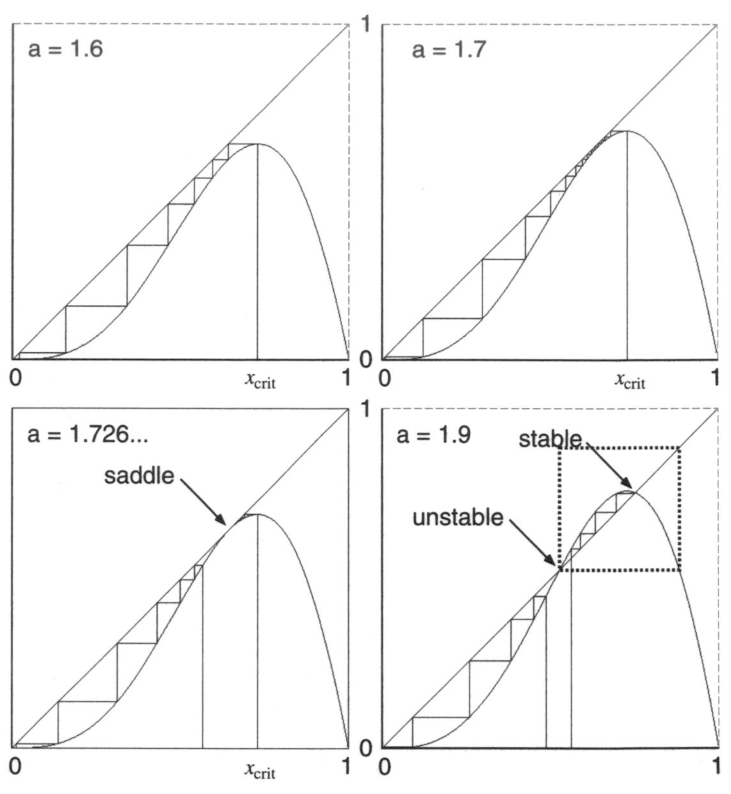

Figure 11.44 : This series of graphs shows the tangent bifurcation for $g_a(x) = ax^2 \sin \pi x$ near $a^* \approx 1.7264$. For parameters below this value (top left and right) all orbits converge to the origin. Right at the bifurcation (lower left) the bisector becomes tangential to the graph of g_a. There is a new fixed point which attracts orbits from the right and repels them on the left. For parameters beyond a^* there are two new fixed points, one is stable and the other is unstable.

bifurcation of fixed points. Let us explain by means of an example which is already familiar from section 11.1. It is the iterator

$$g_a(x) = ax^2 \sin \pi x$$

which we used to illustrate the universality of δ. Now please take another look at figure 11.22, the final-state diagram for g_a. Note that the period-doubling bifurcation tree seems to start out of nothing at about $a \approx 1.7264$. But what is going on for parameters less than 1.7264? Well, let us take a look at the corresponding graphical iteration.

First, figure 11.44 shows the situation for $a = 1.6$ (top left). In this case, the iteration of all initial values will eventually lead to the attractive fixed point 0. Now we increase the parameter a and observe the graph of g_a getting closer and closer to the bisector (top right). Finally, at

$$a = a^* \approx 1.7264289398722975$$

(bottom left) the graph touches the bisector tangentially at

$$x = x^* \approx 0.6457736765434055 \ .$$

If we increase the parameter even further (bottom right), we see that the bisector intersects the graph near x^*. There are two new fixed points, a stable and an unstable one.

Tangent Bifurcation

The new stable fixed point corresponds to the stem of the period-doubling scenario which we have found in the final-state diagram of g_a. When the parameter a increases further, this stable fixed point will become unstable at a certain parameter, giving birth to a stable cycle of period 2. Later stable cycles of period 4 will appear, and so on in the familiar fashion. The square outlined in figure 11.44 (bottom right) encloses a part of the graph which again looks similar to a parabola. Indeed, this part is responsible for the familiar appearance of the final-state diagram. But let us return to the two fixed points. If we reduce the parameter to a^* these two fixed points join into a single point at x^*, called a *saddle point*. If we decrease a even further this fixed point also vanishes. This mechanism is called *tangent bifurcation*.[15]

Note that the point x^* is neither attracting nor repelling. Rather, it attracts values which lie above x^* and repels for points below x^*. In particular if we start the iteration with the initial value $x_0 = x_{\text{crit}}$ (i.e., the x-value belonging to the maximum of g_a), the orbit converges to the fixed point x^* and not to 0 (see figure 11.44, bottom left). But as soon as we decrease the parameter only a little bit, the fixed point disappears. At first only an extremely narrow channel opens (top right). If we again trace the iteration of x_{crit} we observe that the orbit is trapped for quite a while between the graph of g_a and the bisector; it is tacking like a sailboat in a narrow channel. But eventually the orbit leaves this channel and reaches the fixed point 0.

Counting Tacks

If we decrease the parameter a further, the width of the channel becomes larger; and the orbit may pass through the channel with fewer tacks, thus approaching 0 much faster. This behavior is measured in the experiment shown in figure 11.45. We vary the parameter a from 1.6 to a^* and run iterations always starting with $x_0 = x_{\text{crit}} \approx 0.7286$. We count the number of iterations $N(a)$ required for the orbit to reach a small neighborhood of the attractive fixed point 0. As a gets close to a^* the number of iterations

[15] In some texts the term *saddle-node bifurcation* is used instead.

Evaluating Time Series

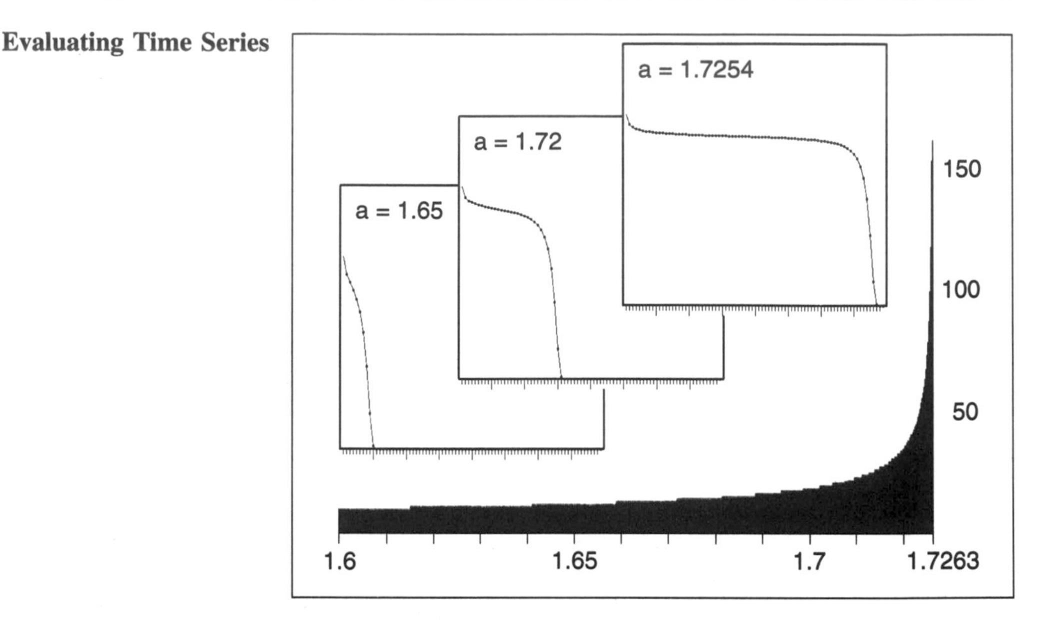

Figure 11.45 : The number of iterates required to bring x_k into a ε-neighborhood of 0 ($\varepsilon = 10^{-8}$), when $x_0 = x_{\text{crit}}$, for $x_{n+1} = g_a(x_n) = ax_n^2 \sin \pi x_n$. The inserts show the corresponding time series for $a = 1.65$, $a = 1.71$ and $a = 1.7254$ illustrating how the iteration is trapped for many iterations near the point x^*, where the graph of g_a touches the bisector when $a^* \approx 1.7264$

increases rapidly. Just how rapidly the number of iterations explode is explored in the experiment which is reported in table 11.46. The last column lists the product $N(a)\sqrt{a^* - a}$ and reveals that it converges to approximately 2.36 as the parameter a approaches a^*. From this we conclude that there is a square root power law which describes the increase of the iteration count,

$$N(a) \approx \frac{2.36}{\sqrt{a^* - a}} \,. \tag{11.8}$$

In other words, as we reduce the distance of a from a^* by a factor of 1/100, the number of iterations goes up by the factor of 10.

What do we learn from this experiment? If an iterator is close to a tangent bifurcation, it is nearly impossible in a numerical study to distinguish between transient and long term behavior. The final state becomes apparent only after many iterations, perhaps many more than the time frame of the experiment and the computer allow.

Let us now return to our quadratic iterator. Figure 11.47 shows the graph of f_a^3 for

$$a = w_3 = 1 + \sqrt{8} \approx 3.828427125$$

Counting Intermittent Iterations

k	$a_k = a^* - 10^{-k}$	$N(a_k)$	$N(a_k)\sqrt{a^* - a_k}$
0	0.7264289398723	4	4.00000
1	1.6264289398723	10	3.16228
2	1.7164289398723	26	2.60000
3	1.7254289398723	77	2.43496
4	1.7263289398723	238	2.38000
5	1.7264189398723	749	2.36855
6	1.7264279398723	2 363	2.36300
7	1.7264288398723	7 467	2.36127
8	1.7264289298723	23 608	2.36080
9	1.7264289388723	74 649	2.36061
10	1.7264289397723	236 056	2.36056
11	1.7264289398623	746 469	2.36054
12	1.7264289398713	2 360 537	2.36054
13	1.7264289398722	7 464 659	2.36053

Table 11.46 : For a sequence of parameters $a_k, k = 0, ..., 13$, approaching the saddle-node bifurcation at a^* from below we compute the number of iterations $N(a_k)$ of the orbit starting at the critical point $x_{\text{crit}} \approx 0.7286$ until it enters a small neighborhood of 0. The last column records the product $N(a_k)\sqrt{a^* - a_k}$ revealing the square root power law governing the exploding numbers of iterations.

Tangent Bifurcation

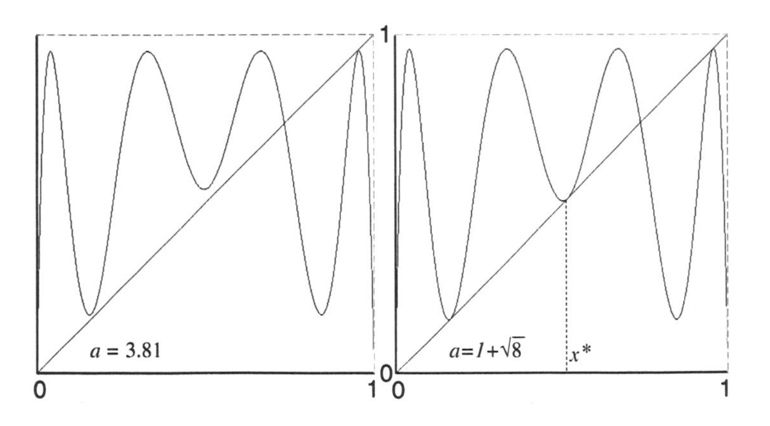

Figure 11.47 : Tangent bifurcation in f_a^3, at $a = 1 + \sqrt{8}$.

and for the slightly smaller parameter value $a = 3.81$. Indeed, we observe a tangent bifurcation at $a = w_3$, with a saddle point at x^*. Note that this point lies slightly to the right of 1/2. Note also that the graph touches the bisector at two more points, namely $f_a(x^*)$ and $f_a^2(x^*)$, which makes up the corresponding 3-cycle for f_a. In

Intermittency in Time Series

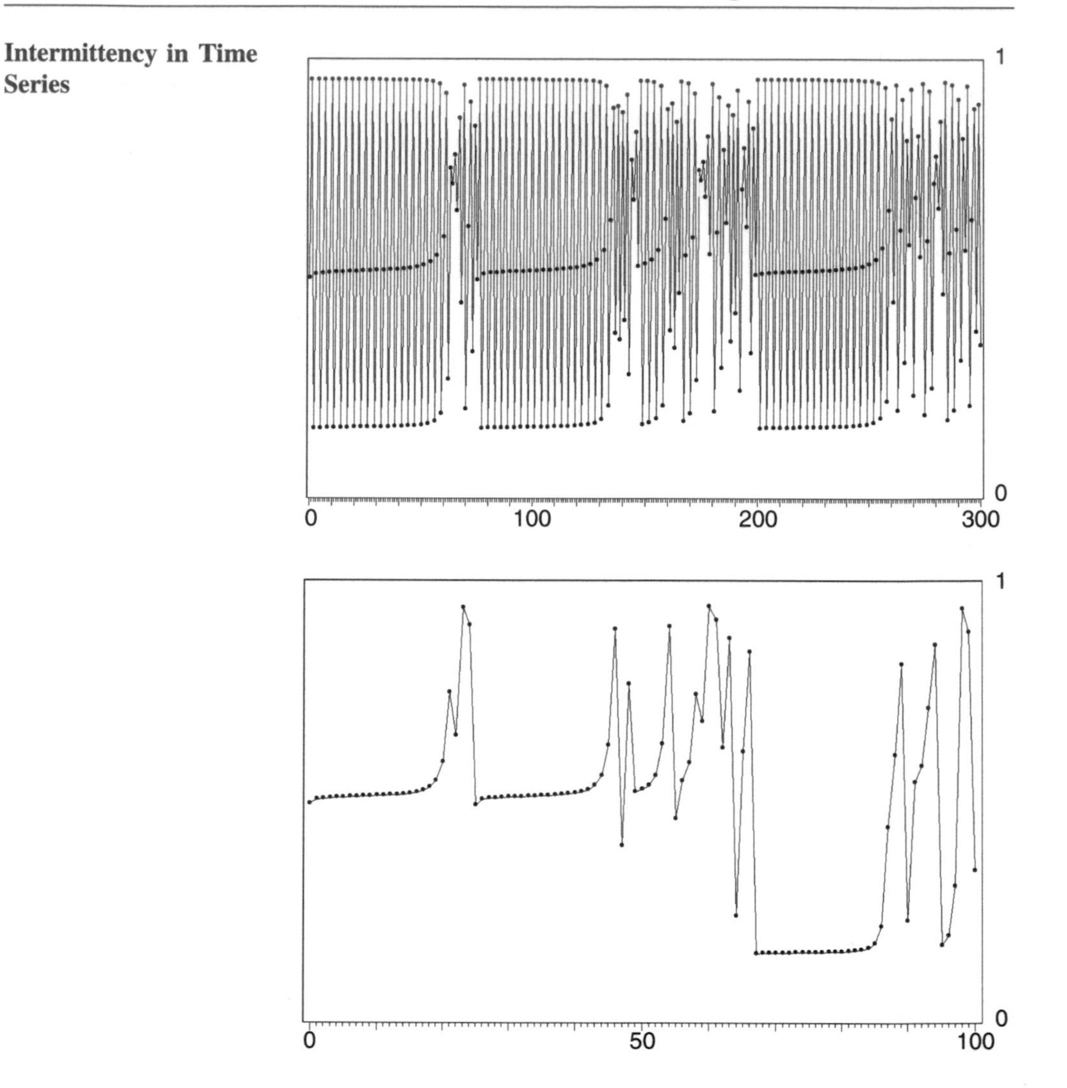

Figure 11.48 : Time series of $x_0 = 0.5$ for $a = 3.82812$: (top) some 300 iterations of f_a, (bottom) the corresponding 100 iterations of f_a^3 (note: time axis is scaled up by 3 to allow a better comparison).

figure 11.42 we have already seen what happens if we increase the parameter a even further: the saddle splits into a stable and an unstable fixed point, and finally we can identify a small copy of the generic parabola. In other words, the tangent bifurcation at $a = w_3$ gives birth to the period-doubling scenario, chaos, and band-merging, which we observe in the period-3 window.

Intermittency

To investigate the behavior of the quadratic iterator for parameters slightly below the tangent bifurcation, we compute a long time series (see figure 11.48). The experiment shows recurrent long phases of almost resting behavior for f_a^3 corresponding to almost perfect cyclic behavior of period 3 for f_a. These phases of 'stability' — also called *laminar phases* — alternate with erratic and chaotic behavior. This dramatic interplay between bursts of chaos and almost periodic behavior is called *intermittency*.

The orderly parts of the time series correspond to phases of the iteration close to x^*, where the orbit is cruising against the wind in a narrow channel. But what happens after the orbit has escaped from these narrows? And what is the mechanism which always forces the iteration back to the orderly phase? This brings us to the second ingredient of intermittency: *homoclinic points*.

Homoclinic Points

Homoclinic points were discovered by the great French mathematician Henry Poincaré in his famous studies of the stability of the solar system at the end of last century. Poincaré already understood very well that homoclinic points generate chaos. However, it was not until the 1960's that Stephen Smale created the appropriate mathematical framework. Homoclinic points can occur in connection with saddle points. We have seen that saddles have unstable and stable parts. Vaguely speaking a homoclinic point is a point which belongs to both parts.

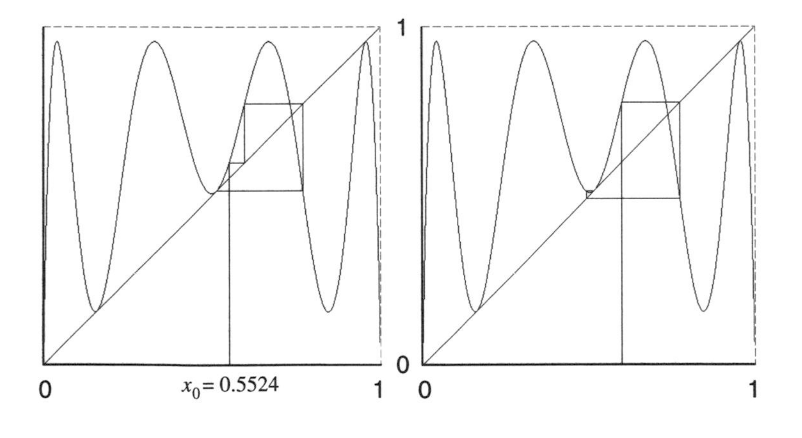

Homoclinic Points

Figure 11.49 : Two homoclinic points for $a = 1 + \sqrt{8}$.

Let us return to our concrete example, the saddle at x^* of f_a^3 for $a = w_3$. In this case a homoclinic point is defined to be any point x_0 on the repelling side of the saddle, which, when iterated, is eventually transformed onto x^* or which gets back to the stable side of the saddle. Formally, either there is a number k such that $x_k = x^*$, or $x_k \to x^*$ as $k \to \infty$. Figure 11.49 shows an example for each case. It is a fact that there exist an infinite number of homoclinic points, and that they densely fill a small interval on the unstable side of the saddle point.

Now consider a homoclinic point as a starting value for the iteration at a slightly reduced parameter $a < w_3$. The saddle point is gone, but the orbit of the formerly homoclinic point does not change very much (at least not for an initial finite number of iterations). This orbit will lead from the formerly unstable side of x^* to the stable side. Now imagine the iteration of an initial value x_0 which has just traveled through the narrow channel. Unavoidably it will get close to a homoclinic point on the unstable side, which then guides it back to the stable side. In other words, we will see intermittent behavior. There is only one possibility for the iteration to escape from this behavior, namely when the orbit hits one of the unstable fixed or periodic points (for example $p_a = (a-1)/a$). But the chance that this will happen is zero.

The Scaling Law of Intermittency

As the parameter a approaches w_3 from below the laminar phases become longer and longer. Eventually, the intermittent chaotic bursts disappear altogether and only the asymptotic periodic behavior remains. The average number of iterations that an orbit spends in the laminar phase thus tends to infinity. Moreover, it obeys a power law, which is of the same quality as the one which we found for the tangent bifurcation for g_a in eqn. (11.8); it is proportional to

$$\frac{1}{\sqrt{w_3 - a}} .$$

This relation can be derived analytically.[16]

Of course the vicinity of w_3 is only one example of the intermittency route to chaos. The same findings can be made at all other periodic windows.

Breakdown of Chaoticity via Crisis

The period-doubling scenario is the primary route to chaos. As a first alternative we have presented intermittency, a backdoor to chaos. There is one more important route which can be regarded as a close relative of intermittency: crisis. At the onset of a crisis — when a parameter of the system is varied appropriately — a chaotic region can be turned into a chaotic repeller. The typical phenomenon is that orbits, started in the previously chaotic region, behave chaotically only for a finite number of iterations. Eventually the chaoticity must break down and the long term fate of the orbit is very definite and predictable. A crisis can be studied for the quadratic system $x \to ax(1-x)$ at the borderline parameter $a = 4$. All our previous work on the quadratic iterator was limited

[16]See Y. Pomeau and P. Manneville, *Intermittent transition to turbulence in dissipative dynamical systems,* Commun. Math. Phys. 74 (1980) 189–197. See also R. W. Leven, B.-P. Koch and B. Pompe, *Chaos in Dissipativen Systemen,* Vieweg, Braunschweig, 1989.

Escaping Orbits

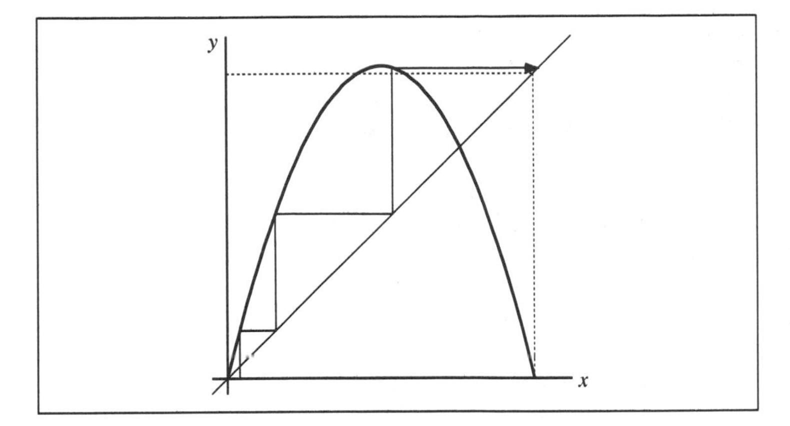

Figure 11.50 : For $a > 4$ in the quadratic iterator $x \to 4x(1-x)$ most orbits escape from the unit interval through the gap at the vertex of the parabola.

Example Orbit

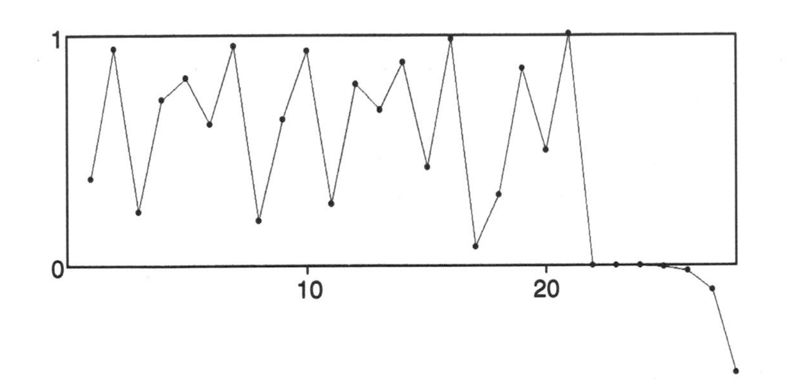

Figure 11.51 : Time series of an orbit started at $x_0 \approx 0.38$ for $a = 4.001$.

right at this point of crisis. But what happens for $a > 4$. A quick look at the corresponding graphical iteration (see figure 11.50) reveals that the unit square which housed the model of pure chaos at $a = 4$ breaks together. The parabola is no longer confined to the unit interval which causes points in a small interval around the critical point $x = 0.5$ to be transformed outside of the unit interval from where orbits rapidly escape to negative infinity. But also all preimages of that interval escape leaving almost no orbits which remain in the unit interval for all times.[17]

Chaotic Transients

Figure 11.51 displays a typical orbit for the parameter $a = 4.001$, slightly above 4. We observe that for a large number of

[17]In fact, only a Cantor set remains. We will pick up this scenario in more detail again in section 13.7.

iterations the orbit looks chaotic until eventually it escapes the unit interval and diverges. Orbits of this type are called *chaotic transients*. In other words, the chaos — which has ceased to exist for this parameter value — still casts a spooky shadow on the orbits. Only in the long term the phantom disappears. But how many iterations are necessary for this to happen? Or what is the life-time of a chaotic transient? This number clearly depends on the choice of the initial point as well as on how close the parameter a is to the value 4. For example, the initial point $x_0 = 0.5$ escapes the unit interval in just one step, while the orbits of the fixed points 0 and 1 as well as their preimages, of course, cannot escape. Moreover, if $a - 4$ is large then large portions of the unit interval escape in only a few iterations since only a small portion of the parabola remains in the unit square. The first deficiency can be removed by considering an average escape time of orbits with initial points uniformly distributed over the entire unit interval. The dependence of this average escape time on the parameter a is established in the form of another interesting power law.

Life-Time of a Chaotic Transient

Let us make a numerical experiment to uncover this power law for parameters a beyond the crisis. We take 10,000 initial points, equally spaced in the unit interval, and compute their orbits until they have escaped the unit interval. Table 11.52 lists the average escape times, obtained for a decreasing sequence of parameters.

Average Escape Times Near Crisis

k	$a = 4 + 10^{-k}$	E_a	$E_a\sqrt{a-4}$
0	5.000000	2.307	2.307
1	4.100000	7.728	2.444
2	4.010000	26.996	2.700
3	4.001000	93.432	2.955
4	4.000100	307.607	3.076
5	4.000010	1017.663	3.218
6	4.000001	3118.888	3.119

Table 11.52 : Escape times E_a averaged over 10000 orbits for a decreasing sequence of parameters $a = 4 + 10^{-k}$, $k = 0, ..., 6$.

Plotted on a doubly logarithmic scale the data reveals the power law (see figure 11.53) represented by a straight line fit. The resulting slope is about $1/2$. Thus, the power law is

$$\text{Average escape time} E_a \propto (a-4)^{-\frac{1}{2}} ,$$

and we say that $\frac{1}{2}$ is the critical exponent of the chaotic transient. The last column of table 11.52 reveals that the product $E_a\sqrt{a-4}$ tends to about 3.1 as the parameter a approaches a = 4. Thus, this is the factor of proportionality in the power law,

$$E_a \approx 3.1(a-4)^{-\frac{1}{2}} .$$

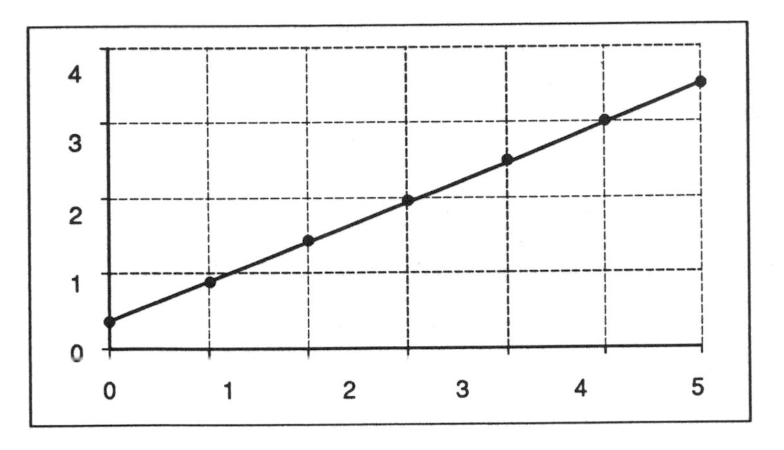

Escape Time Versus Parameter

Figure 11.53 : Plot of the escape times E_a from table 11.52 versus $1/(a-4)$ using a logarithmic scale (base 10) on both axes. The slope of the line fit is about $1/2$.

We can give an argument which supports these numerics.[18] It is based on the expectation that the relative number of orbits which escape the unit interval is approximately proportional to the length of the small interval around $x = \frac{1}{2}$, in which the parabola $ax(1-x)$ surpasses the value 1, i.e. to the fraction of the unit interval which escapes in one iteration. In conclusion, the average escape time should be inversely proportional to the length of the interval. To compute that interval we need to solve the equation $ax(1-x) = 1$ which yields

$$x_{1,2} = \frac{1}{2} \pm \sqrt{\frac{1}{4} - \frac{1}{a}} \,.$$

The sought interval thus is $[x_1, x_2]$. Letting $a = 4 + \varepsilon$ we find that its length is

$$x_2 - x_1 = 2\sqrt{\frac{1}{4} - \frac{1}{a}} = \frac{\sqrt{\varepsilon}}{2} \cdot \frac{1}{\sqrt{1 + \frac{\varepsilon}{4}}}$$

The second factor converges to 1 as ε tends to 0. Thus, the length of the interval is (asymptotically) proportional to $\sqrt{\varepsilon} = \varepsilon^{\frac{1}{2}}$. This yields the same critical exponent $\frac{1}{2}$ as our numerical test above.

[18] Compare with the similar reasoning on page 140 for the life time of an orbit in the mixing experiment.

Boundary Crisis

Let us present another view of the crisis as the parameter a passes through the critical value $a = 4$. Below the critical parameter, orbits are confined to the interval which is bounded by the critical value $v_a = a/4$ and its image $f_a(v_a)$. Orbits started between 0 and 1 outside of this interval rapidly iterate to the final states which are inside the interval. There is one exception, namely the repelling fixed point at 0. It is right on the boundary between initial points whose orbits tend to the invariant set in the unit interval and those which lead to diverging sequences.[19] Precisely at $a = 4$ the final states collide with the repeller at 0 and the chaotic region suddenly disappears. This type of crisis is therefore also called a boundary crisis.[20]

To conclude let us remark that the phenomena of intermittency and crisis are presented here only for the simplest possible model, the quadratic iterator. Of course, they also occur in many other mathematical systems which are far from this simple case. Moreover, intermittency and crises have been observed in physical experiments, for example, in pipe flows, a compass forced by a magnetic field, electronic oscillators, lasers, thermal convection in liquid crystals, and more.[21]

[19]These are *negative* initial values.

[20]There are other types of crises, for example, the *interior crisis* in which a sudden widening of chaotic bands occurs. Such a crisis happens at the parameter which limits the period-3 window. For a review of crises see Grebogi, C., E. Ott, J. A. Yorke, *Crises, sudden changes in chaotic attractors, and transient chaos,* Physica 7D (1983) 181–200.

[21]For references and a review of the theory of chaotic transients see T. Tél, *Transient chaos,* to be published in: *Directions in Chaos III,* Hao B.-L. (ed.), World Scientific Publishing Company, Singapore.

11.6 Program of the Chapter: Final State Diagram

We have discussed the fascinating route from order to chaos as it is captured by the final state diagram of the quadratic iterator. The program of the chapter gives you the opportunity to explore this route yourself. You can reproduce the diagrams shown in this chapter and you can investigate parameter ranges of your own choice.

The program computes the final state diagram (Feigenbaum diagram) for a given parameter range (from `astart` to `aend`) and draws the final states scaled such that a specific range of state values (from `xbottom` to `xtop`) fits to the vertical screen height (given by `h`). You can specify these ranges at the beginning of the program.

Screen Image of Program

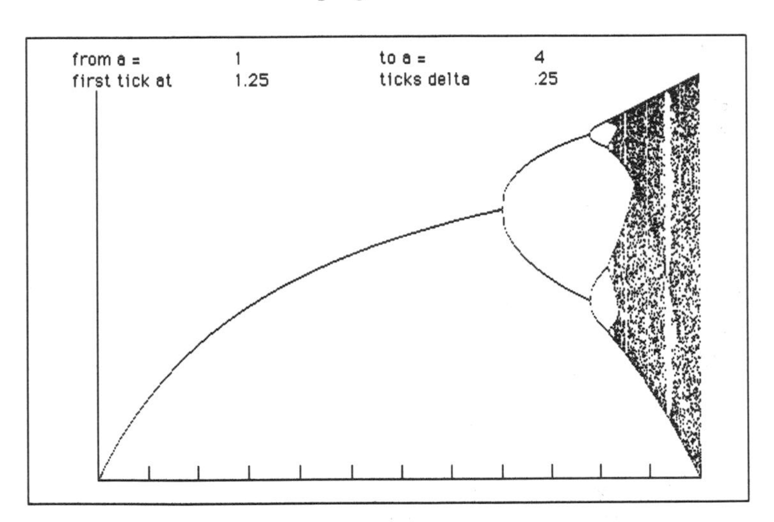

Figure 11.54 : Output of the program 'Feigenbaum'.

First the program draws a coordinate system with tick marks on the horizontal axis. This enables you to better read off the parameter ranges for those parts of the diagram that you want to examine in more detail. The distance between the tick marks is 0.1×10^d, 0.25×10^d or 0.5×10^d depending on the size of the parameter range. It is computed such that 10 to 25 marks are drawn. The program prints the parameter range, the parameter value which is marked by the first tick mark and the difference between succeeding tick marks.

Next the program starts to determine the final state for the given range of parameters. Horizontally it steps forward a number of `dx` pixels and determines the corresponding parameter values starting at `astart`. For each parameter the initial value `xn` is iterated for

BASIC Programm **Feigenbaum**
Title Computation and drawing of a final state diagram

```
left = 20
h = 300
w = 1.5 * h
dx = 2
iter = 200
hide = 100
astart = 1
aend = 4
xbottom = 0
xtop = 1
dw = w/(aend-astart)
dh = h/(xtop-xbottom)

REM Draw coordinates and tick marks
LINE (left,left) - STEP (0,h)
LINE - STEP (w,0)
d = INT(LOG(aend-astart)/LOG(10))
td = 10^d / 10
IF (aend-astart)/td > 50 THEN td = 5*td
IF (aend-astart)/td > 25 THEN td = 2.5*td
ts = td*(INT(astart/td)+1)
FOR i = 0 TO (aend-astart)/td - 1
    a = i*td + ts
    LINE (left+dw*(a-astart),left+h) - STEP (0,-10)
NEXT i
PRINT "from a =", astart," to a =", aend
PRINT "first tick at", ts, " ticks delta ", td

REM Compute iteration and draw final state
xn = .01
FOR i = 1 TO w/dx
    a = astart + i*dx / dw
    FOR k = 1 TO iter
        xn= a*xn*(1-xn)
        IF k < hide GOTO 100
        IF xn > xtop GOTO 100
        IF xn < xbottom GOTO 100
            PSET (left+i*dx,left+(xtop-xn)*dh)
100  NEXT k
NEXT i
END
```

a certain number of steps (given by `iter`). The number `hide` specifies how many of these steps are hidden, i.e., not drawn. The other iteration steps are drawn in the diagram if their value `xn` lies between `xbottom` and `xtop`. You can change the value of `iter` and `hide` at the beginning of the program. The execution time of the program is proportional to values for `iter`. Thus, it is advisable to adjust the number of iterations for each parameter to the speed of the computer. However, if this number is small, the number of points that may actually be plotted, i.e., the difference `iter` – `hide`, may be too small causing the resulting diagram to come out rather incomplete, especially for parameter ranges in the chaotic regime beyond the Feigenbaum point. Also, if the number `hide` (which must be smaller than `iter`) is too small, then the iteration might still be in the transient phase when the plotting of points begins giving a false impression of the Feigenbaum diagram.

If you want to change the range of the `xn` values that are plotted (i.e., `xbottom` and `xtop`) it is also desirable to have tick marks on the vertical axis. This will enable you to read off appropriate values from the diagrams. We have not included the necessary code for this in order to keep the program as short as possible. If you want to improve the program in this way, simply duplicate the part which computes and draws tick marks for the horizontal axis (from the assignment `d = ...` to the statement `NEXT i`). Then change the drawing to

```
LINE(left,left+h-dh*(a-xbottom)) - STEP(10,0)
```

and substitute the variable `astart` by `xbottom` and `aend` by `xtop` for all statements which were duplicated.

You can also use this program for drawing critical lines as shown in figure 11.39. For this purpose you have to reset `xn` to the initial value 0.5 (the critical point) for each new parameter (i.e., just before the loop `FOR k = ...`). Moreover, the variable `hide` should be set to 0, because in this case we want to see all iterates, and the number of iterations `iter` should be small.

Chapter 12

Strange Attractors: The Locus of Chaos

Never in the annals of science and engineering has there been a phenomenon so ubiquitous, a paradigm so universal, or a discipline so multidisciplinary as that of chaos. Yet chaos represents only the tip of an awesome iceberg, for beneath it lies a much finer structure of immense complexity, a geometric labyrinth of endless convolutions, and a surreal landscape of enchanting beauty. The bedrock which anchors these local and global bifurcation terrains is the omnipresent nonlinearity that was once wantonly linearized by the engineers and applied scientists of yore, thereby forfeiting their only chance to grapple with reality.

Leon O. Chua[1]

Having discussed the phenomena of chaos and the routes leading to it in 'simple' one-dimensional settings, we continue with the exposition of chaos in dynamical systems of two or more dimensions. This is the relevant case for models in the natural sciences since very rarely can processes be described by only one single state variable. One of the main players in this context is the notion of *strange attractors*.

To talk about strange attractors we have to consider a particular kind of dynamical systems: dissipative dynamical systems, i.e., systems with some sort of friction. The chief feature of dissipative systems is loss of energy. For example, a real pendulum swinging in air will have dissipation. Energy is lost continuously through the various kinds of friction which the pendulum experiences. In

[1] In: International Journal of Bifurcation and Chaos, Vol. 1, No. 1 (1991) 1–2.

contrast, we speak of conservative dynamical systems when energy is maintained. This is the case in systems without friction. For example, the friction which heavenly bodies sustain is so little that we think of their motion as conservative; no energy is lost.

Guided by mathematical development physicists and mathematicians were led to believe that the long term behavior of dissipative systems would always run into simple patterns of motion such as a rest point or a limit cycle. In contrast, strange attractors are those patterns which characterize the final state of dissipative systems that are highly complex and show all the signs of chaos. They very strongly defy the power of an intuitive understanding, and yet they now are proven to be all around us. It seems as if all of a sudden a whole new world of previously invisible beings is flying around us. Moreover, strange attractors are the point where chaos and fractals meet in an unavoidable and most natural fashion: as geometrical patterns, strange attractors are fractals; as dynamical objects, strange attractors are chaotic. There is now a whole new experimental and theoretical science dealing with strange attractors, their classification, the measurements of their quantitative properties, their reconstruction from physical data, and so on. But undoubtedly the mathematical understanding of strange attractors is just in its infancy and they will be one of the great challenges of future mathematical generations. It is by no means easy to understand even the notion of a strange attractor. In fact, strange attractors still have not received a final mathematical definition. Mathematics is sometimes described as the science which generates eternal notions and concepts for the scientific method: derivatives, continuity, powers, logarithms are examples. The notions of chaos, fractals and strange attractors are not yet mathematical notions in that sense, because their final definitions are not yet agreed upon.

Strange attractors have, however, become a *very* popular topic which has drawn interest not only from physics and mathematics but also from all other natural sciences and even the social sciences. The reason for the overwhelming popularity of chaos and strange attractors lies in the great expectations with which people come to the topic. Scientists hope to be able to crack the mysteries of our planet's climate, or human brain activity, as well as the secrets of turbulence through the metaphor of strange attractors. Fluid turbulence — one of the great unsolved problems in theoretical physics — occurs even in common daily routines, for example, when we open the water tap at the kitchen sink. First a smooth and regular flow of water appears. But as the water flow is increased the fluid starts to forcefully splash out without any regularity: turbulence. More important and with relevant technical applications is the turbulence occurring in the turbo-prop engines of an airplane or at the propeller of a ship or in large water pumps, where turbulence can actually eat away the metal impeller blades. There are also some chemical reactions which are periodic in time. In 1971, David Ruelle, one of the scientific notables in chaos theory, asked a specialist in these periodic reactions if he thought that one would find chemical reactions with chaotic time dependence. Ruelle recalls that he answered that if an experimentalist obtained a chaotic record in the study of a chemical reaction, he would throw away the record, saying that the experiment was unsuccessful.[2] This attitude, of course, was characteristic not only of experimental chemistry, but also of all other natural sciences. But soon after the news of strange attractors had spread around the scientific laboratories of the world in the 1970's, things changed fundamentally. Researchers became aware of the subject and concentrated on the irregular patterns of processes which they previously had dismissed as misfits. Now we know several examples of chaotic behavior even in simple reaction systems. There have also been numerous strange attractors discovered in physics which are similar to the Lorenz attractor (discussed later in this chapter), and the concept of strange attractors is used in the sciences ranging from astronomy almost all the way to zoology.

A part of the drawing power in the concepts of chaos and strange attractors is probably due to the choice of the catchy names for these phenomena. The word 'chaos' was introduced in an article by Tien-Yien Li and James A. Yorke entitled *Period 3 implies chaos*,[3] while the term 'strange attractor' even goes back to 1971. Let us again quote from Ruelle's exposition in the *Mathematical Intelligencer*. "It seems that the phrase 'strange attractor' first appeared in print in a paper by Floris Takens (of Groningen) and myself.[4] I asked Floris Takens if he had created this remarkably successful expression. Here is his answer. 'Did you ever ask God whether he created this damned universe? ... I don't remember anything... I often create without remembering it ...' The creation of strange attractors thus seems to be surrounded by clouds and thunder. Anyway, the name is beautiful, and well suited to these astonishing objects, of which we understand so little."

The first strange attractor ever recognized as such in the natural sciences is the Lorenz attractor, discovered in 1962. However, the work was published in the *Journal of the Atmospheric Sciences* which is not usually read by physicists and mathematicians. So the research on chaos was unnecessarily delayed by a decade or so until the real implications of Lorenz' achievement became clear.

---

[2]From D. Ruelle, *Strange Attractors,* Math. Intelligencer 2 (1980) 126–137.

[3]Li, T. Y. and Yorke, J. A., *Period 3 Implies Chaos,* American Mathematical Monthly 82 (1975) 985–992.

[4]D. Ruelle and F. Takens, *On the nature of turbulence,* Comm. Math. Phys. 20 (1971) 167–192 and 23 (1971) 343–344.

Deterministic Nonperiodic Flow

EDWARD N. LORENZ

Massachusetts Institute of Technology

(Manuscript received 18 November 1962, in revised form 7 January 1963)

ABSTRACT

Finite systems of deterministic ordinary nonlinear differential equations may be designed to represent forced dissipative hydrodynamic flow. Solutions of these equations can be identified with trajectories in phase space. For those systems with bounded solutions, it is found that nonperiodic solutions are ordinarily unstable with respect to small modifications, so that slightly differing initial states can evolve into considerably different states. Systems with bounded solutions are shown to possess bounded numerical solutions.

A simple system representing cellular convection is solved numerically. All of the solutions are found to be unstable, and almost all of them are nonperiodic.

The feasibility of very-long-range weather prediction is examined in the light of these results.

Figure 12.1 : The abstract of Lorenz' paper which pioneered chaotic strange attractors.

Although the Lorenz attractor is one of the 'oldest' known strange attractors, answers to some very basic questions about it are still outstanding. Recently, on the occasion of his sixtieth birthday, the great mathematician Stephen Smale, who has been one of the leading pioneers in dynamical systems and chaos theory for several decades, posed ten major open research problems. One of them asks for a proof that the geometric model of the Lorenz attractor proposed by John Guckenheimer and Philip Holmes[5] is true.

Given a dynamical system such as the Lorenz system, we can see the attractors on our computer graphics screens. This is fine. However, when physicists, for example, make measurements in some real world experiment, they only obtain long and messy sequences of numbers, not equations. Then they must answer the question of what kind of dynamical system is behind the scene or perhaps even whether there is a strange attractor lurking behind their irregular and noisy data. One of the most fascinating achievements of chaos theory is that it has made available a tractable numerical method to attack this problem, the *reconstruction of strange attractors.* It even leads to algorithms which can compute numerical quantities such as dimensions and Ljapunov exponents that specify the degree of strangeness and 'chaoticity' of the attractor.

Strange attractors offer some new understanding of nonlinear effects. Moreover, they can be aesthetically pleasing. Thus, it is

[5] J. Guckenheimer and P. Holmes, *Nonlinear Oscillations, Dynamical Systems, and Bifurcations of Vector Fields,* Springer-Verlag, New York, 1983. See section 5.7 therein.

no wonder that the subject attracts researchers from all disciplines. It has been and continues to be a *hot topic*. The amount of literature that has been written on the topic of chaos surely surmounts any individual's reading capacity. Computers are needed to study dynamics on strange attractors, and computers are equally needed to maintain data bases of that vast literature. An (incomplete) list published by Hao Bai-Lin contains 117 books, conference proceedings, and collections of papers, and an unbelievable number of 2244 technical papers. It is therefore an outstanding achievement for Hao to have compiled the most influential papers, along with some of an introductory type, and the bibliography in one large volume of reprints entitled *Chaos II*.[6] It contains, for example, the original papers by Feigenbaum, Hénon, Lorenz, and May, to name just a few.

[6]Hao, B. L., *Chaos II*, World Scientific, Singapore, 1990. For an even larger bibliography on chaos containing over 7000 references see Zhang Shu-yu, *Bibliography on Chaos — Directions in Chaos Vol. 5*, World Scientific, Singapore, 1991.

12.1 A Discrete Dynamical System in Two Dimensions: Hénon's Attractor

In chapter 10 we built our analysis of chaos for the iteration of the quadratic map starting from the paradigm of the kneading of dough (see section 10.4). The dynamics could be modeled by a stretch-and-fold operation. Of course real dough has some thickness which must be ignored in this approach. Here we learn about a particularly simple transformation which does not neglect this extra dimension. It was suggested by the French astronomer Michel Hénon[7] in 1976 as a simplified model for the dynamics of the Lorenz system which is the topic section 12.4 below. Because of its simplicity, it lends itself to computer studies and numerous investigations followed. Moreover, the gently swirling, boomerang-like shape of the attractor that arises through the dynamics is very appealing aesthetically. This object is now known as the *Hénon attractor*. In fact, it has become another icon of chaos theory next to the Mandelbrot set, the Feigenbaum diagram, and the Lorenz attractor. In a way which we will specify, the Hénon system leads from the one-dimensional dynamics of the quadratic transformation to higher-dimensional strange attractors. It is simple enough to allow an analysis similar to the analysis of chaos in the logistic transformation, yet it possesses features inherent in more complicated attractors such as the Lorenz attractor, about which we do not know nearly as much.

The Model

The stretch-and-fold action of the Hénon system happens in two dimensions, with coordinates denoted by x and y. The transformation, thus, is a transformation in the plane which operates just like *one* of the affine transformations from our paradigm, the Multiple Reduction Copying Machine (MRCM) from the first chapter. Explicitly, Hénon suggested a transformation

$$H(x, y) = (y + 1 - ax^2, bx) \tag{12.1}$$

where a and b are adjustable parameters. An orbit of the system consists of a starting point (x_0, y_0) and its iterated images, i.e.,

$$(x_{k+1}, y_{k+1}) = (y_k + 1 - ax_k^2, bx_k), \quad k = 0, 1, 2, \ldots$$

Similar to the logistic equation, these dynamics depend dramatically on the choice of the constants a and b besides that of the starting point. For some parameters almost all orbits tend to a unique periodic cycle, while chaos seems to reign for other choices. Hénon used the values

$$a = 1.4 \text{ and } b = 0.3 \, .$$

The Hénon Attractor

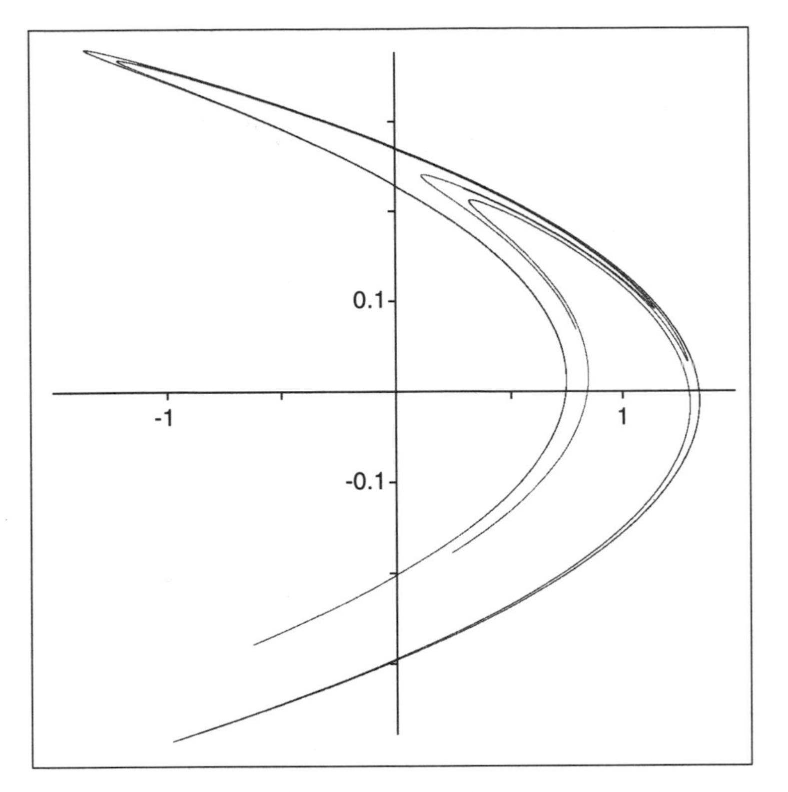

Figure 12.2 : 100,000 computed points of the orbit of the initial point (0,0) of Hénon's system (the first 100 points are omitted). The region shown is $-1.5 \leq x \leq 1.5$ and $-0.4 \leq y \leq 0.4$.

Three Phases of Stretch-and-Fold

Let us study the transformation to see the correspondence to the stretch-and-fold action. We can partition the application of the transformation H into three steps, visualized in figure 12.3.

1. **Bend up.** The first step consists of a nonlinear bending in the y-coordinate given by

$$H_1(x,y) = (x, y + 1 - ax^2) \ .$$

For example, a horizontal line (y = constant) becomes a parabola with vertex at $(0, y+1)$ and opening up at the bottom. In contrast, the remaining two steps are linear transformations.

2. **Contract in x.** Next a contraction in the x-direction is applied,

$$H_2(x,y) = (bx, y) \ .$$

The contraction factor is given by the parameter b, which is 0.3 for Hénon's attractor.

[7]See M. Hénon, *A two-dimensional mapping with a strange attractor,* Comm. Math. Phys. 50 (1976) 69–77.

Decomposition of Hénon's Transformation

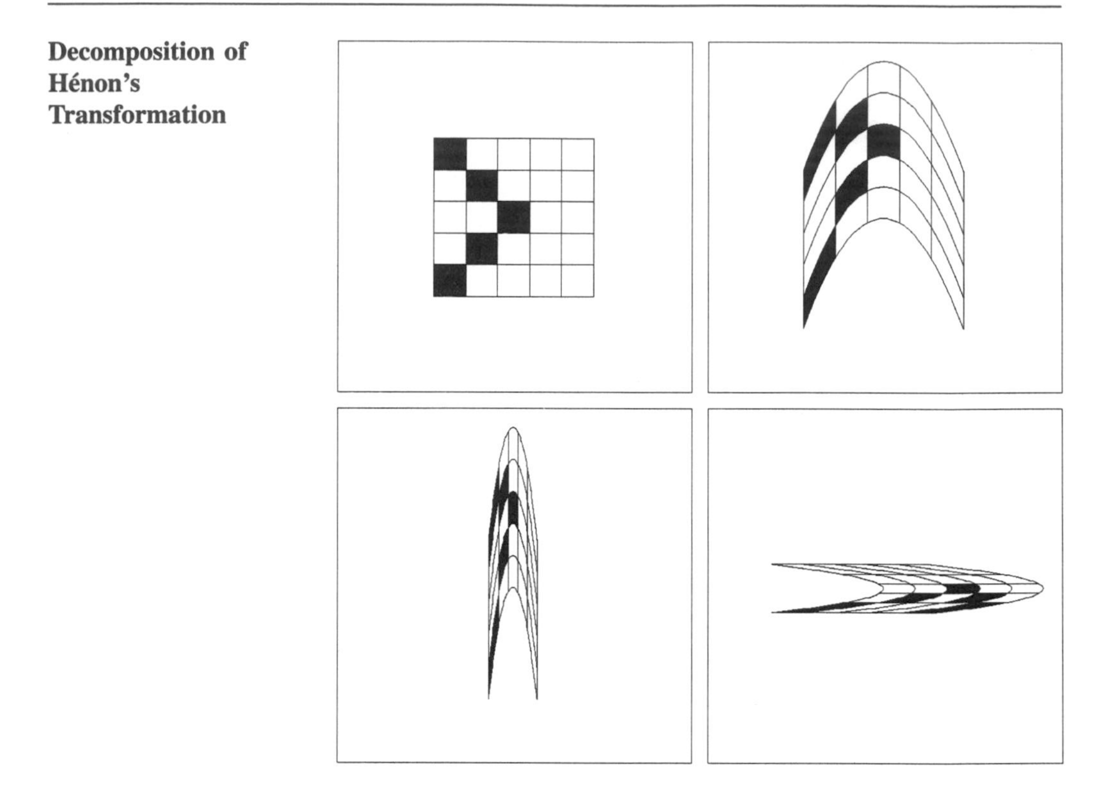

Figure 12.3 : The gridded square in the upper left is transformed in three steps: a nonlinear bending (upper right) in the y-direction, the contraction towards the y-axis (lower left) and a reflection at the diagonal (lower right). The region shown is $-2.2 \le x \le 2.2$ and $-2.2 \le y \le 2.2$.

3. **Reflect.** Finally a reflection at the diagonal,

$$H_3(x, y) = (y, x)$$

is in order.

The result of the compression is the same as applying the original transformation once, i.e.,

$$H(x, y) = H_3(H_2(H_1(x, y))) \ .$$

Speaking in terms of the kneading paradigm, we may say that the first step is the stretching of the dough. One person holds up the dough at the center while another person pulls down at the two ends of the dough. Step 2 folds the dough together, i.e., the second person moves both ends toward each other. Then at the end the dough is put back on the table and turned over.

Plate 1: 3-dimensional cross section of a Julia set in 4-dimensional quarternion space.

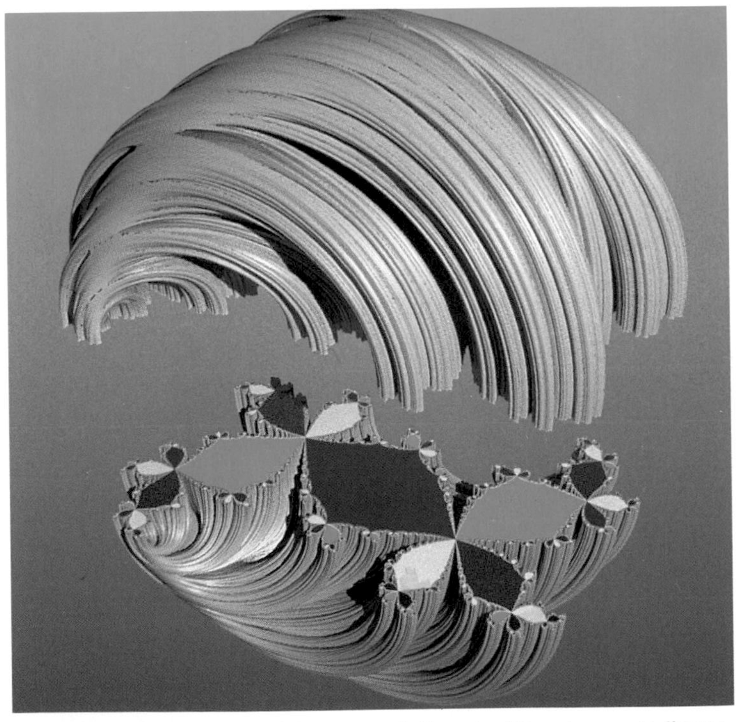

Plate 2: Different view of the same Julia set with cut open 2-dimensional cross section revealing the corresponding Julia set in the complex plane.

Plate 3: High resolution image of the potential of a piece of the Mandelbrot. The book cover shows a similar region viewed from the side.

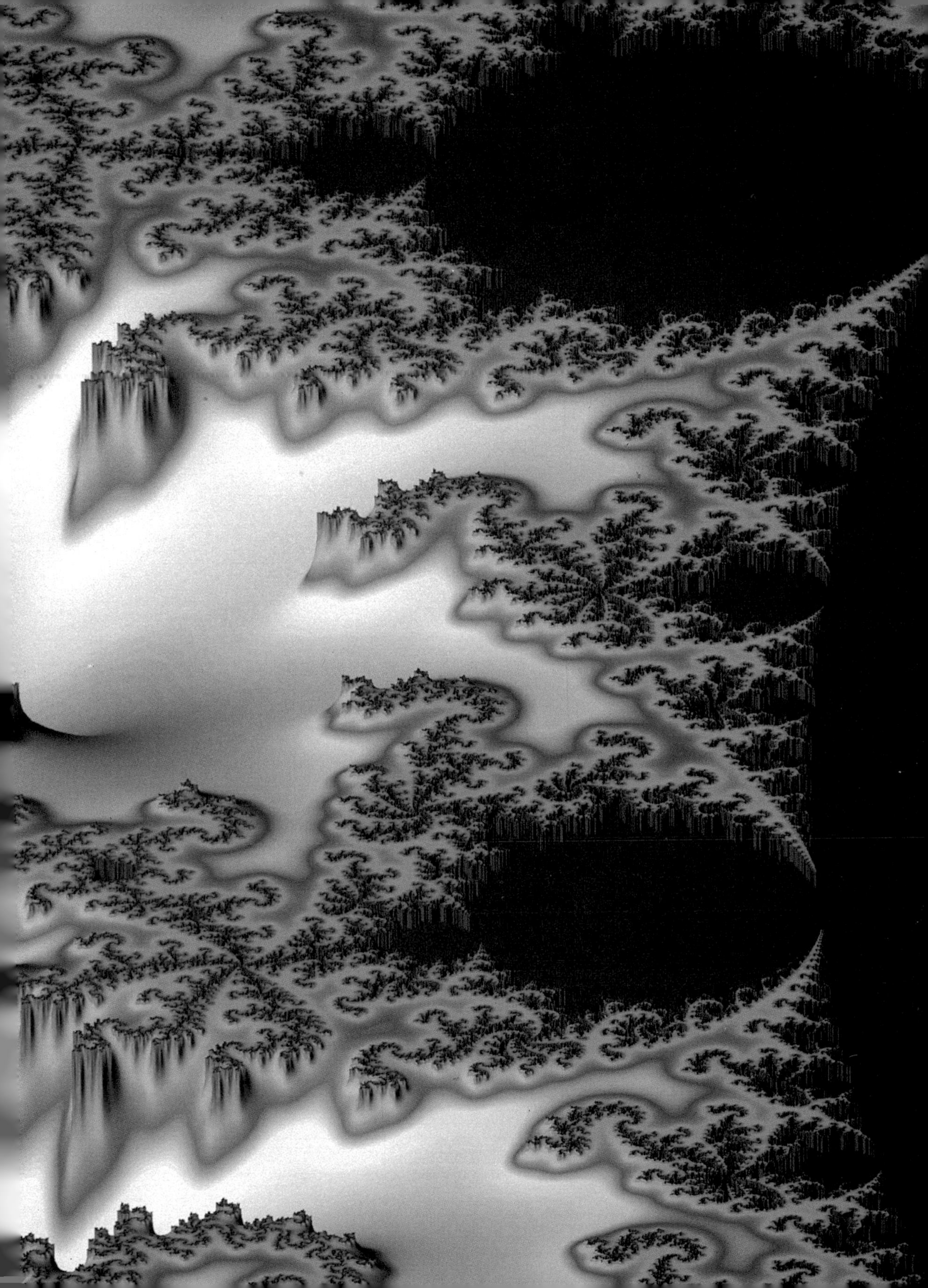

Invariance of the Hénon Attractor

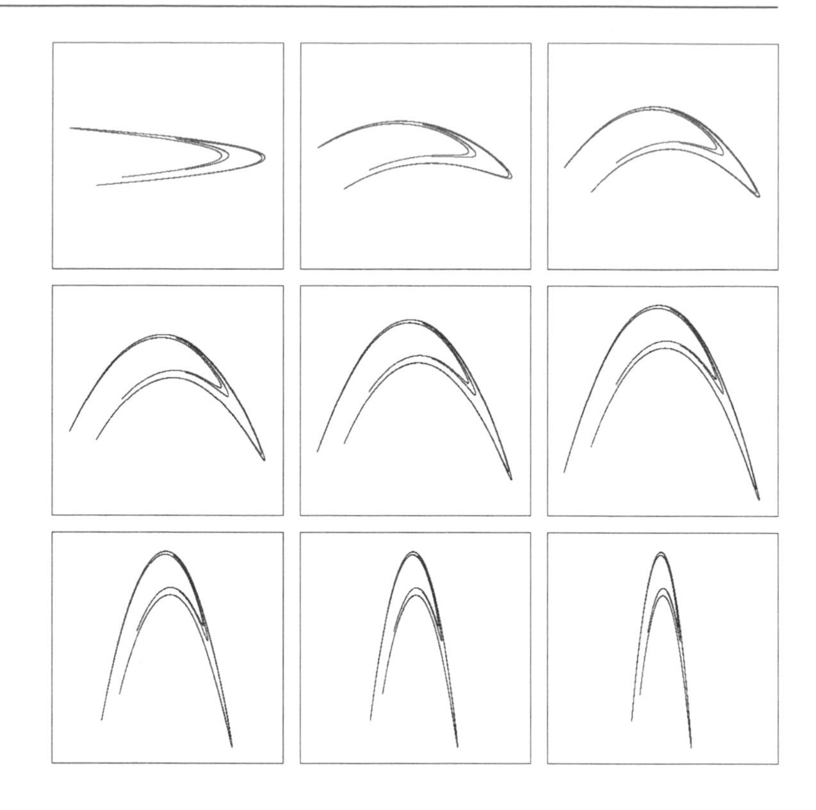

Figure 12.5 : Slow motion action of the Hénon transformation applied to the attractor itself. The upper two rows of figures show how the attractor (top left) is stretched according to $H_1(x, y) = (x, y+1-ax^2)$. The bottom row continues with the folding part of the Hénon transformation ($H_2(x, y) = (bx, y)$). When the bottom right figure is reflected at the diagonal (or turned 90 degrees clockwise and flipped horizontally) the attractor from the top left is exactly reproduced. Figure 12.6 shows how the transformation acts on some example points. The region is $-1.5 \leq x \leq 1.5$, $-1.5 \leq y \leq 1.5$.

the transformation H iteratively (see figure 12.4). Of course, the square is severely deformed and not even recognizable after only a couple of steps. After a few dozen more steps a curious shape emerges — the Hénon attractor. This attractor must be a subset of the plane which is invariant with respect to the kneading. The sequence in figure 12.5 demonstrates how the various parts of the attractor are transformed into each other.

Just as the logistic iterator may have periodic attractors which attract orbits of nearby starting points, the Hénon attractor pulls in nearby orbits. Moreover, these orbits typically fill up the attractor densely.

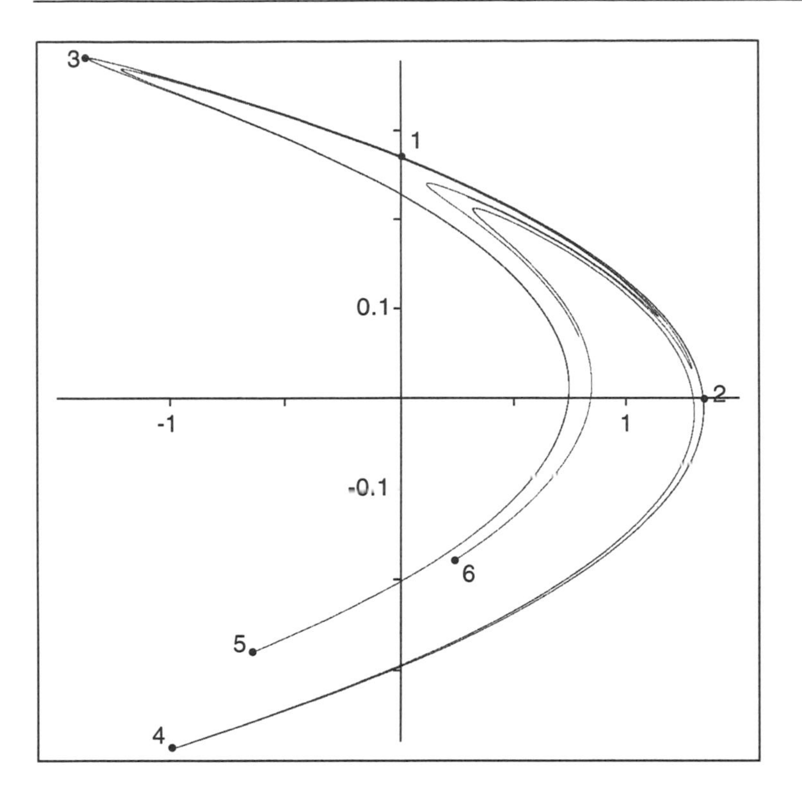

Figure 12.6 : Six points on the Hénon attractor are labeled by numbers 1 to 6. After the stretching, folding, and reflection of the Hénon transformation point 1 arrives at point 2, 2 arrives at 3, and so on. In other words, the initial points are chosen as consecutive points from an orbit. Try to follow these dynamics in figure 12.5. The region is $-1.5 \leq x \leq 1.5$, $-0.4 \leq y \leq 0.4$.

Invariance of the Hénon Attractor

Not All Points are Attracted

But how can we be sure that this phenomenon is not due to our particular choice of the initial point? After all, the transformation given by Hénon is quadratic and if x_0 is large, then x_1 will be much larger, and repeated applications of H drive the orbit beyond all bounds. For example, starting with $x_0 = 10$ and $y_0 = 0$ produces

$$\begin{array}{llll} x_0 = & 10.00 & y_0 = & 0.00 \\ x_1 = & -139.00 & y_1 = & 3.00 \\ x_2 = & -27045.40 & y_2 = & -41.70 \\ x_3 = & -1024035166.32 & y_3 = & -8113.62 \end{array}$$

and clearly the orbit escapes to (negative) 'infinity'. By the way, the initial value of x_0 does not even have to be very large for this effect to take place. The orbit of (1.292, 0.0) also escapes, although the starting point is already quite close to the attractor.

The Trapping Region

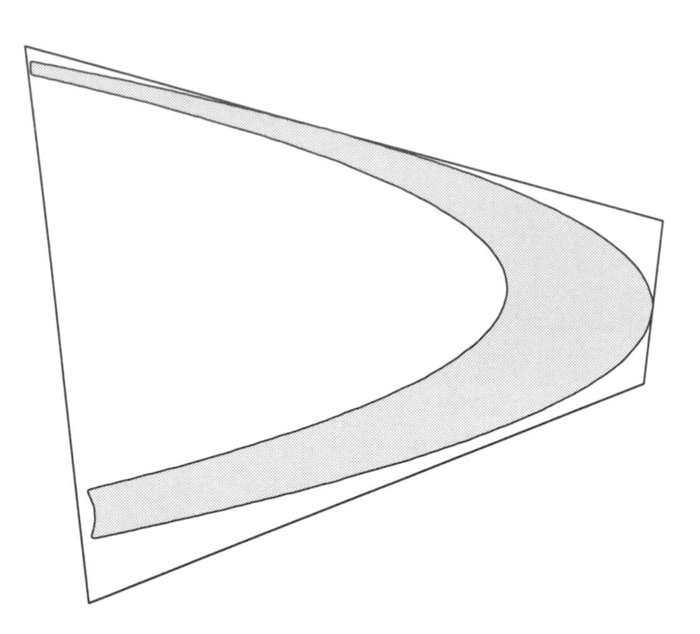

Figure 12.7 : The trapping region is the quadrilateral with vertices $P_1 = (-1.33, 0.42)$, $P_2 = (1.32, 0.133)$, $(P_3 = 1.245, -0.14)$, and $P_4 = (-1.06, -0.5)$. Its image is also shown; it lies entirely inside the trapping region. An orbit of an initial point within the trapping region cannot escape the region.

The Trapping Region

Even though many orbits do escape to infinity we may still speak of an attractor because there is a so called *trapping region* R from which no orbit can escape, thus orbits started within the region must converge to some limit set. The region is a quadrilateral carefully designed by Hénon and shown in figure 12.7. It can be verified using elementary algebra that the image of the region R obtained from one application of Hénon's transformation H does, in fact, lie entirely within the trapping region ($H(R) \subset R$). Thus, repeated application of H must always produce subsets of the region; no orbits can escape. Of course, the Hénon attractor lies in this trapping region. We may now define it as

$$A = \bigcap_{k=0}^{\infty} H^k(R)$$

where H^k means the k-fold composition of H, as usual.

Basin of Attraction

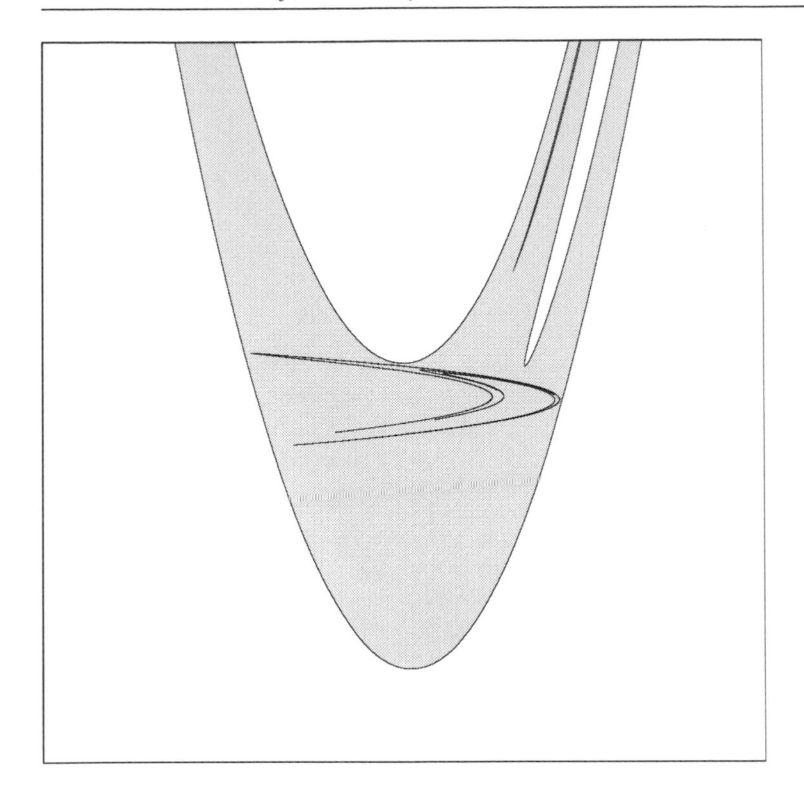

Figure 12.8 : The boundary of the basin of attraction for Hénon's attractor (shaded) which is shown in the center. The region shown is the square $-3 \leq x \leq 3$ and $3 \leq y \leq 3$.

The Basin of Attraction

Now that we know that on the one hand there are escaping points and on the other there is a trapping region for the Hénon attractor, the question arises, which points in the plane have orbits that are eventually caught by the trapping region. The set of all such points is called the basin of attraction of A. Of course, the trapping region itself must be contained in the basin of attraction. Figure 12.8 provides a plot.[8] Basins of attraction will be in the center of interest in the two remaining chapters.

The Area Shrinks

Looking again at figure 12.7 we notice how the area of the quadrilateral shrinks when we apply the transformation. The same holds true for the square in figure 12.4. This observation can be mathematically verified (see the following technical section). The result is that any area considered shrinks by the factor of $b = 0.3$ when iterated once. Thus, taking a region of area 1, we obtain after two iterations an area of only 0.09. After k iterations the area

[8] Pictures of this sort were first published in S. D. Feit, *Characteristic exponents and strange attractors,* Comm. Math. Phys. 61 (1978) 249–260.

has reduced to 0.3^k. When we apply this to the trapping region, we arrive at the conclusion that the attractor A, which must reside in all the iterates of the region, can only cover a subset of the plane with an area equal to 0.

The Area Reduction

Consider a matrix

$$T = \begin{pmatrix} a & b \\ c & d \end{pmatrix}$$

and the parallelogram spanned by its two column vectors. It is a result from basic linear algebra that the corresponding area is

$$A = |\det T| = |ad - bc| ,$$

the absolute value of the determinant of the matrix T. This is also the factor by which an area grows or shrinks when the linear transformation given by the matrix is applied. The Hénon transformation is not linear, but a similar result holds locally, based on the linearization of the transformation. A small area near a point $P = (x, y)$ is reduced by the factor given by the absolute value of the determinant of the derivative (the Jacobian matrix) of the transformation at that point. For the Hénon transformation his is

$$|\det DT(x,y)| = \left|\det \begin{pmatrix} -2ax & 1 \\ b & 0 \end{pmatrix}\right| = |b| .$$

Since $|b|$ is a constant which does not depend on the location of P the area changes uniformly by that factor.

The Strangeness of the Attractor

Let us now come to the chaoticity and the strangeness of the Hénon attractor. We will highlight two corresponding aspects: the sensitive dependence on initial conditions and the fractal structure. Thereafter we present an initial mathematical definition of the concept of a strange attractor.

Sensitive Dependence on Initial Conditions

To obtain a picture of the attractor, it is sufficient to compute just a single orbit of an initial point picked at random somewhere within the trapping region. Picking a different random initial point does not change the visual result in any way if we ignore the first hundred points or so which are needed to bring the orbit sufficiently close to the attractor. However, although two different orbits generate the same limit set, typically there is no correlation between them, even if the initial points are chosen very close to each other (see figures 12.9 and 12.10). Strictly speaking, this is not entirely true. For example, if the second initial point is a point from the first orbit (which may be as close to the first initial value as we want to), then both orbits of course are correlated as demonstrated in

Sensitivity

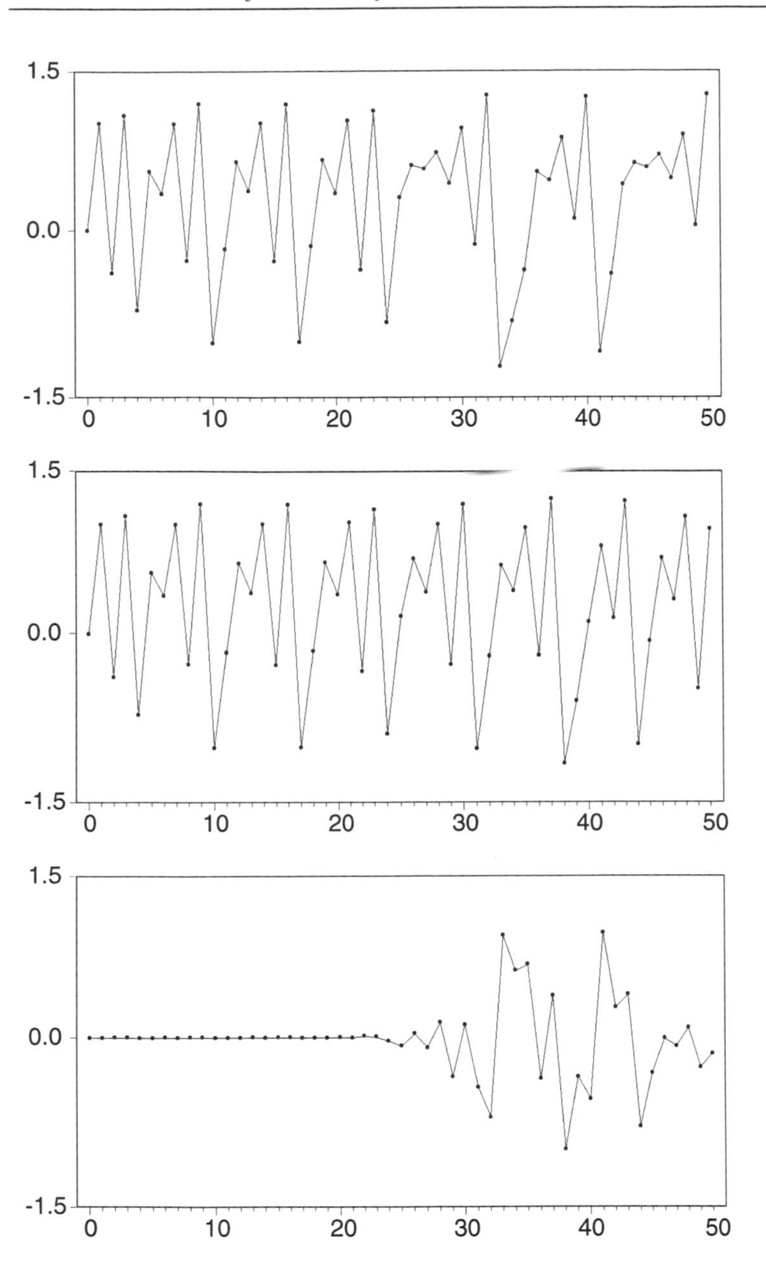

Figure 12.9 : To demonstrate the sensitivity to initial conditions, we compute two orbits $(x_0, y_0), (x_1, y_1), \ldots$ and $(x'_0, y'_0), (x'_1, y'_1), \ldots$ with initial points $(x_0, y_0) = (0, 0)$ and $(x'_0, y'_0) = (0.00001, 0)$. We plot three time series: the values x_k (top), x'_k (center) and the difference $|x_k - x'_k|$ (bottom). In the beginning the time series are undistinguishable, but after a number of iterations, the difference between them builds up rapidly and this 'error' becomes as large as the 'signal' itself — a consequence of sensitivity.

No Correlation Between Orbits

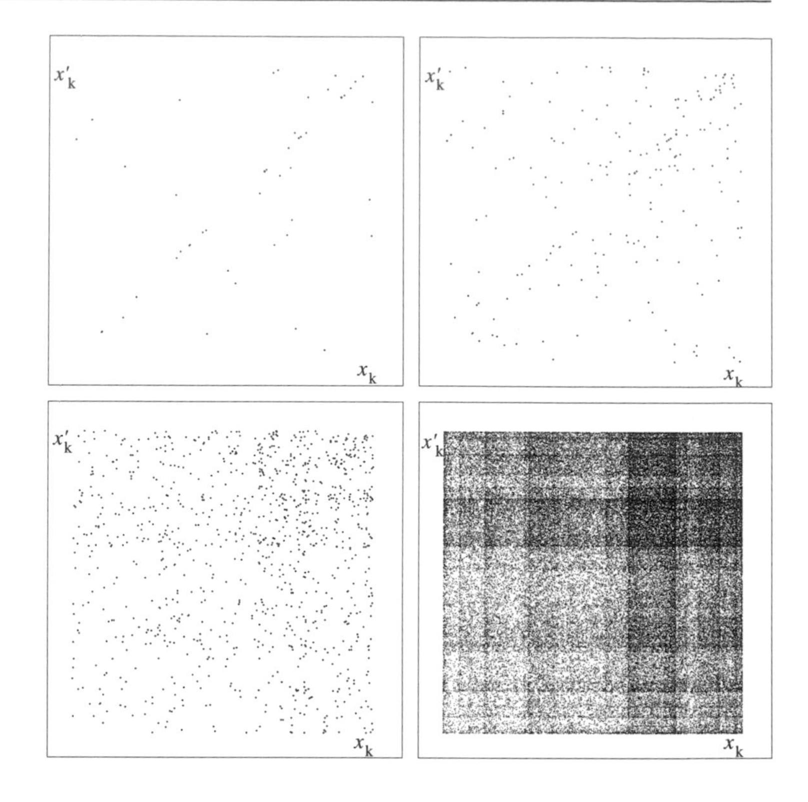

Figure 12.10 : To show another effect of sensitivity to initial conditions, we compare again the two orbits $(x_0, y_0), (x_1, y_1), \ldots$ and $(x'_0, y'_0), (x'_1, y'_1), \ldots$ with initial points as in figure 12.9. We plot points (x_k, x'_k) for the first 50, 200, 1000 and 100,000 iterations (from upper left to lower right). These points densely fill a square. Similar results would be obtained by plotting the second coordinates of the orbits against each other. The region shown is the square $-1.5 \leq x \leq 1.5$ and $1.5 \leq y \leq 1.5$.

figure 12.11. What we mean is that almost all random points chosen arbitrarily close to the first initial value produce orbits which do not correlate with the first one.

Zoom into the Fractal Structure

The first look at the Hénon attractor gives the appearance of a collection of a few curves which look like sections of parabolas. But this impression could not be further from the truth as the enlargements in figure 12.12 shows. The more we magnify a portion of the attractor the more 'curves' become visible. Thus, the Hénon attractor consists of an infinite number of parabola-like layers. When we think about the kneading action of the transformation, this infinitely detailed puff pastry structure becomes, in fact, quite reasonable. Assume that there is at least one parabola-like curve in the attractor. The stretching and folding of this curve

Correlation is Possible

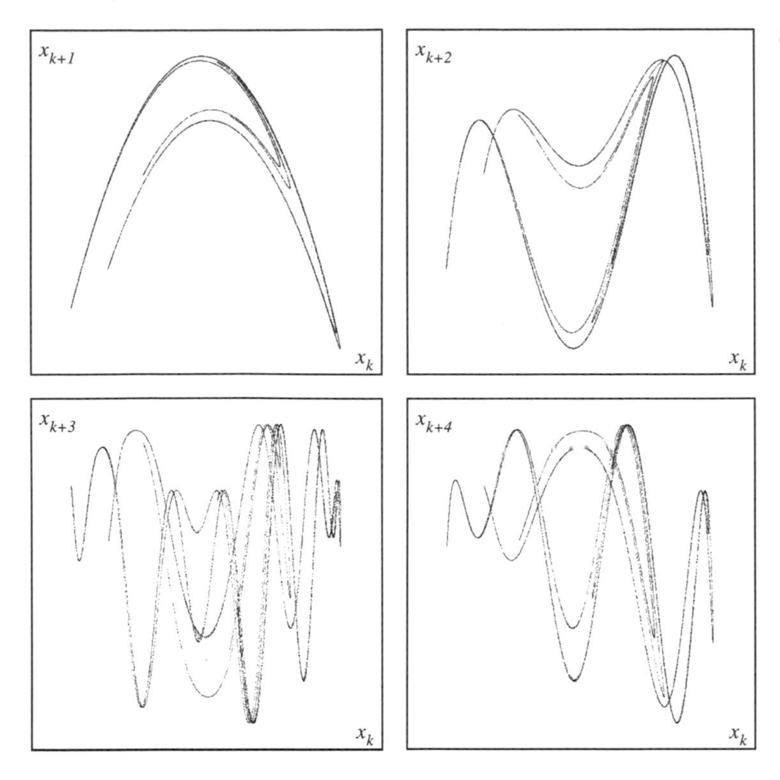

Figure 12.11 : This plot demonstrates that the result from figure 12.10 depends on the choice of initial conditions. Here we chose the initial condition (x'_0, y'_0) of the second orbit to be equal to the m-th iterate of the original orbit, i.e., (x_m, y_m). In effect, we are plotting points (x_k, x_{k+m}), $k = 0, 1, 2, \ldots$ The number m is 1 (upper left), 2 (upper right), 3 (lower left), and 4 (lower right). Note that we can choose the initial points of all orbits considered as close to each other as we wish by placing them in a neighborhood of a (repelling) fixed point. In all graphs there is a clear structure of the collection of points. In fact, they represent just modified versions of the Hénon attractor (compare the account of the reconstruction of strange attractors from time series in section 12.5).

produces a curve composed of a parabola with two layers which must consequently also be part of the attractor.[9] The next iteration generates a structure with four layers, then we get eight, and so on. Thus, after considering all the images of the initial parabola we obtain an infinity of layers, all of which belong to the attractor. This is what we can see in the graphics when we zoom in on a

[9]These two layers are separate, they must not intersect or overlay each other. The reason for this lies in the fact that no two points are transformed to the same image point. Hénon's transformation is one-to-one, which we will work out in detail below.

Enlargements of the Hénon Attractor

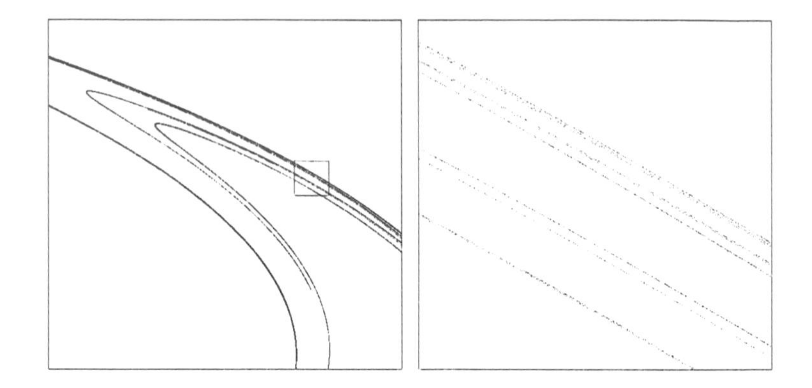

Figure 12.12 : Two successive enlargements of the Hénon attractor. The same 100,000 points as in Figure 12.2 are used. The regions shown are $0 \leq x \leq 1$ and $0 \leq y \leq 0.3$ on the left and $0.7 \leq x \leq 0.8$ and $0.15 \leq y \leq 0.18$ on the right.

point of the attractor. Due to the folding in the transformation, a cross-section of the attractor looks much like a Cantor set. We may say that sections of the Hénon attractor as in figure 12.12 are cross products of an interval with a Cantor set.[10]

The Inverse Transformation and Backward Orbits

In contrast to its one-dimensional cousin, the quadratic iterator, the Hénon transformation has an inverse. This expresses the fact that for any initial point there is not only a unique forward orbit but also a unique sequence of predecessors. The forward iteration of a point is given by

$$\begin{aligned} x_{n+1} &= y_n + 1 - ax_n^2 \\ y_{n+1} &= bx_n \end{aligned}$$

which we now solve for x_n and y_n:

$$\begin{aligned} x_n &= \frac{y_{n+1}}{b} \\ y_n &= x_{n+1} + \frac{a}{b^2}y_{n+1}^2 - 1 \ . \end{aligned}$$

Using these last two equations we can compute the *backward orbit* for any initial point (x_0, y_0), i.e., (x_{-k}, y_{-k}) for $k = 0, 1, 2, ...$ It is not a proven fact, but from numerical studies it seems apparent that all backward orbits must escape to infinity except those started in the attractor or in an unstable invariant set. Since these exceptional points cover a region of area 0, this case should almost never be observed numerically when iterating backwards. Even the smallest roundoff error in the course of the computation will throw such an orbit off the true one, outside of the attractor, and ultimately

[10]This means that we replace each point in the Cantor set by a small vertical line segment.

to infinity. More precisely, orbits diverge such that $y_k \to \infty$ as $k \to -\infty$.

The Fractal Dimension

Again, the result of the magnifications in figure 12.12 shows a Cantor-like structure of parallel lines. The attractor definitely is a fractal. We can compute the box-counting dimension as usual, overlaying the attractor with grids of varying lattice sizes and counting cells which cover part of the attractor (see chapter 4). The result is a dimension of about 1.28, a value well above 1. It may be a surprise that this number is not the same as the dimension of the Cantor set, raised by 1, i.e.,

$$1 + \frac{\log 2}{\log 3} \approx 1.6309 \; .$$

However, $\log 2 / \log 3$ is only the dimension of one particular Cantor set, namely the standard one obtained by recursively deleting the middle thirds of intervals. We can change the construction by subdividing each interval into p equal parts of which we keep the first and last one while deleting all the others. This produces a Cantor set with dimension $\log 2 / \log p$. For example, for $p = 12$ we get that the dimension raised by 1 is

$$1 + \frac{\log 2}{\log 12} \approx 1.2789 \; .$$

which is much closer to the numerically computed dimension of the Hénon attractor.

Characterization of Strange Attractors

Up to this point we have discussed a number of properties of the Hénon transformation. Let us summarize: there is a trapping region within which all initial points have orbits leading to the attractor. These orbits show sensitive dependence on initial conditions, and a single orbit seems to get close to all points of the attractor. Moreover, the attractor exhibits a fractal structure. These are the four chief properties of strange attractors which we note in the list below. Although strange attractors typically exist in spaces of more than two dimensions, we will restrict our presentation for simplicity to the two-dimensional case applicable to the Hénon attractor. Thus, let $T(x, y)$ be a given transformation in the plane with coordinates x and y. A bounded subset A of the plane is a chaotic and strange attractor for the transformation T if there exists a set R with the following properties.[11]

Four Properties

1. **Attractor.** R is a neighborhood of A, i.e., for each point (x, y) in A there is a small disk centered at (x, y) which is contained in R. This implies in particular that A is in R. R is a trapping region, i.e., each orbit started in R remains in R for all iterations.

[11] Except for part 3, which we added here, this definition has been given in D. Ruelle, *Strange Attractors*, Math. Intelligencer 2 (1980) 126–137.

Moreover, the orbit becomes close to A and stays as close to it as we desire. Thus, A is an *attractor.*
2. **Sensitivity.** Orbits started in R exhibit sensitive dependence on initial conditions. This makes A a *chaotic* attractor.
3. **Fractal.** The attractor has a fractal structure and is therefore called a *strange attractor.*
4. **Mixing.** A cannot be split into two different attractors.[12] There are initial points in R with orbits that get arbitrarily close to any point of the attractor A.

We need to point out that the above attempt at a definition is indeed only a first try. The discussion about what should be the most appropriate definition mathematically is still going on, and it seems that we will have to wait for the final clarification until some kind of breakthrough in understanding strange attractors has been achieved.[13]

In fact, the situation is even worse. Up to now no one knows whether the attractor in Hénon's transformation for $a = 1.4$ and $b = 0.3$ really is a strange attractor according to the above or a similar definition even though very extensive numerical checks have been performed which all indicate a positive answer. This underlines the incomplete state of affairs. For example, we could speculate that the experimental observations are due to an attractive periodic orbit with a *very* long period. If that is really the case, then we will never be able to compute that period because rounding errors in the computation will disturb the orbit too much.

Lozi's Piecewise Linear Model

However, the situation is perhaps not as bleak as it may now appear. One first step towards a solution of the problem has been carried out by Michal Misiurewicz.[14] He proved an earlier conjecture by René Lozi,[15] who hypothesized that a simplified version of Hénon's transformation in fact admits a strange attractor. The transformation is given by

$$\tilde{H}(x, y) = (1 + y - a|x|, bx) \ , \tag{12.2}$$

and Lozi suggested the parameter values $a = 1.7$ and $b = 0.5$ (see figure 12.13). The only difference between Lozi's and Hénon's transformation is that the x^2 term is replaced by $|x|$. The fact that this modification makes the transformation linear for $x > 0$ and $x < 0$ allowed Misiurewicz to complete the analysis and the confirmation of a strange attractor.

---

[12] Caution; this does not imply that the attractor must be a connected set.

[13] See the discussion on pages 255–259 in J. Guckenheimer and P. Holmes, *Nonlinear Oscillations, Dynamical Systems, and Bifurcations of Vector Fields,* Springer-Verlag, New York, 1983.

[14] M. Misiurewicz, *Strange Attractors for the Lozi Mappings,* in *Nonlinear Dynamics,* R. H. G. Helleman (ed.), Annals of the New York Academy of Sciences 357 (1980) 348–358.

[15] R. Lozi, *Un attracteur étrange (?) du type attracteur de Hénon,* J. Phys. (Paris) 39 (Coll. C5) (1978) 9–10.

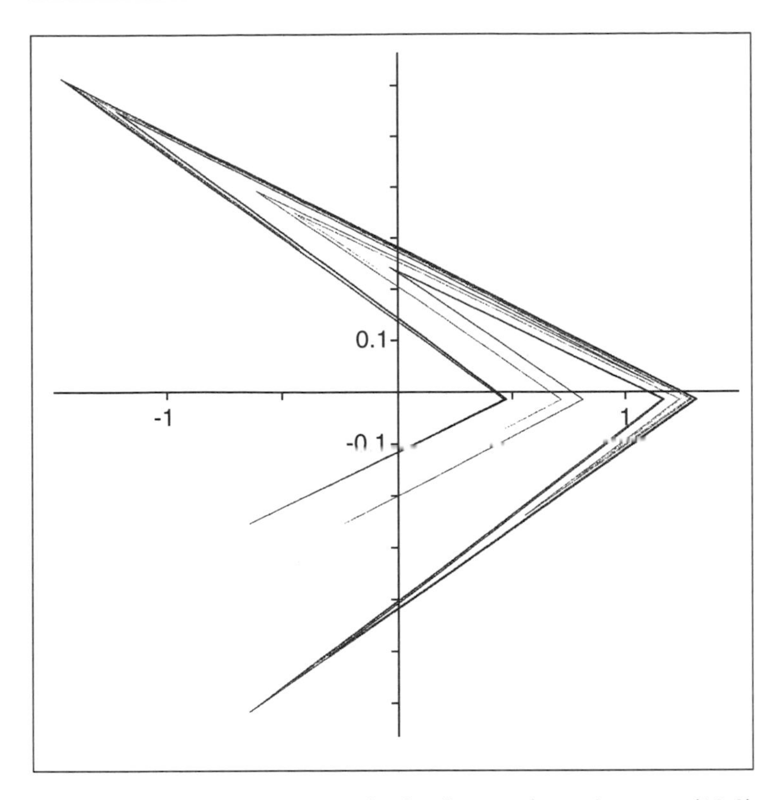

Figure 12.13 : The attractor for Lozi's transformation eqn. (12.2) with $a = 1.7$ and $b = 0.5$. The region shown is $-1.5 \leq x \leq 1.5$ and $-0.75 \leq y \leq 0.75$. 100,000 points are plotted.

The Dynamics

One of the purposes of the Hénon transformation is to provide a generalization of the quadratic iterator to two dimensions. In fact, for the special choice $b = 0$, the Hénon transformation reduces to

$$\begin{aligned} x_{n+1} &= 1 - ax_n^2 \\ y_{n+1} &= 0 \ . \end{aligned}$$

The two coordinates are decoupled, and only the x-coordinate is relevant for the dynamics for $b = 0$. This iterator

$$x \rightarrow 1 - ax^2$$

is quadratic and equivalent to the logistic iterator

$$z \rightarrow rz(1 - z)$$

where we have chosen the symbol r for the parameter to avoid confusion with the parameter a from Hénon's transformation. Therefore, in the case $b = 0$, the Hénon transformation presents just another version of the Feigenbaum scenario (see chapter 11) of the quadratic iterator as the parameter a varies.

Equivalence of $1 - ax^2$ and $rz(1-z)$

In chapter 1 we identified the equivalence of the quadratic iteration $x \to x^2 + c$ with the logistic iterators $p \to p + rp(1+p)$ and $z \to rz(1-z)$ (see pages 63 and 68). Here we proceed along the same lines of argument to show that also the family $x \to 1 - ax^2$, which is nothing but Hénon's transformation eqn. (12.1) for the special choice $b = 0$, is equivalent to the other quadratic iterators. We show that

$$x_{n+1} = 1 - ax_n^2 \tag{12.3}$$

is identical to

$$z_{n+1} = rz_n(1 - z_n) \tag{12.4}$$

when using the setting

$$x_n = \frac{r}{a}(z_n - \frac{1}{2}) \quad \text{and} \quad a = \frac{r(r-2)}{4} \ . \tag{12.5}$$

To verify this statement, we compute x_{n+1} in terms of z_n using equations (12.3) and (12.5) and then compare the result with what we get from equations (12.4) and (12.5). This yields

$$\begin{aligned} x_{n+1} &= 1 - ax_n^2 = 1 - a\frac{r^2}{a^2}\left(z_n - \frac{1}{2}\right)^2 \\ &= -\frac{r^2}{a}z_n^2 + \frac{r^2}{a}z_n + 1 - \frac{r^2}{4a} \end{aligned}$$

and on the other hand

$$\begin{aligned} x_{n+1} &= \frac{r}{a}\left(z_{n+1} - \frac{1}{2}\right) = \frac{r}{a}\left(rz_n(1-z_n) - \frac{1}{2}\right) \\ &= -\frac{r^2}{a}z_n^2 + \frac{r^2}{a}z_n - \frac{r}{2a} \ . \end{aligned}$$

It remains to be shown that

$$1 - \frac{r^2}{4a} = -\frac{r}{2a}$$

which readily follows from $a = r(r-2)/4$. This concludes the proof.

Thus, the case $b = 0$ does not provide anything new. But what happens when $b \neq 0$? Several interesting questions come up. For example, what effect does this choice have on the dynamics? Is there still the Feigenbaum scenario present? If so, what about the universal Feigenbaum constant? How does the Hénon attractor fit into this picture? And are there perhaps other attractors possible besides those indicated by the Feigenbaum scenario? Since the

Hénon transformation is so simple to implement on a computer, these questions can be followed up experimentally to some degree.

The Tangent Bifurcation

We begin by studying the fixed points of the Hénon transformation given by the solutions to the systems of equations

$$x = 1 + y - ax^2$$
$$y = bx \ .$$

There are two solutions (x_1, y_1) and (x_2, y_2), namely

$$x_{1,2} = \frac{b - 1 \pm \sqrt{(b-1)^2 + 4a}}{2a}$$
$$y_{1,2} = bx_{1,2} \ .$$

From now on we will consider only the choice $b = 0.3$, which is the parameter in the attractor suggested by Hénon. The discriminant $(b-1)^2 + 4a$ is negative, if

$$a < a_0 = -\frac{(b-1)^2}{4} = -0.1225 \ .$$

No fixed points exist in this case. Only when a grows above $a_0 = -0.1225$, do both fixed points (x_1, y_1) and (x_2, y_2) exist and the first one of them is attracting. This is exactly the scenario of the tangent bifurcation (or saddle-node bifurcation) discussed in chapter 11, page 255.

The Period-Doubling Cascade

k	Period	Parameter a_k	δ
0	1	−0.122 5	
1	2	0.367 5	
2	4	0.912 5	4.844
3	8	1.026	4.3269
4	16	1.051	4.696
5	32	1.056 536	4.636
6	64	1.057 730 83	4.7748
7	128	1.057 980 893 1	4.6696
8	256	1.058 034 452 15	4.6691
9	512	1.058 045 923 04	4.6691
10	1024	1.058 048 379 80	4.6694
11	2048	1.058 048 905 931	

Table 12.14 : Parameters a_k for the period-doubling cascade in Hénon's transformation (12.1) for $b = 0.3$. This sequence of parameters is approximately geometric. The last column specifies the estimates for the corresponding number δ. In the kth row this estimate is computed by $(a_k - a_{k-1})/(a_{k+1} - a_k)$.

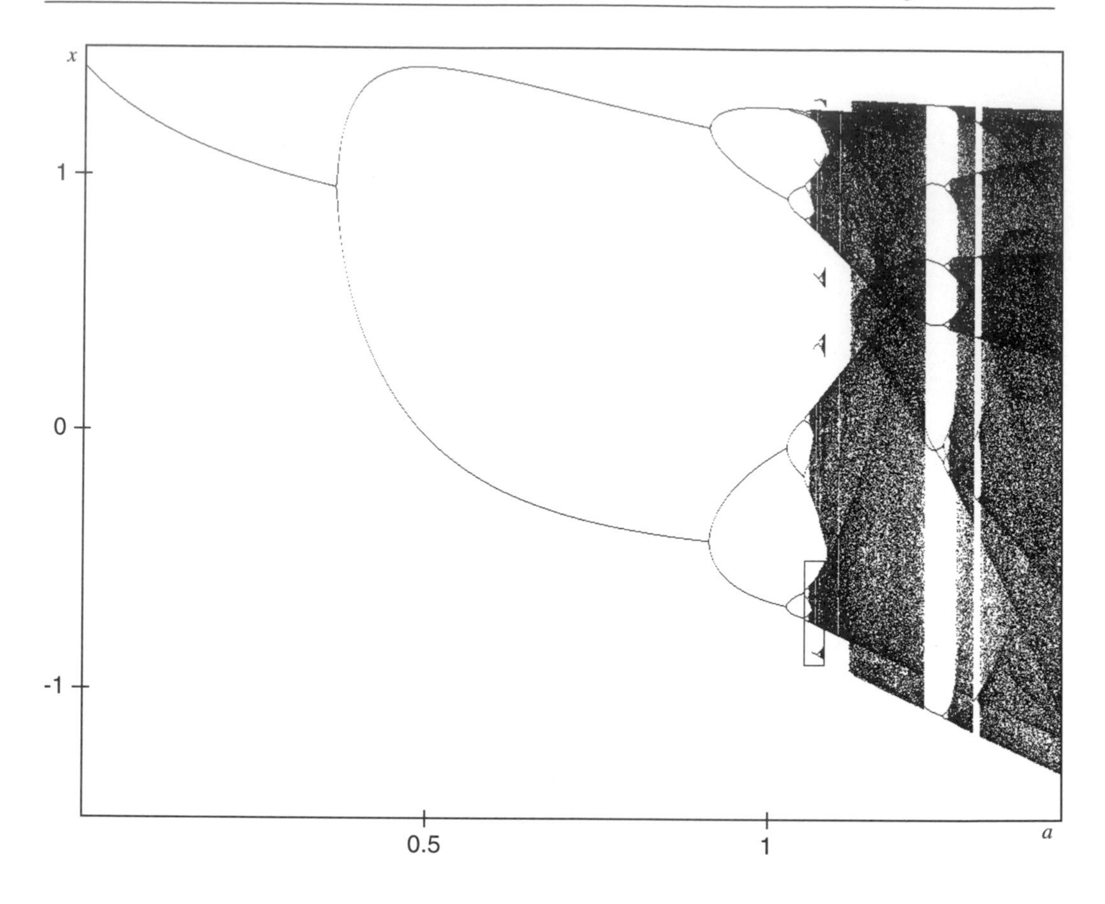

Figure 12.15 : We fix the parameter $b = 0.3$ in Hénon's transformation (12.1). The graph has as horizontal axis the range of parameters $0 < a < 1.5$. For each parameter value of a several initial points were taken and iterated. Note the spurious small specs at parameter value about 1.08. The bottom one is in the the boxed detail. This is enlarged in figure 12.16 revealing that the small spec corresponds to another complete Feigenbaum scenario.

The Feigenbaum Scenario

When the parameter a further increases, the attractive fixed point eventually becomes unstable and gives rise to an attractive cycle of period 2. This transition happens at the value $a_1 = 0.3675$.[16] As expected, the whole sequence of period-doubling bifurcations which is so familiar from the quadratic iterator also appears here in the Hénon transformation for $b = 0.3$ as a increases (see fig-

[16]This value can be derived explicitly using the linearization (i.e., the derivative) of Hénon's transformation at the attractive fixed point (x_1, y_1). Its eigenvalues are given by $\lambda_{1,2} = -ax_1 \pm (a^2x_1^2 + b)^{1/2}$, and precisely at $a_1 = 3(b-1)^2/4 = 0.3675$ the second eigenvalue passes through $\lambda_2 = -1$, which signals a period-doubling bifurcation.

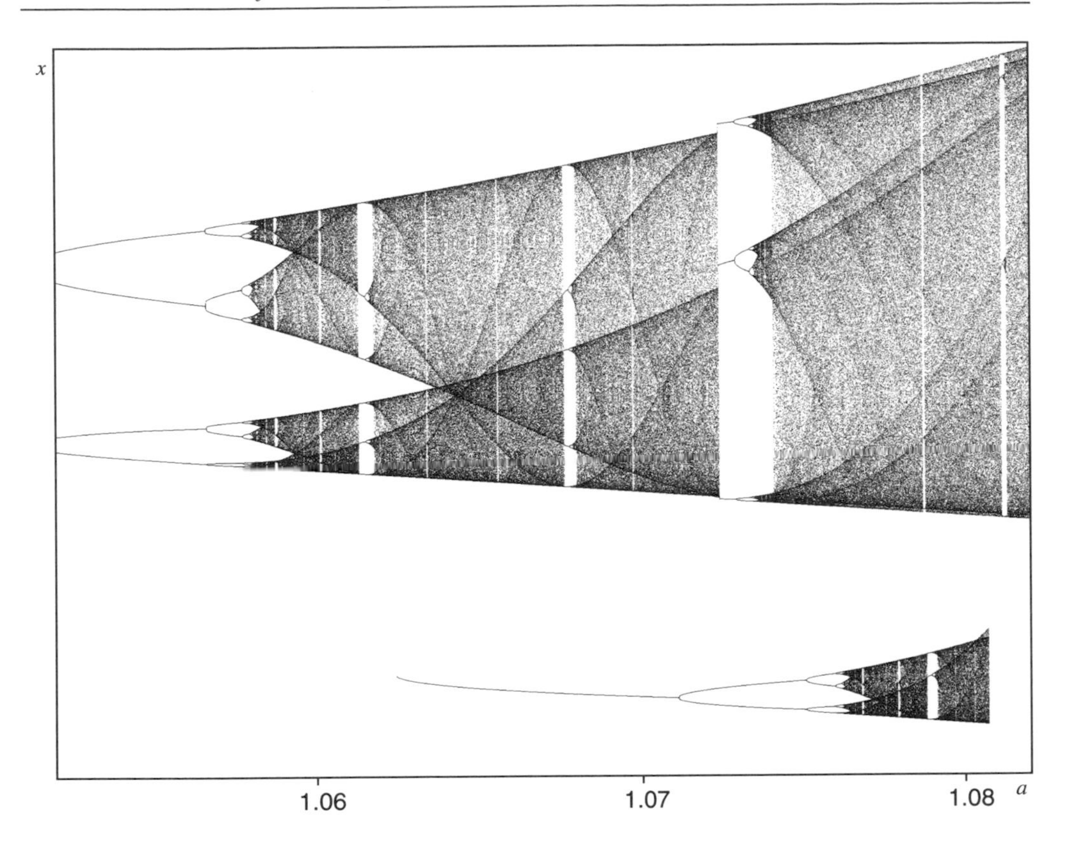

Figure 12.16 : The small boxed region in figure 12.15 is enlarged indicating that there are two coexisting attractors which are shown in the following figure.

ure 12.15 and table 12.14).[17] The sequence of parameters for the period-doubling is almost geometric as for the quadratic iteration. And the table confirms that the 'universal' Feigenbaum constant 4.6692... is also encountered in this two-dimensional system.

Two Attractors

We are tempted to conjecture that everything is the same as in the one-dimensional case. But this is wrong. One important feature which can only arise here is included in the final state diagram. This is the small version of the Feigenbaum tree in the lower part of the diagram in figure 12.16. It corresponds to a separate attractor. Thus, for the parameters about 1.08 there are two attractors with two corresponding basins of attraction (see figure 12.17). This cannot happen in the one-dimensional quadratic transformation.[18]

[17]The table is adapted from B. Derrida, A. Gervois, Y. Pomeau, *Universal metric properties of bifurcations of endomorphisms,* J. Phys. A: Math. Gen. 12, 3 (1979) 269–296.

[18]A very extensive analysis of the scenario of appearing and disappearing strange attractors and periodic orbits has

Coexistence of Two Attractors

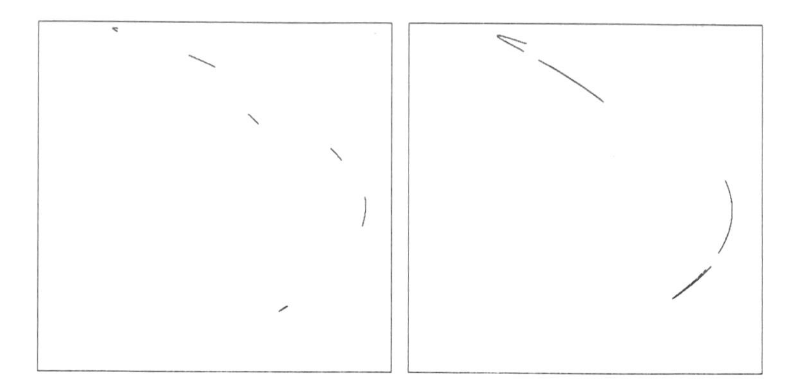

Figure 12.17 : For $a = 1.08$, $b = 0.3$ two attractors coexist. The left attractor belongs to the lower piece of figure 12.16 (initial point $(0.5, 0.0)$) while the right attractor corresponds to the main branch in the final state diagram (initial point $(-0.1111, -0.105555)$).

The Next Step

A spectacular success towards proving what the computer experiments seem to reveal were reported by Michael Benedicks and Lennart Carleson in 1991.[19] They were able to show that the chaotic dynamics which are present in the case $b = 0$ carry over to small values $b > 0$. For each sufficiently small $b > 0$ there are many parameters a such that the dynamics are beyond any doubt chaotic. In fact, when choosing parameter a at random, there is a positive probability that we get chaotic dynamics. The proof, however, does not reveal what these parameter values precisely are. And perhaps we will never know.

been carried out in C. Simó, *On the Hénon-Pomeau attractor,* Journal of Statistical Physics 21,4 (1979) 465–494. See also F. R. Marotto, *Chaotic behavior in the Hénon mapping*, Comm. Math. Phys. 68 (1979) 187–194.

[19]M. Benedicks and L. Carleson, *The dynamics of the Hénon map,* Annals of Mathematics 133,1 (1991) 73–169.

12.2 Continuous Dynamical Systems: Differential Equations

Differential equations have become the language in which modern science encodes the laws of nature. The victories of this approach are numerous. They reach from the laws governing the motion of the planets, to the laws of electromagnetism describing the orbit of an elementary particle in an accelerator, to the air flow carrying an airplane, to the models for the generation of blood cells in the bone marrow, to the mathematical model for numerical weather prediction. The mathematics of differential equations is not elementary. It is one of the great achievements made possible by calculus. Lorenz's discovery of a strange attractor was made in the numerical study of a set of differential equations which he had refined from mathematical models used for testing weather prediction. Although the topic of differential equations is some 300 years old and the results have filled libraries, nobody would have thought it possible that differential equations could behave as chaotically as Lorenz found in his experiments. Moreover, the computer was more than an aid in Lorenz' discovery; it was absolutely crucial, as crucial as for the discovery of Feigenbaum's universality. Up to today a rigorous mathematical understanding of Lorenz' discovery is still open. Lorenz is a meteorologist with a strong mathematical background. In fact one of his teachers at Harvard was George D. Birkhoff who was one of the historical fathers of modern chaos theory.

It is not our aim to introduce a theory of differential equations in this book. After all, we have tried to carefully avoid using methods of calculus wherever we thought it was possible. This is not because we do not appreciate calculus. How could we not. In fact, much of our own research work has been and is in differential equations. But we felt that we should try to explain the major thoughts of fractals and chaos without reference to that more advanced branch of mathematics as far as we could. Here we cannot avoid the subject any more, and we will try to provide a glimpse at what differential equations are about, though without history and any breadth.

Time Steps for the Logistic Equation

Since Lorenz' discovery was made in numerical simulations of differential equations, we think it is a good idea to approach differential equations from the numerical side. We will do that with an old friend: the logistic equation

$$p_{n+1} = p_n + rp_n(1 - p_n) \tag{12.6}$$

with initial value p_0. We rewrite this equation in the form

$$p_{n+1} - p_n = rp_n(1 - p_n) \ . \tag{12.7}$$

Here the left hand side of the equation is the growth of the population from one generation to the next, and p_n denotes the population size at time n, where time is measured in generations. Let us now use a slightly different way to indicate the same dependence substituting $p(n)$ for p_n. When time is measured in discrete steps like $n = 1, 2, 3, \ldots$ we prefer the old notation. We now interpret time as a continuous entity and choose the alternative notation. Let us assume we are given the population sizes $p(t)$ and $p(t + \Delta t)$, where Δt denotes a small increment. Based on these numbers we can estimate the total population change per unit time, i.e., per generation, as

$$\frac{p(t + \Delta t) - p(t)}{\Delta t} \tag{12.8}$$

When substituting t for n in this expression for the left side in eqn. (12.7) we obtain a modified logistic equation

$$\frac{p(t + \Delta t) - p(t)}{\Delta t} = rp(t)(1 - p(t)) \ .$$

Note that in eqn. (12.7) $\Delta t = 1$.

Computing Time Series for Different Step Sizes

Now we can compute the population size $p(t + \Delta t)$ from that at time t by solving the above equation,

$$\begin{aligned} p(t + \Delta t) &= p(t) + \Delta t r p(t)(1 - p(t)) \\ p(0) &= p_0 \ . \end{aligned} \tag{12.9}$$

Given an initial value p_0, a parameter r, and the time step size Δt, this formula generates the complete list of population sizes $p(k\Delta t)$, $k = 0, 1, 2, \ldots$ Figure 12.18 provides graphs for $r = 3$. The plots show population sizes versus time for four different step sizes and three initial points each. As Δt becomes smaller and smaller, the resulting iteration will converge towards 1 in smaller and smaller steps. In fact, the time series which we will observe will look more and more like a smooth curve and there is a deep reason for that. As $\Delta t \to 0$ the iteration in eqn. (12.9) makes a transition into the world of differential equations and in turn can be seen as a numerical approximation of that differential equation. In fact, what we are looking at is the famous Euler's method, which is only one of a whole variety of numerical schemes.

Transition to the Differential Equation

Let us assume that the population is developing along a smooth curve $p(t)$. Then the estimate (12.8) for the population change per unit time can be interpreted graphically as the slope of the secant going through the points $(t, p(t))$ and $(t + \Delta t, p(t + \Delta t))$ (see figure 12.19). As we let $\Delta t \to 0$, the resulting secants appear to approximate the tangent to the curve $p(t)$ in the point $(t, p(t))$. The slope of that tangent would be obtained as we let $\Delta t \to 0$ in

Verhulst Dynamics

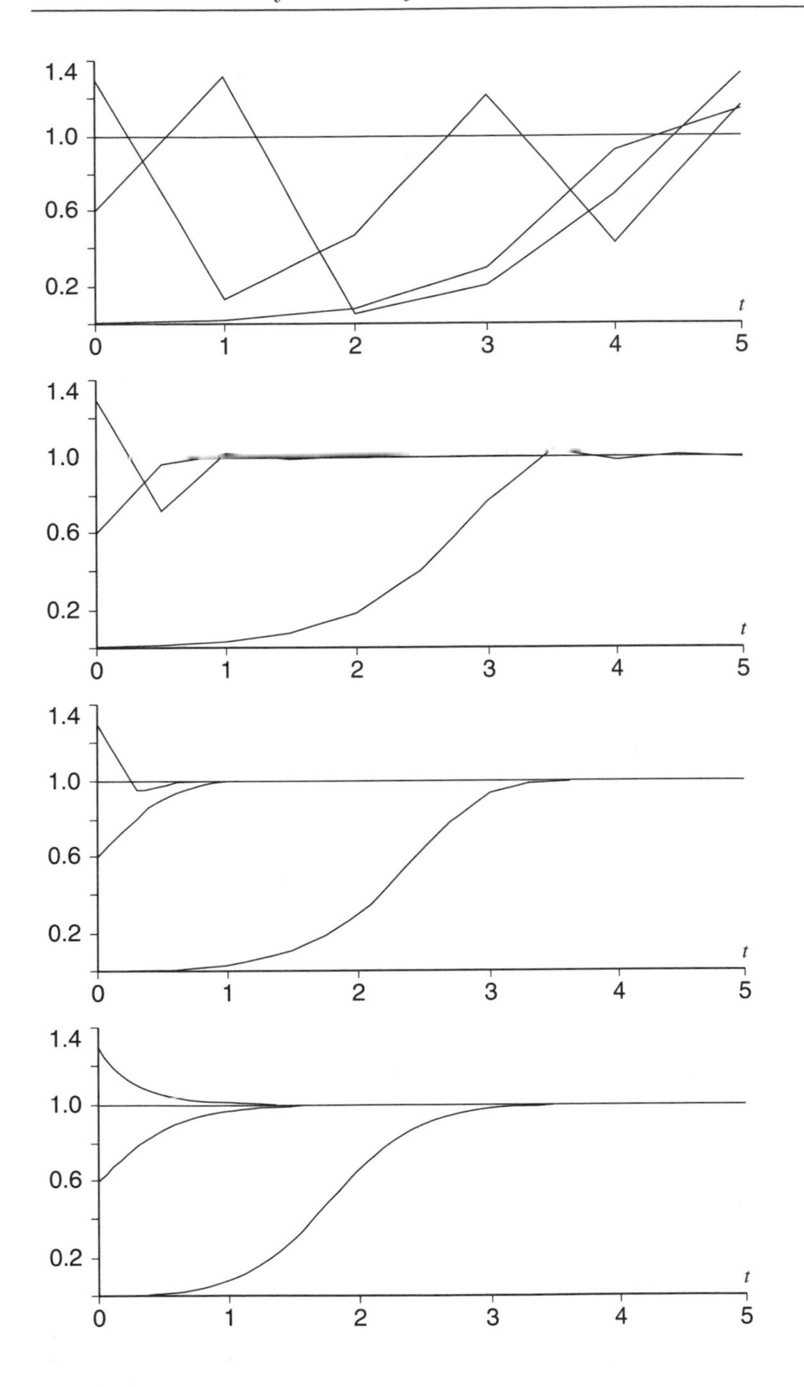

Figure 12.18 : In each diagram, several initial conditions are taken and iterated. The step sizes are $\Delta t = 1.0, 0.5, 0.1, 0.001$ (from top to bottom).

Secant and Tangent

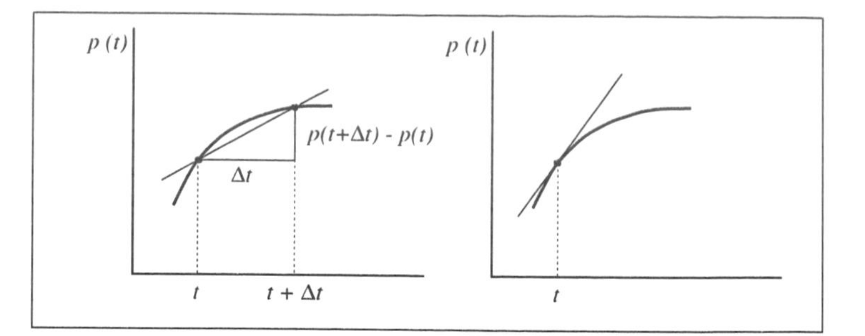

Figure 12.19 : Slope of secant and tangent as limit of secants as $\Delta t \to 0$.

eqn. (12.8). In fact, we say that $p(t)$ has a tangent in $(t, p(t))$ with slope m provided the limit

$$\lim_{\Delta t \to 0} \frac{p(t+\Delta t) - p(t)}{\Delta t} = m$$

exists. This limit is denoted by $p'(t)$ and is called the *derivative* of $p(t)$ at t. Replacing the expression $p(t + \Delta t) - p(t)/\Delta t$ by the derivative $p'(t)$, we arrive at the limit equation, which is a *differential equation* for $p(t)$,

$$\begin{aligned} p'(t) &= rp(t)(1 - p(t)) \\ p(0) &= p_0 \end{aligned} \tag{12.10}$$

In other words, eqn. (12.10) means that we are looking for a function $p(t)$, such that $p(0) = p_0$ and such that for each $t > 0$ the graph $(t, p(t))$ of $p(t)$ has a tangent with slope $rp(t)(1 - p(t))$. Such a function $p(t)$ is called a solution of the differential equation (12.10) with initial value $p(0) = p_0$. A problem like eqn. (12.10) is called an *initial value problem.* This particular equation can be solved analytically. The result is the same regardless of the parameter r; the solution for any given initial value $p_0 > 0$ monotonically[20] tends to the fully saturated population size $p = 1$. This is illustrated in the bottom part of figure 12.18, which shows close approximations to the true solutions.

Solving the Differential Equation

Since the right hand side of eqn. (12.10) is nonlinear, this is an example of a nonlinear differential equation. Usually it is hard or impossible to provide an explicit solution for nonlinear differential equations, while there is a complete theory to solve linear differential

[20]If $p_0 < 1$, then the solution $p(t)$ is a monotonically increasing function. For initial value $p_0 > 1$ it is monotonically decreasing. In both cases $\lim_{t\to\infty} p(t) = 1$.

equations. In our case we are lucky to be able to relate eqn. (12.10) to a linear differential equation by introducing the variable

$$x(t) = \frac{1}{p(t)} .$$

Then formal calculation using eqn. (12.10) gives

$$x'(t) = -\frac{p'(t)}{(p(t))^2} = -\frac{rp(t)(1-p(t))}{(p(t))^2} = -rx(t) + r .$$

When setting $x_0 = 1/p_0$ we obtain the initial value problem

$$\begin{aligned} x'(t) &= -rx(t) + r \\ x(0) &= x_0 \end{aligned}$$

This linear differential equation has the explicit solution

$$x(t) = 1 - (1 - x_0)e^{-rt} .$$

From this formula we read off that if $0 < x_0 < 1$, then $0 < x(t) < 1$ for all times $t > 0$ and $r > 0$. Moreover, the solution approaches the constant 1, $x(t) \to 1$ as $t \to \infty$. Likewise, if $x_0 > 1$, then $x(t) > 1$ for all $t > 0$ and again $x(t) \to 1$ as $t \to \infty$. This translates directly to the populations $p(t) = 1/x(t)$, the solutions of the differential equation (12.10). We have

$$p(t) = \frac{1}{x(t)} = \frac{1}{1 - (1 - \frac{1}{p_0})e^{-rt}} = \frac{p_0 e^{rt}}{p_0 e^{rt} - p_0 + 1}$$

and in either case the population will go into saturation, $p(t) \to 1$. The various choices of $r > 0$ will only affect how fast saturation is reached.

True Solution Versus Approximation

This is dramatically different from the discrete model eqn. (12.9), where we see that the numerical solution substantially depends on the choice of the step size Δt. If the step size is large, as in the upper plot of figure 12.18, the solution is not momotonic and does not even approach the saturation $p = 1$. Let us explain why this is the case. Compare the numerical solution

$$p(t + \Delta t) = p(t) + \Delta t r p(t)(1 - p(t))$$

with the original logistic equation

$$p_{n+1} = p_n + rp_n(1 - p_n)$$

Note that these formulas coincide for $\Delta t = 1$. Or interpreted differently, that means that the growth law for arbitrary time steps Δt reduces to the original Verhulst model by replacing the expression $\Delta t r$ by the parameter r. We know quite well from chapters 10

Graphical Iteration

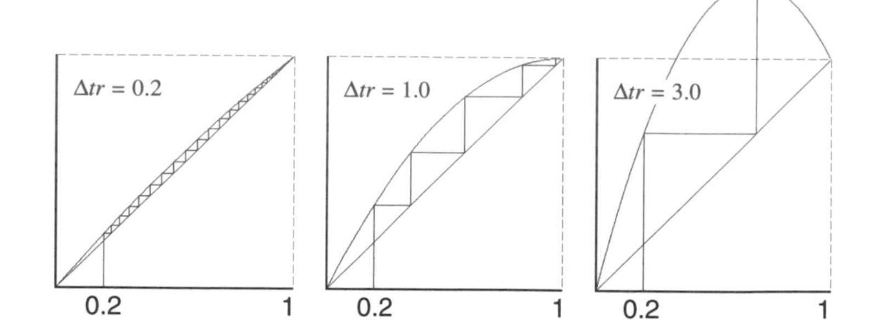

Figure 12.20 : Iteration of $p(t + \Delta t) = p(t) + \Delta trp(t)(1 - p(t))$ for different values of Δtr which correspond to the values used in the upper three plots in figure 12.18.

and 11 that changing the parameter r may dramatically effect the behavior of the corresponding orbits. In other words, if we let $r = 3$, for example,[21] then we have chaotic orbits for $\Delta t = 1$, while we have orbits converging to 1 for $0 < \Delta t < 2/3$. Look at figure 12.20 to understand this change in behavior in terms of graphical iteration.

Stability Condition

In terms of the numerical approximation for the differential equation this means that for a given parameter r we have to restrict the step size Δt so that

$$\Delta tr < 2 \quad \text{or} \quad \Delta t < \frac{2}{r} \tag{12.11}$$

in order to achieve at least the convergence to the saturation $p = 1$. A condition like this is called a *stability condition* for the numerical approximation. Conditions of this sort were first observed by Richard Courant, Hans Lewy, and Kurt Otto Friedrichs and were the initial impulse in the development of the field of *numerical stability analysis*.

The relation of the differential equation (12.10) to its numerical approximation in eqn. (12.6) is very delicate. The stability condition eqn. (12.11) is a reflection of that. Another one lies in the fact that the continuous model eqn. (12.10), though formally nonlinear, is actually related to a linear problem, much unlike the discrete approximation eqn. (12.6) which is really nonlinear. This shows dramatically that passing to limits — or passing to discrete approximations — may change the nature of a problem significantly, a fact which has only entered the conscience of numerical analysts quite recently. This is another merit of chaos theory.

[21]Recall that the logistic equation is equivalent to the quadratic iterator $x_{n+1} = ax_n(1 - x_n)$ using $x_n = rp_n/(r + 1)$ and $a = r + 1$ (see chapter 1, page 68).

Numerical Methods

Differential equations such as $x'(t) = f(x(t))$ are one of the most important tools for modeling processes in the natural sciences, particularly in physics. Thus, there is a continuously growing body of mathematical research on the various different types of differential equations. However, for most equations considered, there is no known solution that we could write down in terms of a common formula. Only the numerical approximation with the help of the computer is possible; and, in fact, widely used. There are plenty of numerical methods available for this task. The first one, given by

$$x_{k+1} = x_k + \Delta t f(x_k) ,$$

which we have already introduced above. It is called *Euler's method*, and the step from one point x_k to the following, x_{k+1} is called an *Euler step*. The smaller the step size Δt the more accurate is the solution. Euler's method is of first order, meaning that the error of the numerical solution relative to the true solution of the differential equation is proportional to the step size Δt. In a second order method, the error is proportional to the square Δt^2. Such methods are much superior to Euler's method because here the error decreases by a factor of $1/4$ each time the step size is halved. In other words, in order to achieve the same precision we may choose a much larger step size and thus may save a lot of computational effort in comparison to Euler's method. As an example for a second order method we mention

$$x_{k+1} = x_k + \frac{\Delta t}{2}(f(x_k) + f(x_k + \Delta t f(x_k))) ,$$

which is called the *trapezoidal* or *Heun's method.* Both methods may also be used when the variable x is a vector of several components, which is the case considered in the following. Almost all text books on numerical methods present an analysis of such methods and some others. There exist many computer codes for higher order methods which even adjust the step size to the local properties of the solution.[22]

No Chaos in Dimensions 1 and 2

One of the crucial differences between the discrete system eqn. (12.6) and the continuous counterpart eqn. (12.10) is the fact that it is plainly impossible for the dynamics of the differential equation to be chaotic. The reason is that no two trajectories (as shown in the lower part of figure 12.18 when $\Delta t \to 0$) can cross each other. This is important.

Life in the real world, however, is not so simple. In almost any physical system the state cannot be described by a single variable such as a population in the Verhulst model. Usually there are two, three and more (sometimes many more) variables necessary.

[22]For most figures and computations presented in this book, we have used the code in *Numerical Recipes in C* by W. H. Press, B. P. Flannery, S. A. Teukolsky, W. T. Vetterling, Cambridge University Press, 1988.

For example, with a pendulum which is allowed to swing in a plane, we need to know angle and angular momentum (besides the configuration of the pendulum) in order to be able to forecast the motion. Or to compute the orbit of an asteroid, we need to know position and velocity, each one of which has three components for each of the spatial directions, giving a total of six variables. It turns out that in order to find chaos in such continuous dynamical systems we need to consider at least three-dimensional systems. As in the one-dimensional case, trajectories must not cross; and this implies that trajectories in a plane cannot act chaotically. They typically converge to a point, or escape to infinity, or perhaps spiral around closing in on some loop.[23] But in three dimensions chaos may reign; and it is the general consensus that in dynamical systems in nature, chaos is typical rather than the exception.

Approximation Method in 3 Dimensions

Let us therefore now briefly introduce a more general viewpoint, which is necessary to comprehend Lorenz' work. Consider the differential equation

$$\begin{aligned} x'(t) &= f(x(t)) \\ x(0) &= x_0 \end{aligned}$$

where $f(x)$ is some function. How would we think about this equation? Just like we did in the above special case. We would use the numerical Euler approximation

$$\begin{aligned} \frac{x(t+\Delta t) - x(t)}{\Delta t} &= f(x(t)) \\ x(0) &= x_0 \end{aligned}$$

for Δt small and pass to the limit $\Delta t \to 0$. With this in mind we can easily go one step further and look at systems of differential equations. Let us take real-valued functions $f(x, y, z)$, $g(x, y, z)$, and $h(x, y, z)$ and consider the system

$$\begin{aligned} x'(t) &= f(x(t), y(t), z(t)), \quad & x(0) &= x_0 \\ y'(t) &= g(x(t), y(t), z(t)), \quad & y(0) &= y_0 \\ z'(t) &= h(x(t), y(t), z(t)), \quad & z(0) &= z_0 \,. \end{aligned}$$

As for one equation, we would just set up a system of numerical

[23] The basic tool for understanding planar dynamical systems is the Poincaré-Bendixson theory (see chapter 11 in M. W. Hirsch and S. Smale, *Differential Equations, Dynamical Systems, and Linear Algebra,* Academic Press, New York, 1974).

approximations

$$\begin{aligned}
\frac{x(t+\Delta t)-x(t)}{\Delta t} &= f(x(t),y(t),z(t)), \quad x(0)=x_0 \\
\frac{y(t+\Delta t)-y(t)}{\Delta t} &= g(x(t),y(t),z(t)), \quad y(0)=y_0 \\
\frac{z(t+\Delta t)-z(t)}{\Delta t} &= h(x(t),y(t),z(t)), \quad z(0)=z_0 \ ,
\end{aligned}$$

and construct solutions by letting $\Delta t \to 0$.

12.3 The Rössler Attractor

The Rössler Model

In 1976 Otto E. Rössler found a particularly simple system, which is probably the most elementary geometric construction of chaos in continuous systems. Thus, before we start to discuss the Lorenz attractor in detail, let us follow Rössler's ideas.[24] His system of differential equations is

$$\begin{aligned} x' &= -(y+z) \\ y' &= x + ay \\ z' &= b + xz - cz \end{aligned} \tag{12.12}$$

where the three coefficients a, b, c are adjustable constants. In this section, we will fix the parameters a and b and change only c.

$$a = 0.2, \quad b = 0.2.$$

A Trajectory

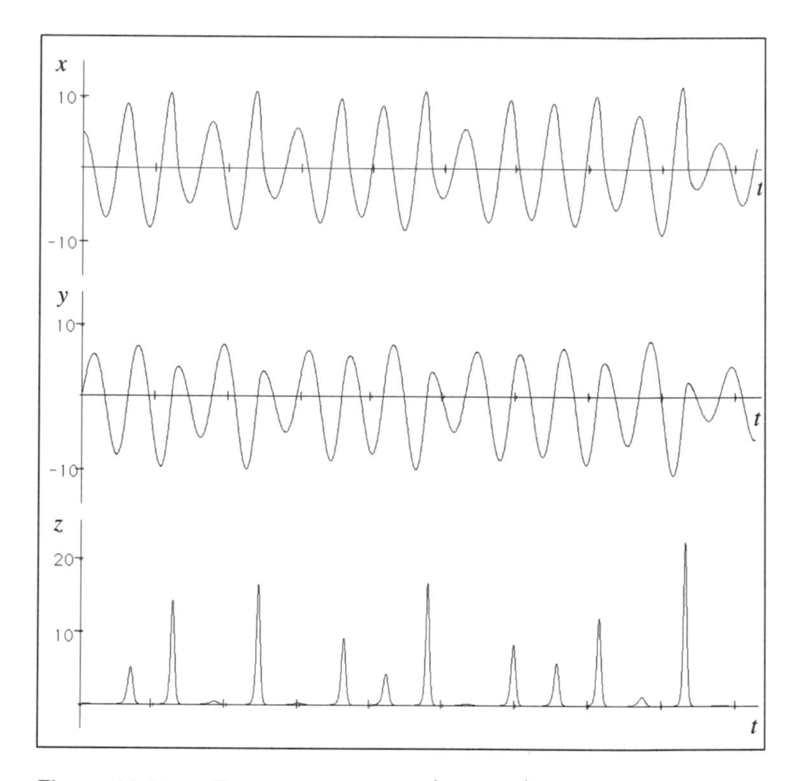

Figure 12.21 : The initial condition $(-1, 0, 0)$ for Rössler's system eqn. (12.12) produces these plots of $x(t)$, $y(t)$ and $z(t)$ versus time t. The parameter is $c = 5.7$.

[24]Rössler, O. E., *An equation for continuous chaos,* Phys. Lett. 57A (1976) 397–398.

The Rössler Attractor

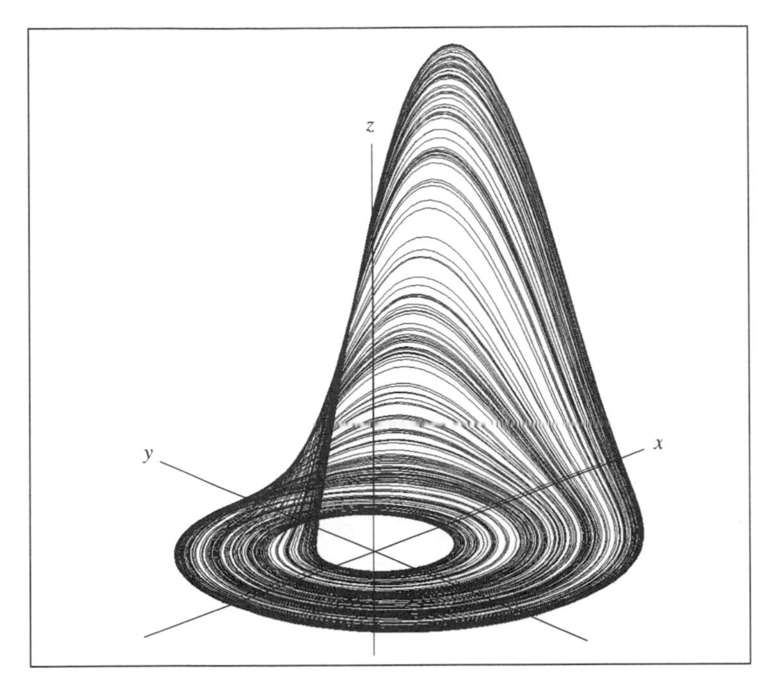

Figure 12.22 : The trajectory from figure 12.21 is plotted in three-dimensional space revealing a first picture of the Rössler attractor. Two projections are given in figure 12.23.

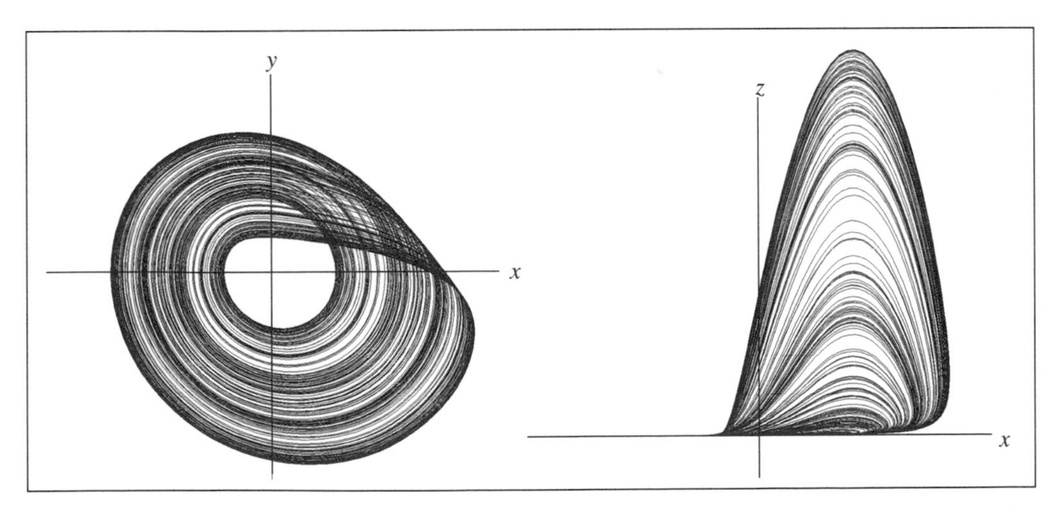

Figure 12.23 : The Rössler attractor, top view (left) and side view (right).

This system is the same type as that of the continuous Verhulst system in eqn. (12.10), with the difference that here we are given

Dynamics on the Attractor

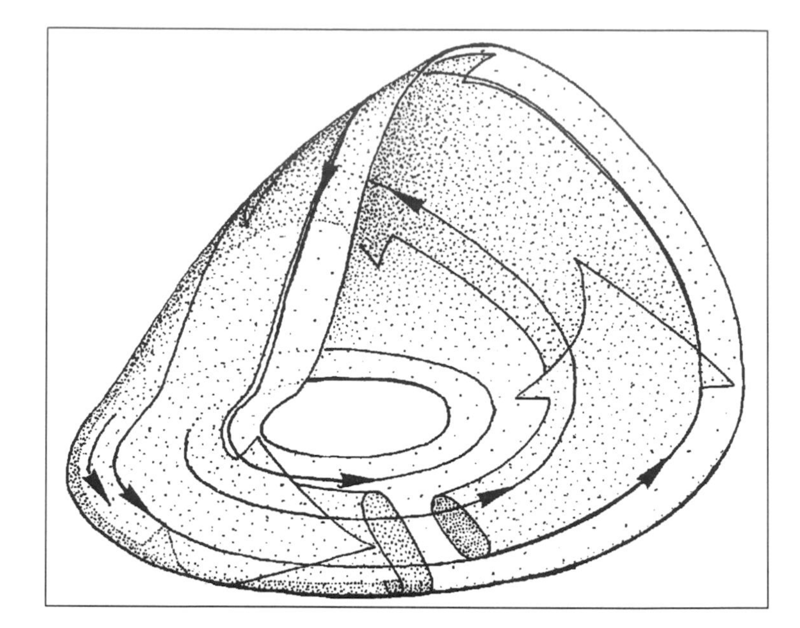

Figure 12.24 : Schematic graph of the dynamics on the Rössler attractor. Reprinted with permission from R. H. Abraham, C. D. Shaw, *Dynamics, The Geometry of Behavior*, Part Two: *Chaotic Behavior*, 1983, Aerial Press, Santa Cruz, second edition Addison-Wesley, 1992.

three variables x, y and z instead of the one for the population, p. Thus, we can interpret the system in eqn. (12.12) as a collection of laws of motion for a point at coordinates (x, y, z) in three-dimensional space. For any given initial coordinates (x_0, y_0, z_0) the system defines a unique trajectory which is parametrized by time t and satisfies the equations at all times. Denoting the coordinates of this trajectory by $(x(t), y(t), z(t))$ for time $t \geq 0$ this means

$$\begin{array}{ll} x'(t) = -(y(t) + z(t)) & x(0) = x_0 \\ y'(t) = x(t) + ay(t) & y(0) = y_0 \\ z'(t) = b + x(t)z(t) - cz(t) & z(0) = z_0 \ . \end{array}$$

Figure 12.21 shows a plot of the three components $x(t)$, $y(t)$, $z(t)$ versus time t. However, it is more instructive to plot this trajectory in three-dimensional space with coordinates (x, y, z) and consecutive points connected by short line segments; we obtain a first picture of the Rössler attractor (see figure 12.22). Orbits on the attractor spend most of their time near the xy-plane spiraling out from the origin. When an orbit has attained some critical distance from the origin it is first lifted away from the xy-plane. Then, after reaching some maximal z-value, it is reinserted into the spiraling

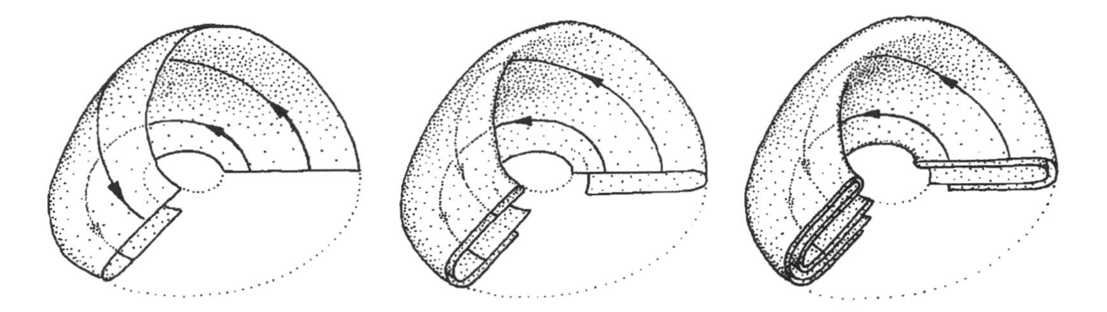

Figure 12.25 : Three iterations of the stretch-and-fold action are illustrated here for an initial line segment. Reprinted with permission from R. H. Abraham, C. D. Shaw, *Dynamics, The Geometry of Behavior,* Part Two: *Chaotic Behavior,* 1983, Aerial Press, Santa Cruz, second edition Addison-Wesley, 1992.

piece of the attractor close to the plane. The larger the z-amplitude of this excursion has been, the closer to the origin the orbit will land, and the spiraling process followed by ejection and reinsertion repeats (see figure 12.24).

The Attractive Property

Let us first take a look at the phenomenology of the attractor by means of some simple numerical experiments before we try to understand how the equations provide the foundations for these effects. First of all we note that we are indeed dealing with an attractor. When we start the solution of the differential equation at some other initial point somewhere in the vicinity of the structure shown in figure 12.22, we get essentially the same result. Only the first part of the trajectory is noticeably different stemming from the transitional period necessary for the solution to get close to the attractor.

The Kneading Transformation

In order to understand the chaotic behavior of the dynamics in Rössler's attractor, we begin by showing how the nonlinear stretch-and-fold operation is hidden in the system. From the pictures, it appears that the attractor has the structure of a *folded band.* Starting from a section across the band near the negative x-axis we can observe two effects (see figure 12.25).

1. As the band winds around the center for about half of a turn its width increases. This corresponds to the stretching in the dough analogy.
2. Near the positive x-axis the band is more than twice as wide and the outer part begins to fold over and eventually covers the inner part and also part of the 'hole' at the center. This folding action is completed after about another half of a turn and the process then repeats.

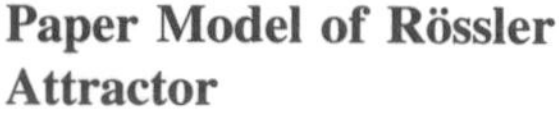

Paper Model of Rössler Attractor

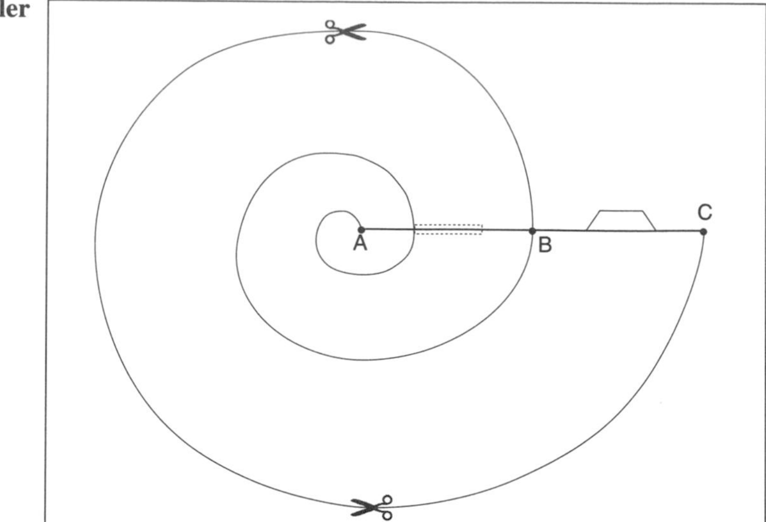

Figure 12.26 : To attain a three-dimensional paper model of the Rössler attractor follow these steps: 1. Make a photo copy of the figure enlarging it by a factor of two. 2. Using scissors cut the figure along the outline. 3. Cut a small slit at the position indicated between points A and B. 4. Fold at point B so that point C comes to lie on point A. 5. Insert the little flap between points B and C through the slit and fix it to the underside of the figure with some glue. Now we can follow around the trajectories of the system. The one already drawn is a periodic one which circles around the center 3 times before repeating. Hint: the lines are appearing only on one side of the paper, it may be more instructive to also draw them on the reverse side before starting the paper construction.

The One-Dimensional Return Map

In other words, what we see here in three dimensions is essentially the same as the iteration of stretch-and-fold operations. Therefore, there must also be a direct relation to the quadratic iteration[25]

$$x \to rx(1-x)$$

because its dynamics are also nothing but stretch-and-fold operations. Let us uncover this relation and make it explicit. Recall that the relation between the quadratic transformation and the kneading is very elementary: the value x specifies a particle in the dough and $rx(1-x)$ is its position after one stretch-and-fold operation. Here we can proceed in the same manner. We define a cross-section of the Rössler band as a reference frame — for example, the part of the attractor which lies along the negative x-axis where

[25] Here we have called the parameter r in order to avoid confusion with the parameter a of the Rössler system.

The Return Map for Rössler's Attractor

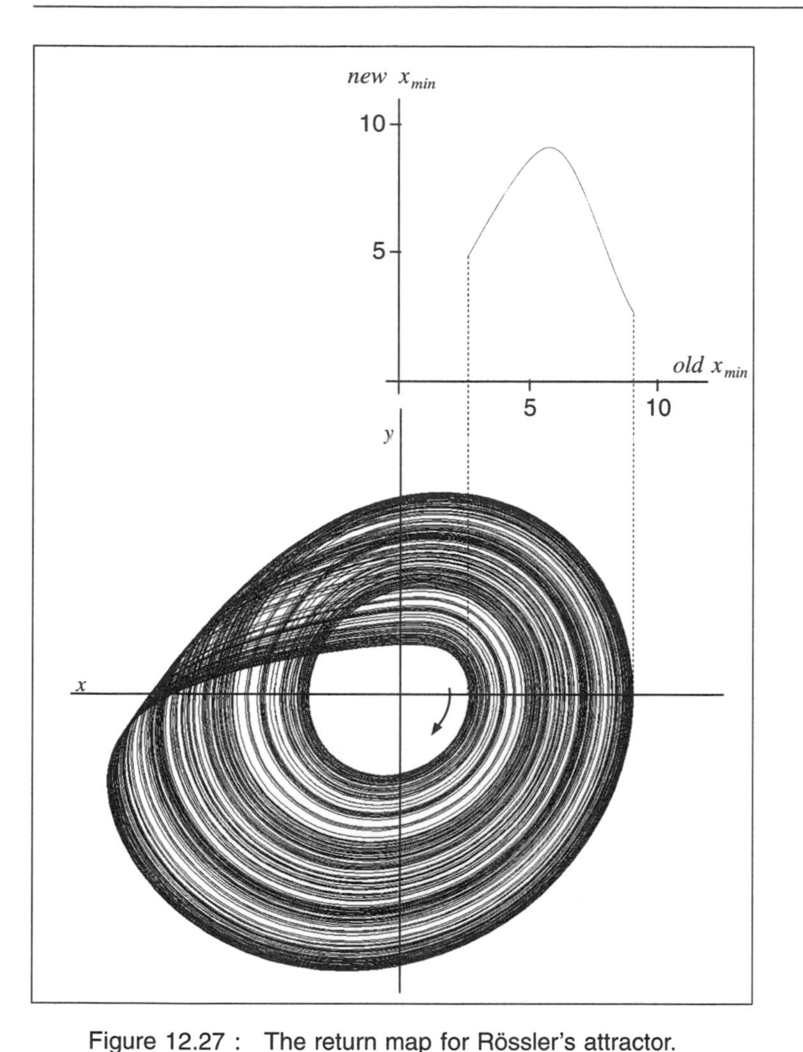

Figure 12.27 : The return map for Rössler's attractor.

the x-values of trajectories are at their minimum.[26] The band is quite flat, and therefore we can identify points in the intersection of the attractor and the half plane simply by the absolute values of their x-coordinates. We regard such points as initial points and follow their trajectories around the band for one complete turn until they reenter the half plane. We arrive at a new position with a new x-coordinate. In order to record the stretch-and-fold operation we can mark in a diagram the absolute value of this new x-value compared to the old one. When performing this procedure

[26]We consider the plane given by the equation $y + z = 0$. This implies that at the points where a trajectory pierces through the plane the value of $x'(t)$ is zero. Thus, the plane cuts the trajectories precisely at their minima and maxima of the x-value.

for a long trajectory on the Rössler band, we should obtain a good representation of the stretch-and-fold dynamics, which is in the same spirit as the parabola in the graphical iteration method (see figure 12.27). And moreover, the figure indeed reveals a shape which closely resembles a parabola! A function modeling the plot is called a *Lorenz map*. It provides a link between the dynamics of a continuous system and the discrete dynamics of transformations of an interval.[27] Once this link is known, it provides a shortcut for computing the dynamics of the underlying system. Instead of following a trajectory of the differential equation with possibly many steps and a large computational effort, we may simply evaluate the Lorenz map once (or perform one step of graphical iteration) and arrive at the same result. What is more, it is usually much easier to analyze the properties of a Lorenz map than the dynamics for a differential equation. Thus, the results of chapters 10 and 11 regarding the chaos for transformations of intervals and the routes leading to it help to understand the dynamics of the more complicated continuous system.

We may conclude that the chaos present in the iteration of the quadratic iterator carries over to the Rössler band by means of the Lorenz map. Of course, when we change the parameters a, b or c in Rössler's system the parabola-like graph of the Lorenz map must change accordingly. In particular, we expect that the band can be replaced by periodicity, implying a periodic attractor in the Lorenz map corresponding to a periodic loop in phase space (x, y, z) for the differential equation (see figure 12.28).

The Feigenbaum Experiment

Let us continue these experiments with the next logical step: drawing a Feigenbaum type diagram. Does the analogy between the quadratic iterator and the Rössler system go as far as to reproduce the period-doubling cascade or even the Feigenbaum constant 4.6692...? We do not claim to have performed this experiment with utmost numerical care, but the result shown in figure 12.29 seems to suggest that it is true. We retrieve the fundamentals of the spectrum of dynamics from the quadratic iterator.

The Strangeness of the Attractor

This may seem like the end of the story of the chaotic attractor, but we have omitted one crucial limitation of its one-dimensional model, the Lorenz map; *it is false*. If it were exact, then this would imply that initial points in the cut through the band come in pairs which would land at precisely the same point when iterated once with the Lorenz map. (A parabola transforms two points into one.[28]) This would violate the fundamental property of the continuous systems discussed here, namely that trajectories must not

[27] The discovery of this quasi-one-dimensional character of the dynamics of a system of differential equations was made by Lorenz (see the following section 12.4).

[28] When using $rx(1-x)$ these points are $x_0 = 0.5 + d$ and $x_0 = 0.5 - d$, where d can be any number.

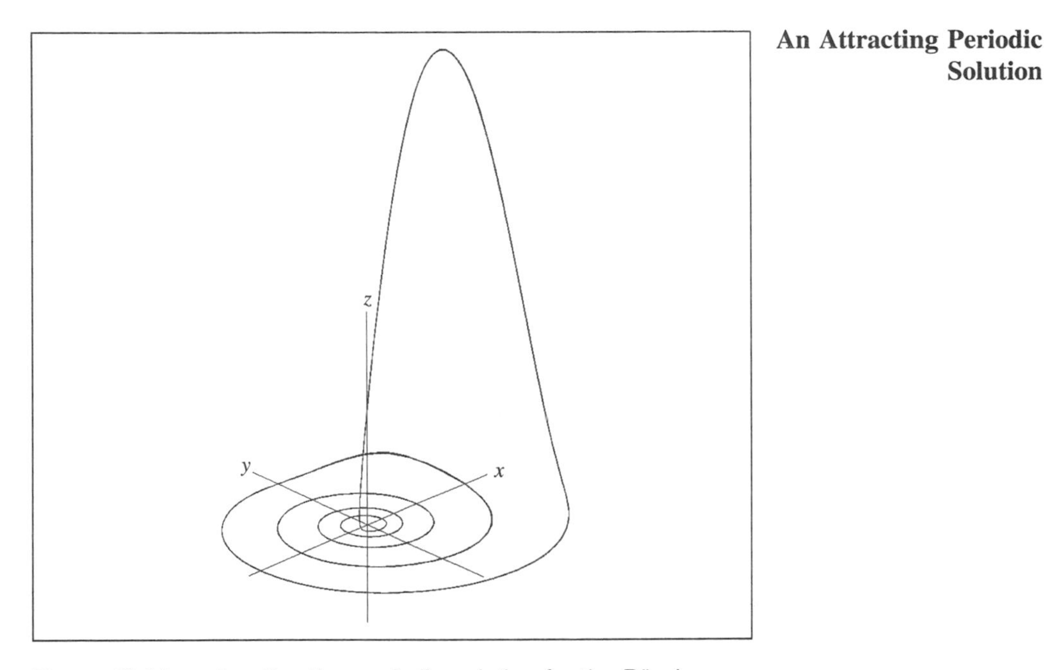

Figure 12.28 : An attracting periodic solution for the Rössler system. $c = 8.0$.

An Attracting Periodic Solution

cross. If they share a single point then they must be identical for all times from that point on. And this applies in both directions of a trajectory forward and backward in time. So what is really happening? Our experiments are certainly not completely wrong. Similar studies which yield the same qualitative results have, however, been carried out by numerous researchers with more precision and effort. Thus, we are forced to conclude that since the folded part of the Rössler band cannot exactly merge with the other part, it can at least come close to it. Indeed, the two parts of the band can come *very* close to each other, as close as we want them to. Thus, what we see in the pictures after one turn around the attractor is not one band but two tightly packed layers of the band. After another rotation, both of these layers (which still must be separate) fold over and form four layers. Then we get eight, then sixteen and so on. Thus, in effect we should see an infinite number of layered bands stacked up somewhat like a pile of strangely intertwined extra thin pancakes which, moreover, have been subjected to a compactor. Expressed more academically, it is a *Cantor set of sheets*. And because these sheets are so close to each other, we cannot see them in the phase plots or in the figure 12.27, from which we assumed the Lorenz map. Only if we work with extraordinary precision, do we have a chance to see a glimpse of

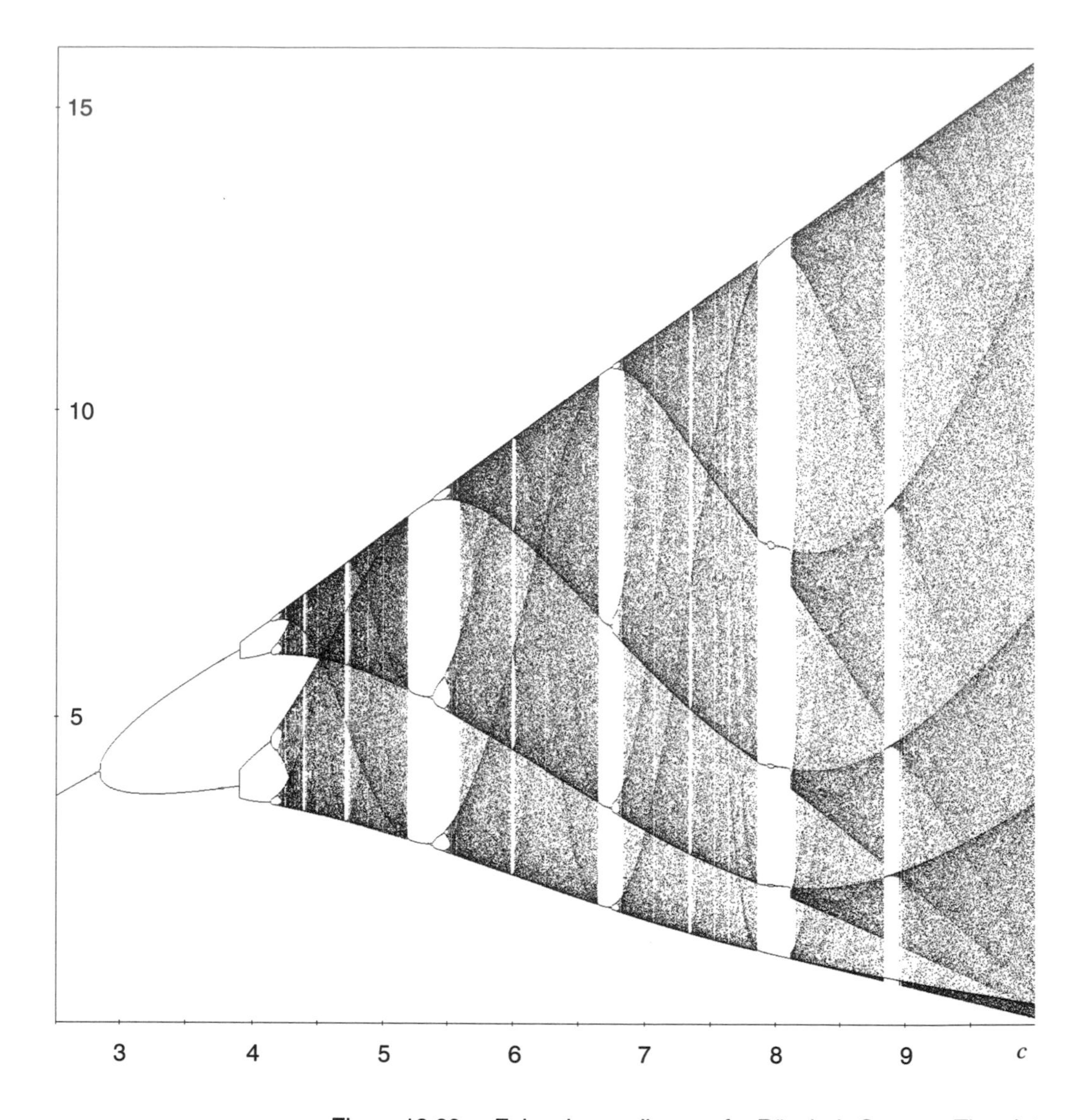

Figure 12.29 : Feigenbaum diagram for Rössler's System. The plot shows the parameter range $2.5 < c < 10.0$. Vertically the absolute values of the minimal x-values of the corresponding trajectories from the attractor are shown. This corresponds to a projection of a Lorenz map diagram such as given in figure 12.27 on the vertical axis. Initially, for small values of c, the attractor consists of a periodic orbit which has only one local minimum of x-values, i.e., it is a single loop. As the parameter c increases, this periodic orbit undergoes a period-doubling bifurcation. Check that for $c = 8$ there are 5 points in the diagram corresponding to the periodic solution shown in figure 12.28 which has 5 loops.

these layers. In other words, the fractal dimension of a chaotic attractor from the Rössler family must be $2+\varepsilon$ where ε is a very small number.[29] It is this fractal character that qualifies Rössler's attractor as a *strange* one.

Poincaré Sections

The one-dimensional Lorenz map is a simplification of the real dynamics happening in the Rössler system. The dots which accumulate in the diagram of a section of the Rössler band in figure 12.27 do *not* exactly lie on a curve. They are very close, almost indistinguishably close, to a curve which looks like a parabola. To create a discrete transformation which reflects the true dynamics let us consider a rectangle containing the Rössler band in the section given by the z-axis and the negative part of the x-axis. This rectangle is *transversal* to the flow of the system which means that trajectories pierce right through it and do not approach it tangentially. A surface with this property is also called a *Poincaré section*.[30] The dynamics of the system can be described with a transformation T defined on such a section. Given an initial point p on the section we follow the corresponding trajectory until the section is entered again. That point is taken as $T(p)$ and by the same procedure an image point $T(p)$ is defined for all initial points p of the section. The transformation T is called a *Poincaré map* (see figure 12.30). In place of a 'complete' trajectory, we may consider the corresponding iterates of the Poincaré map. For example, a periodic trajectory matches up with a fixed point of the Poincaré map or one of its iterates.

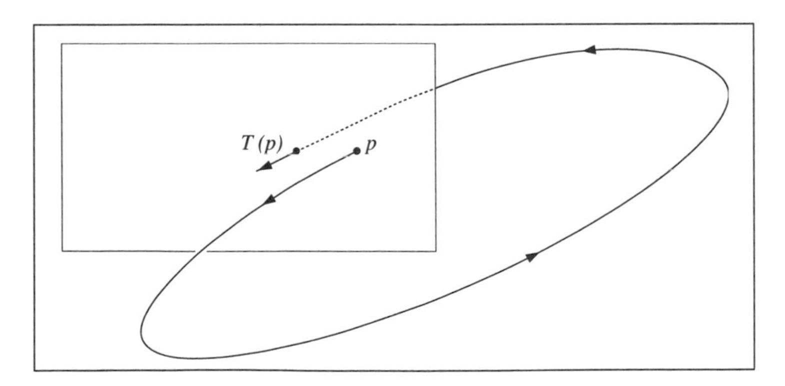

Figure 12.30 : The Poincaré map transforms the point p into $T(p)$ defined by the first reentry point of the trajectory starting at p in the shown cross-section.

[29] The dimension can be estimated between 2.01 and 2.02.

[30] After Jules-Henri Poincaré (1854–1912) who laid the mathematical foundations of modern dynamical systems theory.

How the Differential Equation Operates

To explain how the differential equations of the Rössler system

$$x' = -(y+z)$$
$$y' = x + ay$$
$$z' = b + xz - cz$$

can generate trajectories of the type discussed, we reproduce here the explanation of J. M. T. Thompson and H. B. Stewart from their book *Nonlinear Dynamics and Chaos,* Wiley, Chichester, 1986, p. 235–236. This simple, autonomous system has only a single nonlinear term, the product of x and z in the third equation. Each term in these equations serves its function in generating the desired global structure of trajectories. Considering the first two equations, let us for the moment suppose that z is negligibly small. Then the subsystem

$$x' = -y$$
$$y' = x + ay$$

can be transformed to the second-order linear oscillator

$$x'' - xx' + x = 0\ .$$

With positive a, this oscillator has negative damping, and the origin is an unstable focus for $0 < a < 2$. Thus in the full system of three first-order equations, trajectories near the (x, y) plane spiral outwards from the origin. This produces a spreading of adjacent trajectories, which is the first ingredient in the mixing action of chaos.

This spreading is achieved with only linear terms. But if the full system of three equations were linear, the spreading would merely continue as all trajectories diverge to an infinite distance from the origin. To confine the spreading action within a bounded attractor, the nonlinear term is required. The constant c in the third equation acts as a threshold for switching on the nonlinear folding action. Considering the third equation alone, whenever the value of x is less than the constant c, the coefficient of z is negative, and the z subsystem is stable, tending to restore z to a value near $-b/(x-c)$. However, if x should exceed c, then z will appear in the third equation multiplied by a positive factor, and the previously self-restoring z subsystem diverges. Choosing $b > 0$ ensures that this divergence will be towards positive z.

The effect of this is shown in figure 12.22. A trajectory spirals outwards while appearing to remain in a plane near to and parallel to the (x, y) plane. When x becomes large enough, the z subsystem switches on and the trajectory leaps upwards. Once z becomes large, the z term in the first equation comes into play, and x becomes large and negative, throwing the trajectory back towards smaller x. Eventually x decreases below c, the z variable becomes self-restoring, and the trajectory lands near the (x, y) plane again. Through the feedback of z to the x equation, trajectories are folded back and reinserted closer to the origin, where they begin an outward spiral once more.

12.4 The Lorenz Attractor

The Rössler system is an artificial system designed solely with the purpose of creating a model for a strange attractor which uses only the simplest chaos generating mechanism, stretch-and-fold. Of course, Rössler knew about the Lorenz system, which had been published 13 years before. In fact, we may say the Rössler attractor is a model of the Lorenz model.

The Lorenz System

The system of equations that Lorenz proposed does not look any more complicated than that of Rössler. Here it is:

$$\begin{aligned} x' &= -\sigma x + \sigma y \\ y' &= Rx - y - xz \\ z' &= -Bz + xy \end{aligned} \qquad (12.13)$$

The numbers σ, B, and R are the system's physical parameters, which Lorenz fixed at

$$\sigma = 10, \quad B = \frac{8}{3}, \quad R = 28 \; .$$

Figure 12.31 shows the corresponding attractor, which is now called the Lorenz attractor. Clearly the geometry is more involved than for the Rössler band. There are two sheets in which trajectories spiral outwards. When the distance from the center of such a spiral becomes larger than some particular threshold, the solution is ejected from the spiral and attracted by the other spiral, where it again begins to spiral out and the game is repeated. The number of turns that a trajectory spends in one spiral and then in the other is not specified. It may wind around one spiral two times, then three times around the other, then ten times around the first and so on. In fact, we believe that for any sequence of positive numbers which are not too large, for example 3, 11, 7, ... there exists a trajectory on the Lorenz attractor with precisely these numbers as turns around the spirals. Thus, there is a solution that turns 3 times around the right spiral, then 11 times around the left, then 7 times around the right again and so on.

Lorenz's Physical Model

What is the connection between these wildly spinning solutions and weather forecasting which is what Lorenz was interested in? Certainly, the trajectories should not be mistaken for the paths of air currents! If this were the case then the Lorenz attractor would act similar to a black hole in astrophysics, sucking in all the atmosphere — leaving nothing but emptiness around it and laying waste to the whole planet Earth. But we are not far from the truth. The Lorenz system is in fact a model of thermal convection which, however, includes not only a description of the motion of some viscous fluid or atmosphere but also the information about distribution of heat, the driving force of thermal convection.

The Lorenz Attractor

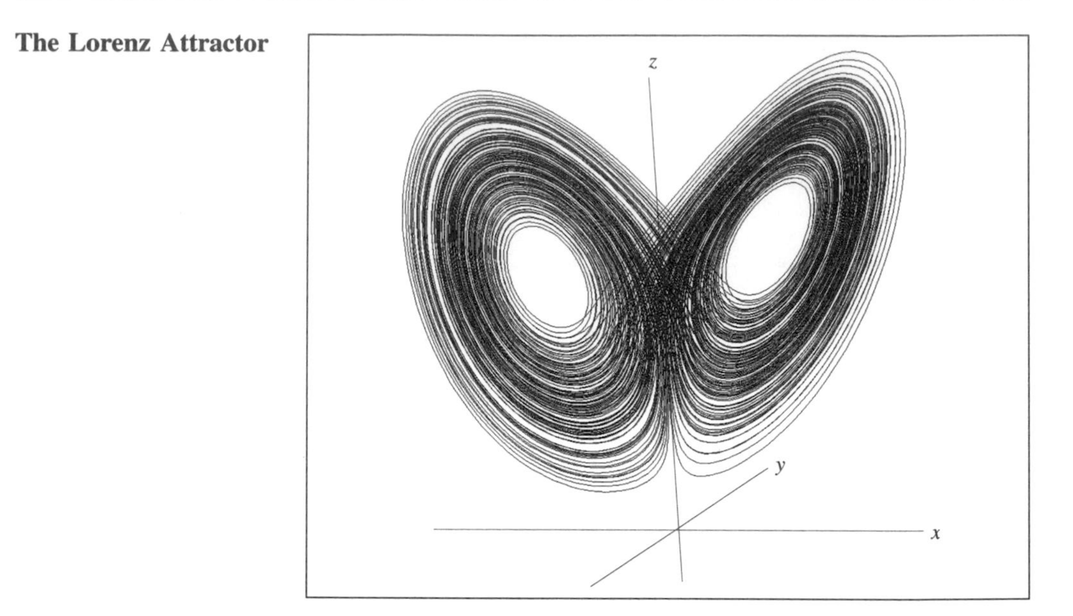

Figure 12.31 : Some trajectories from the Lorenz attractor.

When air is warmed near the Earth's surface it rises. This is an important factor in the atmospheric weather factory. Convection air currents may accumulate and give rise to convection cells of several types and, when forced more vigorously, may produce very turbulent motion in the atmosphere. Examples of convection cells are cylindrical rolls and structures, which are called *Bénard cells*, resembling a honeycomb from above, In these hexagonal cells the warmed portions of the fluid rise in the center, get colder near the top and sink back down to the surface around the boundary of the cell. The Lorenz system is related more to the cylindrical roll type of fluid motion in which one of the dimensions can be disregarded pretending that these rolls extend to infinity. The mathematical model of the fluid motion had originally been developed by Lord Rayleigh[31] in 1916. It assumes that all convection happens in a rectangular region whose bottom is heated such that the temperature difference between bottom and top remains constant. With certain parameter configurations in this model it turns out that the solutions to the model equations have a rather special form which was already known to Rayleigh. Lorenz took these special solutions, regarded their amplitudes as time-dependent, inserted them in the Rayleigh model, disregarded all terms that are not in this

[31]Lord Rayleigh, *On convective currents in a horizontal layer of fluid when the higher temperature is on the under side,* Phil. Mag. 32 (1916) 529–546.

special form,[32] and arrived at a system of differential equations for the time-dependent amplitudes, eqn. (12.13).[33]

The Meaning of x, y and z

It is almost impossible to find out what the variables x, y, and z in the Lorenz system precisely stand for without consulting Lorenz' original paper. Let us give some technical details and visualizations of the convection process which was investigated by Lorenz. As already stated we deal with convection currents and temperature distribution in a rectangular region where the temperature difference ΔT between the bottom and the top is kept constant. The dynamics are assumed to be identical in all slices parallel to the rectangular region. The governing equations for the more general three-dimensional problem were worked out by Lord Rayleigh. The simplification to the two-dimensional case considered here is by B. Saltzman.[34] We do not include these formulas but instead, present the type of approximation of a solution used by Lorenz. It involves a so-called stream function Ψ and a temperature function. The variables in these functions are the spatial coordinates u and v, and time t. In place of the actual temperature, the difference Θ with a temperature profile belonging to the state of no convection is used, i.e., where the temperature decreases linearly from some value T_0 at the bottom to $T_0 - \Delta T$ at the top.

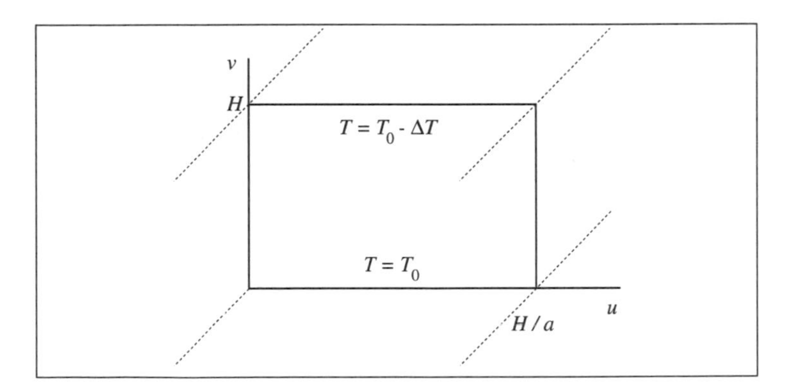

Figure 12.32 : The coordinate system in the cross-section of a bar where the Lorenz equations present a model for fluid flow and temperature.

[32] Except for one particular term related to the temperature distribution.

[33] Another interpretation is that the solutions of the Rayleigh equations can be written as Fourier series with time-dependent coefficients. Using these series in place of the original variables produces a system containing infinitely many equations. When keeping only the three most significant of these we again obtain the Lorenz system given in eqn. (12.13).

[34] B. Saltzman, *Finite amplitude free convection as an initial value problem — I*, J. atmos. Sci. 19 (1962) 329–341.

Let us give these complicated looking equations and then explain.

$$\begin{aligned}
\frac{a}{(1+a^2)\kappa}\Psi(u,v,t) &= x(t)\sqrt{2}\sin\left(\tfrac{\pi a}{H}u\right)\sin\left(\tfrac{\pi}{H}v\right)\\
\frac{\pi R_a}{R_c \Delta T}\Theta(u,v,t) &= y(t)\sqrt{2}\cos\left(\tfrac{\pi a}{H}u\right)\sin\left(\tfrac{\pi}{H}v\right) \qquad (12.14)\\
&\quad -z(t)\sin\left(\tfrac{2\pi}{H}v\right)
\end{aligned}$$

The symbols used have the following interpretation.

$\Psi = \Psi(u,v,t)$	stream function
$\Theta = \Theta(u,v,t)$	local temperature difference
u	horizontal spatial coordinate
v	vertical spatial coordinate
$x(t), y(t), z(t)$	time-dependent coefficients (amplitudes)
t	time
H	depth of fluid layer (maximum of v)
a	parameter of geometry (fixed at $a = 1/\sqrt{2}$)
κ	thermal conductivity
R_a	Rayleigh number
R_c	critical value of R_a ($R_c = \pi^4(1+a^2)^3/a^2$)
ΔT	total temperature difference

The rectangular region has coordinates u ranging from 0 to $H/a = \sqrt{2}H$ and v ranging from 0 to H. The stream function Ψ is interpreted in the following sense. Ψ is a scalar field and the fluid motion at a time t_0 occurs along the isolines $\Psi(u,v,t_0) = const.$ Thus, to obtain a picture of the fluid motion we can simply plot these isolines (see figure 12.33). More precisely, the velocity V of the fluid at a given point (u,v) in space and t in time is

$$V(u,v,t) = \begin{pmatrix} -\dfrac{d}{dv}\Psi(u,v,t) \\ \dfrac{d}{du}\Psi(u,v,t) \end{pmatrix}$$

The corresponding temperature profile is given by $\Theta(u,v,t)$ and can be read directly from the formula (12.14).

Using this form of solution, Saltzman's equations reduce to the system of Lorenz which contains only the unknown time-dependent coefficients. Recall

$$\begin{aligned}
x' &= -\sigma x + \sigma y\\
y' &= Rx - y - xz\\
z' &= -Bz + xy
\end{aligned}$$

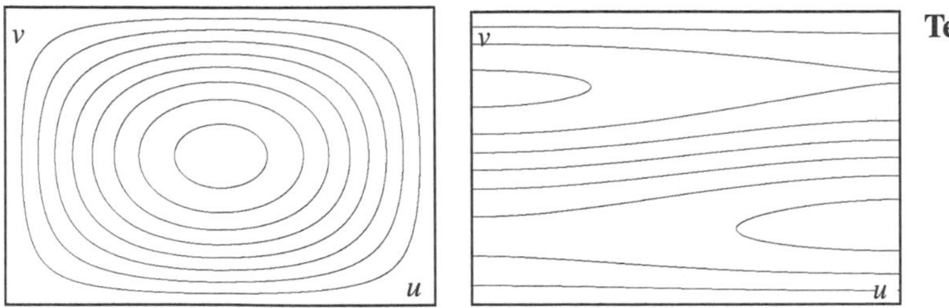

Temperature Profile at Steady State

Figure 12.33 : Streamlines of convection currents (left) and corresponding temperature profile (right) at steady state.

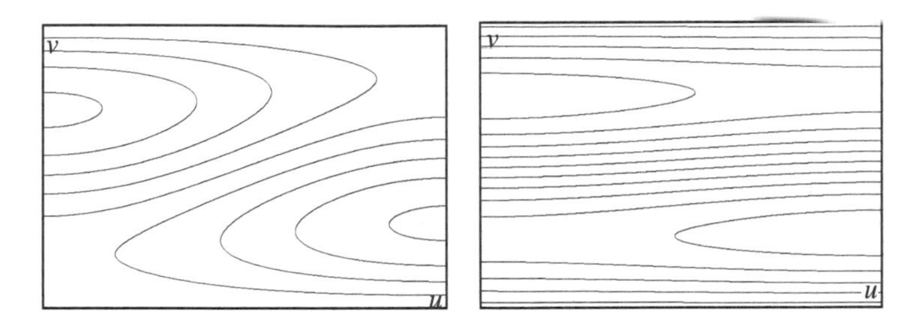

Other Temperature Profiles

Figure 12.34 : Two more temperature profiles. The corresponding state variables (x, y, z) are $(2.403, 4.892, 4.673)$ (left) and $(16.610, 7.428, 45.428)$ (right).

which includes three parameters: σ, the Prandtl number, $R = R_a/R_c$ and $B = 4/(1+a^2) = 8/3$.[35] Lorenz explains: 'In these equations x is proportional to the intensity of the convective motion, while y is proportional to the temperature difference between the ascending and the descending currents, similar signs of x and y denoting that warm fluid is rising and cold fluid is descending.'

Lorenz chose the parameters $\sigma = 10$ and $R = 28$. For this setting there are two steady states of the differential equation, i.e., values of x, y, z which remain constant (besides the origin $(0,0,0)$). These are

$$(6\sqrt{2}, 6\sqrt{2}, 27) \text{ and } (-6\sqrt{2}, -6\sqrt{2}, 27)$$

and correspond to the centers of the two 'holes' in the attractor shown in figure 12.31. Associated with these solutions are steady states of convection. We illustrate the first of these in figure 12.33. Other points in phase space correspond to other convection currents and temperature profiles. Two more examples for points on the Lorenz attractor are given in figure 12.34. Thus, when we follow a point along its trajectory on the Lorenz attractor, we must

[35] The derivatives x', y', z' are with respect to reparametrized time τ, namely $\tau = \pi^2(1+a^2)\kappa t/H^2$.

interpret its coordinates in this sense as given by the formulas in eqn. (12.14) and the figures. With these remarks we conclude the discussion of the modeling aspect of the Lorenz system.[36]

As outlined, there are several severe simplification steps before we get to the final set of equations, and we may rightfully say that the solutions of the system may not bear any significance for the real convection process. But it was not Lorenz' intention to be as precise as possible in the modeling. On the contrary, after having discovered a strange attractor (in a more complex system) he strived for the most elementary system that can be derived from the convection equations and that would still demonstrate the extreme sensitivity to initial conditions, which since has become the trademark of chaos.

A Model for the Lorenz Dynamics

The chaos generating mechanism in the Lorenz system is a bit more involved than the one in the Rössler system. Rather than featuring a stretch-and-fold action, we have a stretch-split-and-merge operation as shown in the model in figure 12.35. Around the two spirals a stretching takes place. Both stretched bands split near the horizontal line in the center, one half of them returning to the left spiral, the other to the right. During the subsequent turn, the two bands on each part of the attractor merge, and the cycle is completed. Similar to the Lorenz map for the folded band, a Lorenz map for points on the central line segment can now be defined. The result is shown in the lower part of the figure.

The graphical iteration using this graph corresponds to the dynamics of points (x, y, z) on the attractor. The left half of it belongs to the left spiral of the attractor, the other to the right. There is a connection between this one-dimensional model and the shift transformation in chapter 10, $x \to \mathrm{Frac}\,(2x) = 2x \bmod 1$, from which we deduced the essential properties of chaos: sensitive dependence on initial conditions, dense periodic points, and mixing.

Another Lorenz Map

Originally Lorenz had proposed a different one-dimensional model. He observed that a trajectory "leaves one spiral only after exceeding some critical distance from its center. Moreover, the extent to which this distance is exceeded appears to determine the point at which the next spiral is entered; this in turn seems to determine the number of circuits to be executed before changing spirals again." He then concluded that the maximum z-value alone suffices to predict the maximum for the following circuit. To check this idea, he plotted many points with coordinates being two

[36]For more details and references see the original paper of E. N. Lorenz, *Deterministic nonperiodic flow,* J. Atmos. Sci. 20 (1963) 130–141. The book by J. M. T. Thompson and H. B. Stewart, *Nonlinear Dynamics and Chaos,* Wiley, Chichester, 1986, contains a broad introduction and much more material about the geometry underlying the attractor and its route to chaos.

Model of the Dynamics with Lorenz Map

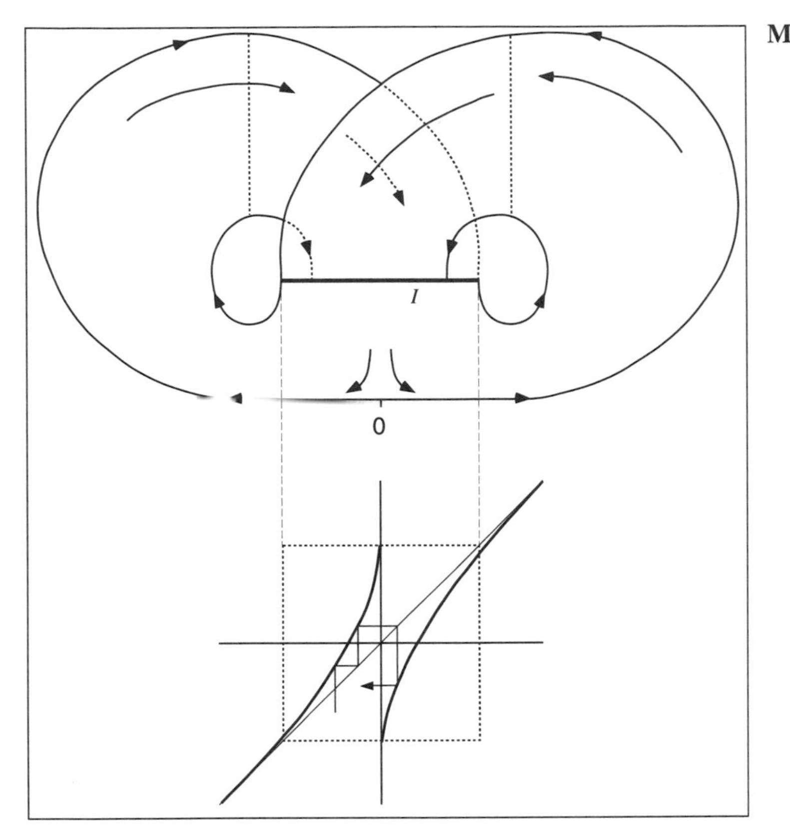

Figure 12.35 : A schematic diagram of the stretch-split-and-merge process in the dynamics on the Lorenz attractor is shown at the top. The Lorenz map of the system shown below models the stretch-split-and-merge process as observed at the interval I. Each (graphical) iteration corresponds to a turn around one of the lobes of the attractor.

consecutive maximum z-values of a trajectory as in figure 12.36; thus was born the first Lorenz map.[37] The points appear to lie on a curve and graphical iteration can be used to predict the maximum z-value of the following spiral turns if the current one is given. The similarity of the graph to the tent function is quite apparent and Lorenz' paper continues with a short study of the chaotic dynamics associated with the tent function and concludes that there must be an infinity of trajectories corresponding to what he called nonperiodic deterministic flow, i.e., that one now calls chaos.

[37]There is a surface which contains all points (x, y, z) with a maximum z-value of a corresponding trajectory. At such a maximum the derivative z' must necessarily vanish. Thus, from eqn. (12.13), $xy - Bz = 0$, or $z = xy/B$. This equation describes a surface, the graph of the function xy/B. A portion of this surface may be interpreted as a proper Poincaré section of the Lorenz system.

Lorenz Map

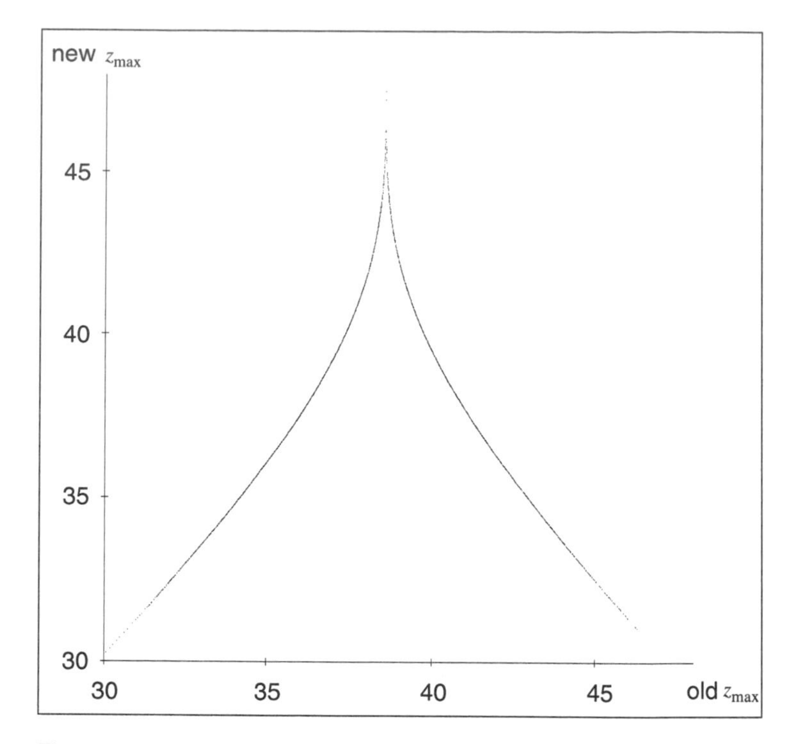

Figure 12.36 : This Lorenz map for the Lorenz system models the dynamics of the attractor as observed at the dotted vertical lines pictured in figure 12.35. From this point of view it can be described by a stretch-and-fold process.

Cantor Set Structure and Fractal Dimension

Again, it is true that the Lorenz maps from figures 12.35 and 12.36 are only models for a truely two-dimensional Poincaré map. If such one-dimensional models were exact, this would imply a perfect merging of the two 'spiral surfaces' which contradicts the uniqueness of solutions of Lorenz' system of differential equations. Similar to the bands in the Rössler attractor, these surfaces come very, very close to each other, indistinguishable to the eye, but they cannot completely merge. Again, there is a Cantor set structure of these surfaces leading to a fractal dimension only slightly above 2.[38]

Changing the Parameter R

Let us conclude this section by looking at phenomena brought to light by changing the parameter R in the Lorenz system. There are a large number of periodic solutions, some of which are stable and attracting, while others are unstable and repelling. For some parameters, chaos and stable equilibria coexist. The whole palette of features worth discussing cannot be included here; it would take

[38]The dimension has been estimated as 2.073. See E. N. Lorenz, *The local structure of a chaotic attractor in four dimensions,* Physica 13D (1984) 90–104.

A Periodic Solution

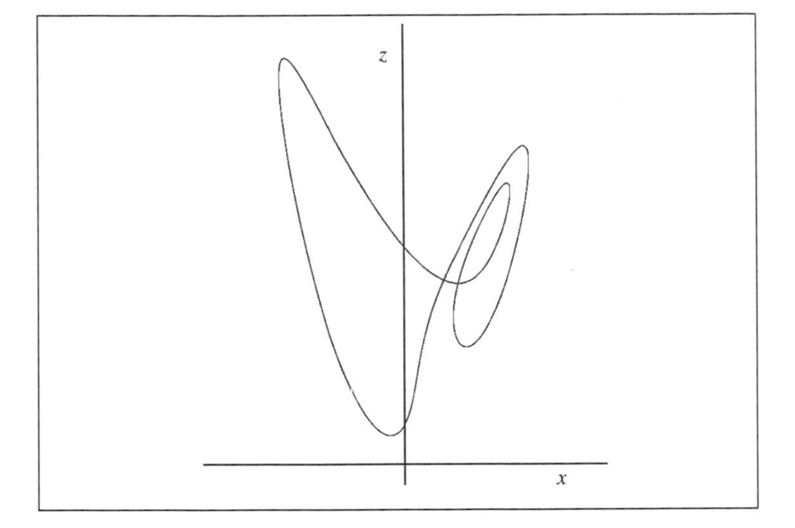

Figure 12.37 : A stable, periodic solution to Lorenz' system at parameter $R = 100.5$, projected onto the xz-plane.

up too much space. Here we can only highlight a couple of aspects; for further details we can only point to the literature.[39]

Period Doubling Window and Feigenbaum Number

We begin by looking at the range of parameters $99.524 < R < 100.795$. Here we can observe a period-doubling scenario similar to the one in the Rössler attractor. This cascade of bifurcations occurs as we decrease the parameter R. Figure 12.37 shows an attractive periodic orbit obtained for $R = 100.5$. It spirals around twice in the 'positive' half space $x > 0$, then one time in the negative half space $x < 0$ before it repeats. This solution would be called an x^2y-solution following the naming conventions in some of the literature.[40] When we lower the parameter below $R \approx 99.98$ this solution doubles up to a periodic solution of twice the period and type $x^2yx^2y = (x^2y)^2$. This happens in the same spirit as in the bifurcation of periodic solutions in Rössler's system. Further bifurcations of period-doubling can be observed when we continue to lower the parameter, as listed in the following table.

[39]For example, see the book by C. Sparrow, *The Lorenz Equations: Bifurcations, Chaos, and Strange Attractors*, Springer-Verlag, New York, 1982.

[40]In the code x^ky^l an x stands for a turn around the fixed point in the half space $x > 0$, while the symbol y denotes a turn in the other half space. Thus, in this example there are k turns of type x followed by l turns of type y. The notation can be extended to longer symbol strings such as xy^2xy^3 and so on.

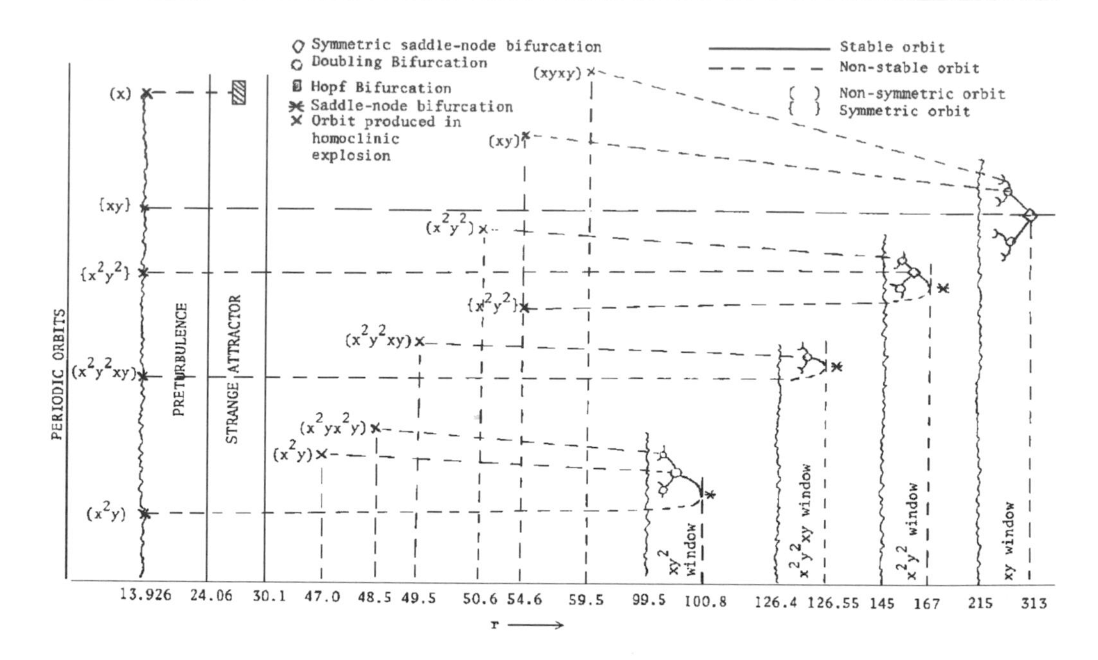

Figure 12.38 : Bifurcation diagram showing a few of the shorter periodic trajectories. Reproduced with permission of the publisher from C. Sparrow, *The Lorenz Equations: Bifurcations, Chaos, and Strange Attractors,* Springer-Verlag, New York, 1982, page 99.

Parameter R	Type of Solution
100.795	x^2y
99.98	$(x^2y)^2$
99.629	$(x^2y)^4$
99.547	$(x^2y)^8$
99.529	$(x^2y)^{16}$
99.5255	$(x^2y)^{32}$

Based on more precise calculations the Feigenbaum number δ belonging to this sequence has been estimated.[41] The result is $\delta \approx 4.67$, a number which at this precision is indistinguishable from Feigenbaum's famous ratio 4.6692... There are other windows in the parameters R with period-doubling bifurcations. To give an impression of the complicated scenarios, we reproduce a chart (figure 12.38) compiled by Colin Sparrow.

[41] V. Franceschini, *A Feigenbaum sequence of bifurcations in the Lorenz model,* Jour. Stat. Phys. 22 (1980) 397–406.

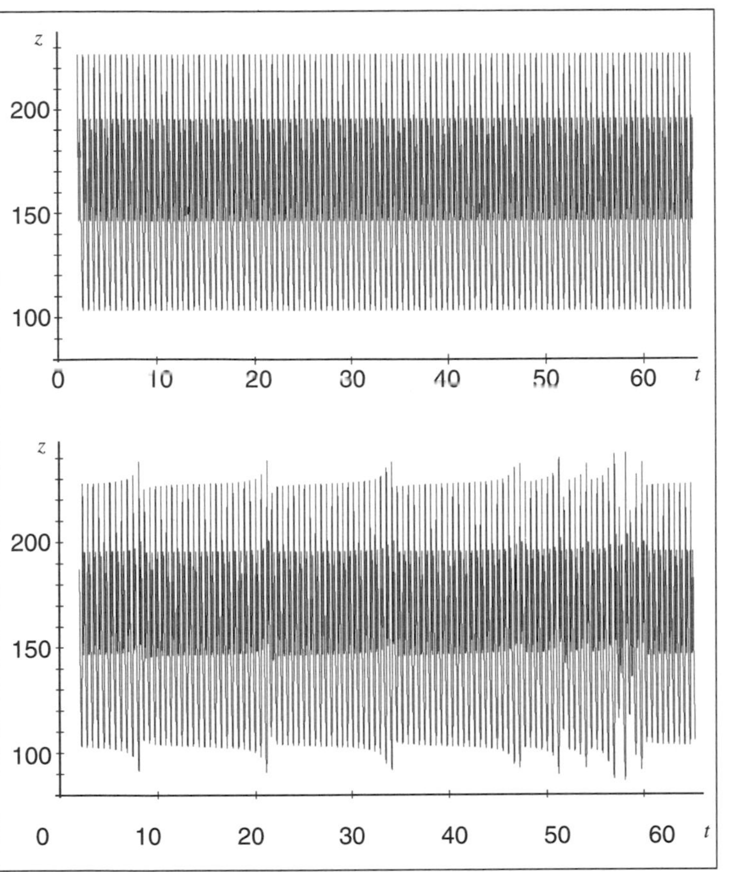

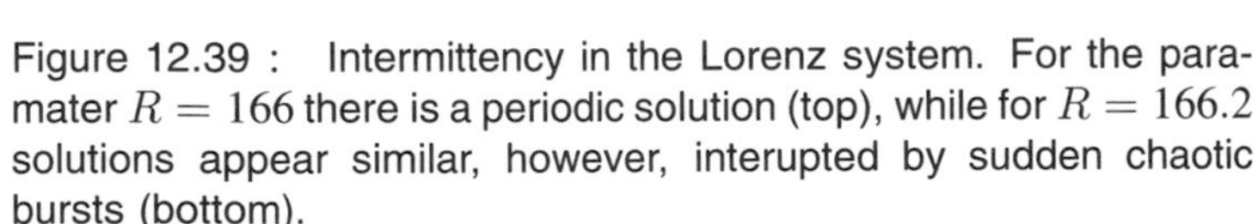
Figure 12.39 : Intermittency in the Lorenz system. For the parameter $R = 166$ there is a periodic solution (top), while for $R = 166.2$ solutions appear similar, however, interupted by sudden chaotic bursts (bottom).

Intermittency

Knowing that the dynamics of Lorenz' and Rössler's systems have a strong connection to the one-dimensional iteration of quadratic transformations, it is no longer a surprise to again find the back doors to chaos that we discussed in chapter 11: intermittency and crises. Recall that intermittency means that a solution spends most of its time near a periodic solution but is interrupted by sudden and erratic chaotic bursts. For an only slightly perturbed parameter value, these bursts disappear and only the periodic behavior remains. Exactly the same can be observed, for example, in the Lorenz system for the parameters $R = 166.2$ and $R = 166.0$ (see figure 12.39).

A Crisis of the Lorenz Attractor

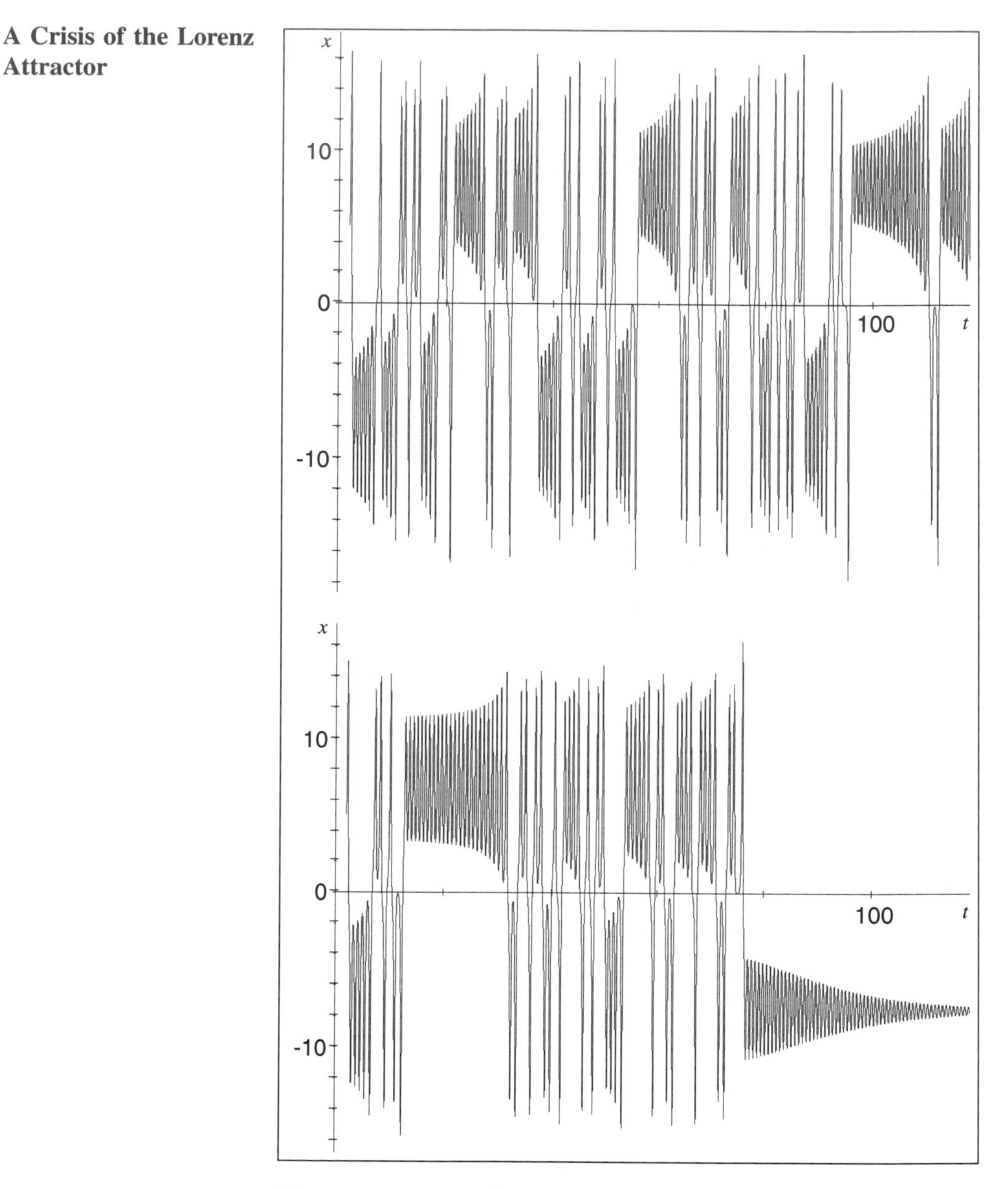

Figure 12.40 : For $R = 25$ (top) there is a strange attractor for the Lorenz system. When the parameter is lowered to $R = 22.4$ (bottom) the attractor undergoes a crises. Trajectories appear chaotic only in an initial transient phase. After that one of the attractive rest points is approached.

Crises

When a periodic solution 'disturbs' the chaos, an intermittent trajectory is produced. In this case, the chaos prevails. However, the chaotic attractor deteriorates, when a periodic trajectory steals the attractivity from the attractor. What remains of the chaos, is the long transient chaotic behavior of solutions, all of which approach the periodic one asymptotically. The chaotic attractor is said to be in a crisis. And again this can also be discovered in the Lorenz system (see figure 12.40). These phenomena are by no means restricted to the quadratic iterator and the Lorenz system; they have been identified in many other mathematical and physical dynamical systems.[42]

[42]See T. Tél, *Transient chaos, Directions in Chaos III,* Hao B.-L. (ed.), World Scientific Publishing Company, Singapore.

12.5 The Reconstruction of Strange Attractors

Understanding natural processes does not start with a set of equations for a dynamical system. On the contrary such models are usually obtained at the end of a long course of action consisting of the identification of the phenomena to be studied, conducting series of often difficult and elaborate experiments, running trial and error computer simulations, and finally making a mathematical analysis. Somewhere in this process the question about order or chaos arises. How can we determine from measured data whether there is some underlying deterministic governing equation for the phenomena observed, or whether the data is merely noise without any structure? In other words, we want to know from a given sequence of numbers whether they come from a strange attractor.

Let us imagine a somewhat simpler situation; we have a black box in which some continuous dynamical system is running. We may only probe the system at discrete time intervals and obtain the value of one of the state variables of the system. For example, choose one of the variables, calling it $z(t)$, and a time interval τ. Our examination of the black box would yield the sequence of numbers

$$z_0 = z(0), z_1 = z(\tau), z_2 = z(2\tau), z_3 = z(3\tau), ... \qquad (12.15)$$

Given such data, can we reconstruct some meaningful picture of some underlying attractor? At first thought this seems a rather hopeless undertaking.

Distinguishing Random from Deterministic Behavior

The prospects, however, are not as bleak as they seem. Here is an example. Figure 12.41 shows three time series all of which look more or less random. However, if one of them really is deterministic, then the numbers must follow some rule. In other words, z_k may be determined from the past of the sequence, and we may hope to be able to put it in the form of

$$z_k = \Psi(z_{k-1}, z_{k-2}, z_{k-3}, ...)$$

where Ψ denotes some (so far) unknown transformation. Let us be even more optimistic and assume that z_k strongly depends on its predecessor, z_{k-1}, and only mildly (or not at all) on all earlier predecessors. To check this assumption we produce plots of z_k versus z_{k-1} as shown in figure 12.42. The result is quite clear. There is no evident structure in the first set of data. The points obtained from the second series clearly lie on a section of what appears to be a parabola. This tells us that this set of data can be generated by means of graphical iteration of a function whose graph is the parabola pictured in the figure. In fact, we used the formula $4x(1-x)$ of the generic parabola to produce the data.

Random or not Random Data?

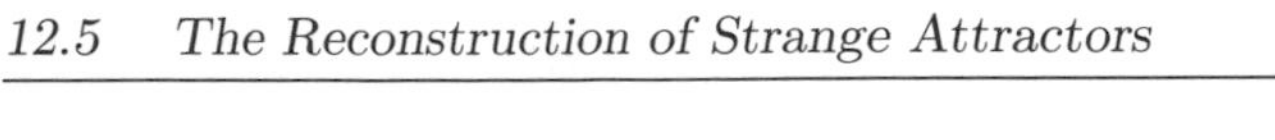

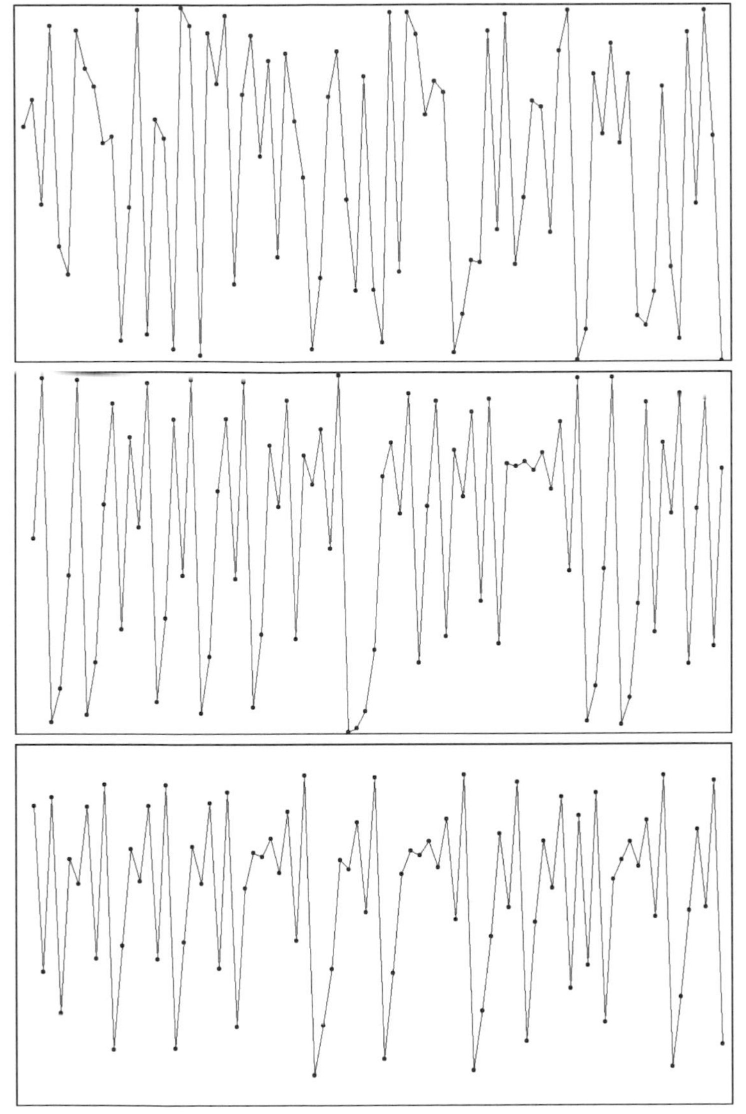

Figure 12.41 : Three series obtained from a 'black box'. Which one is random, which is deterministic? Or are they all random?

Thus, this simple procedure already enables us to completely unravel the random looking data and to uncover its deterministic quadratic generation process. But will such a cheap trick work in real applications where other variables are hidden in the 'black box' or the dependence of the presence on the past is more complicated? The third data set presents such a case. The corresponding

Reconstruction Attempts

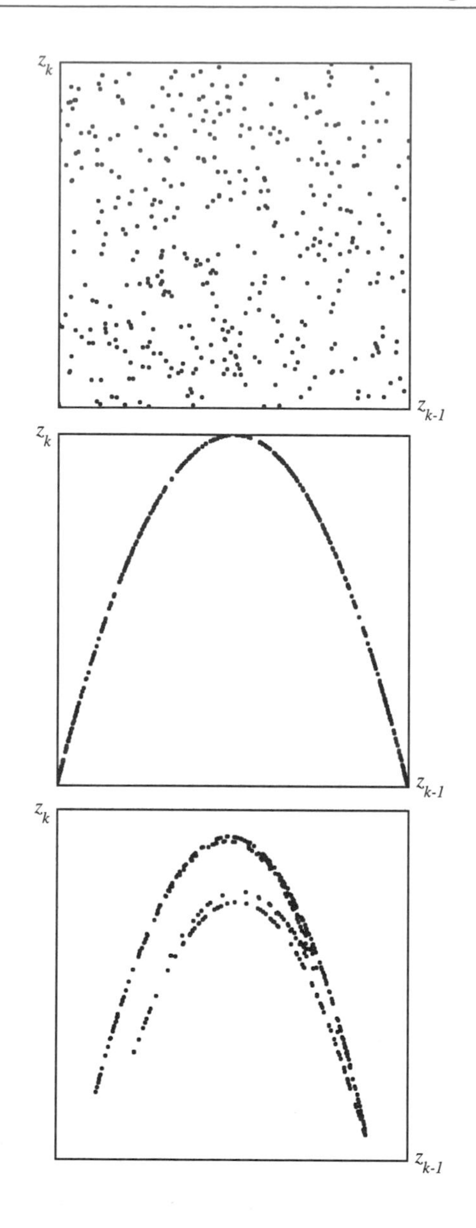

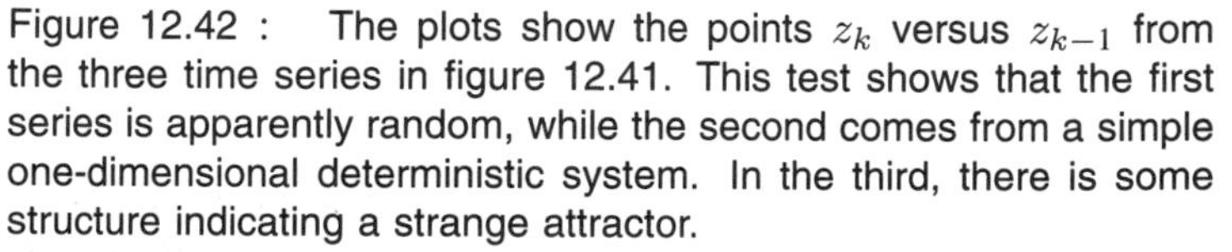
Figure 12.42 : The plots show the points z_k versus z_{k-1} from the three time series in figure 12.41. This test shows that the first series is apparently random, while the second comes from a simple one-dimensional deterministic system. In the third, there is some structure indicating a strange attractor.

plot of z_k versus z_{k-1} shows a collection of points which are not distributed throughout the entire square in contrast to the points

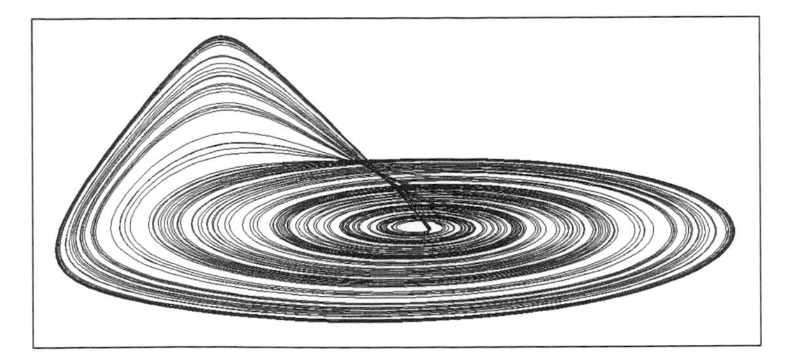

Reconstruction of the Rössler Attractor

Figure 12.43 : Using a time delay of $T = 0.5$ we obtain a picture of phase space with a reconstruction of the Rössler attractor.

of the first random data set. They form a clear structure which, however, cannot be obtained as the graph of a function. But this structure can be interpreted as an attractor of some underlying system which is a crucial insight opening the door to further numerical investigations.[43]

The Phase Space Reconstruction

Thus, this method of analysis of time series may lead to useful results. The fact is that a straightforward extension of this very simple procedure allows us to retrieve the geometric structure of *any* underlying attractor.[44] Let us hide the Rössler system in the 'black box', run the system, and again assume that the sequence in eqn. (12.15) is extracted from the machine. We now choose a time delay T (a multiple of τ) and look at the following sequence of vectors

$$\begin{array}{l}(z(0), z(T), z(2T))\\ (z(\tau), z(\tau+T), z(\tau+2T))\\ (z(2\tau), z(2\tau+T), z(2\tau+2T))\\ \vdots\\ (z(k\tau), z(k\tau+T), z(k\tau+2T))\ .\end{array}$$

Plotting these points in three-dimensional space with connecting line segments, we obtain figure 12.43. Clearly, the essential features of the Rössler attractor are apparent.

The Reconstruction Principle

This is not an accident! Strange attractors theoretically can always be faithfully reconstructed using the above procedure. However, working in three dimensions, we cannot expect the procedure to perform when the dimension of the attractor surpasses 3. In

[43]The structure seen in the bottom plot of figure 12.42 seems to be related to the Hénon attractor. In fact, the underlying sequence $z_0, z_1, z_2, \ldots$ has been generated using the Hénon system $(x_{k+1}, y_{k+1}) = H(x_k, y_k)$ (see section 12.1) and setting $z_k = x_k, k = 0, 1, 2, \ldots$

[44]This procedure was suggested by David Ruelle, see N. H. Packard, J. P. Crutchfield, J. D. Farmer, R. S. Shaw, *Geometry from a time series,* Phys. Rev. Lett. 45 (1980) 712–716.

Different Time Lags

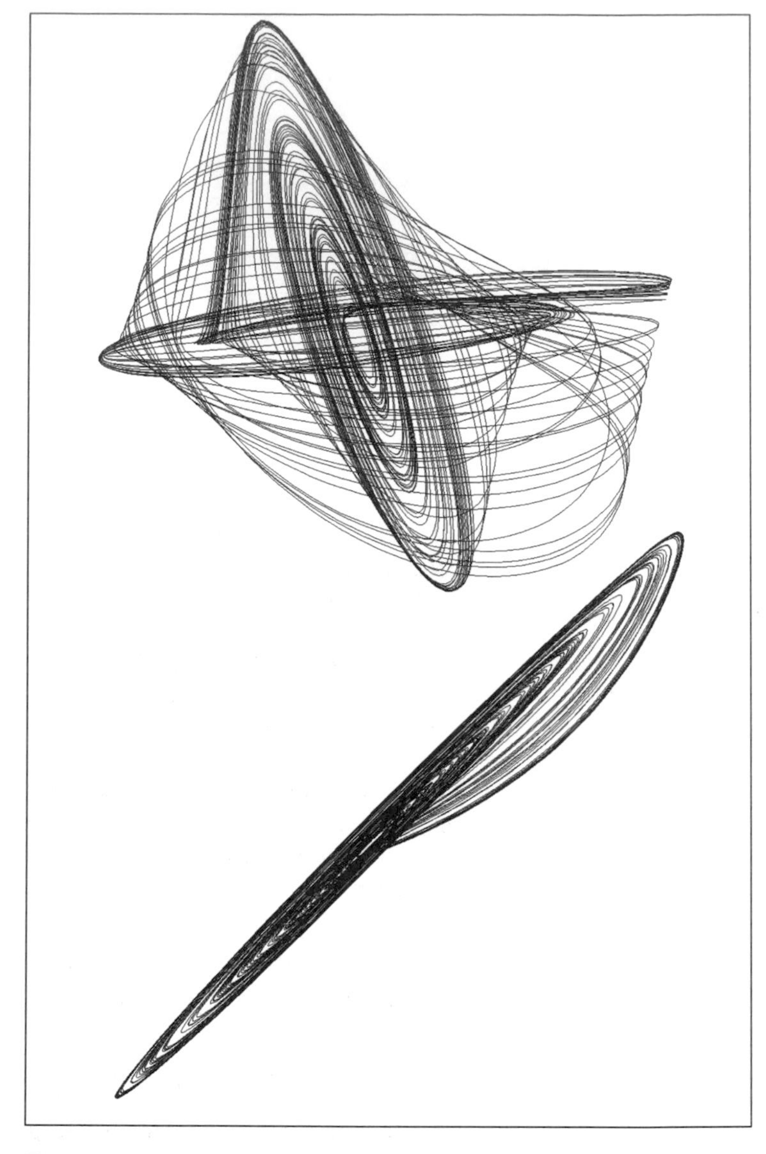

Figure 12.44 : The time lag in the reconstruction process should not be too small or too large. Here $T = 10$ (top) and $T = 0.1$ (bottom) for the Rössler attractor.

such a case a dense subset of the three-dimensional space would be filled. We may, however, simply work in higher-dimensional spaces using vectors

$$u(t) = (z(t), z(t+T), ..., z(t+2NT)) \tag{12.16}$$

with $2N+1$ components. If N is chosen large enough the attractor

Reconstruction of the Lorenz Attractor

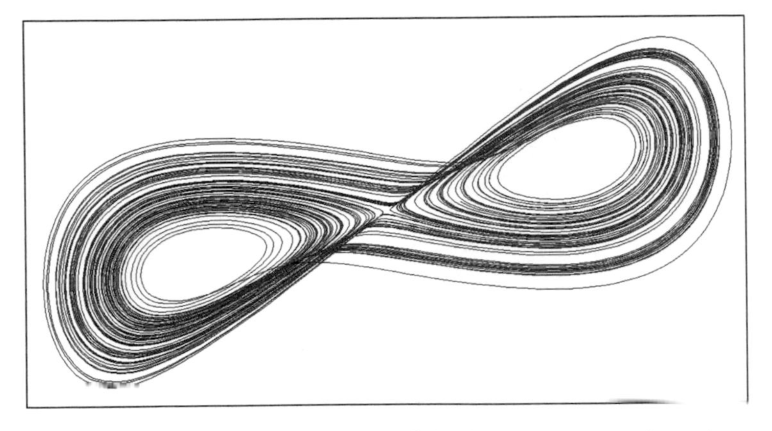

Figure 12.45 : A reconstruction of the Lorenz attractor based on time series of the x-coordinate. We have used $T = 0.05$.

will 'fit' in the chosen space. Following some theorems derived by Ricardo Mañé and Floris Takens, this can be guaranteed if the dimension of the attractor is not larger than N.[45] Here the choice of the time lag T is almost arbitrary. However, in practice there are limitations. If T is quite small, then the vectors to be plotted will have components which are almost identical, resulting in a reconstructed attractor which will be very close to the 'diagonal' of the space. On the other hand if T is very large, then there is only very little correlation between the components of the vectors, and trajectories on the attractor appear to wander all around phase space such that the structure is hard to detect. In figure 12.44 we have chosen two different values of the time lag T to illustrate these points.

The reconstruction of strange attractors can be interpreted as a change of coordinates. Often the attractor is defined in some infinite dimensional space (e.g., a space of functions).[46] In this case the reconstruction amounts to a projection of the original to a finite dimensional Euclidean space. Choosing the dimension $2N + 1$ of the embedding space large enough guarantees that the projection is injective. This means that each point in the projected attractor corresponds to one and only one point in the original attractor. In other words, we see a truthful representation and not some image

[45] Mañé, R., *On the dimension of the compact invariant set of certain nonlinear maps,* in: *Dynamical Systems and Turbulence, Warwick 1980,* Lecture Notes in Mathematics 898, Springer-Verlag (1981) 230–242. Takens, F., *Detecting strange attractors in turbulence,* in: *Dynamical Systems and Turbulence, Warwick 1980,* Lecture Notes in Mathematics 898, Springer-Verlag (1981) 366–381.

[46] Doyne Farmer presents an in-depth study of reconstructions of attractors in an infinite dimensional space with calculations of dimensions and Ljapunov exponents. See D. Farmer, *Chaotic attractors of an infinite-dimensional system,* Physica 4D (1982) 366–393. This study is continued in P. Grassberger and I. Procaccia, *Measuring the strangeness of strange attractors,* Physica 9D (1983) 189–208.

Reconstruction of the Lorenz Attractor

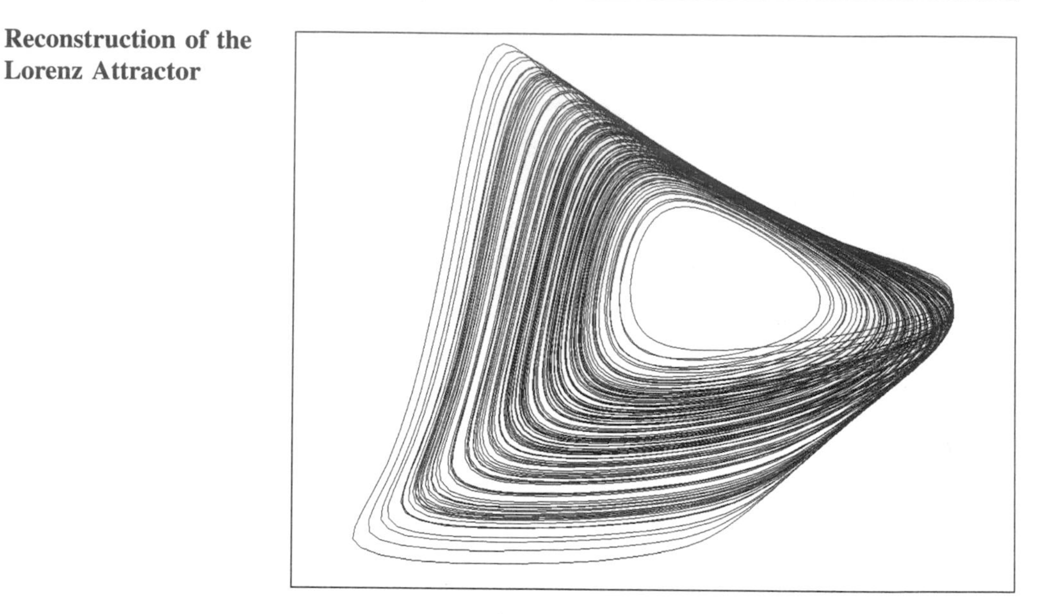

Figure 12.46 : Another reconstructions of the Lorenz attractor based on time series of the z-coordinate. We have used $T = 0.2$. This image reveals that the folded band structure of the Rössler attractor is hidden in the Lorenz attractor.

where parts of the attractor are collapsed onto each other. Thus, the reconstructed attractor is not identical to the original but a more or less distorted copy (see figures 12.45 and 12.46 for reconstructions of the Lorenz attractor). Changing coordinates moderately does not effect the dimension. Thus, we should be able to extract that information from a time series of a single variable.

Michael Faraday's Pioneering Experiment

We conclude this section with an impressive application of the time delay reconstruction method to a physical system termed *acoustic turbulence*. This example is very interesting from several perspectives. On the one hand, it provides a convincing reconstruction along with a successful calculation of dimensions. On the other, there is a link to the first historical observation of the period doubling route to chaos. This is an experiment by Michael Faraday[47] with a periodically driven nonlinear system and dates back to the year 1831.[48] Faraday constructed a large, 18 foot long vibrating plate which held a shallow layer of water. He observed how 'heaps' of liquid were oscillating in a sloshing motion. In particular he reported that "each heap recurs or is re-formed in two complete vibrations of the sustaining surface." In other words, he

[47] Michael Faraday (1791–1867), British physicist, pioneered electromagnetism and invented the dynamo.

[48] M. Faraday, *On a peculiar class of acoustical figures, and on certain forms assumed by groups of particles upon vibrating elastic surfaces,* Phil. Trans. Roy. Soc. London 121 (1831) 299–340.

Setup for Acoustic Chaos

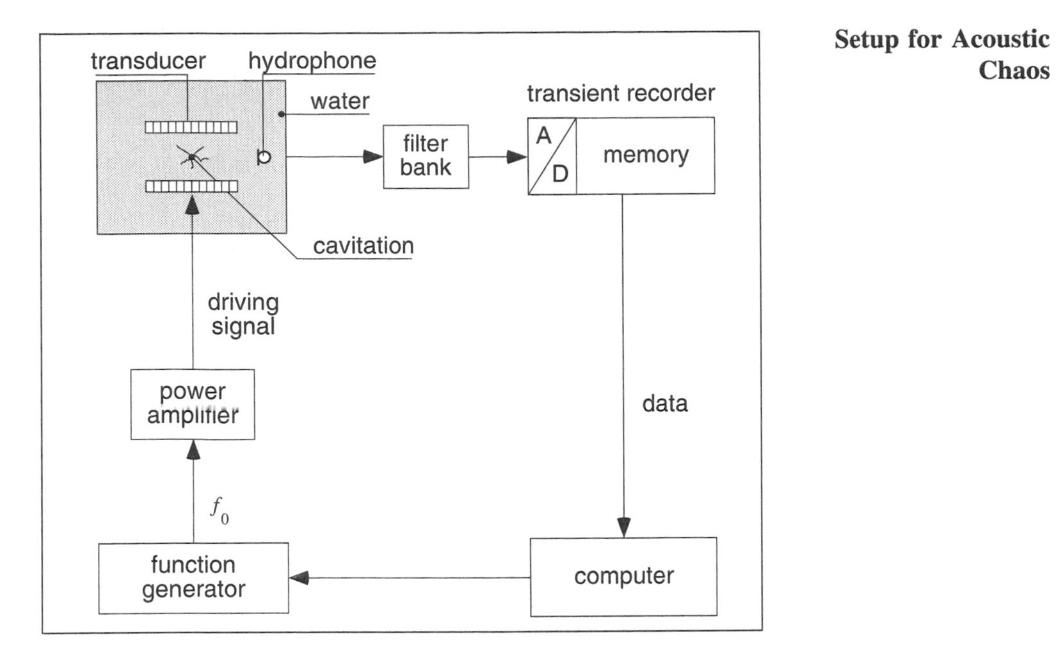

Figure 12.47 : Experimental arrangement for acoustic measurement of driven liquids.

observed the first step in a period-doubling cascade. Later, in 1883, Lord Rayleigh confirmed these results, and today the sophisticated methods of chaos theory and modern computer-controlled experiments are performed to study these transitions from order to chaos in similar systems.[49]

Acoustic Chaos in Ultrasonic Cavitation

For display in this section we have chosen one of these experiments reported by Werner Lauterborn and Joachim Holzfuss. Its purpose is to study acoustic chaos in ultrasonic cavitation, which appears as a hissing noise when a liquid is bombarded with sound of high intensity. This noise stems from the rupture of the liquid structure which results in an organized cloud of small bubbles undergoing complicated dynamics. The driving signal is of a constant frequency and the intensity increases from zero to a high value within a fraction of a second. A spectral analysis of the emitted noise shows strong subharmonic spectral lines at 1/2, 1/4, and also 1/3 of the driving frequency before the broadband hissing noise occurs at high intensity values of the driving sound near the end of the experiment.[50] These spectral lines are a sign of a period-doubling cascade and the Feigenbaum scenario.

[49]For references see the historical notes in W. Lauterborn and J. Holzfuss, *Acoustic chaos,* International Journal of Bifurcation and Chaos 1,1 (1991) 13–26.

[50]The first observations of this sort go back to R. Esche in 1952.

Reconstruction

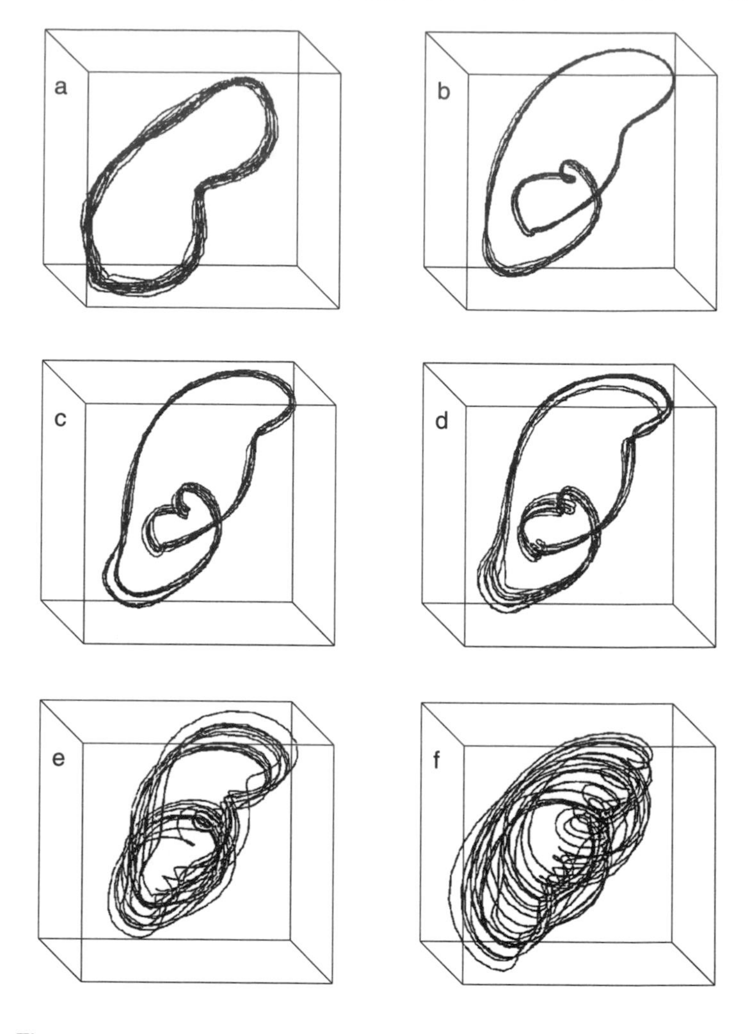

Figure 12.48 : Phase space reconstruction of periodic trajectories and a chaotic attractor.

The experimental setup is shown in figure 12.47.[51] The water in the container must be specially prepared with certain additions. The high intensity sound which drives the experiment is produced in a cylindrical transducer of about 3 inches in length and diameter. The voltage applied to the transducer is computer controlled. Finally, the sound produced is picked up by the hydrophone, appropriately 'cleaned' by the filter bank and stored in the computer memory for later processing. The duration of such an experiment typically is about the order of a quarter of a second.

[51] Our exposition is based on the above mentioned paper by W. Lauterborn and J. Holzfuss. The figures 12.47, 12.48, and 12.49 are reproduced with their kind permission.

Dimension

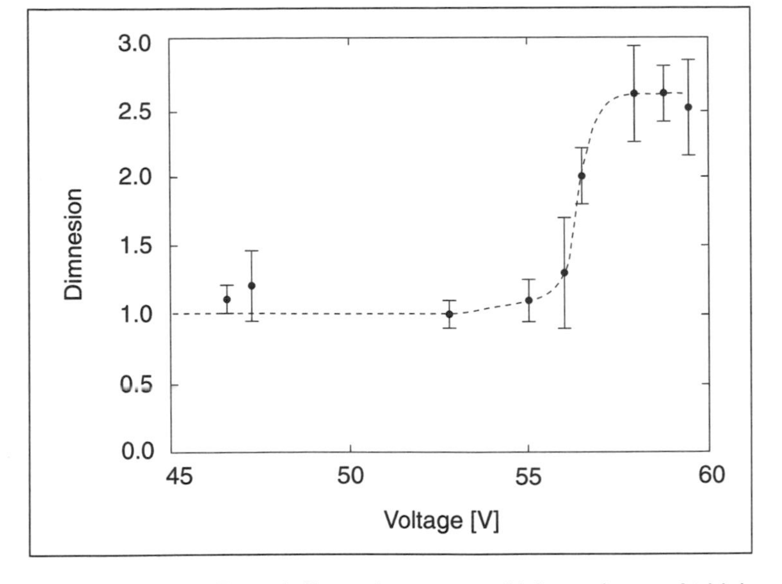

Figure 12.49 : Fractal dimension versus driving voltage. At high voltages the dimension is between 2 and 3.

For the phase space reconstruction of the pressure data $p(t)$, it is sufficient to consider only a three-dimensional setting, i.e., vectors $(p(t), p(t+T), p(t+2T))$, where the time delay T is about one tenth of the period of the driving signal. Figure 12.48 shows six such reconstructions for increasing driving intensities. Following the interpretation of Lauterborn and Holzfuss, the periodic trajectory in part (a) splits into two bands in (b) and four bands in (c). This structure breaks up when the voltage is further increased in parts (d) to (f), where a strange attractor appears. Thus, the hissing sound in the experiment clearly is not some stochastic noise but linked to pronounced effects in nonlinear dynamics. This has been further supported by the computation of dimensions (see figure 12.49) for the reconstructions.

Acoustic chaos is only one of the many examples which convincingly demonstrate that the methods and notions of chaos theory can be successfully applied to problems which only a couple of decades ago would have been discarded as intractable.[52] Chaos theory has opened our eyes and shown us where to look to find those rich structures which indicate once hidden laws of nature.

[52]To give one example, reconstructions of chaotic attractors in hydrodynamics have been reported by Tom Mullin, in: *Chaos in physical systems,* in: *Fractals and Chaos,* A. J. Crilly, R. A. Earnshaw, H. Jones (eds.), Springer-Verlag, New York, 1991. Several examples for chaos in mechanical sytems is described in the introductory text F. C. Moon, *Chaotic Vibrations,* John Wiley & Sons, New York, 1987.

12.6 Fractal Basin Boundaries

Sensitive dependence on initial conditions is one of the central properties in chaos. In this last section of this chapter we present a different kind of sensitivity, namely the so called *final state sensitivity*. This phenomenon may occur whenever there are several coexisting attractors. We have seen such a case already for the Hénon transformation in certain parameter ranges.[53] These may be strange attractors or perhaps simply attractive fixed points of a transformation. With such a transformation on hand, the orbit (or trajectory) for a given initial point will typically converge to either one of the attractors. Therefore, there must be a *boundary* of the corresponding basins of attractions. Such boundaries often are fractals. Physically, and also numerically, an initial point can only be specified up to some precision s. If all orbits started within the distance of s from the initial point converge to the *same* attractor, then there is no problem regarding the prediction of the final state. However, if some of these orbits converge to one attractor and the rest of them to the other attractor, we have a problem. No longer can we safely predict the final state belonging to the given initial point. Clearly, the severity of this problem becomes worse when the fractal dimension of the basin boundaries gets larger. Thus, we have that fractal basin boundaries with a large fractal dimension present an increased obstruction to the predictability in nonlinear systems with several attractors. In other words, in this situation the fractal dimension obtains an immediate dynamical interpretation.

How Large is the Unsafe Region?

Given an uncertainty s, we call an initial point *unsafe*, if there is another initial point at a distance less than s, which converges to a different attractor. We may now ask how large the area (or the volume) $A(s)$ formed by unsafe initial points is. In the simple situation pictured in figure 12.50, this area is given by a strip of width $2s$ around the (smooth) basin boundary and is thus proportional to s. This means that when we improve our precision of initial points by a factor of 2, the region of unsafe initial points reduces by the factor 1/2.

However, if the basin boundary is fractal, then the scaling law should be more complicated. To see this, recall the box-counting dimension (see chapter 4). The size of the region of unsafe initial points is about the same as the area (or volume) $A(s)$ of the $N(s)$ boxes of width s needed to cover the basin boundary. If D_f denotes the fractal dimension and d the Euclidean dimension of the embedding space, we get that the number $N(s)$ is proportional to $1/s^{D_f}$ and

$$A(s) \approx N(s) \cdot s^d \propto s^{d-D_f} = s^{\alpha} .$$

[53] See figure 12.15.

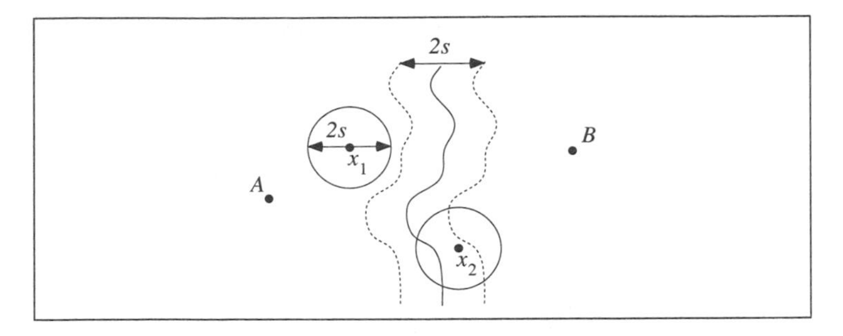

Figure 12.50 : Safe and unsafe initial points. Two attractors (A and B) with basins bounded by a smooth curve. The initial point x_1, which has an uncertainty of amount s, is safe, while the point x_2 is unsafe. The region of unsafe initial points is the strip of width $2s$ around the basin boundary.

The exponent

$$\alpha = d - D_f$$

is called the *uncertainty exponent*, and because $D_f \geq d - 1$, we get that $0 < \alpha \leq 1$.

Let us present another interpretation of this exponent. Assume that we intend to reduce the size of the region of unsafe initial points by a factor of 1/10 by improving the precision with which we measure and approximate the initial points. How many more significant digits do we need? If r is the necessary precision then $r^\alpha = s^\alpha/10$. Taking logarithms (with base 10) we get

$$\log_{10} r = \log_{10} s - \frac{1}{\alpha}\ .$$

Thus, we need $1/\alpha$ additional digits of precision to achieve our goal. If the uncertainty exponent α is small then this increase in precision can become very expensive.

The Pendulum Experiment

In the remainder of this section we present a study of such fractal basin boundaries arising in a physical experiment. Imagine a metal ball tied to a string which is attached to the ceiling. This pendulum may swing in all directions, and we adjust the length of the string so that the ball is close to the ground, where we place three strong magnets near the resting position of the pendulum (see figure 12.51). Holding the ball close to each of the three magnets, we can feel the attracting forces; when we let go of the ball close to one of the magnets, it will stay there. When we release the ball somewhere else, we can observe how the pendulum swings back and forth; and when the course of the ball comes close to one of the magnets, the direction of movement may drastically change due to

The Pendulum Setup

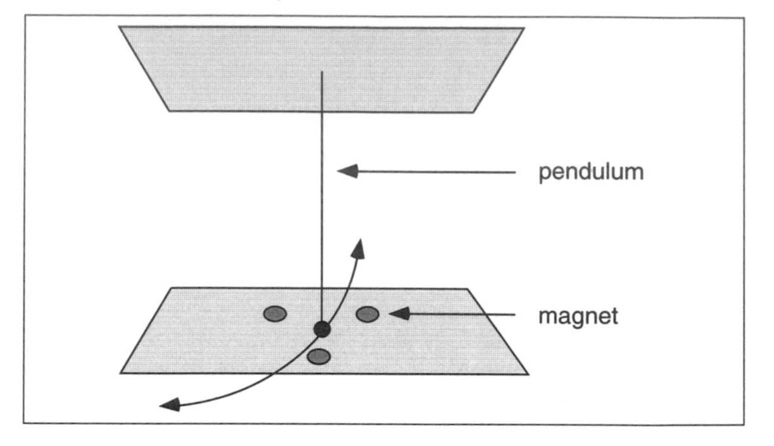

Figure 12.51 : The metal ball of the pendulum swings over three magnets.

the attracting force. In general, the pendulum's swings seem erratic and quite unpredictable. But eventually, the ball comes to rest due to the friction involved (air resistance and internal friction in the string, the attachment, and so on). You can play a little game to forecast which of the magnets the ball will finally be attracted to. For some initial positions it is easy to make a reliable prediction; for others it is nearly impossible. In fact, the experiment is one exhibiting the final state sensitivity described above.

A Computer Model is Necessary

In spite of this inherent difficulty, we may try to make many experiments with the pendulum in order to draw a map depicting the final rest point of the ball for all the chosen initial positions. Of course, it would be a very long and tedious physical experiment to do this for an array of, say, 100 by 100 initial conditions. Therefore, we prefer to employ a computer simulation for the experiment. This involves setting up a mathematical model for the pendulum motion and the implementation of routines for the solution of the underlying equations. Our central assumptions are the following:

- The pendulum length is long compared to the spacing of the magnets. Thus, we may assume for simplicity that the ball moves about on a plane rather than on a sphere with a large radius.
- The magnets are point attractors positioned a short distance below the pendulum plane at the vertices of an equilateral triangle.
- The force applied to the ball by a particular magnet is proportional to the inverse of its squared distance from the magnet.

System Parameters

The conclusion of the premises is given by a differential equation of second order in two variables. There are several parameters

Trajectories and Friction

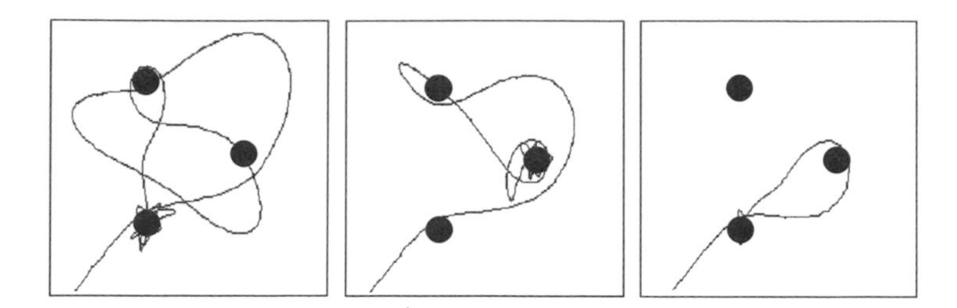

Figure 12.52 : Top view of the paths of the pendulum. Same initial point in the lower left, but different friction parameters (low friction on the left and high friction on the right). Note that the final rest point of the pendulum is not the same in the three experiments.

Trajectories and Gravitational Force

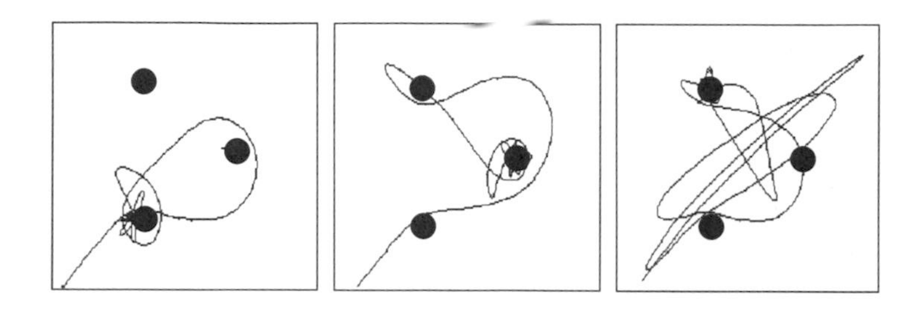

Figure 12.53 : Top view of the paths of the pendulum. Same initial point and different force parameters with low force pulling the pendulum to its mid-point position (left), and strong force (right).

in the model. For example, the amount of friction in the pendulum is a parameter that can easily be changed in a computer program even while it is running, whereas it would be very hard, if not impossible, to change the friction in a physical experiment (we might have to consider building a pendulum suspended in a liquid). Another parameter is the strength of the force pulling the pendulum to its mid-point position. The figures 12.52 and 12.53 list the outcomes of some trajectories of solutions for different parameter settings.

Basins of Attraction

We now come to the central experiment with the pendulum, plotting the basins of attraction. Here we can produce three diagrams (one for each of the three magnets) collecting all the initial points which lead the pendulum to one of the three resting positions. These sets of points are called the basins of attraction. The three magnets are in competition over all points in the plane, and it is not at all clear how the plane is divided up between them. Figure 12.54 shows the outcome of the experiment. Note, that the union of the three shaded regions essentially fills up the whole

Basins of Attraction

Figure 12.54 : Basins of attraction for the pendulum over three magnets. For each of the three magnets, one of the above figures shows the basin shaded in black. The fourth picture displays the borders between the three basins. This border is not a simple line; but within itself it has a Cantor-like structure, as the enlargement in figures 12.55 and 12.56 show (see also the color plates 10 and 11).

square.[54] The boundaries of the basins are shown in the fourth plot of the figure. Although it looks as if the boundaries are made up of just a few line segments, it is demonstrated in a close-up that they really have a very complicated structure (see figure 12.56) which has similarity to a Cantor set. In other words, wherever two basins seem to meet, we discover upon closer examination that the third basin is there in between them, and so ad infinitum.

[54]Theoretically it is also possible that the pendulum comes to rest at another fourth position. For example, if we choose very weak magnets and place them at a certain distance from the natural resting position of the pendulum, then the ball may come to a full stop close to this resting position as if the magnets did not exist. In the experiments discussed here, the magnets are assumed to be so strong that this fourth resting position is not stable. Any arbitrarily small deviation will be magnified, and the ball will be driven to one of the other three resting positions above the magnets. Therefore, this unstable fourth resting position is not numerically observable. It is located exactly in the centers of the plots in figure 12.54.

Figure 12.55 : Same as the lower right picture in figure 12.54, however at the fourfold resolution of 2048 by 2048 pixels.

We can also observe that the fractal structure of the basin boundaries becomes more apparent when we reduce the friction parameter in the computer simulation (see figure 12.57).

The Equation of Motion for the Pendulum Over Three Magnets

The pendulum point mass moves about in the xy-plane, and the magnets are positioned below the xy-plane, say at a distance of d from the plane. Thus, assume the pendulum at position $(x, y, 0)$ and a magnet at $(x_1, y_1, -d)$. According to the second law of Coulomb, we assume that the force that the magnet applies to the pendulum

Blowup of Basin Boundaries

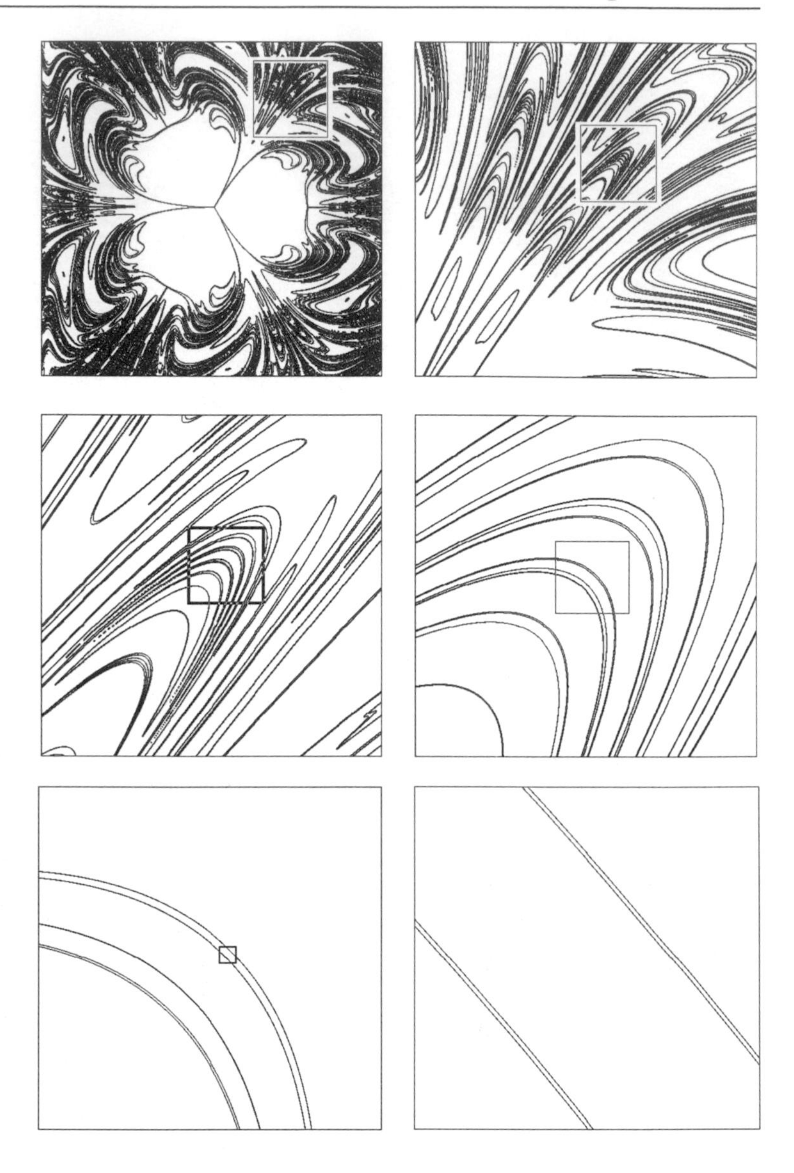

Figure 12.56 : This enlargement of a portion of figure 12.54 reveals the Cantor-set like structure of the boundaries of the basins of attraction in the pendulum experiment.

is inversely proportional to the square of the distance between the two given points, i.e., proportional to

$$\frac{1}{(x_1 - x)^2 + (y_1 - y)^2 + d^2} \,.$$

However, the pendulum motion is restricted to the xy-plane, and

Basin Boundaries With Low Friction

Figure 12.57 : A sequence depicting the fractal basin boundaries for increasing friction parameters (0.2, 0.3, 0.4, 0.5, from upper left to lower right). Increasing the friction results in smaller areas of uncertain initial points. Systems with high friction are more predictable than those with little friction

therefore we should multiply the force by the cosine of the angle α, which is indicated in the figure 12.59.

After some transformations using the elementary expressions for the cosine, we arrive at a force in the xy-plane proportional to

$$\frac{1}{\left(\sqrt{(x_1-x)^2+(y_1-y)^2+d^2}\right)^3}\begin{pmatrix} x_1-x \\ y_1-y \end{pmatrix}.$$

This is a vector in the xy-plane. There are two other forces that must be considered, namely the gravitational force that pulls the pendulum ball back to the center of the xy-plane, and the friction force. The gravitational force can be modeled simply as a force proportional to

$$-\begin{pmatrix} x \\ y \end{pmatrix}$$

Basin Boundaries With Low Spring Constant

Figure 12.58 : The fractal basin boundaries change with the choice of the spring constant C. Here we have used the values 0.3, 0.2, 0.1, 0.0 (from upper left to lower right.) The pendulum introduces complexity in the system. The more pronounced the self-restoring force of the pendulum the larger the region of unsafe initial points becomes.

whereas the friction force acts in opposition to the direction of movement and is proportional to the speed. Thus, the force vector is taken to be proportional to

$$-\begin{pmatrix} x' \\ y' \end{pmatrix}.$$

The above forces can be summed up in a differential equation using Newton's law, which relates the total force to the acceleration of the mass. With a setup of three magnets at positions (x_1, y_1), (x_2, y_2), (x_3, y_3), the force from the magnets becomes a sum over three terms. Thus, after moving all terms onto the left side of the equation

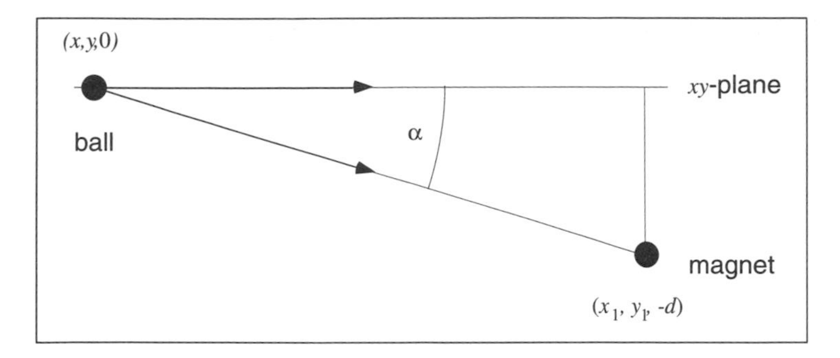

Figure 12.59 : The force of a magnet applied to the pendulum.

we obtain

$$x'' + Rx' - \sum_{i=1}^{3} \frac{x_i - x}{\left(\sqrt{(x_1 - x)^2 + (y_1 - y)^2 + d^2}\right)^3} + Cx = 0$$

$$y'' + Ry' - \sum_{i=1}^{3} \frac{y_i - y}{\left(\sqrt{(x_1 - x)^2 + (y_1 - y)^2 + d^2}\right)^3} + Cy = 0$$

which is a system of two ordinary differential equations of second order. The solution of the corresponding initial value problem requires specification of the position (x, y) and velocity (x', y'). In our experiment the position varies over a square region around the magnet positions and the initial velocity is always chosen to be zero.

The numerical procedure used to solve such an initial value problem can take many different forms. Any standard textbook on numerical analysis will have a number of methods to choose from, starting from the simple Euler scheme, which is not recommended here because of a serious lack of stability, up to more complex algorithms such as adaptive Runge-Kutta methods.[55]

---

[55] A good source of reference and algorithms, including codes in three programming languages, is given in *Numerical Recipes*, W. H. Press, B. P. Flannery, S. A. Teukolsky, W. T. Vetterling, Cambridge University Press, 1986.

12.7 Program of the Chapter: Rössler Attractor

We have introduced the Rössler attractor as a fundamental example of a strange attractor. The program of the chapter implements its generation as a small BASIC program which you can modify to investigate other examples, like the Lorenz attractor.

The parameters of the program are set up to generate the classical Rössler attractor. But you can change any of the parameters. For example, if you change the parameter `c` (see the diagram of figure 12.29) to `c = 8`, you should get a periodic attractor.

Screen Image of Program

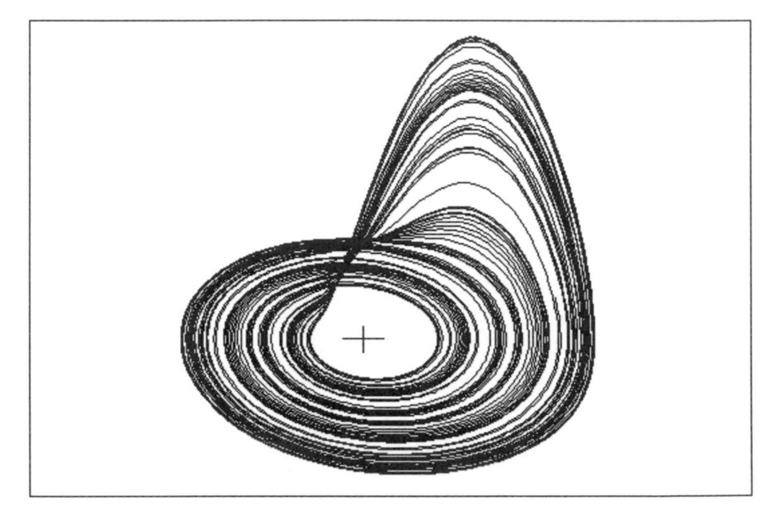

Figure 12.60 : Output of the program 'Attractor'.

The program uses Heun's method to integrate the equation.[56] The computation starts at a given initial point, which need not be on the attractor. The initial transient part of the trajectory is not drawn in order to allow the solution to approach the attractor sufficiently well. The number of the initial hidden points is specified by the variable `hide`. During the time when points are not drawn, the size of the attractor is being computed, which is needed for the final screen projection of the data. In other words, the program keeps track of appropriate minimum and maximum values of the solution.[57] Finally, a coordinate system is drawn (from $x = -1$ to $x = 1$ and $y = -1$ to $y = 1$). The entire drawing is scaled such that the range `xmin` to `xmax` and `ymin` to `ymax` fits into a screen area of width and height equal to `w`. Observe that in the vertical direction we draw the sum of the y- and z-coordinates. This simulates a view of the attractor from an angle of 45 degrees.

[56]See page 301.

[57]At label 100 you find the code which tests for the maximal and minimal values of `xn` and the sum `yn + zn`.

BASIC Programm **Attractor**
Title Computation of a strange attractor

```
left = 20 : w = 300
a = .2 : b = .2 : c = 5.7
dt = .04 : dt2 = dt/2
xmin = 1000 : xmax = -1000 : ymin = 1000 : ymax = -1000
yn = 1
hide = 2000

FOR i = 1 TO 100000!
    REM integration by Heun's method
    GOSUB 500
    xn0 = xn : yn0 = yn : zn0 = zn
    xfn0 = xfn : yfn0 = yfn : zfn0 = zfn
    xn = xn0 + dt*xfn0
    yn = yn0 + dt*yfn0
    zn = zn0 + dt*zfn0
    GOSUB 500
    xn = xn0 + dt2*(xfn + xfn0)
    yn = yn0 + dt2*(yfn + yfn0)
    zn = zn0 + dt2*(zfn + zfn0)
    IF i < hide GOTO 100
        LINE -(left+xfac*(xn-xmin), left+w-yfac*(yn+zn-ymin))
        GOTO 200
        REM find the range of x and y+z
100     IF xn > xmax THEN xmax = xn
        IF xn < xmin THEN xmin = xn
        IF yn+zn > ymax THEN ymax = yn+zn
        IF yn+zn < ymin THEN ymin = yn+zn
        IF i < hide-1 GOTO 200
            REM Draw coordinate axis
            xfac = w/(xmax-xmin)
            yfac = w/(ymax-ymin)
            PSET (left-xfac*(1+xmin), left+w+yfac*ymin)
            LINE -(left+xfac*(1-xmin), left+w+yfac*ymin)
            PSET (left-xfac*xmin, left+w-yfac*(1-ymin))
            LINE -(left-xfac*xmin, left+w+yfac*(1+ymin))
            PSET (left+xfac*(xn-xmin), left+w-yfac*(yn+zn-ymin))
200 NEXT i
END

REM Evaluate function at current point
500 xfn = - yn - zn
    yfn = xn + a*yn
    zfn = b+xn*zn-c*zn
    RETURN
```

At the end of the program a subroutine contains the definition of the system of differential equations under consideration. These are the equations for the Rössler system. To experiment with the Lorenz system change this part to

```
zfn = a * (yn - zn)
yfn = c*zn - yn - zn*xn
xfn = -b*xn + zn*yn
```

The classical Lorenz attractor[58] is obtained for parameters

a = 10 (for σ)
b = 2.667 (for B)
c = 28 (for R) .

However, you should decrease the step size of the integration (dt = 0.02) to obtain a smoother image of the attractor.

[58] If you compare this with the original equations, you will notice that we have exchanged the x- and z-coordinates. Using the original equation would require changing the program to plot $x + y$ in the horizontal direction and z in the vertical direction in order to obtain a nice view of the attractor.

Chapter 13

Julia Sets: Fractal Basin Boundaries

I must say that in 1980, whenever I told my friends that I was just starting with J. H. Hubbard a study of polynomials of degree 2 in one complex variable (and more specifically those of the form $z \to z^2 + c$), they would all stare at me and ask: Do you expect to find anything new? It is, however, this simple family of polynomials which is responsible for producing these objects which are so complicated — not chaotic, but on the contrary, rigorously organized according to sophisticated combinatorial laws.[1]

Adrien Douady

The goal of this chapter is to demonstrate how genuine mathematical research experiments open a door to a seemingly inexhaustible new reservoir of fantastic shapes and images. Their aesthetic appeal stems from structures which are beyond imagination and yet, at the same time, look strangely familiar. The ideas we present here are part of a world wide interest in so called *complex* dynamical systems. They deal with chaos and order, both in competition and coexistence. They show the transition from one condition to the other and how magnificently complex the transitional region generally is. One of the things many dynamical systems have in common is the competition of several centers for the domination of the plane. A single boundary between territories is seldom the result of this contest. Usually, an unending filigree

[1] Adrien Douady, *Julia sets and the Mandelbrot set*, in: The Beauty of Fractals, H.-O. Peitgen, P. H. Richter, Springer-Verlag, Heidelberg, 1986.

entanglement and unceasing bargaining for even the smallest areas results. We studied the quadratic iterator in chapters 1, 10 and 11 and learned that it is the most prominent and important paradigm for chaos in deterministic dynamical systems. Now we will see that it is also a source of fantastic fractals. In fact the most exciting discovery in recent experimental mathematics, i.e., the Mandelbrot set, is an offspring of these studies. Now, about 10 years after Adrien Douady and John Hamal Hubbard started their research on the Mandelbrot set, many beautiful truths have been gained about this 'most complex object mathematics has ever seen'. Almost all of this progress stems from their work.

This chapter begins with an informal discussion of basin boundaries (section 13.1) and a short introduction to complex numbers (section 13.2) and methods for complex quadratic equations (section 13.3). Thus, readers who are already familiar with these notions and tools may want to only briefly scan through these sections and then start reading section 13.4 where Julia sets finally come in.

13.1 Julia Sets as Basin Boundaries

Pixel Game Rules

In section 12.6 we have presented the pendulum over three magnets as a physical example of competition between centers of attraction. The corresponding dynamical system is given by means of certain physical laws, which manifest themselves in the form of a differential equation. We now consider a sort of game where the dynamical laws are much simpler. They are given as a table of rules. Imagine a large square board where the fields are assigned labels as in the game of chess. For example A1 denotes the lower left square, and C8 is the eighth square of the third row. In each square we write the coordinates of a follow-up square as shown in figure 13.1. The game simply consists in following the instructions: we place a peg on an initial square at pleasure, read the coordinates from that square and move the peg to the indicated destination. Then the procedure is repeated.

Pixel Game Scenarios

Having understood the simple rules of the game, we now may ask what the possible dynamical patterns in this game are. The answer is quite easy. Assuming that the player is never required to leave the domain given by the board, i.e., all destinations are coordinates on the board, the journey of the peg has the following alternatives:

- The path comes to a halt, i.e., at some point the peg arrives on a square whose destination is itself. Such a square may be called a fixed point.
- The above is not true. In that case the sequence of visited squares must be periodic from some point on. This is true because there are only finitely many squares on the board; and thus, after a finite number of moves, the peg will arrive on a square that was already visited before. From then on the sequence becomes periodic.

Let us carry out this procedure for all squares on the board. The result is shown in figure 13.2. In summary, the figure shows:

- There are three fixed squares, namely B3, F10 and K3.
- All trajectories, i.e., all sequences starting anywhere on the board, terminate at one of the three above fixed squares. There are no cycles.
- For each fixed square (for example, B3), there is a *basin of attraction* which consists of all initial squares which eventually lead to the fixed square. The basins of attraction can have several components.

We may add one additional piece of information on the entries of the board, namely the number of moves necessary to advance the peg from an initial square to the final fixed square. This is also

Pixel Game Transitions

	1	2	3	4	5	6	7	8	9	10	11
L	K 2	K 3	K 3	K 4	K 4	I 5	I 6	I 7	I 8	I 8	I 9
K	K 3	K 3	×	K 3	I 3	H 4	H 5	G 7	H 8	H 9	H 9
I	I 3	I 3	K 4	K 3	I 2	G 3	F 5	F 7	F 8	G 9	G 10
H	H 3	I 4	K 5	L 3	K 1	D 2	B 5	D 7	F 9	F 9	G 10
G	G 4	H 5	K 7	K 6	L 2	A 2	C 7	C 10	E 10	F 10	F 10
F	F 4	F 6	F 9	F 10	F 11	F 11	F 11	F 11	F 11	×	F 10
E	E 4	D 5	B 7	B 6	A 2	L 2	I 7	I 10	G 10	F 10	F 10
D	D 3	C 4	B 5	A 3	B 1	H 2	K 5	H 7	F 9	F 9	E 10
C	C 3	C 3	B 4	B 3	C 2	E 3	F 5	F 7	F 8	E 9	E 10
B	B 3	B 3	×	B 3	C 3	D 4	D 5	E 7	D 8	D 9	D 9
A	B 2	B 3	B 3	B 4	B 4	C 5	C 6	C 7	C 8	C 8	C 9

Figure 13.1 : Each square contains the coordinates of a next square. Pick an initial square at pleasure and follow the sequence of squares. Does this journey ever end; and if so, where does it end?

shown in figure 13.2, see the numerical entries in the squares. It seems to be suggested here that from squares near the boundary of a basin it takes more moves.

Julia Sets and the Pixel Game

This simple game already shows the basic procedure used in the computation of Julia sets. The squares correspond to the finite, though possibly very large, number of points in the plane that can be represented in the computer. The rules that prescribe the transition from one square to the next are given in terms of a formula, and not as a table as in the figure. In fact, as the reader probably has already guessed, the transition rules of our pixel game have been derived from a formula corresponding to a Julia set with three basins of attraction.

Complex Polynomial Transition Functions

The mathematical analysis of the dynamical properties analogous to our simple pixel game is carried out in dynamical systems theory. There are many different classes of dynamical systems. In section 12.6, for example, we looked at a mathematical model of

Results of the Pixel Game

2	1	1	2	2	5	6	4	4	4	4
1	1	○	1	3	3	3	5	6	4	4
3	3	2	1	4	5	3	3	3	3	2
5	2	4	2	2	3	4	5	3	3	2
5	3	4	4	2	2	4	4	2	1	1
2	3	3	1	2	2	2	2	2	●	1
5	3	4	4	2	2	4	4	2	1	1
5	2	4	2	2	3	4	5	3	3	2
3	3	2	1	4	5	3	3	3	3	2
1	1	●	1	3	3	3	5	6	4	4
2	1	1	2	2	5	6	4	4	4	4

Figure 13.2 : This figure summarizes the results for the Pixel Game of the previous figure. There are three fixed squares (marked by a large dot) with three corresponding basins of attraction (black, white and grey) around them. The numbers in the squares indicate the number of moves necessary to advance to the final fixed square.

a pendulum — in mathematical terms, a so called ordinary differential equation. Population models in biology have led to another example, namely discrete dynamical systems based on the iteration of a real function (a function with one real variable and real values, e.g., $z \longrightarrow z + rz(1-z)$, such as those discussed in chapter 1). When the transition function from one state of the system to the next is given as a complex polynomial or complex rational function, then the corresponding background is provided by the theory of Julia sets.

Gaston Julia and Pierre Fatou

The name Julia set stems from the French mathematician Gaston Julia (1893–1978), who developed much of the theory while he was recovering from his war wounds in an army hospital. During World War I he served as an officer. On 25 January 1915 the German headquarters decided to celebrate the Kaiser's birthday by organizing an attack on the French front with the intention of taking

Cayley's Paper

Desiderata and Suggestions.

BY PROFESSOR CAYLEY, *Cambridge, England.*

No. 3.—THE NEWTON-FOURIER IMAGINARY PROBLEM.

THE Newtonian method as completed by Fourier, or say the Newton-Fourier method, for the solution of a numerical equation by successive approximations, relates to an equation $f(x) = 0$, with real coefficients, and to the determination of a certain real root thereof a by means of an assumed approximate real value ξ satisfying prescribed conditions: we then, from ξ, derive a nearer approximate value ξ_1 by the formula $\xi_1 = \xi - \frac{f(\xi)}{f'(\xi)}$; and thence, in like manner, $\xi_1, \xi_2, \xi_3, \ldots$ approximating more and more nearly to the required root a.

In connexion herewith, throwing aside the restrictions as to reality, we have what I call the Newton-Fourier Imaginary Problem, as follows.

Take $f(u)$, a given rational and integral function of u, with real or imaginary coefficients; ξ, a given real or imaginary value, and from this derive ξ_1 by the formula $\xi_1 = \xi - \frac{f(\xi)}{f'(\xi)}$, and thence $\xi_1, \xi_2, \xi_3, \ldots$ each from the preceding one by the like formula.

A given imaginary quantity $x + iy$ may be represented by a point the coordinates of which are (x, y): the roots of the equation are thus represented by given points $A, B, C\ldots$, and the values $\xi, \xi_1, \xi_2 \ldots$ by points $P, P_1, P_2, \ldots$ the first of which is assumed at pleasure, and the others each from the preceding one by the like given geometrical construction. The problem is to determine the regions of the plane, such that P being taken at pleasure anywhere within one region we arrive ultimately at the point A; anywhere within another region at the point B; and so for the several points representing the roots of the equation.

The solution is easy and elegant in the case of a quadric equation, but the next succeeding case of the cubic equation appears to present considerable difficulty.

CAMBRIDGE, *March 3d, 1879.*

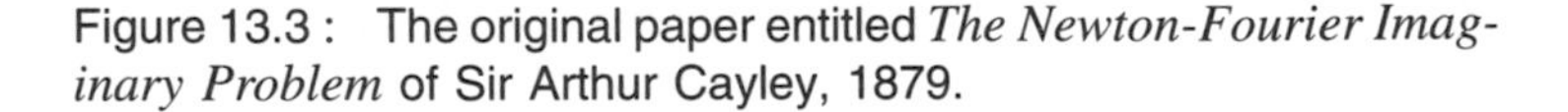

26 97

Figure 13.3 : The original paper entitled *The Newton-Fourier Imaginary Problem* of Sir Arthur Cayley, 1879.

1000 prisoners. The attack took place and was extremely violent, but the goal was not achieved. Many on both sides were killed or wounded. Julia was one of those who were badly wounded. He lost his nose during the aforementioned attack, and thereafter wore a black leather strap across his face. Simultaneously Julia's competitor Pierre Fatou (1878–1929) created another huge volume of results in the area of complex iteration.[2] These early results were almost forgotten for many years and became popular again only in the 1980's through Mandelbrot's work. The immense progress that Julia and Fatou were able to make must be valued all the more because in those days there were no computers to aid in the

[2]In addition to his work in astronomy at the Paris observatory Pierre J. L. Fatou was very productive in mathematics delivering numerous results, in particular in complex analysis.

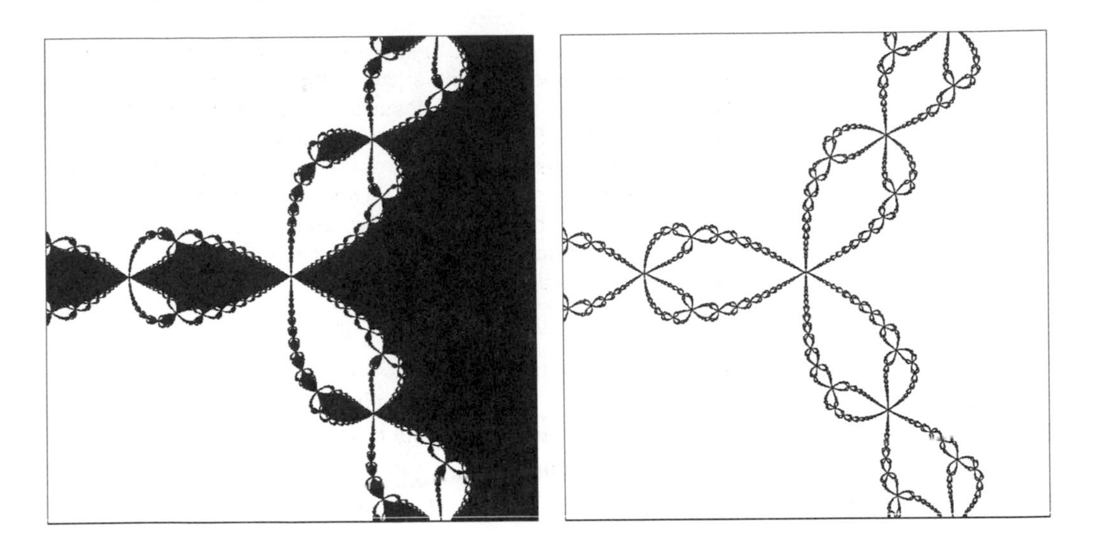

Figure 13.4 : The figure shows the basins of attraction for Newton's method applied to the equation $z^3 - 1 = 0$. Left, the basin for the solution $z = 1$ is shown in black, while right only the basin boundaries are pictured. It is an example of a Julia set.

understanding of the complicated matter; instead they had to rely completely on their imagination. Much of Julia's work was motivated by a one page article by the famous British mathematician Sir Arthur Cayley, who published a problem in 1879, the 'Newton-Fourier Imaginary Problem' (see figure 13.3).[3] When computing the roots of $z^3 - 1 = 0$ using Newton's method one is led to a competition situation which is very similar to our pendulum experiment.

Cayley's Problem

Cayley's problem is related to the pendulum over three magnets from the last chapter in a very nice way. We considered the question which magnet the pendulum would come to rest over. Cayley considered Newton's method and asked, which zero of the equation $z^3 - 1 = 0$ in the complex plane[4] would the method converge towards if one starts with an arbitrary initial guess. The equation $z^3 - 1 = 0$ has three solutions, $\{1, e^{2\pi i/3}, e^{4\pi i/3}\}$. Newton's method for this equation is given by the feedback iteration

$$z_{n+1} = z_n - \frac{z_n^3 - 1}{3z_n^2} \,. \tag{13.1}$$

Its basic properties are designed so that the three solutions of the cubic equation behave like magnets for the iteration. Julia understood

[3]Arthur Cayley, *The Newton-Fourier imaginary problem,* American Journal of Mathematics 2, 1879.

[4]Complex numbers will be introduced in the next section.

that the three corresponding basins of attraction have a common boundary. Any boundary point is like a three-corner point between three countries, i.e., arbitrarily close to any boundary point we can find points in each of the three basins of attraction.[5]

Actually, our pixel game of figures 13.1 and 13.2 is based on a variant of Newton's method. The pixels correspond to squares in the complex plane, their centers being taken as points z_n, and the coordinates of the target square being derived from the complex number z_{n+1} according to equation 13.1.

Before being able to understand and compute Julia sets you must feel comfortable with complex numbers. On the following few pages we provide the reader with a short introduction that contains the most important facts. Readers already basically familiar with complex numbers may skip to the following section dealing with complex square roots.

[5]A recent collection of papers discussing Newton's method as dynamical systems is in *Newton's Method and Dynamical Systems,* H.-O. Peitgen (ed.), Kluver Academic Publishers, Dordrecht, 1989. See also H.-O. Peitgen P. H. Richter, *The Beauty of Fractals,* Springer-Verlag, Heidelberg, 1986, chapters 6 and 7.

13.2 Complex Numbers — A Short Introduction

Complex numbers are an extension of real numbers. The historical motivation for their invention stems from the desire to be able to solve algebraic equations that normally, i.e., by using traditional real numbers, have no solution. For example, $x^2 + 1 = 0$ has no real solution and, thus, a symbolic solution was created and called the *imaginary unit* i with the postulated property

$$i^2 = -1 \ .$$

A *complex number* z has two components which are called the *real* and the *imaginary* part. We write

$$z = x + yi \ ,$$

where $\text{real}\,(z) = x$ denotes the real, $\text{imag}\,(z) = y$ the imaginary part, and i the imaginary unit.

The arithmetic of complex numbers is a straight forward extension of the arithmetic of real numbers. We add two numbers $z = x + yi$ and $w = u + vi$ by

$$z + w \ = \ (x + yi) + (u + vi) = (x + u) + (y + v)i \ ,$$

i.e., by adding the real and imaginary parts separately. We multiply the two numbers by

$$\begin{aligned} z \cdot w &= (x + yi)\ (u + vi) = xu + xvi + yui + yvi^2 \\ &= (xu - yv) + (xv + yu)i \ . \end{aligned}$$

Note that the term yvi^2 contributes to the real part of the product, since $i^2 = -1$.

Embedding of the Real Numbers

Let us look at a special case, namely choosing z and w with imaginary component being zero, i.e.,

$$z = x + 0i = x \quad \text{and} \quad w = u + 0i = u \ .$$

For these numbers the addition and multiplication rules say that

$$\begin{aligned} z + w &= (x + u) + (0 + 0)i = x + u \ , \\ z \cdot w &= (xu - 0 \cdot 0) + (x \cdot 0 + u \cdot 0)i = xu \ . \end{aligned}$$

This indicates that the real numbers are embedded in the complex numbers as those having imaginary components being zero. All the arithmetic laws for real numbers also apply for the real numbers interpreted as complex numbers. However, there is an important distinction between real and complex numbers. There are positive real numbers, and this idea induces an ordering of all real numbers. In other words, given two distinct real numbers, x and y, we have either $x < y$ or $y < x$. Such an ordering does not exist in the complex numbers.

The Complex Number

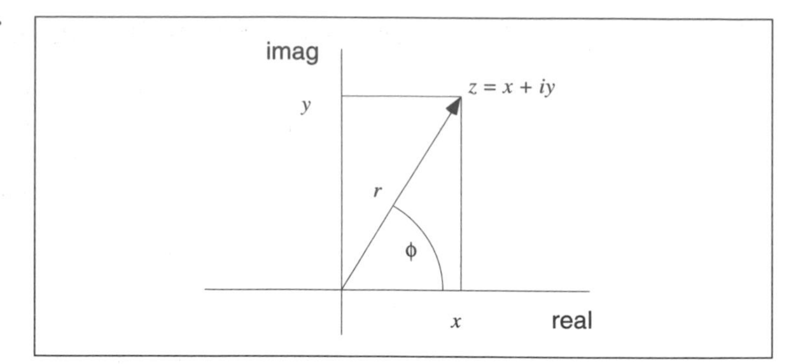

Figure 13.5 : The complex number z corresponds to a point in the plane with coordinates x and y. The length of the vector from the origin to the point is $r = |z| = \sqrt{x^2 + y^2}$ and is called the absolute value or modulus of z. The counterclockwise angle ϕ that the vector makes with the real positive axis is called the argument $\arg z$.

Complex Addition

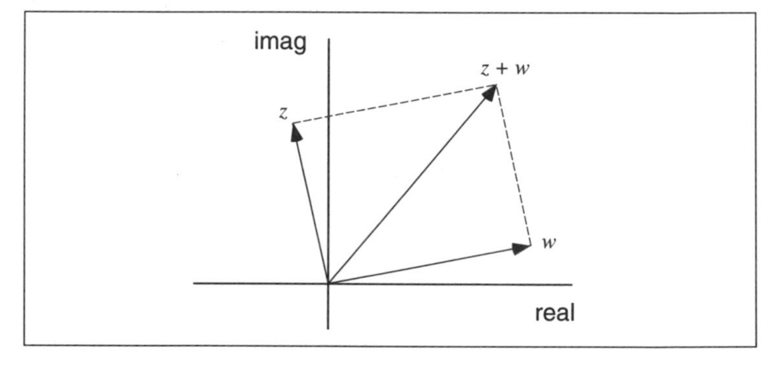

Figure 13.6 : The addition of two complex numbers is identical to the addition of two point vectors from the origin to the numbers.

Complex Numbers as Points in the Plane

While real numbers can be geometrically interpreted as points on a line, it is common to identify complex numbers as points in a plane. The coordinates of a complex point are its real and imaginary parts (see figure 13.5). Thus, we may write a complex number $z = x + yi$ as a vector

$$z = \begin{pmatrix} x \\ y \end{pmatrix} = \begin{pmatrix} \text{real part} \\ \text{imaginary part} \end{pmatrix}$$

and x and y are also called the *Cartesian coordinates* of z.

Polar Coordinates

The length of the point vector corresponding to a complex number z is called its *absolute value* or *modulus* $r = |z|$. For $z = x + yi$ it is given by

$$r = |z| = \sqrt{x^2 + y^2} \ .$$

The counterclockwise angle ϕ that the vector makes with the real positive axis is called the argument

$$\phi = \arg z \ .$$

Thus, there are two ways to specify a complex number z: either by the Cartesian coordinates $x = \text{real}(z)$ and $y = \text{imag}(z)$, or using its so called *polar coordinates*, i.e., its modulus $r = |z|$ and argument $\phi = \arg z$. Using the trigonometric function sin and cos we may write

$$z = r(\cos\phi + (\sin\phi)i)r \ .$$

Thus, the conversion from polar to Cartesian coordinates is given by

$$x = r\cos\phi$$
$$y = r\sin\phi \ .$$

The conversion of polar to Cartesian coordinates, however, is more complicated.

Converting to Polar Coordinates

A point in the xy-plane can be specified as usual by its x- and y-coordinates. An alternative way is given by the *polar coordinates* which characterize a point (x, y) (respectively a complex number $z = x + yi$) by the absolute value $r = |z|$ and the angle ϕ between the vector from the origin to the point (x, y) and the x-axis. This angle is also called the *argument* of z. Thus,

$$z = r(\cos\phi + i\sin\phi) \ .$$

Care must be taken since the argument is not uniquely defined: if the point z is zero, all angles can be used, if z is not zero, then the argument is defined only up to integer multiples of 2π. Thus, ϕ and $\phi + 2\pi$ describe the same. As a formula, the *polar coordinate transformation* point is given by

$$r = \sqrt{x^2 + y^2}$$
$$\sin\phi = \frac{y}{r} \quad \text{and} \quad \cos\phi = \frac{x}{r} \ ,$$

The second formula is not explicit, it defines the argument ϕ in terms of $\sin\phi$ and $\cos\phi$. In practice we would use the arctan function applied to y/x to obtain an angle between $-\pi/2$ and $\pi/2$ (assuming $x \neq 0$). Then it is left to decide whether the argument is this angle or the angle plus π. Explicitly, the algorithm is:

```
if (x > 0) then
   phi = arctan(y/x)
else if (x < 0) then
   phi = pi + arctan(y/x)
else if (y > 0) then
```

```
   phi = pi/2
else if (y < 0) then
   phi = -pi/2
else
   print "Error: x = y = 0, argument undefined."
end if
```

Complex Multiplication

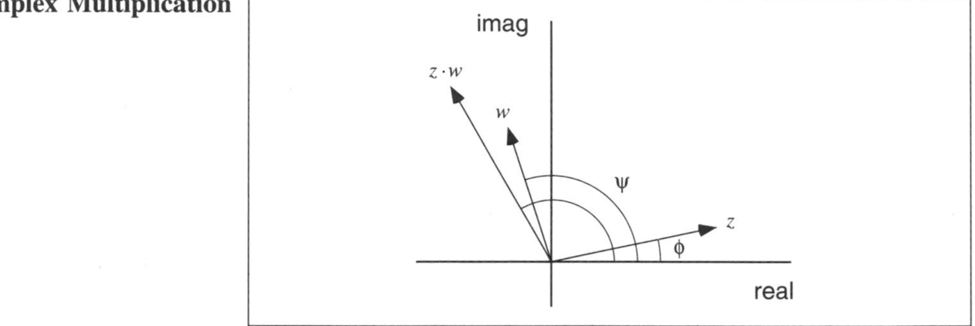

Figure 13.7 : Multiplication of two complex numbers. The product is given by a point whose argument is the sum of the arguments of the factors and whose absolute value, i.e., its distance to the origin of the plane, is given by the product of the absolute values of the factors.

Geometry of Multiplication

Using polar coordinates we can present an elegant geometrical interpretation of multiplication of complex numbers. Consider two complex numbers

$$\begin{aligned} z &= r(\cos\phi + i\sin\phi)\ , \\ w &= s(\cos\psi + i\sin\psi)\ . \end{aligned}$$

and their product zw. Then the polar coordinates of the product are easy to obtain. The modulus of the product is equal to the product of the moduli r and s, and its argument is simply the sum of the arguments of the factors, $\phi + \psi$. Thus,

$$z \cdot w = rs(\cos(\phi + \psi) + i\sin(\phi + \psi))\ . \tag{13.2}$$

In other words, multiplying two complex numbers means adding the corresponding angles and multiplying the lengths of the associated vectors (see figure 13.7).

Derivation

It remains to be shown that the geometric interpretation agrees with the initial definition of the product of two numbers. We let

$$\begin{aligned} z &= x + yi = r(\cos\phi + i\sin\phi) \\ w &= u + vi = s(\cos\psi + i\sin\psi) \end{aligned}$$

and make use of the double angle identities

$$\begin{aligned} \cos(\phi + \psi) &= \cos\phi\cos\psi - \sin\phi\sin\psi \\ \sin(\phi + \psi) &= \sin\phi\cos\psi + \sin\psi\cos\phi \ . \end{aligned}$$

We compute

$$\begin{aligned} zw &= (xu - yv) + (xv + uy)i \\ &= r\cos\phi \cdot s\cos\psi - r\sin\phi \cdot s\sin\psi \\ &\quad + (r\cos\phi \cdot s\sin\psi + s\cos\psi \cdot r\sin\phi)i \\ &= rs(\cos\phi\cos\psi - \sin\phi\sin\psi) \\ &\quad + rs(\sin\phi\cos\psi + \sin\psi\cos\phi)i \\ &= rs(\cos(\phi + \psi) + \sin(\phi + \psi))i \ . \end{aligned}$$

Thus, eqn. (13.2) is shown.

Euler Notation

Closely related to the polar representation is the Euler notation of complex numbers. In this notation a complex number with modulus 1 and argument ϕ can be written as

$$e^{i\phi} = \cos\phi + i\sin\phi \ .$$

The polar representation of a number with modulus r and argument ϕ is conveniently expressed as $re^{i\phi}$. With this definition the laws for the exponential function with real exponents carry over to complex exponents. For example,

$$\begin{aligned} e^{a+bi} &= e^{a}(\cos b + i\sin b) \\ re^{i\phi} \cdot se^{i\psi} &= rs\ e^{(\phi+\psi)i} \ . \end{aligned}$$

The remarkable fact about complex numbers is that we can compute with them in almost the same fashion as with ordinary real numbers. For example, the common laws such as the commutative and the distributative laws apply.

Subtraction and Division

Let us finally state the arithmetic rules for subtraction and division of complex numbers. Subtraction is straightforward just like addition

$$z - w = (x - u) + (y - v)i$$

while division is somewhat more complicated. Division, of course, is the inverse of multiplication. Thus, geometrically, dividing by a complex number with modulus $r > 0$ and argument ϕ should

The Complex Conjugate

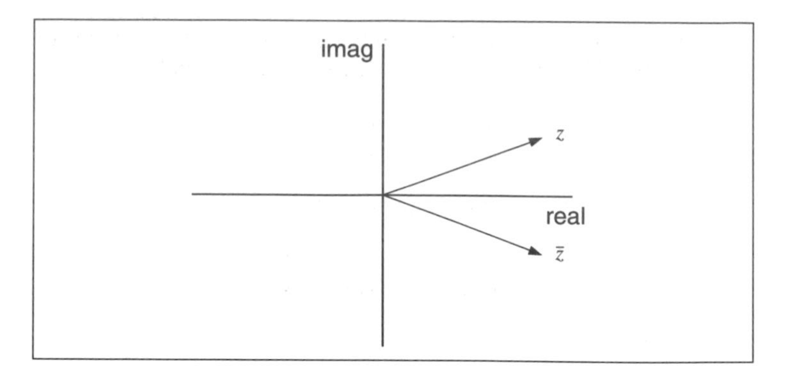

Figure 13.8 : For a complex number $z = x + yi$ the complex conjugate is $\bar{z} = x - yi$, and their product is $z \cdot \bar{z} = x^2 + y^2 = |z|^2$, the squared modulus of z.

correspond to a scaling by the inverse factor $1/r$ and a *clockwise* rotation by the angle ϕ. Thus, in polar coordinates,

$$\frac{z}{w} = \frac{r(\cos\phi + i\sin\phi)}{s(\cos\psi + i\sin\psi)}$$
$$= \frac{r}{s}\left(\cos(\phi - \psi) + i\sin(\phi - \psi)\right) .$$

To derive the formula for division in terms of Cartesian coordinates, it is helpful to introduce the so-called *complex conjugate* number $\bar{z}$. If $z = x + yi$, then $\bar{z} = x - yi$. Thus, in the complex plane $\bar{z}$ is the point z mirrored at the real axis (see figure 13.8). To be specific, we have that $z = \bar{z}$ if and only if z is a real number. Also,

$$z \cdot \bar{z} = x^2 + y^2 = |z|^2 .$$

Now, we compute for $w \neq 0$

$$\frac{z}{w} = \frac{z\bar{w}}{w\bar{w}} = \frac{z\bar{w}}{|w|^2} = \frac{xu + yv}{u^2 + v^2} + \frac{-xv + uy}{u^2 + v^2} i .$$

Thus, a division is computationally more expensive than a multiplication.

On the Cost of a Division

In the complex division scheme presented above, a total of 8 multiplication and division steps are used. It is possible to reduce this number to 6 as follows. First compute

$$q = \frac{v}{u}$$

and then multiply nominator and denominator by $1 - qi$, i.e., we have

$$\frac{z}{w} = \frac{x + yq}{u + vq} + \frac{y - xq}{u + vq} i \ .$$

This procedure is numerically stable for $|u| \geq |v|$. In the other case, for $|u| < |v|$, multiplication of nominator and denominator by $u/v - i$ will do the job. It has been proven that we cannot further reduce the number of multiplications and divisions in a complex division operation. Whether this improvement is of practical value or not depends very much on the computer hardware used and on the precision desired in the calculation. Note that we have not counted additions, which on some computers are about as costly as a multiplication.

Stereographic Projection

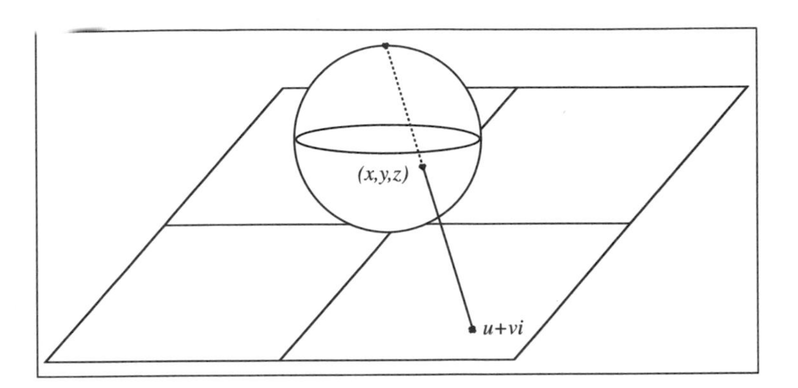

Figure 13.9 : Stereographic projection is a method to create a planar map of a sphere. Here we use a sphere of radius 1 and the complex plane which touches the south pole of the sphere at the origin. For each point (x, y, z) on the sphere there is a corresponding projected point $u + vi$ in the plane. To construct this projection consider the ray from the north pole through (x, y, z). The intersection of the ray with the complex plane defines the stereographic projection $u + vi$ of (x, y, z).

The Point at Infinity

Mathematicians like to think about the complex plane as a sphere, called the *Riemannian sphere*. This is justified if we carry out an identification of a punctured sphere with a plane, much like when we create a map of our planet (see figure 13.9). In this identification every point in the plane corresponds to a unique point on the sphere, and every point on the sphere, except the north pole, corresponds to a point in the plane. Observe that points in the plane which are further and further away from the origin correspond to points on the sphere located closer and closer to the north pole. In this way mathematicians regard the north pole on the sphere as the *point at infinity*.

The Mathematics of Stereographic Projection

There is a transformation that defines the complex number $u + vi$ corresponding to a point (x, y, z) on the sphere as shown. To be specific, we choose the radius of the sphere equal to 1 and fix the south pole of the sphere at the origin of the complex plane. The z-axis points upwards through the poles of the sphere. Then the transformation is

$$P(x, y, z) = \frac{2}{2-z}(x + yi) .$$

Hence,

$$u = \frac{2}{2-z}x \text{ and } v = \frac{2}{2-z}y .$$

The inverse of the transformation required for the computation of the point (x, y, z) on the sphere when $u + vi$ is given. The formula is

$$P^{-1}(u + iv) = \frac{1}{u^2 + v^2 + 4}(4v, 4u, 2u^2 + 2v^2) .$$

Let us take as an example the complex number $u+vi = \sqrt{2}(1-i)$. The last formula for the inverse specifies the corresponding point on the sphere

$$\begin{aligned} x &= \sqrt{2}/2 \\ y &= -\sqrt{2}/2 \\ z &= 1 . \end{aligned}$$

It is located on the equator. The transformation P applied to this point produces the original complex number $\sqrt{2}(1 - i)$, of course.

The z-component of the inverse transformation P^{-1} is

$$z = \frac{2(u^2 + v^2)}{u^2 + v^2 + 4} .$$

When $u + vi$ tends to infinity, i.e., $u^2 + v^2$ grows without bounds, then the z-component tends to 2,

$$\lim_{u^2+v^2 \to \infty} \frac{2(u^2 + v^2)}{u^2 + v^2 + 4} = 2 .$$

Thus, the point on the sphere slides up to the north pole, which therefore is called the point at infinity.

13.3 Complex Square Roots and Quadratic Equations

Complex Square Roots With the above explanation of complex multiplication, it is also geometrically clear how to compute a square root of a complex number $z = r(\cos\phi + i\sin\phi)$. We must divide the argument by 2 and take the square root of the absolute value:

$$\sqrt{z} = \sqrt{r}\left(\cos\frac{\phi}{2} + i\sin\frac{\phi}{2}\right) .$$

Except for $z = 0$ there are always two square roots, $\sqrt{z}$ as in the formula, and $-\sqrt{z}$,

$$\begin{aligned}-\sqrt{z} &= -\sqrt{r}\left(\cos\frac{\phi}{2} + i\sin\frac{\phi}{2}\right)\\ &= \sqrt{r}\left(\cos\left(\frac{\phi}{2}+\pi\right) + i\sin\left(\frac{\phi}{2}+\pi\right)\right) .\end{aligned}$$

However, in most applications relevant to Julia sets we will work with the Cartesian coordinates (x, y) rather than the polar coordinates (r, ϕ). It is possible to first make a change of coordinates to obtain the polar coordinates, then carry out the square root as described and finally make another change of coordinates to get back the usual Cartesian coordinates. This is rather cumbersome. Therefore, let us derive a more efficient algorithm for complex square root computation. Let

$$w = u + vi$$

be the complex number for which the square root

$$z = \sqrt{w} = x + yi$$

is desired. We have that

$$z^2 = (x^2 - y^2) + (2xy)i$$

and to satisfy $z^2 = w$ the two equations

$$\begin{aligned}u &= x^2 - y^2\\ v &= 2xy\end{aligned}$$

must hold. We now solve this system of equations for x and y. The second equation solved for y yields

$$y = \frac{v}{2x}$$

and, substituting this into the first one we obtain

$$u = x^2 - \frac{v^2}{4x^2} .$$

Sorting terms in this equation then gives rise to a biquadratic equation for x,

$$x^4 - ux^2 - \frac{v^2}{4} = 0 \ .$$

We now solve this for x^2 using the standard formula for quadratic equations. There is a special case, namely when $v = 0$, which we postpone until later, so let us assume for now that v is not zero. Then we have

$$x^2 = \frac{u + \sqrt{u^2 + v^2}}{2} \ .$$

Notice that the right hand side is strictly positive since the square root of u^2+v^2 is always greater than $|u|$. Here we use only the plus sign before the square root because a minus sign would give us a negative right side, which does not lead to a real solution x. Taking the square root again and inserting the result into the equation for y yields the final result

$$\begin{aligned} x &= \pm\frac{\sqrt{2}}{2}\sqrt{u + \sqrt{u^2 + v^2}} \ , \\ y &= \frac{v}{2x} \ . \end{aligned}$$

Note that this denotes two symmetric solutions, a pair (x, y) for each sign in the first expression.

The special case $v = 0$ is very simple, of course. It corresponds to the root of a real (not necessarily positive) number u. In this case, when $u \geq 0$ we have

$$\begin{aligned} x &= \sqrt{u} \ , \\ y &= 0 \ , \end{aligned}$$

while

$$\begin{aligned} x &= 0 \ , \\ y &= \sqrt{-u} \ , \end{aligned}$$

when $u < 0$.

Errors in the Computation

The derived formulas, although being mathematically correct, may exhibit a severe limitation once implemented on a computer. Suppose that v is very small in magnitude as compared to a negative value of u, say, for example $u = -10^6$, and $v = 10^{-6}$. Then $u^2 + v^2 = 10^{12} + 10^{-12}$ and although u^2 and v^2 are computed exactly on the machine, their sum cannot be represented correctly as a 32 bit single-precision or even as a 64 bit double-precision floating point number. This is so because $10^{12} + 10^{-12}$ would require 24 significant decimal digits in the machine representation of numbers. Thus, the machine rounds off to obtain the result $u^2 + v^2 = 10^{12}$. As a conclusion

$$x = \frac{\sqrt{2}}{2}\sqrt{-10^6 + \sqrt{10^{12} + 10^{-12}}}$$

evaluates to zero. This is the wrong answer, since

$$\sqrt{10^{12} + 10^{-12}} \approx 10^6 + 0.5 \cdot 10^{-18}$$

and, thus,

$$x \approx \sqrt{\frac{10^{-18}}{4}} = 0.5 \cdot 10^{-9} \ .$$

The relative error in the computed value of x is therefore 100 percent. This may not sound so bad, since the absolute error is indeed very small. But the worst is yet to come. The evaluation of $y = v/(2x)$ yields a division by zero — an error which usually terminates the program rather unexpectedly. The problem encountered here is a consequence of what has been termed by numerical analysts *loss of significance* or *cancellation of significant digits*.

Fortunately, in our case there is a remedy for the problem. The solution to the square root problem $u = x^2 - y^2$, $v = 2xy$ can also be written as

$$y = \pm\frac{\sqrt{2}}{2}\sqrt{-u + \sqrt{u^2 + v^2}} \ ,$$
$$x = \frac{v}{2y} \ .$$

This is derived in much the same way as our first solution, and we skip these details. Let us again assume that $|v|$ is small in comparison to $|u|$. The second solution also exhibits loss of significance, namely when $u > 0$, while in the previous solution cancellation occurs for $u < 0$. Thus, we may effectively eliminate loss of significance by choosing one or the other method depending on the sign of u. You can check that for our specific example $u = -10^6$, $v = 10^{-6}$ we indeed get $y = 10^3$ and $x = 0.5 \cdot 10^{-9}$.

The Complete Algorithm

Summarizing, a complete algorithm for the complex square root which also properly takes into account the special cases $u = 0$ and $v = 0$ is outlined below. Here only one of the roots is computed, namely the one with a positive real component (if the real component is not zero). The other root is identical except for the signs of x and y. This algorithm is very fast. It essentially involves just substituting two real square roots for one complex square root while avoiding loss of significant digits.

Algorithm for the Complex Square Root of $u + vi$

The algorithm below computes the complex square root which has nonnegative real part.

```
if (u > 0) then
   x = sqrt((u + sqrt(u*u + v*v))/2)
   y = v / (2*x)
else if (u < 0) then
   y = sign(v) * sqrt((-u + (u*u + v*v))/2)
   x = v / (2*y)
else
   x = sqrt(abs(v)/2)
   if (x > 0) then
      y = v / (2*x)
   else
      y = 0
   end if
end if
```

The functions `sqrt`, `sign`,`abs` are the real positive square root, the sign, and the absolute value of real numbers. The sign of v is $+1$ if $v \geq 0$ and -1 otherwise. The factor `sign(v)` ensures that the root computed has a non-negative real component.

Solving Complex Quadratic Equations

With this tool for calculating the complex square root we can solve any quadratic equation in complex numbers using the same formula as for real ones. If

$$az^2 + bz + c = 0 ,$$

where all numbers are complex, must be solved, then we can write the solutions as

$$z_{1,2} = \frac{-b \pm \sqrt{b^2 - 4ac}}{2a} .$$

Similar to the case of a complex square root, cancellation may occur when $|4ac|$ is small compared to $|b^2|$.

As demonstrated, polynomial equations of degree two have two complex roots. In general, any polynomial of degree $n \geq 1$ has exactly n complex zeros, where the zeros must be counted with proper multiplicities. For example, if $b^2 = 4ac$ in the quadratic equation, the root $z_1 = z_2$ is counted as a double root.

Let us present an example that will come up again in a section further below. Up to three decimal places, solve the quadratic equation

$$z^2 - z + c = 0$$

where $c = -1/2 + (1/2)\ i$. The recipe formula prescribes the computation (set $a = 1$ and $b = -1$)

$$z_{1,2} = \frac{1 \pm \sqrt{1 - 4c}}{2} = \frac{1 \pm \sqrt{3 - 2i}}{2} .$$

Now the algorithm for the computation of the square root of $3-2i$ can be applied (here we have $u=3$ and $v=-2$). Thus,

$$x = \sqrt{\frac{u+\sqrt{u^2+v^2}}{2}} = \sqrt{\frac{3+\sqrt{13}}{2}} \approx 1.817$$
$$y = \frac{v}{2x} \approx \frac{-2}{2\cdot 1.817} \approx -0.550$$

and thus

$$\sqrt{3-2i} \approx \pm(1.817 - 0.550i) \ .$$

Substituting this result in the expression for z_1 and z_2, we obtain

$$z_{1,2} \approx \frac{1 \pm (1.817 - 0.550i)}{2} \ ,$$

i.e.,

$$z_1 \approx 1.408 - 0.275i \ ,$$
$$z_2 \approx -0.408 + 0.275i \ .$$

13.4 Prisoners versus Escapees

We are now well prepared to come to the real stuff of this chapter: Julia sets. The simplest example of a nonlinear iteration procedure in the complex numbers is given by the transformation

$$z \longrightarrow z^2 .$$

Geometrically the squaring of a complex number means that the corresponding length of z is squared in the ordinary sense and that the corresponding angle $\arg z$ of z is doubled (mod 2π). The following table lists three examples. We take an initial point inside the unit circle, i.e., a complex number with absolute value less than 1, another initial point on the unit circle, and finally one outside the unit circle.[6] Squaring z gives z^2, squaring it again gives z^4, and so on as listed.

Dynamics of $z \rightarrow z^2$

	length	angle	length	angle	length	angle
z	0.8	10°	1.0	10°	1.5	50°
z^2	0.64	20°	1.0	20°	2.25	100°
z^4	0.4096	40°	1.0	40°	5.06	200°
z^8	0.1678	80°	1.0	80°	25.63	40°
z^{16}	0.0281	160°	1.0	160°	656.90	80°
z^{32}	0.0008	320°	1.0	320°	431439.89	160°

Table 13.10 : The iteration of three initial points using the simple squaring operation $z \rightarrow z^2$.

Table 13.10 shows the dynamical behavior of the generated points. In all cases we observe that the sequences of points circle around the origin. However the initial point inside the unit circle leads to a sequence which converges to the origin, the point exactly on the unit circle leads to a sequence which remains there forever, and the point outside leads to a sequence which escapes to infinity, i.e., the absolute value of the iterates become larger and larger (see also the figure 13.11). In fact, if the initial point has absolute value of 10, say, then the absolute values of the iterates are $100 = 10^2$ for the first, $10,000 = 10^4$ for the second, $100,000,000 = 10^8$ for the third, $10,000,000,000,000,000 = 10^{16}$ for the fourth, $100,000,000,000,000,000,000,000,000,000,000 = 10^{32}$ for the fifth, and so on. In other words, the absolute values of the iterations literally explode within a few steps.

Prisoner and Escape Set

This leads us to an important dynamic dichotomy: the complex plane of initial values is subdivided into two subsets. The first one

[6]The unit circle is defined as the set of points in the complex plane having distance 1 to the origin, i.e., $\{x + yi \mid x^2 + y^2 = 1\}$.

Iterating $z \to z^2$

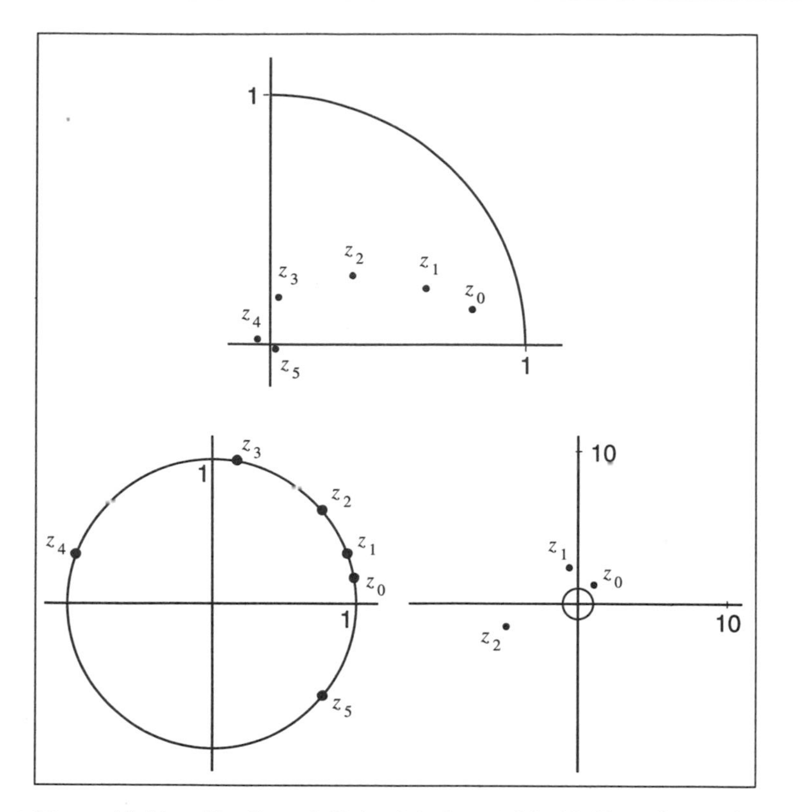

Figure 13.11 : The three initial points from table 13.10 are iterated.

collects points for which the iteration escapes. It is called the *escape set* E. The iteration for all other initial values remains in a bounded region forever, and we collect these points in the so-called *prisoner set* P. Note that P is the disk around zero with radius 1 and that E is the outside of that disk.[7] The boundary between E and P is the unit circle. In this context it is called the *Julia set* of the iteration.

For initial values in the Julia set, i.e., $|z| = 1$, the iteration generates only points which again lie on the unit circle. In other words, the Julia set is *invariant* under iteration.

Two Basins of Attraction

Note that the interior of P can be interpreted as a basin of attraction, the attractor being the point 0. If we just restrict attention to real initial points z_0, then the behavior can be visualized by graphical iteration (see figure 13.12).

Fixed Points and Basins of Attraction

Note that the iteration has two fixed points; 0 and 1. However, 0 is attracting and 1 is repelling, i.e., if we start near 1 as, for example with $z_0 = 1 + \varepsilon$, then z_1 is about twice as far from 1 than

[7] Compare the section about the Cantor set as a prisoner set in chapter 2.

Graphical Iteration for $z \to z^2$

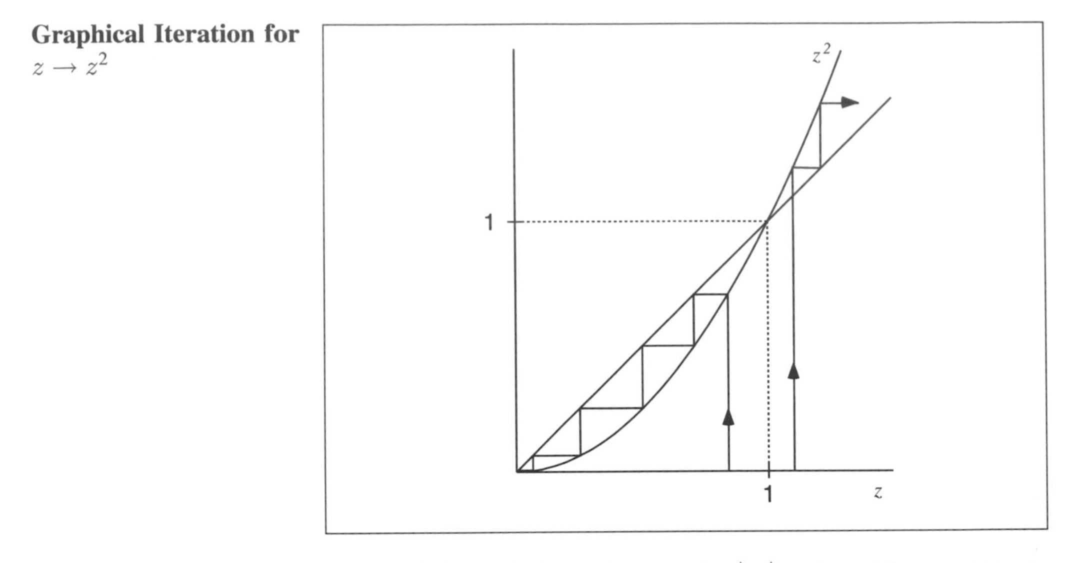

Figure 13.12 : Initial points z_0 with $|z_0| < 1$ rapidly converge to the origin, while points with $|z_0| > 1$ escape to infinity.

z_0, $z_1 = (1+\varepsilon)^2 \approx 1 + 2\varepsilon$. If we consider the iteration of z^2 on the sphere as explained in the previous section, the north pole is an attractor for the dynamics of $z \to z^2$. Likewise the escape set E can be interpreted as a basin of attraction, the basin of the point at infinity. Thus the dynamic dichotomy is that we have orbits $z_0, z_1, z_2, \ldots$ which escape to infinity, and orbits which do not.

We have two attractors, the origin 0 and the point at infinity, whose basins of attraction are the open unit disk and everything outside the unit disk respectively, while the boundary points, which are all located on the unit circle, wander around on the circle forever. It is the boundary, which is shared between the basins of attraction, that is called the Julia set. Here in this simple example, the Julia set is a circle and, thus, a geometrical object from classical Euclidean geometry. However, it is very special because it is not a fractal, and most Julia sets are. Besides that property, however, it shares the important typical properties of many Julia sets: it is the boundary of basins of attraction; and the dynamics on it are chaotic.

Julia Sets for the Quadratic Family

The special case $z \to z^2$ is the entrance door into an amazing zoo of beautiful fractal Julia sets. One department of that zoo of Julia sets is built on the iteration of $R_c(z) = z^2 + c$, where c is some complex parameter. The Julia set of $z \to z^2$ is right in the center of this class; set $c = 0$. As the unit circle is the boundary of the escape set for $R_0(z) = z^2$, the other Julia sets are the boundaries of the escape set of $z \to z^2 + c$. This we can

Three Escaping Points

	Orbit 1		Orbit 2		Orbit 3	
	x	y	x	y	x	y
z_0	1.00	0.00	0.50	0.25	0.00	0.88
z_1	0.50	0.50	-0.31	0.75	-1.27	0.50
z_2	-0.50	1.00	-0.96	0.03	0.87	-0.77
z_3	-1.25	-0.50	0.43	0.44	-0.34	-0.85
z_4	0.81	1.75	-0.51	0.88	-1.12	1.07
z_5	-2.90	3.34	-1.01	-0.39	-0.41	-1.90
z_6	-3.26	-18.91	0.37	1.30	-3.93	2.04
z_7	-347.46	123.68	-2.04	1.46	10.79	-15.52
z_8			1.53	-5.46	-124.77	-334.49
z_9			-28.01	-16.27		

Table 13.13 · The iteration of three initial points for $z \to z^2 + c$, $c = -0.5 + 0.5i$. All three orbits escape to infinity.

view as their definition. To see these other Julia sets we need a special viewing device. Let us therefore learn how to build this apparatus. In reality it is only a visualization of the escape set E for an arbitrary choice of c, because whatever remains will then be the prisoner set P, and the common boundary between E and P is the Julia set.

The Definition

Let us first make some more precise definitions. The escape set for the parameter c is

$$E_c = \{z_0 \ : \ |z_n| \to \infty \text{ as } n \to \infty\} \ .$$

In this definition the orbit $z_0, z_1, z_2, ...$ of the initial point z_0 of course is given by

$$z_{n+1} = z_n^2 + c, \ n = 0, 1, 2, ...$$

The prisoner set for parameter c is

$$P_c = \{z_0 \mid z_0 \notin E_c\} \, ;$$

it is the complement of E_c. The Julia set for parameter c is the boundary of the escape set E_c. A point is a boundary point if arbitrarily close to it we can find initial points with escaping orbits as well as points from the prisoner set.

A First Fractal Julia Set

For a first example let us choose $c = -0.5 + 0.5i$ and a few initial points (see the tables 13.13 and 13.14). We can observe two basic behaviors. In the first table, the iterated points escape to infinity. In the second table, the iterated points do not escape but eventually settle down to a certain point, namely $z \approx -0.408 + 0.275i$. This indicates that we again have two basins of attraction, but zero is no longer one of the attracting points. Further below,

Three Prisoner Points

	Orbit 1		Orbit 2		Orbit 3	
	x	y	x	y	x	y
z_0	0.000	0.000	0.500	-0.250	-0.250	0.500
z_1	-0.500	0.500	-0.313	0.250	-0.688	0.250
z_2	-0.500	0.000	-0.465	0.344	-0.090	0.156
z_3	-0.250	0.500	-0.402	0.180	-0.516	0.472
z_4	-0.688	0.250	-0.371	0.355	-0.456	0.013
z_5	-0.090	0.156	-0.488	0.237	-0.292	0.488
z_{100}	-0.473	0.291	-0.393	0.290	-0.438	0.217
z_{200}	-0.394	0.279	-0.411	0.271	-0.409	0.290
z_{300}	-0.411	0.273	-0.409	0.276	-0.407	0.272
z_{400}	-0.408	0.276	-0.409	0.275	-0.409	0.276
z_{500}	-0.409	0.275	-0.409	0.275	-0.409	0.275

Table 13.14 : The iteration of three initial points for $z \to z^2 + c$, $c = -0.5 + 0.5i$. All three orbits do not escape to infinity. Rather they seem to converge to a point $z \approx 0.41 + 0.28i$.

in section 13.6, we will compute this fixed point of the iteration as a solution of an equation.

Figure 13.15 shows the result of an extended experiment which tests all points in the complex plane. The prisoner set is pictured in black, while everything white belongs to the escape set. Consequently, the boundary of the black region is the Julia set. This bordering curve is obviously a typical fractal. No matter how much we magnify a region near the Julia set, there is always detail which looks similar (see the next figure 13.16).

The Point of No Return

The key to the computation of the escape set E_c is the observation that points z_k from an orbit will escape to infinity with certainty once their absolute value is large enough. This seems clear because the square of a large number is much larger and adding the constant c will be rather insignificant. Then in the next step of the iteration of $z \to z^2 + c$ this effect is even more pronounced. As a result we can see that in the iteration for large z, the constant c can be neglected, and we are left with $z \to z^2$, which we understand very well. But how large must an iterate be so that we can decide that the orbit will definitely escape to infinity? Fortunately, there is an optimal answer to this question. It is given by a computable number $r(c)$ which depends on the parameter c. We can show that one may choose the number $r(c)$ as the maximum of the absolute value $|c|$ and 2:

$$r(c) = \max(|c|, 2) \ .$$

Thus, if $|z_k|$ exceeds $r(c)$ in absolute value, we can be confident that the iteration escapes to infinity. The algorithm which classifies

Prisoner Set for $c = -0.5 + 0.5i$

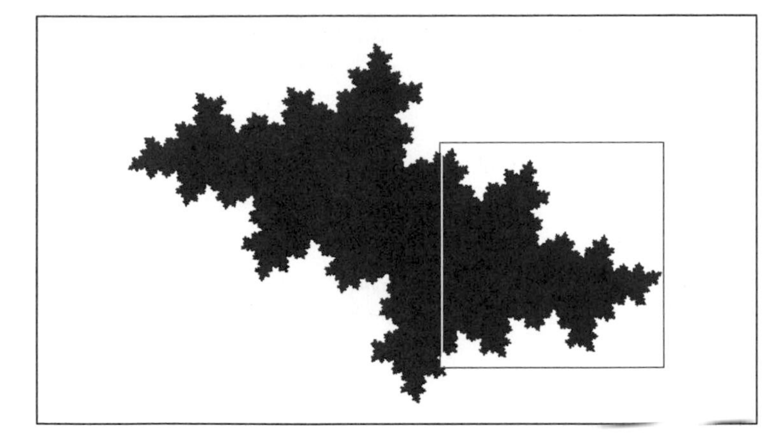

Figure 13.15 : The prisoner set for $z \to z^2 + c$, $c = -0.5 + 0.5i$ is shown in black. Points outside escape to infinity. The framed region is enlarged in figure 13.16.

Blowups of Prisoner Set for $c = -0.5 + 0.5i$

Figure 13.16 : The prisoner set for $z \to z^2 + c$, $c = -0.5 + 0.5i$ from figure 13.15 is successively enlarged near a boundary point. Each picture (from left to right) is a computation of the small framed region in the previous one.

an initial point as a member of the escape set or the prisoner set may therefore proceed as follows. If the absolute value exceeds $r(c)$ at some iteration, the algorithm terminates and returns the result that the initial point is in the escape set. Clearly, the iterates of a point that escapes to infinity must exceed $r(c)$ in magnitude at some iteration. Otherwise, the points from the orbit would never be able to move far away, and the initial point would have to be a prisoner. Therefore, the criterion really catches all escaping points. However, in practice it may take a very long time until the orbit escapes a disk of radius $r(c)$. Thus, we prescribe a maximal number of iterations; and if the iterated point does not exceed $r(c)$ in absolute value during these iterations, then we must assume — up to the precision of the algorithm — that the initial point does not belong to the basin of infinity, but to the prisoner set.

The Threshold Radius

Here we demonstrate the following proposition:

Let z be a complex number not less than c and greater than 2 in absolute value. Then z is an escaping point for the iteration $z \to z^2 + c$.

Let a parameter c be given and set $r(c) = \max(|c|, 2)$. For the starting point z of the iteration we assume that

$$|z| \geq |c| \text{ and } |z| > 2$$

holds. Then there exists a (possibly) small but positive number $\varepsilon > 0$ with $|z| = 2 + \varepsilon$. The triangle inequality for complex numbers implies

$$|z^2| = |z^2 + c - c| \leq |z^2 + c| + |c| \ .$$

Solving this inequality for $|z^2 + c|$ and continuing, we derive

$$\begin{aligned} |z^2 + c| &\geq |z^2| - |c| \\ &= |z|^2 - |c| \\ &\geq |z|^2 - |z| \\ &= (|z| - 1)|z| \\ &= (1 + \varepsilon)|z| \ . \end{aligned}$$

Thus, if we iterate once, the absolute value will increase by at least a factor $1 + \varepsilon$. The kth iterate of z will thus be at least $(1 + \varepsilon)^k$ times as large as z in magnitude. Therefore, the absolute values clearly tend to infinity and z is in the basin of attraction of infinity.

From this proposition it immediately follows that the orbit must escape to infinity if any one point in an orbit for $z \to z^2 + c$ is larger than $r(c) = \max(|c|, 2)$ in magnitude; the initial point of that orbit is in the escape set.

Encirclement of the Prisoner Set

Let us demonstrate how an algorithm makes use of this important fact in order to encircle the prison closer and closer. To begin with we choose a parameter c and compute the threshold radius $r(c) = \max(|c|, 2)$. We also define a region of interest in the complex plane and place it on a grid of pixels for viewing on a computer graphics screen. The pixels correspond to initial points for the orbits which we will test next. First we check if there are some initial points z_0 which are greater in magnitude than the threshold radius $r(c)$. All these can be discarded; they have already been identified as points of the escape set E_c. Thus, our initial approximation of the prisoner set P_c is simply the disk of radius $r(c)$ centered at 0. For this approximation, we introduce the notation $Q_c^{(0)}$,

$$Q_c^{(0)} = \{z_0 \ : \ |z_0| \leq r(c)\} \ .$$

We now allow one iteration for each of the remaining points or pixels, obtaining a complex number z_1 for each pixel. Again we

Encirclement for $c = -0.5 + 0.5i$

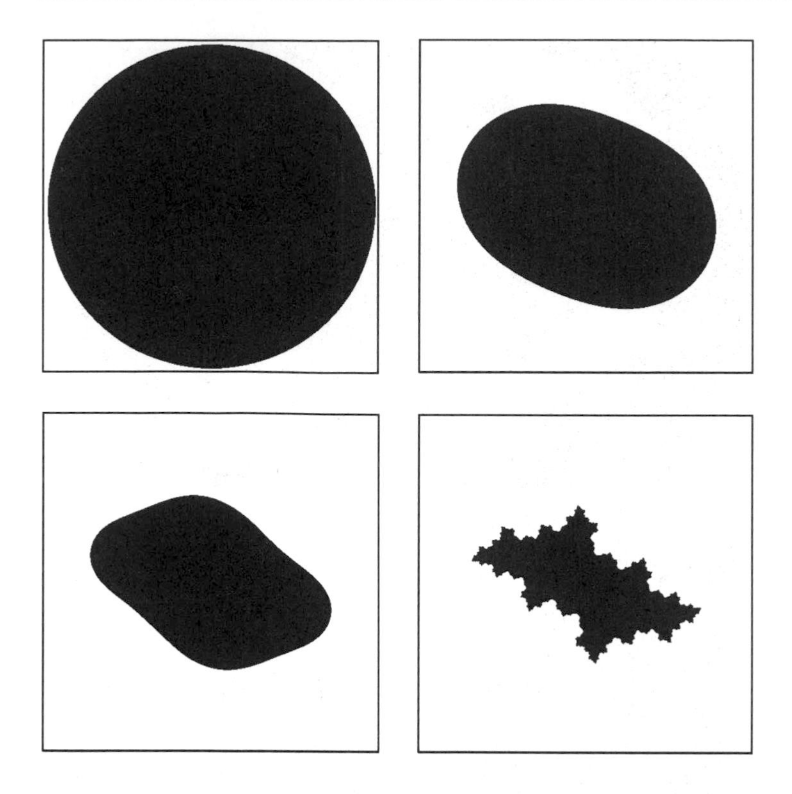

Figure 13.17 : Encirclement for $c = -0.5 + 0.5i$. The initial approximation $Q_c^{(0)}$ is shown in the upper left (a disk of radius 2). The other approximations are $Q_c^{(-k)}$ for $k = 1, 2$, and 10.

can apply the criterion for escaping orbits. If the absolute value $|z_1|$ is larger than $r(c)$, then the corresponding point is identified as an escapee and can be removed from further consideration. In effect we obtain an improved approximation containing the prisoner set. It consists of all points which after one iteration of R_c are still in $Q_c^{(0)}$. We say that they form the preimage of $Q_c^{(0)}$ and use notation $Q_c^{(-1)}$.

$$Q_c^{(-1)} = \{z_0 \ : \ |z_1| \leq r(c)\} \ .$$

Next we repeat the procedure for all remaining points, which yields the next approximation, and so on. In each iteration we remove points and thus move in on the prisoner set more and more closely. Formally, we write

$$Q_c^{(-k)} = \{z_0 \mid |z_k| \leq r(c)\} \ , k = 0, 1, 2, ...$$

Encirclement in Alternating Colors

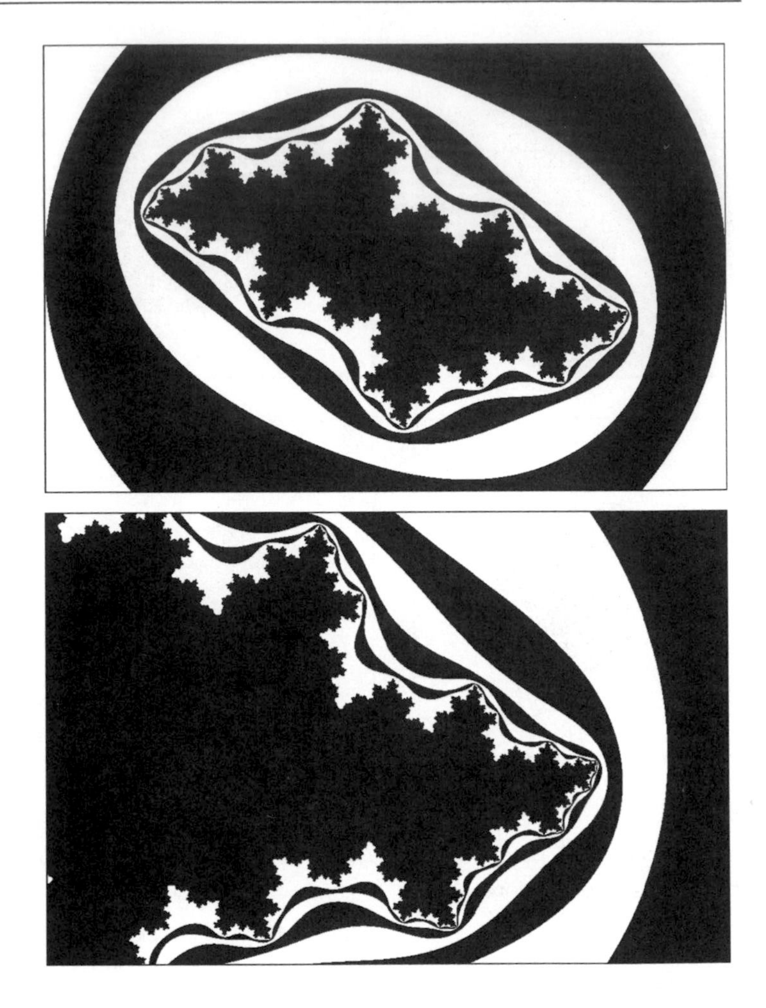

Figure 13.18 : In the upper figure an overlay of encirclements for $c = -0.5 + 0.5i$ is drawn in alternating colors. In black: $Q_c^{(0)}, Q_c^{(-2)}, Q_c^{(-4)}$, and $Q_c^{(-20)}$, and in white: $Q_c^{(-1)}, Q_c^{(-3)}$ and $Q_c^{(-5)}$. The lower figure shows a detail of the encirclement. In black: $Q_c^{(0)}, Q_c^{(-2)}, Q_c^{(-4)}, Q_c^{(-6)}$, and $Q_c^{(-20)}$, and in white: $Q_c^{(-1)}, Q_c^{(-3)}, Q_c^{(-5)}$ and $Q_c^{(-7)}$.

and

$$\lim_{k\to\infty} Q_c^{(-k)} = P_c \ .$$

The Case $c = 0$

For the case $c = 0$ we can write down explicitly what these approximating sets are. We have $r(c) = 2$ and the initial approximation $Q_c^{(0)}$ is the disk with radius 2. The following approximations are also disks. For example, $Q_c^{(-1)}$ is the collection of all

Encirclements

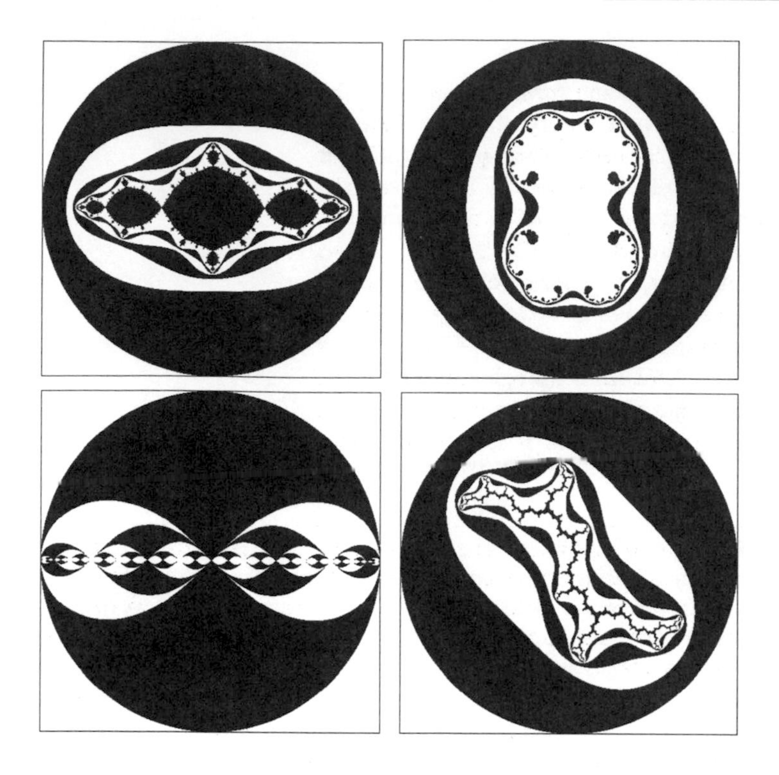

Figure 13.19 : Encirclements for $c = -1, 0.4, -2, i$ (from upper left to lower right).

points z that satisfy $|z^2| \leq 2$. This is the set of points z with $|z| \leq \sqrt{2}$. Thus, $Q_c^{(-1)}$ is a disk of radius $\sqrt{2}$. The radius of the next approximation, $Q_c^{(-2)}$, is $\sqrt{\sqrt{2}} = 2^{1/4}$. A general formula is

$$Q_c^{(-k)} = \left\{ z_0 \ : \ |z_0| \leq 2^{1/2^k} \right\} .$$

This sequence of radii rapidly approaches 1, the radius of the prisoner set P_c. Here is a list of the first few radii: 2, 1.414, 1.189, 1.090, 1.044, 1.022, 1.011, 1.005, ...

The Case $c = -0.5 + 0.5i$

For the case $c = -0.5 + 0.5i$ from figures 13.15 and 13.16, we cannot give explicit formulas for the approximations of the prisoner set, but we show a sequence of figures illustrating how the prisoner set is encircled more and more closely (see figure 13.17).

Stacking Encirclements

It is possible to stack encirclements on top of each other, allowing a better comparison and insight into how the prisoner set is approximated. For this purpose it is necessary to shade the encirclements differently, for example, using alternating black and white sets. In our next figure 13.18 we show encirclements drawn on top of each other in this way. $Q_c^{(0)}, Q_c^{(-2)}$, and $Q_c^{(-10)}$ are black while $Q_c^{(-1)}$ and $Q_c^{(-5)}$ are white.

In figure 13.19 this is done for some further choices of the parameter c. Indeed, we can observe a variety of rather different results. For $c = -1$ we have a connected (one piece) Julia set which is the common boundary of two basins of attraction (the basin of the infinite attractor and the basin of a finite attractor, which in this case is the period-2 cycle: $-1, 0, -1, 0, ...$). In the other cases there is no finite attractor, the prisoner set has no interior points and is equal to the Julia set. For $c = 0.5$ the prisoner set has dissolved into a dust of points, while for $c = -2$ and $c = i$ we observe boundary cases: the Julia set is a single connected set. We are about to discover an important dichotomy: prisoner sets are either connected or a dust of points. This is explained further in section 13.7.

Implementation Details for the Pixel Game

Pixel game algorithms to compute Julia sets (and the Mandelbrot set) have become very popular through an article in the mathematics column in the *Scientific American*.[8] Here we want to visualize the prisoner set approximation $Q_c^{(k)}$ by coloring an array of pixels like the one shown in figure 13.2. A pixel with coordinates x and y is colored black if $z = x + yi$ is an initial point belonging to $Q_c^{(k)}$. At the beginning we present an overview of the algorithm using complex notation for the variables c and z. In a computer implementation the complex addition and multiplication rules must be properly used.

```
R = max(|c|,2)
i = 0
while (i < k)
   if (|z| > R) then
      return (z belongs to the escape set)
   end if
   z = z*z + c
   i = i + 1
end while
return (z belongs to Q_c^(k))
```

The algorithm requires the iteration of `z*z + c`. Here some care must be taken. Let x and y be the real and imaginary part of z, and cr and ci those of c. People have often erroneously written

```
x = x*x - y*y + cr
y = 2*x*y + ci
```

The mistake is that when the new value of y is computed, the old value of x has already been lost (overwritten by its new value); and thus `2*x*y` is not what one expects. The correct version, of course, uses a temporary variable to hold the new value of x as in

[8] A. K. Dewdney, *Computer Recreations: A computer microscope zooms in for a look at the most complex object in mathematics,* Scientific American (August 1985) 16–25.

```
temp = x*x - y*y + cr
y = 2*x*y + ci
x = temp
```

One hint for an efficient implementation is in order here. It is far more efficient to check whether $|z|^2 > R^2$ than to check if $|z| > R$, because $|z|^2 = x^2 + y^2$ and the absolute value of z requires a costly square root for the computation. Moreover, the numbers x^2 and y^2 can be reused for the next iteration, resulting in an additional speed-up. Here are the details:

```
R2 = max(|c|,2) * max(|c|,2)
i = 0
x2 = x*x
y2 = y*y
while (i < k)
   if (x2 + y2 > R2) then
      return (z belongs to the escape set)
   end if
   y = 2*x*y + ci
   x = x2 - y2 + cr
   i = i + 1
   x2 = x*x
   y2 = y*y
end while
return (z belongs to P_c^(k))
```

A typical implementation of the pixel game is shown in the program of the chapter, however for the Mandelbrot set (see section 14.4).

13.5 Equipotentials and Field Lines for Julia Sets

Capturing the prisoner set P_c by the encirclements $Q_c^{(-k)}$ has turned out to be a very fruitful idea. We will now see how we can refine it to an even more powerful tool. This leads us to the work of Douady and Hubbard. More than 60 years had passed since the outstanding work of Julia and Fatou at the beginning of the century before Douady and Hubbard developed new methods to continue the unveiling of the secrets of Julia sets. They found a beautiful way to do this using an analogy from electrostatics.

The Electrostatic Field

Think of the prisoner set as a piece of metal charged with electrons. This charge produces an *electrostatic field* in the surrounding space, resulting in an attracting force on any small test charge of the opposite polarity. The field is given by vectors indicating the direction and the strength of the force per unit test charge (see figure 13.20). The lines which follow the vectors from any given point to the charged prisoner set are those that infinitesimally small test particles would travel when exposed to the field. These lines are called *field lines*.

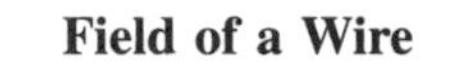
Field of a Wire

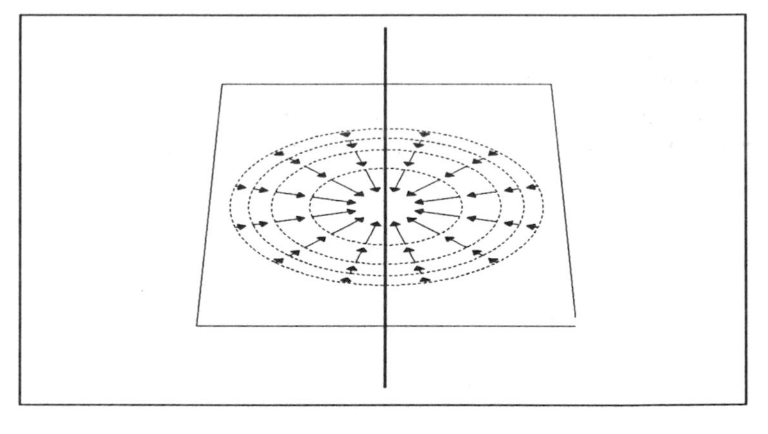

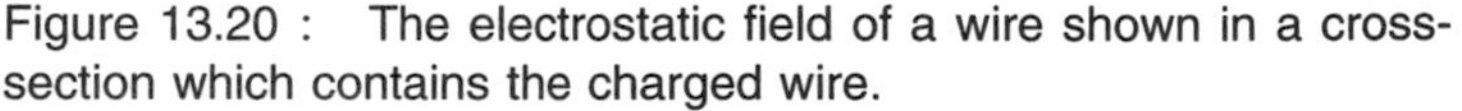
Figure 13.20 : The electrostatic field of a wire shown in a cross-section which contains the charged wire.

Now the theory of electrostatics applies to objects and charges in three-dimensional space, while prisoner and Julia sets are embedded in the two-dimensional space of complex numbers. However, a flat prisoner set placed in three-dimensional space has an electrostatic field which varies in all three directions of space. To get back to a theory which is completely two-dimensional in nature, we extend the prisoner set in the third dimension; we consider an infinitely long cylinder-like set whose cross-section is the prisoner set. On it we imagine an infinite amount of charge uniformly distributed along its entire length. The electrostatic field of this extended prisoner set is identical in all cross-sectional slices, any one of which completely describes the whole field.

The Field of an Infinite Cylinder

In principle we can compute the electrostatic field from Coulomb's law, which states that the force between two charged particles is proportional to the product of the two charges and inversely proportional to the square of their distance. Thus, the field of a single point charge is spherical: all field vectors point to the center, and the field strength is the same for all points on a sphere centered at the point charge (inversely proportional to the square of the radius). From this basic principle it follows that the electrostatic field of an infinite straight is cylindrical and the strength is inversely proportional to the distance to the wire (and not the square of the distance). Moreover, the electrostatic field is the same when we consider a wire which has some thickness, for example with a unit disk as the cross-section, and when the distance is measured to the center of the wire.

The Potential Function ...

For a given force field such as an electrostatic field it is possible and useful to discuss the *potential energy* of an object in terms of the external work necessary to move it from place to place against the forces of the field. The idea is familiar from Newtonian mechanics and the gravitational force field. In this case the potential is proportional to height; the work necessary to lift an object from height h_0 to height h_1 is proportional to the height *difference* $h_1 - h_0$. In an electrostatic field there is also a potential function which allows us to compute the work required to move a test charge from one point to another as the difference of the potentials at the two points. This implies, of course, that this work is independent of the particular path chosen to move the test charge. Moreover, we can retrieve the energy by allowing the particle to return to its initial position. We say that the electrostatic field is *conservative*.

... and its Equipotential Surfaces

Of special importance are the equipotential surfaces. These are defined as surfaces on which the potential is constant.[9] For example, the equipotentials of a point charge are spheres, those of the infinite wire are cylinders. Equipotential surfaces are perpendicular everywhere to the direction of the electrostatic field. In many cases, a system of equipotentials and field lines can be regarded as a very special system of polar coordinates. By the definition of equipotentials it is clear that there is no external work required to move a test charge to any other place as long as the destination point is on the same equipotential surface as the initial point. But more than this, the equipotential surfaces give an idea of the intensity of force: the intensity of the field is inversely proportional to the distance between equipotential surfaces when they are drawn

[9] In the following we often call these surfaces just 'equipotentials'.

for equally spaced values of the potential. Crowded equipotentials mean relatively high force, and sparse equipotentials, relatively low force.

The computation of the electrostatic potential requires advanced techniques from calculus. The work to move a particle in a field is proportional to the distance traveled and the active field force. However, because the field force is not a constant, this calculation must be done in practice as a summation over many steps, each one representing the work performed to move a particle a small distance in which the field is approximately constant. The background for this computation is the theory of line integrals; and we have chosen not to present these details here.[10] In general the complexity of the computation is too great to allow for explicit formulas for the electrostatic potential. But there are some special cases, namely the potential of a point charge is proportional to the inverse of the distance, and the potential of the infinitely long wire is proportional to the logarithm of the distance.[11]

Potential of the Unit Disk

Let us now return to prisoner sets and how their potential functions can aid in the understanding of the dynamics of the quadratic iteration. This ground-breaking connection between dynamics and potentials was made in Douady and Hubbard's theory. It applies to cases where the prisoner and Julia sets are *connected*. This implies that all field lines converge to the prisoner set; there are no rest points where the field force is zero. We begin by discussing the important example of the unit disk, the prisoner set of R_c, $c = 0$, $z \to z^2$, corresponding to an infinite wire with a unit disk cross-section. The electrostatic field and the potential are not affected by the third (vertical) dimension. Thus, in the following we will ignore this third dimension of the physical setting and concentrate on the two-dimensional complex plane. Of course, the field lines in this plane cannot leave the plane. Moreover, the equipotential surfaces intersect the complex plane in curves, which we call equipotential curves or simply equipotentials. As already explained above the potential function is logarithmic, we may write

$$p(z) = \log |z| \ ,$$

for the potential, where the base of the logarithm does not matter.[12] The equipotential curves are concentric circles (given by $|z| = r$,

[10] Any college level physics text book on electricity and magnetism should have a section on the electrostatic potential and some examples for its computation.

[11] The reader familiar with calculus will see a connection with the corresponding force fields. If r denotes the distance to the point charge or the center of the infinite wire, then the electric fields are proportional to $1/r^2$ and $1/r$ respectively. Integrating these functions gives factors $1/r$ and $\log r$.

[12] In any case the formula ignores the factor of proportionality which depends on the density of charge considered for the infinite wire, among other things. Changing the base of the logarithm is the same as introducing a different factor which affects neither the equipotentials nor the field lines.

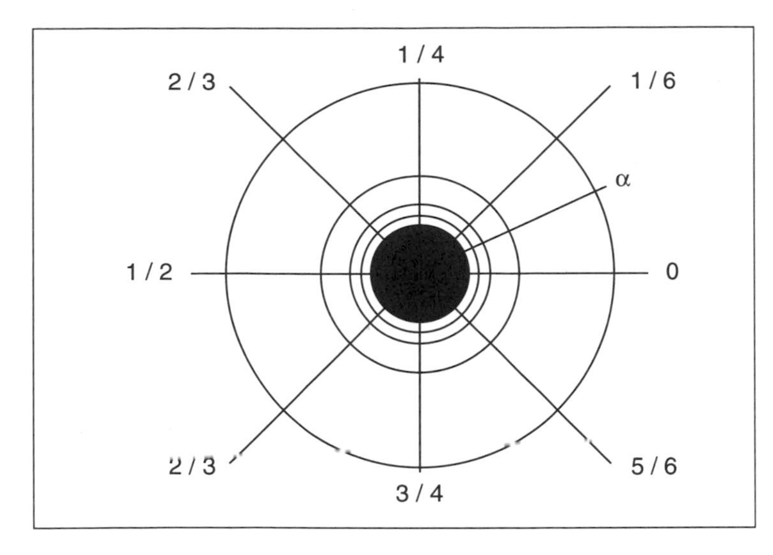

Figure 13.21 : Field lines and equipotential lines for the unit disk. The angles of the field lines are given in multiples of 2π (i.e., $\phi = 2\pi\alpha$, $0 \le \alpha \le 1$).

Potential of Unit Disk

where r is the radius with $r > 1$) and the field lines are straight lines (given by $\arg z = \phi$, where ϕ is an angle between 0 and 2π). In other words, we choose angle zero for the radial line attached to the disk at $z = 1$ and going counterclockwise we identify each field line by an angle between 0 and 2π. In figure 13.21 we also show some equipotential lines. These are circles with $\log|z| = 2^k \log 2$. Stepping from circle to circle outward the potential $p(z)$ becomes twice as large. Note that if we use logarithms to the base 2, the equation for the circles becomes $\log_2|z| = 2^k$. In the following we will therefore use base-2 logarithms.

Connected versus Disconnected

A set is called *connected* provided it cannot be decomposed into two disjoint, non-empty subsets (which are both open and closed in the topology of the set). There are several other mathematical notions of connectedness. For example, a set is called *pathwise connected* provided any two of its points can be connected by a continuous path which is entirely within the set. The property of being connected is not the same as pathwise connected. For example, take as a set A the graph of the function $f(x) = \sin 1/x$ together with the line segment $\{(0, y) \mid -1 \le y \le +1\}$, see figure 13.22. A is connected but not pathwise connected, because a point on the graph and a point on the line segment cannot be connected by a continuous path which is entirely in A! On the other hand, a set which is not connected can be decomposed into disjoint parts. In particular, a set is called *totally disconnected* provided

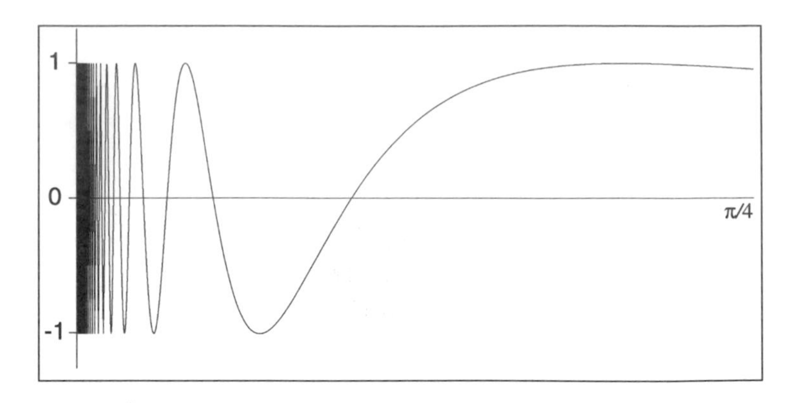

Figure 13.22 : The graph of the function $\sin 1/x$ together with a piece of the y-axis form a set that is connected, but not pathwise connected.

its connected components (i.e., the maximal connected subsets) are single points. Any finite set of points is totally disconnected. Infinite point sets can also be totally disconnected, for example, the set $\{1/n \mid n = 1, 2, 3...\}$ or the set of rational numbers within all real numbers. These are examples where the sets are still countable. The Cantor set is an example of a set which is both totally disconnected and uncountable (see chapter 2).

Iteration of Field Lines

The field lines of the unit disk are closely related to the dynamics of R_c, $c = 0$, $z \to z^2$. From our introduction to complex numbers, we know that squaring a number z means doubling its argument ϕ. In other words, the iteration of all initial points from a field line with argument ϕ produces results which are all on the field lines with twice that argument. Thus, field lines are transformed into field lines. This transformation is given by

$$\phi \to 2\phi \bmod 2\pi = \begin{cases} 2\phi \text{ if } \phi < \pi \\ 2\phi - 2\pi \text{ otherwise} \end{cases}$$

When defining $\phi = 2\pi\alpha$ with $0 \leq \alpha \leq 1$, this is nothing else but our familiar *shift transformation*

$$\alpha \to \text{Frac}\,(2\alpha)$$

which we studied extensively in chapter 10. There it turned out that it was fruitful to use the binary expansion of numbers to reveal properties of the iteration. Let us apply this to our situation. For example, we have the binary expansion $1/3 = 0.\overline{0101}$. Applying the shift transformation means that we have to shift digits:

$$\begin{aligned} &\text{step 1:} \quad 0.1\overline{01} = 2/3 \\ &\text{step 2:} \quad 0.\overline{01} = 1/3\ . \end{aligned}$$

After two iterations an initial point is back on the same field line. This corresponds to a 2-cycle in the transformation of field lines

$$\alpha_1 = \frac{1}{3} \quad \rightarrow \quad \alpha_2 = \frac{2}{3} \quad \rightarrow \quad \alpha_1 = \frac{1}{3} \quad \rightarrow \quad \cdots,$$

and this periodicity is apparent in the binary expansion of the angle. From the expansion $1/6 = 0.0\overline{01}$ we read off that $\alpha = 1/6$ iterates in one step to the cycle just discussed. Indeed, all kinds of iterative behavior of the shift transformation can also be found in the transformation of field lines: periodic angles (like 1/3), pre-periodic ones (like 1/6), but also sensitive dependence on initial conditions, and so on. These dynamics also apply for points on the unit circle (the Julia set), where the field lines terminate.

Periodic and Pre-Periodic Angles

Rational numbers $\alpha = p/q$ in the unit interval $(0 < \alpha < 1)$ can be written as infinite periodic decimal or binary fractions. In other words, we can write α as a decimal of the form

$$\alpha = 0.d_1 d_2 ... d_l \overline{d_{l+1} d_{l+2} ... d_{l+m}},$$

where the d_i are decimal or binary digits and the overlining denotes periodic repetition. Here we consider only binary digits.

What happens if we iterate $\alpha \rightarrow 2\alpha \bmod 1$? This can be read off directly from the binary expansion. In the case $l = 0$ we obtain an orbit of period m:

$$\begin{aligned} \alpha_0 &= 0.\overline{d_1 d_2 ... d_m} \\ \alpha_1 &= 0.\overline{d_2 ... d_m d_1} \\ &\vdots \\ \alpha_{m-1} &= 0.\overline{d_m d_1 ... d_{m-1}} \\ \alpha_m &= 0.\overline{d_1 d_2 ... d_m} = \alpha_0 \, . \end{aligned}$$

If $l > 0$, then α is pre-periodic; l iterations will lead to the periodic cycle. We can also write α in the form

$$\alpha = \frac{k}{2^l (2^m - 1)}$$

where k is an integer less than $2^l(2^m - 1)$. Let us briefly verify this notation. First, we again consider the case $l = 0$, thus $\alpha = 0.\overline{d_1 d_2 ... d_m}$. We let

$$k = d_1 d_2 d_3 ... d_m = \sum_{i=1}^{m} d_i 2^{m-i}$$

and obtain $2^m \alpha = k + \alpha$. Solving this equation for α yields $\alpha = k/(2^m - 1)$. The case $l > 1$ can be verified in just the same way (let k be the integer $k = d_1 ... d_{l+m} = \alpha 2^l (2^m - 1)$).

Riemann Mapping Theorem

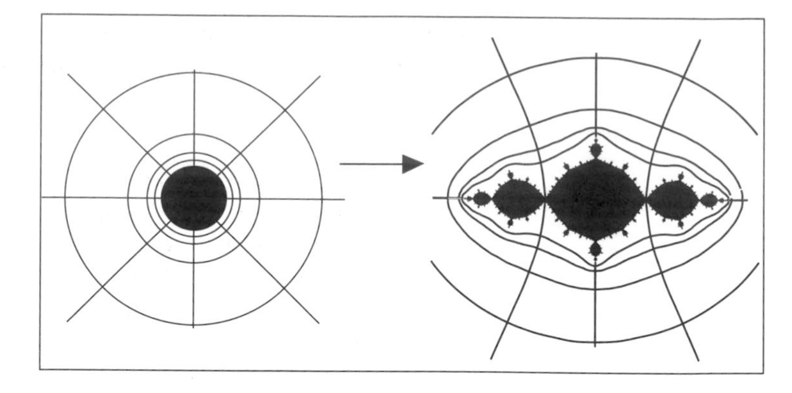

Figure 13.23 : A one-to-one correspondence between the potential of the unit disk and the potential of any connected prisoner set.

From the Potential of P_0 to that of P_c

The electrostatic potential has turned out to be the crucial tool for the mathematical analysis of prisoner sets. This is the fundamental work of Adrien Douady and John H. Hubbard. If the prisoner set P_c is connected, then the escape set E_c carries a system of field lines and equipotentials (see figure 13.24 for $c = -1$). Recall that the potential of the prisoner set P_0, the unit disk, can be interpreted as an ordinary polar coordinate system equipped with a dynamics given by $z \to z^2$. The beauty of Douady and Hubbard's work[13] lies in the fact that the potential of any connected prisoner set P_c can be interpreted as a particular polar coordinate system for the escape set E_c equipped with dynamics given by $z \to z^2 + c$. Underlying this work is a famous result of the German mathematician Bernhard Riemann, the Riemann mapping theorem, which allows to relate the potential of the unit disk to that of any connected prisoner set: equipotentials and field lines of P_0 and P_c can be brought into a one-to-one correspondence. Moreover, the dynamics of field lines of P_c can be played back to the dynamics of the field lines of the unit disk P_0, which is governed by the binary arithmetic for the transformation $\alpha \to \text{Frac}\,(2\alpha)$.

Field Lines for $z \to z^2 + c$

Let us begin to work out the procedure for labeling the field lines for the prisoner set P_c corresponding to $z \to z^2 + c$. There are many possible choices but there is only one which relates the dynamics of $z \to z^2$ in E_0 to the dynamics of $z \to z^2 + c$ in E_c. The goal is a labeling of field lines by angles $\alpha \in [0, 1)$ such that a field line with angle α is transformed to another with angle $\text{Frac}\,(2\alpha)$ under $z \to z^2 + c$. The crucial field line is the one corresponding to the line $\alpha = 0$ of the unit disk that lands at the point $z = 1$ on the boundary of the disk. The property of that

[13] A. Douady, J. H. Hubbard, *Étude dynamique des pôlynomes complexes,* Publications Mathematiques d'Orsay 84-02, Université de Paris-Sud, 1984.

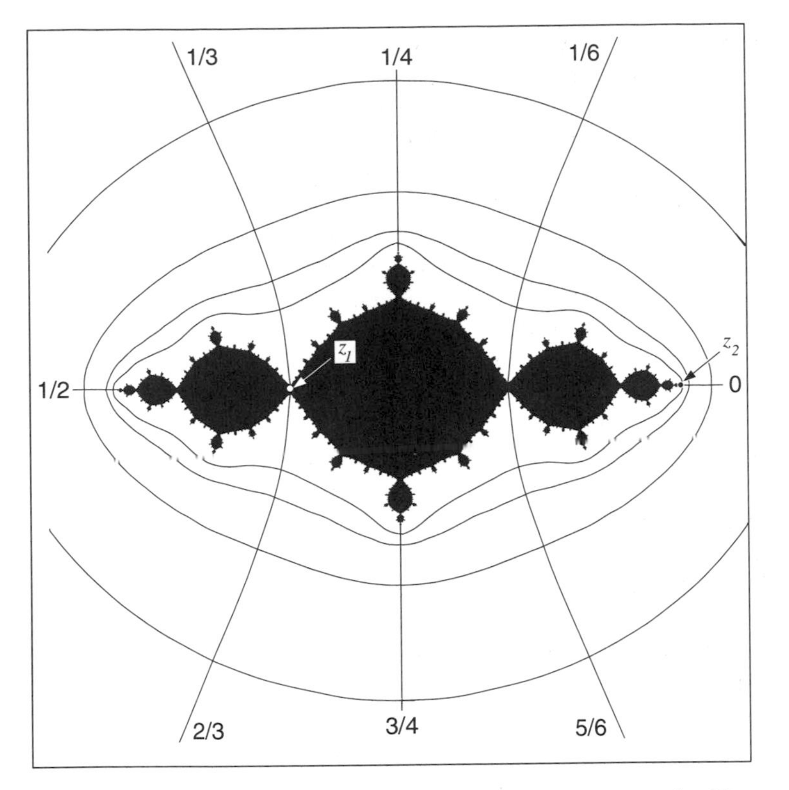

Figure 13.24 : Equipotentials and field lines for $c = -1$. The angles of the field lines are given in multiples of 2π.

Potential and Field Lines

point in this context is that it is a fixed point of $z \to z^2$. Thus, the entire field line landing at $z = 1$ is transformed onto itself under the dynamics of $z \to z^2$. Moreover, note that $z = 1$ is a repelling fixed point. Now we turn to $z \to z^2 + c$ using the example $c = -1$ (see figure 13.24). There are two fixed points,

$$z_1 = \frac{1 - \sqrt{5}}{2}, \text{ and } z_2 = \frac{1 + \sqrt{5}}{2} .$$

They are located at the left far end of P_{-1} and at the pinching point marked in figure 13.24. The derivatives of $f_c(z) = z^2 + c$ at z_1 and z_2 are $|1 \mp \sqrt{5}| > 1$. Thus, both fixed points are repelling and consequently points of the Julia set, which is the boundary of P_{-1}. Therefore, we expect that each will identify a field line. It would be natural to define angles for the field lines of P_{-1} so that angle $\alpha = 0$ corresponds to either z_1 or z_2. But which one? Well, if we want to be consistent with the dynamics of $z \to z^2 - 1$ we have to be careful. There is one field line landing at z_2. That is the set $\{z = x + 0i \mid x \geq z_2\}$. We verify that if z is from that line then $z^2 - 1$ is from that line as well.[14] In other words, we label that field line with $\alpha = 0$, because it remains invariant under $z \to z^2 - 1$, exactly as the corresponding field line of $z = 1$ under $z \to z^2$. But now there seems to be a problem. What happens with the field line(s) landing at z_1? First of all, if there was only one line landing at z_1, then it would also remain invariant under $z \to z^2 - 1$, following the dynamics associated with field lines. But that would trouble our choice of labeling the field line for z_2, because there is only one angle α such that $\alpha = \text{Frac}(\alpha)$, namely $\alpha = 0$ which we have already used for the field line which lands at z_2. However, there is a way out of the dilemma allowing two field lines landing at z_1, which is just what we see in figure 13.24. This means that P_{-1} must be pinched at z_1. Now $z \to z^2 - 1$ sends field lines to field lines by simply doubling the angle. Thus, the two field lines at z_1 must be transformed to each other. There are only two angles α_1 and α_2 having the property that $\alpha_1 = \text{Frac}(2\alpha_2)$, and $\alpha_2 = \text{Frac}(2\alpha_1)$, namely $\alpha_1 = 1/3$ and $\alpha_2 = 2/3$. Thus, we label the two lines landing at the fixed point z_1 by $1/3$ and $2/3$. Continuing in this way one can understand all the pinching points which give rise to an interpretation of the prisoner set P_{-1} itself. The same procedure applies to other parameter values c as long as the associated prisoner set is connected.

Let us summarize. The comprehension of connected Julia sets requires a study of the potential and an appropriate analysis of the dynamics of field lines under $z \to z^2 + c$. That dynamics is just given by an angle-doubling. To work out the details it is necessary to understand where on the Julia sets field lines land. A priori it is not even clear that field lines land at all. It turns out that for a special class of Julia sets[15] all field lines land. For all other cases it is only known that field lines corresponding to *rational* angles α land on the boundary of the prisoner set.

Towards the Potential of Prisoner Sets

The computation of a potential function for connected prisoner sets is strongly related to encirclements of the prisoner sets. Recall that the encirclements of Julia sets from the previous section are approximations of the prisoner sets given by

$$Q_c^{(-k)} = \{z_0 \mid |z_k| \le r(c)\} \ .$$

which also can be written as

$$Q_c^{(-k)} = \left\{z_0 \mid z_k \in Q_c^{(0)}\right\} \ .$$

In other words, the $Q_c^{(-k)}$ are the iterated *preimages* of $Q_c^{(0)} = \{z \mid |z| \le r(c)\}$, which can be called a reference or *target set*.

[14] Let $x \ge (1+\sqrt{5})/2$. Then $x^2 - 1 = (6 + 2\sqrt{5})/4 - 1 = (1+\sqrt{5})/2$.

[15] The class of locally connected Julia sets, see section 14.2.

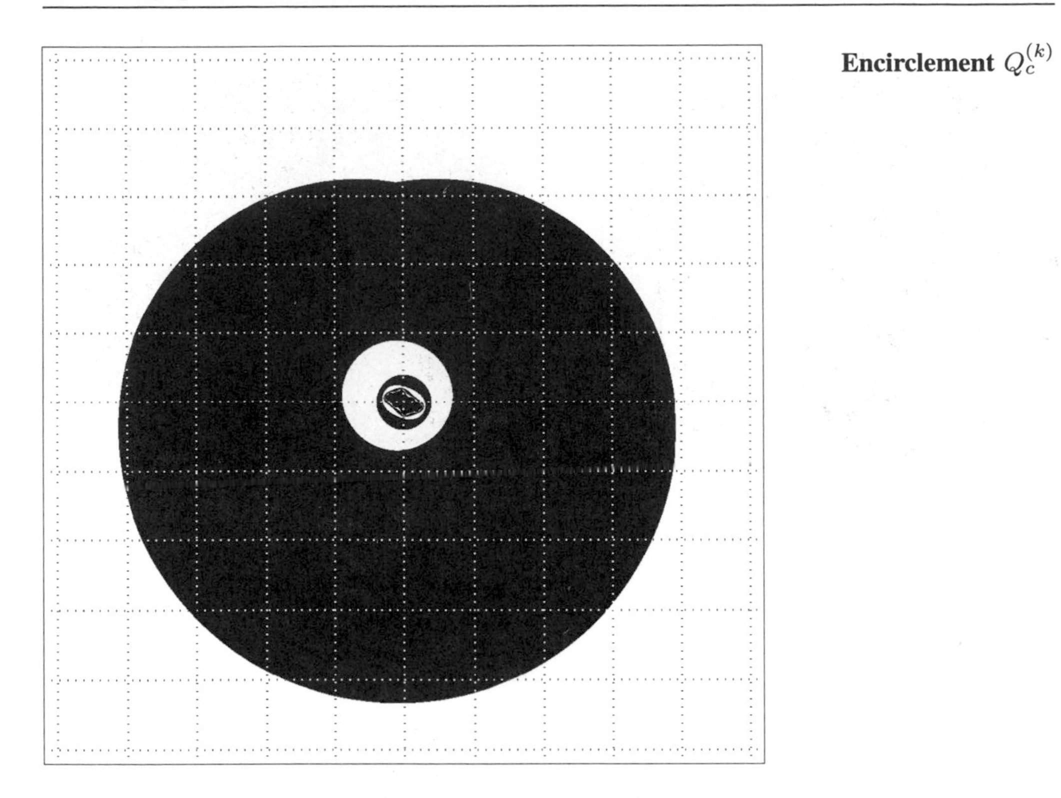

Encirclement $Q_c^{(k)}$

Figure 13.25 : Overlay of encirclements for $c = -0.5+0.5i$ drawn in alternating colors: $Q_c^{(k)}$, for $k = 2, 1, 0, -1, -2$ and -3. Observe that only $Q_c^{(0)}$ is a circle.

Now we can also look at the *images* of $Q_c^{(0)}$, denoted by

$$Q_c^{(k)} = \left\{ z_k \mid z_0 \in Q_c^{(0)} \right\} .$$

The same construction can be carried out using an arbitrary target set T, setting

$$Q_c^{(k)}(T) = \{z_k \mid z_0 \in T\}$$

for all integers k. When k is negative, z_k denotes a preimage of z_0, i.e., k iterations started at z_k produce z_0. If we look at the special case $c = 0$ and the target set $T = Q_c^{(0)}$, then the images and preimages $Q_c^{(k)}(T)$ are just the disks which are bound by the equipotential lines shown in figure 13.21, denoted by

$$D^{(k)} = \{z \mid \log_2 |z| \leq 2^k\} .$$

Choosing the Right Target Set

If, however, we choose $c \neq 0$, then things become a bit more complicated and interesting. We can observe that in this case only

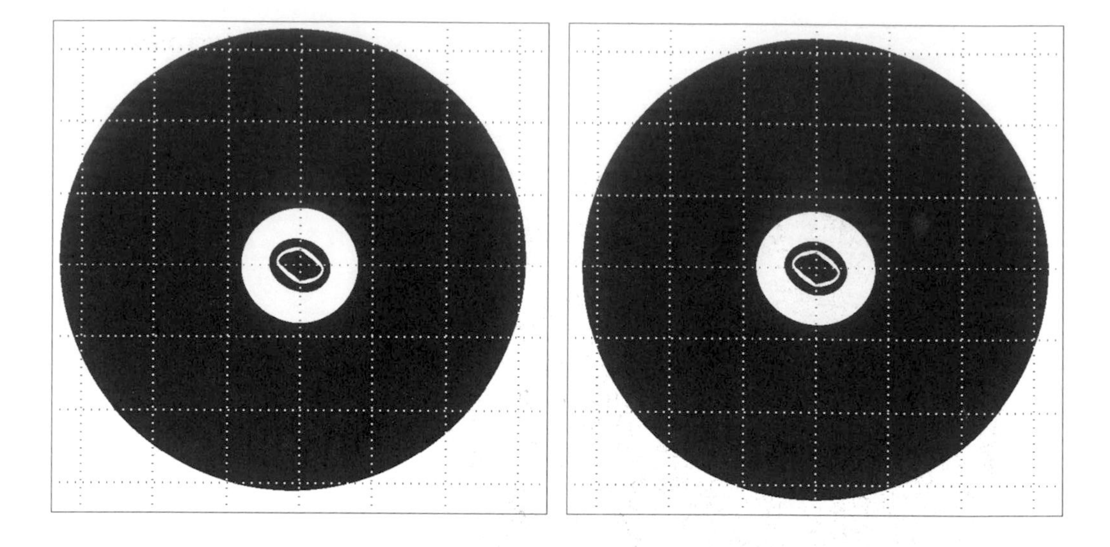

Figure 13.26 : Approximation of $P_c^{(k)}$ by $Q_c^{(k-1)}(D^{(1)})$ and $Q_c^{(k-2)}(D^{(2)})$.

the target set $T = Q_c^{(0)}$ is a disk (see figure 13.25). Indeed most published images of Julia sets show the sets $Q_c^{(k)}(D^{(0)})$ in black and white (or with an appropriate color coding). In other words, they use the disk with radius $r = 2$ as the target set. Such images may look nice, but they have a major deficiency. We cannot interpret the boundaries of $Q_c^{(k)}$ as approximations of equipotential lines when $c \neq 0$. This is clear because we cannot expect that the circle with radius 2 is an equipotential curve for all prisoner sets. There is only a single case where this holds, namely for $c = 0$. For all other cases the equipotentials are not circles. Only when we consider equipotentials distant from the prisoner set, it acts approximately like the unit disk and the equipotentials are very close to circles. The defect is visually apparent. For example, observe that in figure 13.18 the boundaries of the $Q_c^{(k)}$ come closer in some places and diverge in other places in a rather artificial way which is not related to the approximated prisoner set. Equipotential lines should not exhibit such a behavior. The problem arises from the choice of the target set T. Let us again look at the case $c = 0$. For this parameter the target set $T = Q_c^{(0)} = D^{(0)}$ is not very special. We could also take $T = Q_c^{(1)} = D^{(1)}$ or $T = Q_c^{(2)} = D^{(2)}$ and obtain essentially the same result. In fact one has the equality

$$Q_c^{(k-l)}(D^{(l)}) = Q_c^{(k)}(D^{(0)})$$

for the special case $c = 0$. In other words, we can say that the

Encirclements

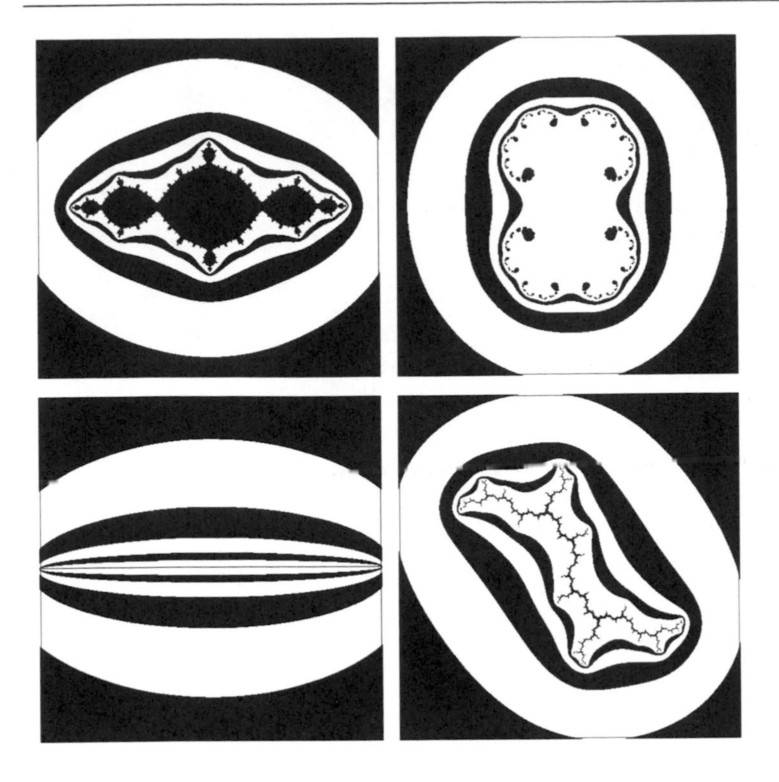

Figure 13.27 : Encirclements by $P_c^{(k)}$ for $c = -1, 0.4, -2, i$ (from upper left to lower right). Compare these images with figure 13.19. Note, that for $c = 0.4$ the Julia set is not connected. Thus the image has to be valuated with care. In this case field lines are not even defined.

boundary of $Q_c^{(k-l)}(D^{(l)})$ is a potential curve independent of the value of l.

For $c \neq 0$ this is obviously not true. But what we might consider to be next best in fact holds true. The limit, denoted by

$$P_c^{(k)} = \lim_{l \to \infty} Q_c^{(k-l)}(D^{(l)})$$

does converge to a set with an equipotential as its boundary.[16] Figure 13.26 shows the approximations of a prisoner set by $Q_c^{(k-l)}(D^{(l)})$ for $l = 1$ and $l = 2$. In fact, even for $l = 3$ it is hard to see any differences as compared to $l = 2$. Note, that the radius of the disk $D^{(k)}$ is 2^{2^k}. Thus the radius of $D^{(3)}$ is already 256. In practice we can use $Q_c^{(k-3)}(D^{(3)})$ as a rather good approximation

[16] A. Douady, J. H. Hubbard, *Étude dynamique des pôlynomes complexes,* Publications Mathematiques d'Orsay 84-02, Université de Paris-Sud, 1984.

of $P_c^{(k)}$. More explicitly, we can write

$$P_c^{(k)} = \left\{ z_0 \;\middle|\; \lim_{l \to \infty} \frac{\log_2 |z_l|}{2^l} \leq 2^k \right\} . \tag{13.3}$$

In fact, the boundary of $P_c^{(k)}$ is an equipotential curve, and the potential function is given by

$$p_c(z_0) = \lim_{l \to \infty} \frac{\log_2 |z_l|}{2^l} .$$

Using this function has another advantage; it allows us to drop the special considerations with respect to the threshold $r(c)$ (which is 2 if c is small, but is c if $|c| > 2$). It is no longer required in the definition of the sets $P_c^{(k)}$ which also provide an encirclement of the prisoner set P_c, with the additional effect of simultaneously approximating the equipotentials.

13.6 Chaos Game and Self-Similarity for Julia Sets

The method for the computation of the prisoner sets and the Julia sets as their boundaries using the pixel game is quite slow. Even on modern PCs and workstations, the resulting image will not appear in a matter of seconds. To obtain a faster, but usually sufficiently detailed, picture of the Julia set we can use the chaos game as introduced in chapter 6. For this purpose we need a set of transformations that when applied iteratively to a point will generate an approximation of the Julia set. In contrast to chapter 6, however, these transformations cannot be affine transformations because one look at our figures reveals that Julia sets do not possess the affine self-similarity properties characteristic of iterated function systems.

Julia Sets as Attractors

The idea of the chaos game for Julia sets is not as far fetched as we might think. When points from the escape set close to the Julia set are iterated using the rule $z \to z^2 + c$, they move away from the Julia set and approach infinity. Thus, we may call the Julia set a repeller with regard to the transformation $z \to z^2 + c$. Now, we can think about the inverted transformations, i.e., the transformations that take a point w to the point z, where $w = z^2+c$, thus, effectively iterating backwards. With respect to such inverse transformations the character of the Julia set must be different, it cannot be a repeller anymore but now plays the role of a new kind of attractor. The situation is completely analogous to the Multiple Reduction Copy Machine (MRCM), where the attractor in the chaos game is also a more or less complicated geometric point set in the plane. Now the remaining question is: what are the correct inverse transformations, that must be applied for this kind of chaos game?

Two Nonlinear Transformations

The problem to be solved is the following: given a complex number w, what are its preimages z such that $w = z^2 + c$? This is an exercise in equation solving for

$$z^2 - w + c = 0 \ .$$

We need not go back to the general formula for quadratic equations from section 13.3 since the above equation does not contain a linear z-part. We simply rewrite it as $z^2 = w - c$ and apply the complex square root to obtain

$$z_{1,2} = \pm\sqrt{w - c} \ .$$

Taking the plus-minus sign into account we have two solutions to the equation. This means there are generally two preimages z for each point w (except for $w = c$). Thus, there are two transforma-

MRCM for Julia Set

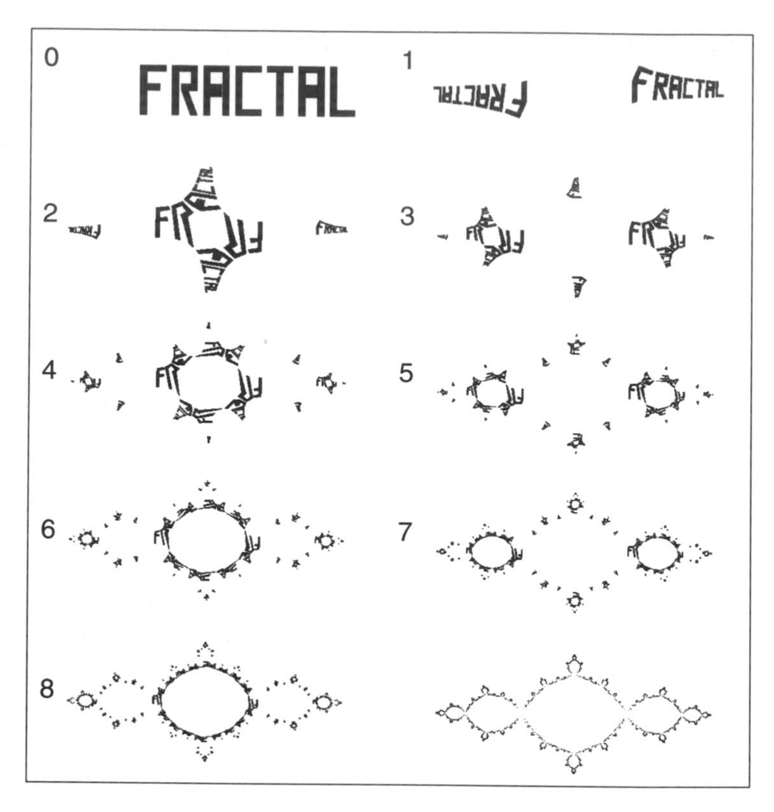

Figure 13.28 : The MRCM with the two nonlinear lenses $\pm\sqrt{w-c}$ with $c=-1$ is applied to an initial image consisting of the sequence of letters 'FRACTAL'. In each step two deformed copies of the input image are composed which rapidly converge to the corresponding Julia set.

tions in the MRCM:[17]

$$w \;\rightarrow\; +\sqrt{w-c} \quad\text{and}\quad w \;\rightarrow\; -\sqrt{w-c}$$

The chaos game for these two transformations then works as usual. First, we pick an arbitrary initial point w. Then one of the two preimages according to the above formulas is selected at random and w is replaced by this preimage and shown on the monitor. The process is repeated until enough points have been collected. The first few computed points should not be displayed as they come from the transitional period necessary for the random initial point to be attracted close enough to the Julia set. We can skip this phase if we start directly on the Julia set. Fortunately there is one point

[17]Heuristically, the situation is quite clear. Taking square roots, for example, $\sqrt{100}$, means to reduce, at least for large numbers.

of the Julia set which can be easily determined: a repelling fixed point.[18]

A fixed point can be computed directly from the equation $z = z^2 + c$, without iterating, so

$$z^2 - z + c = 0$$

must be solved. Let us do this for $c = -0.5 + 0.5i$. This is exactly the exercise from section 13.3. The result was two solutions

$$\begin{aligned} z_1 &= 1.408 - 0.275i, \\ z_2 &= -0.408 + 0.275i \ . \end{aligned}$$

The second solution z_2 is an attracting fixed point and certainly not a point of the Julia set. However, the other fixed point z_1 is a repeller (i.e., points nearby are pushed away by the iteration); and z_1 is in the Julia set.

A Derivative Criterion for Repelling and Attracting Fixed Points

The distinction between attractive and repelling fixed points can be made mathematical so that one does not have to test numerically for the attractive or repelling property. However, this test involves a tool from calculus because the essence of the method lies in the derivative of the transformation $z \to z^2 + c$ at the fixed point z. This derivative is computed the same way as for real functions; it is $2z$. Given this number the fixed point can be classified: the fixed point is *attractive* if the absolute value of the derivative at that fixed point is *less than* 1; it is *repelling* if the derivative is *greater than* 1 in absolute value. The remaining case, in which the absolute value of the derivative is exactly 1 is undecided; such fixed points are called *indifferent* and have been the source of rather deep mathematical research. The criterion is also applicable for all other underlying transformations besides the quadratic $z \to z^2 + c$. The proof of this criterion is not difficult, and we omit it.

In our example of the quadratic transformation with $c = -0.5 + 0.5i$ and its two fixed points $z_1 = 1.408 - 0.275i$ and $z_2 = -0.408 + 0.275i$, we compute the absolute values of the derivatives up to three decimals of precision

$$|2z_1| = |2.816 - 0.550i| = \sqrt{2.816^2 + 0.550^2} = 2.869$$

and

$$|2z_2| = |-0.816 + 0.550i| = \sqrt{0.816^2 + 0.550^2} = 0.984 \ .$$

The result is clear; the first fixed point z_1 is repelling, and the other one is attracting.

[18] It is already known from the work of Julia and Fatou that any repelling periodic point belongs to the Julia set. In fact, the repelling periodic points are dense in the Julia set.

Progress of the Chaos Game

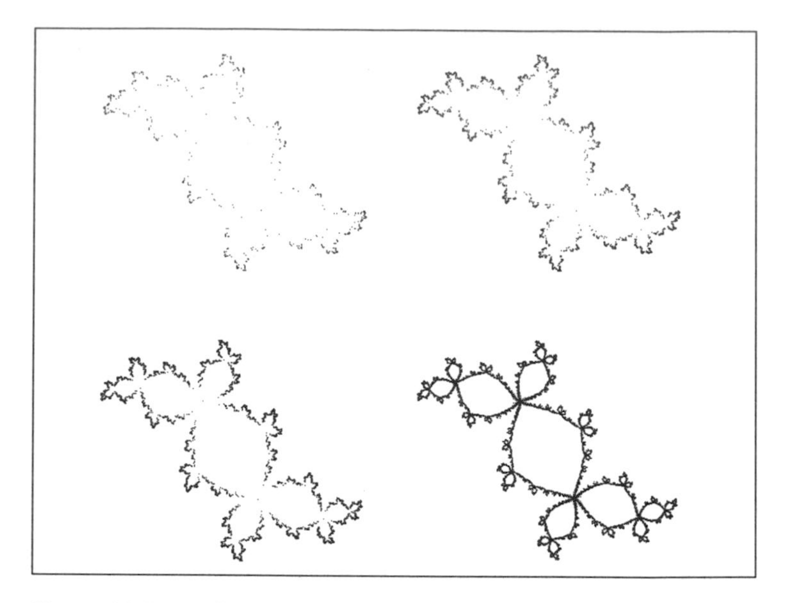

Figure 13.29 : Computation of the Julia Set for $c = 0.12 + 0.74i$ (termed the *rabbit* by Douady and Hubbard). Although the performance of the Chaos Game is in this case not too satisfactory, a first overview of the Julia set appears rather rapidly. The top left image shows 1,000 points, the top right one shows 10,000. In the bottom left one even 100,000 points of the Chaos Game are plotted. For the bottom right image the Modified Inverse Iteration Method was used. It requires only 4,750 points.

Depending on the choice of the parameter c, the performance of the chaos game algorithm may be more or less satisfactory. In some cases there are regions in the Julia set that are hard to get to. Then the Julia set may look as if it is composed of several parts, while it really is connected. There are modifications possible to improve the method, called the *Modified Inverse Iteration Method*.[19] The next figure illustrates the progress of the algorithm as more and more points are computed.

The Invariance of Julia Sets

Julia sets can be seen as attractors in the chaos game, and this reveals an important fact about the self-similarity of Julia sets. Recall that for the attractor of an MRCM the whole attractor is covered by small copies of itself. These small copies are nothing but the images of the whole attractor under the transformations in the iterated function system.[20] The same property should be true here. This says specifically that as we apply one of the two

[19]See page 178 in *The Science of Fractal Images*, H.-O. Peitgen, D. Saupe (eds.), Springer-Verlag, New York, 1988.

[20]Recall from chapter 5 that if $w_1, w_2, \ldots, w_n$ are the contractions and A is the attractor, then $A = \bigcup_{k=1}^{n} w_k(A)$.

Self-Similarity of a Julia Set

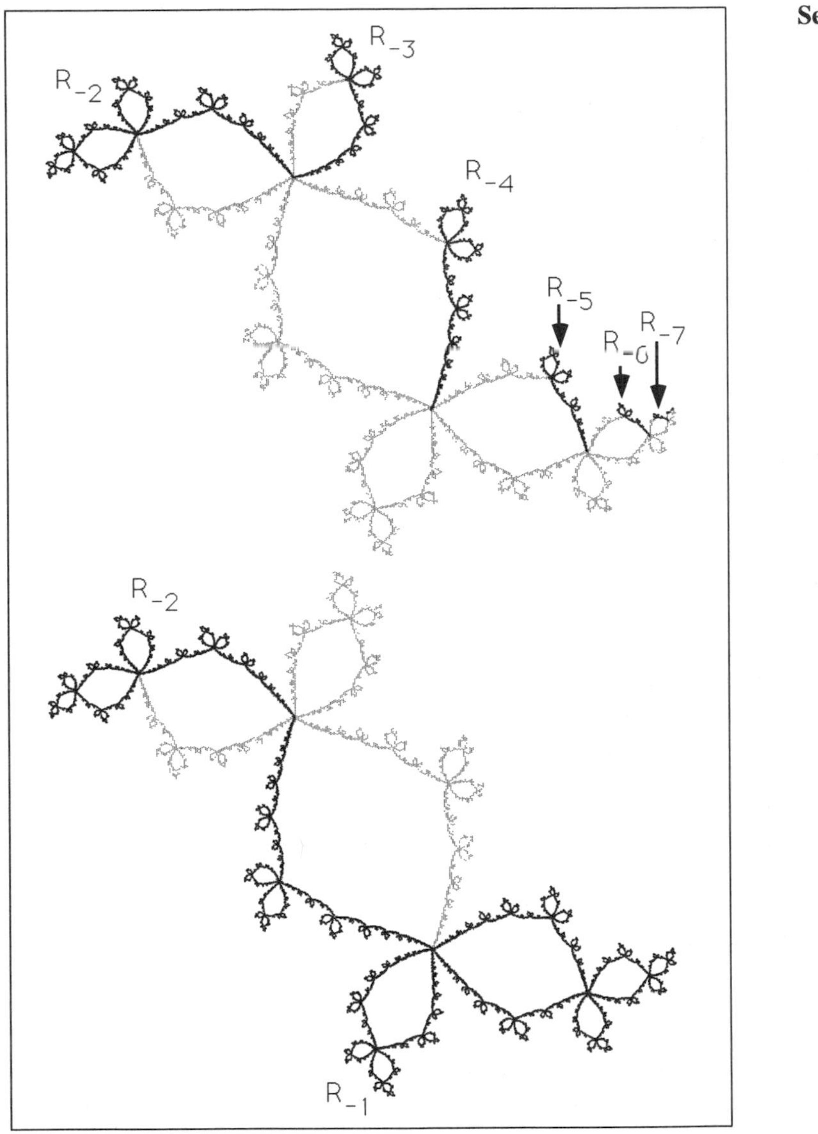

Figure 13.30 : The self-similarity of the Julia sets. These two pictures show how a very small section of the Julia set, denoted by R_{-7}, is transformed several times. In each transformation the covered portion of the Julia set indicated by the bold black parts labeled R_{-6} to R_{-1}, increases. After six iterations the result R_{-1} is already one half of the Julia set; one more application of $z \to z^2 + c$ yields the whole set R_0.

transformations of our iterated function system for the Julia set

$$w \to +\sqrt{w-c}$$
$$w \to -\sqrt{w-c}$$

to any point w of the Julia set, then we obtain yet another point of the Julia set. Therefore, the Julia set is called an *invariant set* with respect to the inverse transformations of $z \to z^2 + c$.

Moreover, if again z is a point from the Julia set, we may ask the question: What kind of point is the image of z, i.e., $z^2 + c$? It cannot be in the basin of attraction of infinity because then the initial point z would also have to be an escaping point, but z was chosen in the Julia set. On the other hand, $z^2 + c$ cannot be in the interior of the prisoner set; it must be on the boundary. The reason for this lies in the continuity of the quadratic transformation: arbitrarily close to z there are escaping initial points, and continuity of $z \to z^2+c$ implies that this neighborhood relation must also hold for the set of transformed points. Summarizing, this observation is expressed in the statement that the Julia set is invariant not only with respect to the inverse transformations $w \to +\sqrt{w-c}$ and $w \to -\sqrt{w-c}$, but also with respect to the transformation $z \to z^2 + c$ itself.

The Self-Similarity of Julia Sets

Thus, the Julia set remains invariant under forward iteration (using $z \to z^2 + c$) as well as under backwards iteration (using either of the two inverse transformations). This property is called *complete invariance* and describes one of the key properties of Julia sets. Therefore, the global structure of the Julia sets must also appear in the images and preimages of the Julia set, which explains the apparent self-similarity. This self-similarity is not to be confused with the strict or affine self-similarity that is discussed in earlier chapters of Part One of this book. The similarity here is based on a *nonlinear* transformation, and thus, the smaller copies of the Julia set contained in itself are not exact copies but strongly distorted ones, which are even folded back on themselves.

Nevertheless, the following amazing property has been shown to be true (see figure 13.30). Take any small section of the Julia set, e.g., intersect a small disk with the Julia set and assume that this intersection is not empty. Then we apply the iteration $z \to z^2 + c$ to every point in this set. We obtain a new, typically larger, subset of the Julia set. Iterating this procedure a *finite* number of times will result in the complete Julia set! This says that the immensely complicated global structure of the Julia set is already contained in any arbitrarily small section of it.

13.7 The Critical Point and Julia Sets as Cantor Sets

We have visualized Julia sets using encirclements $Q_c^{(k)}$ and $P_c^{(k)}$ leading to the powerful tools of potential functions and field lines. These, however, apply only to connected Julia and prisoner sets. Let us now develop an understanding of the reasons for this limitation.

For this purpose we return for a moment to the discussion of the iterative behavior of $x \to ax(1-x)$ for real numbers a and x. You will recall that the dynamics of this iteration is equivalent to that of $z \to z^2 + c$, again for real numbers z and c. If $c \leq 1/4$ we can translate one into the other by

$$x_n = \frac{1}{2} - \frac{z_n}{a} \quad \text{and} \quad a = 1 + \sqrt{1-4c}\,. \tag{13.4}$$

We will use this equivalence later to interpret our following discussion. What are the prisoner sets for $x \to ax(1-x)$, and how do they change with the parameter a. Let us adapt our encirclement experiments to the case of interest. We start with some large interval $Q_a^{(0)}$ guaranteed to enclose the prisoner set P_a. Then we look at the iterated preimages $Q_a^{(-k)}$, which in the limit tend to the prisoner set P_a. This would not be difficult by explicit calculation, but we prefer to carry out the evaluation graphically to make the issue visually apparent. For this task we introduce *backward graphical iteration.* We are given a value, say y, and we want to find x so that $ax(1-x) = y$. This amounts to solving a quadratic equation which we do graphically as follows.

Backward Iteration — One Stage

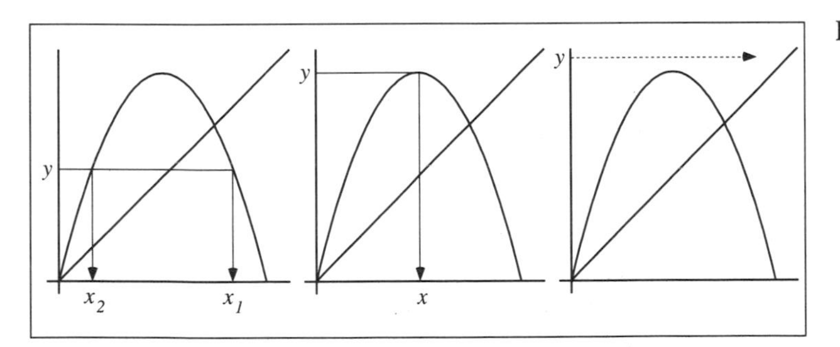

Figure 13.31 : Three cases for the backward iteration: two solutions, one solution, and no solution.

Backward Graphical Iteration

We pick y on the y-axis and draw the horizontal line through this point. It may intersect the graph of $g(x) = ax(1-x)$ in two points, one point, or no point at all. Then we draw vertical lines from these intersection points to the x-axis. There we can read off the solutions to the equation. These are the preimages of y. The

Backward Iteration — Two Stages

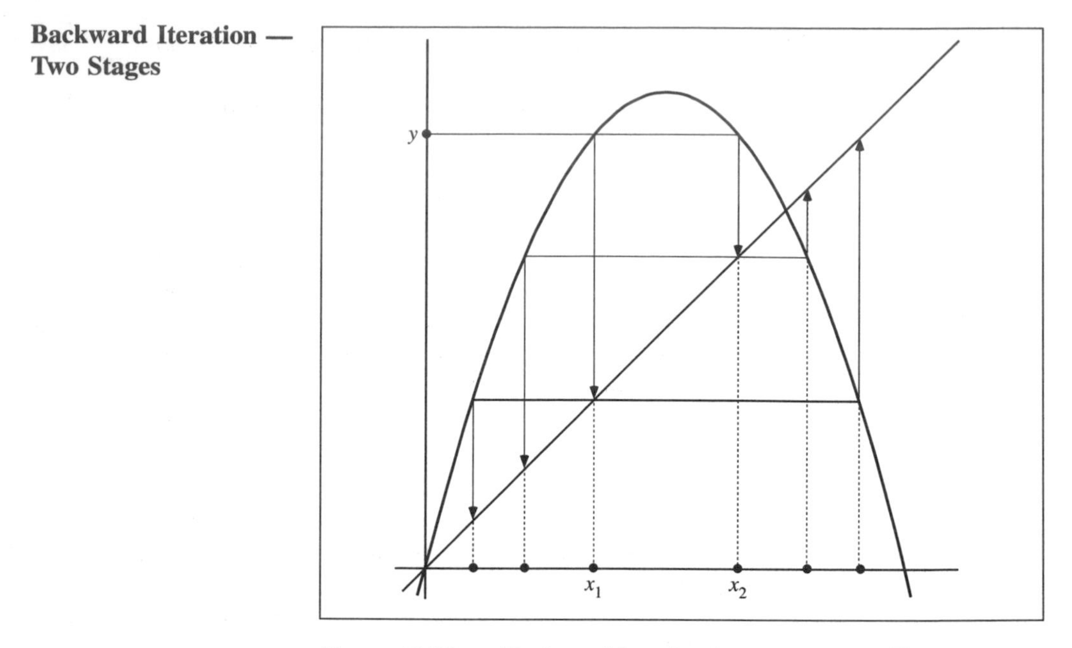

Figure 13.32 : Backward iteration for two stages. There are two preimages of y and there are four points which, when iterated twice, land in y.

procedure really is nothing but carrying out the usual graphical iteration backwards. If we want to repeat this procedure, we draw the vertical lines only up to the diagonal and then draw horizontal lines again to meet the graph (see figure 13.32). In this way we generate backward orbits, which can be described by a tree: given y, we may find two, one, or no preimages. For each of those we may again find two, one, or no preimages, and so on. When there is no preimage, then the tree is pruned at the corresponding branch.[21]

Now we use this method to find $Q_a^{(1)}$ when $Q_a^{(0)}$ is given as an initial large interval.[22] Figure 13.33 shows the result for $a = 3.5$. We observe that the $Q_a^{(-k)}$ are a nested sequence of intervals which shrink towards the unit interval; and in fact, if we restrict attention to real numbers only, then we may conclude that the prisoner set is $P_a = [0, 1]$.

Indeed, for this example ($a = 3.5$) we could have obtained the result much more cheaply. Observe first, that any $x_0 < 0$ leads to an orbit escaping to infinity. Then also orbits starting at

[21]Using complex numbers in place of real ones, branches are never pruned since we always have two preimages (except for $y = c$ where only one preimage $x = 0$ exists). Thus, when computing the backward iteration in the (complex) chaos game it is always possible to choose randomly between two options.

[22]In fact, one can take any interval which encloses the unit interval $[0, 1]$.

Backward Iteration of an Interval

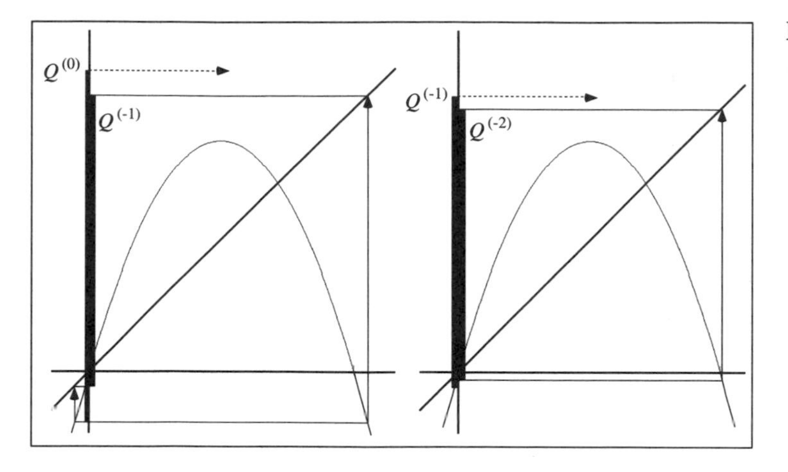

Figure 13.33 : Encirclement by backward iteration. Here an entire interval indicated by the thick line on the y-axis labeled $Q^{(0)}$ is iterated backwards once (left) and again (right). ($a = 3.5$)

Encirclement by Backward Iteration

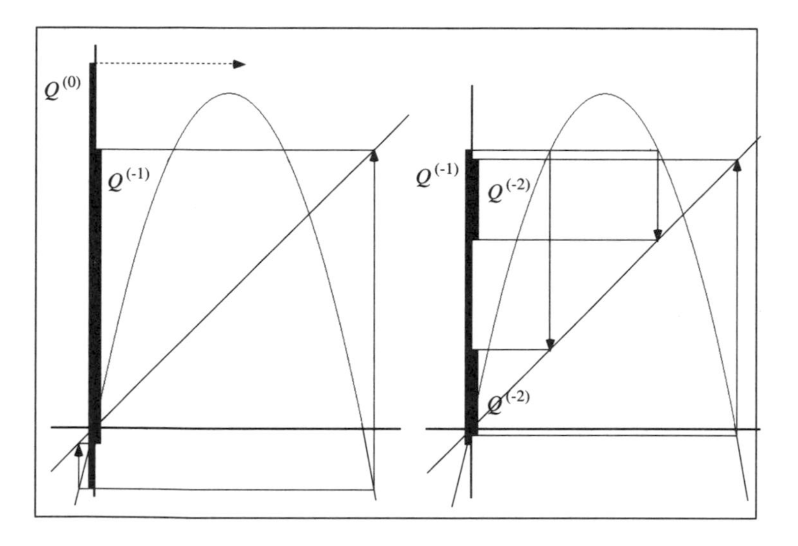

Figure 13.34 : Two cycles of backward iteration of the interval $Q^{(0)}$ yields two intervals. ($a = 4.5$).

$x_0 > 1$ escape because the first iterate is $x_1 < 0$, and thus the following iterates grow negatively without bound as in the first case. Moreover, an orbit for $0 \leq x_0 \leq 1$ cannot escape the unit interval because $0 \leq x_1 = ax_0(1 - x_0) \leq 1$.

However, if we take $a = 4.5$, for example, we cannot argue that way, although it is true that orbits starting outside the interval $[0, 1]$ do escape to infinity as before. Let us look at the backward graphical iteration of a large interval. Its preimage consists of *two*

Backward Iteration Generating Cantor Set

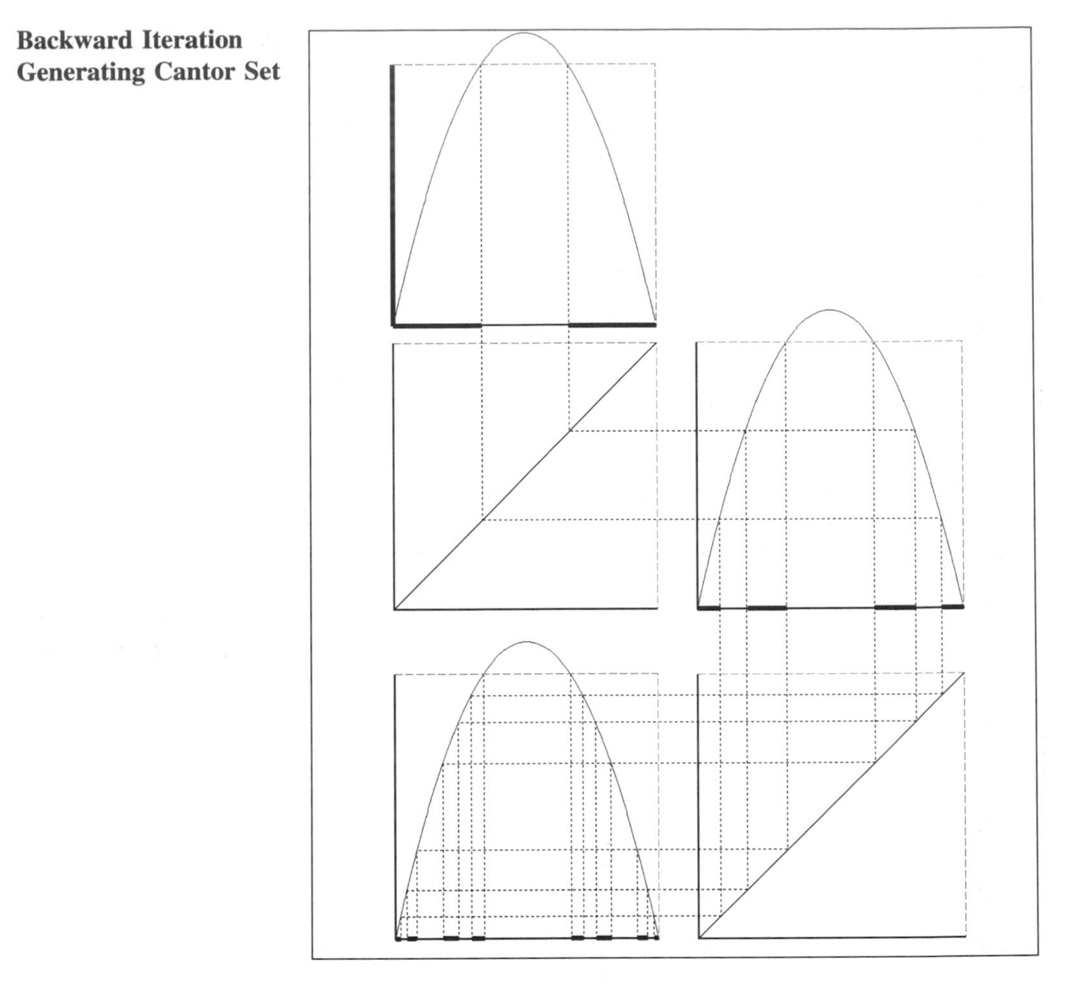

Figure 13.35 : Encirclement of Cantor set through backward iteration of the unit interval. ($a = 4.5$)

disjoint parts (refer to figure 13.34). Thus, we are led to conjecture that the prisoner set is not connected.

To investigate further let us iterate backwards a few times (see figure 13.35). We position several graphs of $g(x) = ax(1 - x)$ for $a = 4.5$ in such a way that the result of the first step feeds into the second step and so on. We make a very important observation: the resulting encirclements $Q_a^{(-i)}$ correspond to a Cantor set construction. In other words, P_a will be a Cantor set. Why do we call P_a a 'Cantor set'? Usually one refers to the Cantor set as an interval from which the (open) middle thirds are recursively removed. Thus, in the usual Cantor set all pieces in a given construction stage have the same length; and the resulting limit object is strictly self-similar. In our construction of encirclements we obtain something very similar, but the pieces of a given step have different sizes and the limiting object is not self-similar in the strict sense. It is a Cantor set — but slightly distorted.

The Fate of the Critical Point is Decisive

Let us summarize. We have just seen that for $a = 3.5$ the prisoner set on the real line is an interval, while for $a = 4.5$ the prisoner set is a Cantor set. In other words, as we increase a from 3.5 to 4.5 something comparable to a phase transition occurs. Exactly at which parameter a is the transition? To answer this question we observe that the disintegration of the preimage of the unit interval into two parts is caused by the fact that the vertex of the parabola of the graph of $g(x) = ax(1-x)$ is *above* the unit square $[0,1] \times [0,1]$. The coordinates of the vertex are $(x, y) = (1/2, a/4)$. The x-coordinate is usually called the *critical point*, while the y-coordinate is referred to as the *critical value* of the function $g(x) = ax(1-x)$. Both are characterized by the fact that the critical value has only one preimage, namely the critical point (compare figure 13.31). The fate of the critical point, i.e., the long term behavior of the orbit of $x_0 = 1/2$, called the *critical orbit*

$$\frac{1}{2} \to \frac{a}{4} \to \frac{4a^2 - a^3}{16} \to \cdots,$$

determines whether P_a is an interval (one connected piece) or a Cantor set. When $a \le 4$ then the critical value does not exceed 1, $a/4 \le 1$, and thus the critical orbit remains in the unit interval for all iterations; P_a is an interval. However, if $a > 4$ then $a/4 > 1$, the critical orbit goes to $-\infty$, and P_a is a Cantor set.

Translation to $z^2 + c$

So far we have verified only the real case and the iteration $x_{n+1} = ax_n(1-x_n)$. Let us now begin to translate this into the iteration $z_{n+1} = z_n^2 + c$ for real z_n. Is there a square which corresponds to the unit square for the x_n-iteration? Indeed, this is easy to locate, using either the explicit transformation rules in eqn. (13.4) or by an analysis of the graphical iteration (see figure 13.36). The square here is determined by the far right intersection of the graph of $f(z) = z^2 + c$ with the diagonal, which is at the positive solution of $z^2 + c = z$, i.e.,

$$z = \frac{1 + \sqrt{1 - 4c}}{2}$$

as long as $c < 1/4$. This number is equal to $a/2$ according to eqn. (13.4). Thus, the square extends from $-a/2$ to $a/2$; and we call it the *essential square*.

What is the critical orbit here? Again we can use the transformation rules in eqn. (13.4) and compute the corresponding initial point: $z_0 = a(1/2 - x_0)$ for $x_0 = 1/2$, which is $z_0 = 0$. Alternatively, we can observe that the prisoner set P_c is the interval

Essential Squares

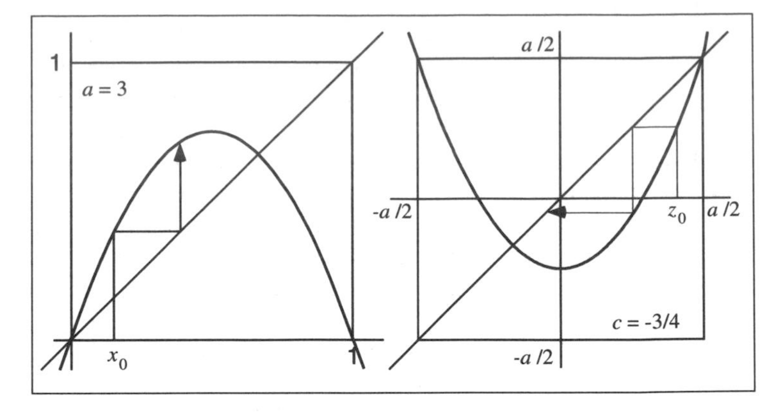

Figure 13.36 : Graphical iteration of $x \to ax(1-x)$ (left) and $z \to z^2 + c$ (right).

$[-a/2, a/2]$ as long as the minimum of the graph of $f(z) = z^2 + c$ does not exceed the essential square in figure 13.36. In other words, the critical orbit is now given by the orbit generated by the critical point $z_0 = 0$,

$$0 \to c \to c^2 + c \to \cdots$$

And again we have to check whether the critical orbit goes to infinity or not. Also note that if the critical value c leaves the essential square we will have a Cantor set for P_c as before.

This observation allows us to make a conjecture about all real c-values corresponding to connected prisoner sets. First, we observe that the minimum of the graph of $f(z)$ touches the base of the essential square when $c = -a/2$. Using the transformation rules, this equation is the same as

$$c = -\frac{1 + \sqrt{1 - 4c}}{2} \tag{13.5}$$

and also

$$\sqrt{1 - 4c} = -2c - 1 \ .$$

Squaring the last equation yields

$$1 - 4c = 4c^2 + 4c + 1$$

and

$$4c^2 + 8c = 0$$

or, equivalently,

$$4c(c + 2) = 0 \ .$$

At $c = 1/4$

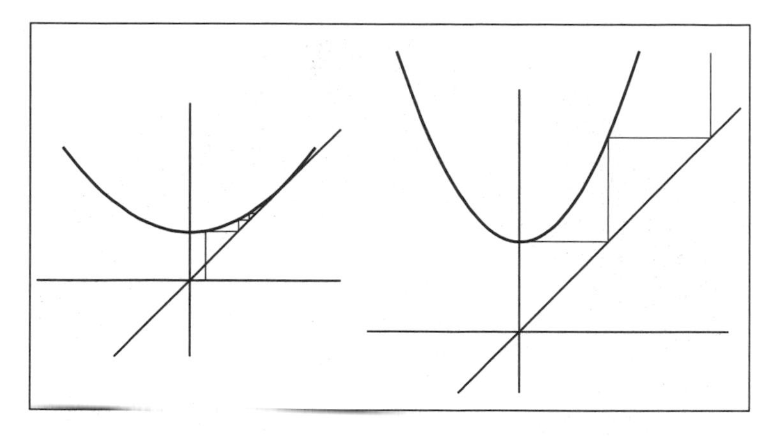

Figure 13.37 : When c is large, all orbits escape to infinity (right). Precisely at $c = 1/4$ the parabola touches the bisector and some orbits do not escape (left).

Clearly, there are two solutions to this last equation, $c = 0$ and $c = -2$. But only the latter one solves the original eqn. (13.5).[23] Thus, the vertex of the parabola exceeds the essential square when $c < -2$. On the other hand, if c is very large then the parabola is completely above the bisector as in figure 13.37 (left), and then the critical orbit definitely goes to infinity. The transition between this case and the case where the critical orbit will remain in the prisoner set occurs at a certain parameter c. Precisely at this parameter c the parabola just touches the bisector (see the right hand graph in figure 13.37). This configuration occurs when the two points of intersection of the graph of $f(z)$ with the diagonal coincide, i.e., when

$$\frac{1+\sqrt{1-4c}}{2} = \frac{1-\sqrt{1-4c}}{2} .$$

This is the case when $4c = 1$, thus $c = 1/4$. In summary, we expect that the prisoner set P_c is connected provided c is in the interval $[-2, 1/4]$.

Complex Parameters c

Let us now take the last step and allow the parameter c to be complex and see how the critical orbit determines which Julia sets are Cantor sets.

The prisoner set P_c is a subset of $P_c^{(m)}$, which, for sufficiently large m, is a disk (or at least as close to a disk as we like). It is a connected set, and it is symmetric with respect to 0. Now, what is the shape of $P_c^{(k)}$ for $k < m$? Well, for k close to m it is certainly still very much like a disk, but is this true for all values of k?

[23] The other solution $c = 0$ is an artefact of the squaring of the equation.

A Cantor Set in the Complex Plane

Figure 13.38 : For $c = -2.2$ the prisoner set is a Cantor set.

The crucial observation is that if $P_c^{(k)}$ is disk-like (i.e., a deformed disk), then there are exactly three possible cases for the next approximation $P_c^{(k-1)}$ of the prisoner set: $P_c^{(k-1)}$ is either

- disk-like and contains the critical point,
- made up of two disjoint disk-like subsets and does not contain the critical point,
- made up of two disk-like subsets which touch exactly at the critical point, i.e., its boundary forms a figure-eight.

The critical point is decisive for these cases. If we now assume that all sets $P_c^{(k)}$ are disk-like for $k < m$, then the enclosed prisoner set is connected. On the other hand, if $P_c^{(k-1)}$ is made of two sets then $P_c^{(k-2)}$ is made of 4 sets, and $P_c^{(k-l)}$ is made of 2^l disjoint subsets (see figure 13.38). This observation leads to the dichotomy:

- The Julia set is a Cantor set if, and only if, the iteration of the critical point 0 leads to infinity (in absolute value).
- The Julia set is one piece (connected) if, and only if, the iteration of the critical point 0 is bounded.

This dichotomy will be a major theme in the last chapter.

13.8 Quaternion Julia Sets

Complex numbers are a two-dimensional extension of real numbers. It is possible to extend the space of complex numbers further. However, the attempt by the Irish physicist and mathematician William R. Hamilton to create a space of numbers with three components failed. Instead he had to resort directly to a space of numbers with four components. This space, invented in the year 1843, is called the space of quaternions H. A quaternion $x \in H$ can be represented by the symbol

$$x = x_0 + x_1 i + x_2 j + x_3 k$$

where j and k denote two additional imaginary units and x_0 to x_3 denote the four components of the quaternion. Thus, x_0 is the real part of x and $x_0 + x_1 i$ is the complex part of the quaternion x. Almost all the rules for real and complex number hold also in the space of quaternions. The only exception is the multiplication, which is not commutative, i.e., if x and y are two quaternions, then generally $xy \neq yx$.

Rules for Quaternions

Let x and y be two quaternions

$$x = x_0 + x_1 i + x_2 j + x_3 k$$
$$y = y_0 + y_1 i + y_2 j + y_3 k \ .$$

Then the sum is

$$x + y = (x_0 + y_0) + (x_1 + y_1)i + (x_2 + y_2)j + (x_3 + y_3)k \ .$$

The product xy is computed by formally multiplying x and y by use of the distributive law and the conventions

$$i^2 = j^2 = k^2 = -1$$
$$ij = -ji = k$$
$$jk = -kj = i$$
$$ki = -ik = j \ .$$

Division of quaternions is also defined but not relevant to this section.

With these definitions we can interpret the iteration $x \to x^2 + c$ for quaternions. Moreover, we can extend the notion of prisoner and escape sets from the complex to the quaternion space. A quaternion Julia set consequently is defined as the boundary of a quaternion prisoner set. However, the visualization of quaternion Julia sets is significantly more demanding compared to the complex case, because we have to deal with fractal objects in four dimensions. Rendering techniques have been pioneered by V. Alan

Norton in 1982, who produced a collection of amazing images.[24] In the color section of this book we have included two images of this type.[25]

Let us explain some more details which are needed to understand what these images are displaying. First we note that a simplification can be achieved by choosing the quaternion parameter c in the iteration of $x \to x^2+c$ as a complex number, i.e., $c = x_0+x_1 i$. If the initial point $x \in H$ is complex (i.e., with third and fourth component being 0) then the orbit remains in the complex plane and we have the same outcome as if we had worked in the complex space from the beginning. In other words, the quaternion Julia set, restricted to the subset which corresponds to the complex plane, is identical to the traditional Julia set. Moreover, the set

$$H_0 = \{x \in H \mid x = x_0 + x_1 i + x_2 j\} ,$$

which contains all quaternions with the last component $x_3 = 0$, is invariant under the iteration. In other words, if $x \in H_0$ then $x^2+c \in H_0$ also. Thus, in a first attempt, we may ignore the fourth component of quaternions and work only with the first three. The result is a Julia set in three dimensions. An example is shown in color plate 2 for the case $c = 0.2809 + 0.53i$. The object is cut open at the complex plane revealing a familiar looking rabbit-type Julia set (compare figure 13.29). In the third dimension the Julia set seems to have a very complicated structure, which, however, also reveals some elements of regularity.[26] This picture shows only a three-dimensional section of the entire Julia set which is a subset of the four-dimensional quaternion. To get a feeling for the whole structure of the Julia set we may display a different section (see color plate 1). Note that again the rabbit-type complex Julia set is contained as a cross-section of the object displayed.

[24]See B. B. Mandelbrot, *The Fractal Geometry of Nature,* W. H. Freeman and Co., New York, 1982, and V. A. Norton, *Generation and display of geometric fractals in 3-D,* Computer Graphics 16, 3 (1982) 61–67. A more recent reference with an advanced rendering technique is J. C. Hart, D. J. Sandin, L. H. Kauffman, *Ray tracing deterministic 3-D fractals,* Computer Graphics 23, 3 (1989) 91–100.

[25]The pictures stem from the computer science diploma thesis (1991) of our student Ralph Lichtenberger, which is based on the ray tracing technique introduced by Hart, Sandin, and Kauffman mentioned in the above footnote.

[26]The Julia set contains circles. For a discussion of this effect see V. A. Norton, *Julia sets in the quaternions,* Computers and Graphics 13, 2 (1989) 267–278.

13.9 Program of the Chapter: Julia Sets

We have discussed several algorithms to visualize Julia sets. For the program of the chapter, we have chosen the chaos game algorithm using inverse iterations. It is rather fast in providing a first impression of the shape of the Julia set, although for some parameter choices it takes a very long time to obtain all the details.

The program uses the complex square root algorithm discussed in section 13.3 to compute the root which has a non-negative real part. The chaos game iteration starts at the repelling fixed point as suggested in section 13.6. This is a point of the Julia set. A fixed point for the iteration of $z^2 + c$ is explicitly given by

$$z = 1/2 \pm \sqrt{1/4 - c}\,.$$

The fixed point is repelling if the modulus $|z|$ is greater than $1/2$.[27] Therefore, we obtain a repelling fixed point if we use the root $\sqrt{1/4 - c}$ having a positive real part in the above formula.

Screen Image of Program

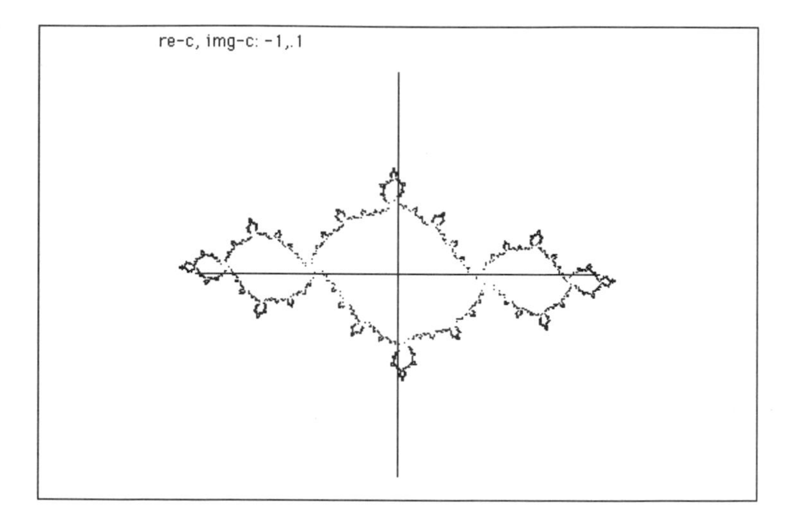

Figure 13.39 : Output of the program 'JuliaSets'.

The program starts by setting `xn = 0.25` (real part) and `yn = 0` (imaginary part) representing the complex number $z = 1/4$. Then the square root of $z - c$ (with a positive real component) is computed. At the first step of the iteration we add $1/2$ to the result and obtain the repelling fixed point which is the initial point for the following steps of the iteration.

After having computed the square root of $z - c$, the program chooses one of the two possible roots $\pm\sqrt{z - c}$ at random (using

[27] Then the derivative $2z$ of $z^2 + c$ at the fixed point is greater than 1 in modulus.

BASIC Programm **JuliaSets**
Title Computation of Julia sets by inverse iteration

```
left = 30
w = 300
s = w/3
orig = left + w/2
INPUT "re-c, img-c:", xc, yc
REM draw coordinate system
LINE (left,left+w/2) - (left+w,left+w/2)
LINE (left+w/2,left) - (left+w/2,left+w)

REM set up to start iteration with fixed point computation
xn = .25
yn = 0
FOR i = 1 TO 5000
    a = xn - xc
    b = yn - yc
    REM compute square root with re >= 0
    IF a > 0 GOTO 100
    IF a < 0 GOTO 150
        xn = SQR(ABS(b)/2)
        IF xn > 0 THEN yn = b/(2*xn) ELSE yn = 0
        GOTO 200
100     xn = SQR((SQR(a*a + b*b) + a)/2)
        yn = b / (2 * xn)
        GOTO 200
150     yn = SQR((SQR(a*a + b*b) - a)/2)
        IF b < 0 THEN yn = -yn
        xn = b / (2 * yn)
        REM first step is computation of repelling fixed point
200 IF i = 1 THEN xn = xn + .5
    REM choose one of the two symmetric root at random
    IF RND < .5 GOTO 300
        xn = -xn
        yn = -yn
300 PSET (xn*s + orig,-yn*s + orig)
NEXT i
END
```

the random number generator RND). Finally the result is plotted as a point using PSET. The drawing is scaled such that a unit circle (i.e., the Julia set for $c = 0$) covers 2/3 of the width w of the drawing area.

Chapter 14

The Mandelbrot Set: Ordering the Julia Sets

In the Mandelbrot set, nature (or is it mathematics?) provides us with a powerful visual counterpart of the musical idea of 'theme and variation': the same shapes are repeated everywhere, yet each repetition is somewhat different. It would have been impossible to discover this property of iteration if we had been reduced to hand calculation, and I think that no one would have been sufficiently bright or ingenious to 'invent' this rich and complicated theme and variations. It leaves us no way to become bored, because new things appear all the time, and no way to become lost, because familiar things come back time and time again. Because this constant novelty, this set is not truly fractal by most definitions; we may call it a borderline fractal, a limit fractal that contains many fractals. Compared to actual fractals, its structures are more numerous, its harmonies are richer, and its unexpectedness is more unexpected.

Benoit Mandelbrot[1]

The Mandelbrot set is certainly the most popular fractal, probably the most popular object of contemporary mathematics at all. Some people claim that it is not only the most beautiful but also the most complex object which has been *seen*, i.e., made visible. Since Mandelbrot made his extraordinary experiment in 1979, it has been duplicated by tens of thousands of amateur scientists around the world.[2] They all like to delve into the unlimited variety of pictures

[1]Edited from an interview in the video film: *Fractals, An Animated Discussion,* by H.-O. Peitgen, H. Jürgens, D. Saupe, C. Zahlten, W. H. Freeman and Company, New York, 1990.

[2]See B. B. Mandelbrot, *Fractal aspects of the iteration of* $z \to \lambda z(1-z)$ *for complex* λ *and* z, Annals New

Benoit B. Mandelbrot

Figure 14.1 : Benoit B. Mandelbrot.

which can develop on a computer screen. Sometimes many hours are required for their generation; but this is the price you have to pay for the adventure of finding something new and fantastic where nobody has looked before.

Is this wealth just a generous gift from mathematics to those who like to marvel at beautiful pictures, or does this apparent beauty and complexity have a deeper meaning? Do the apparent pictorial features of the Mandelbrot set have an equal counterpart in its mathematical beauty? In other words, does the Mandelbrot set present a glimpse of what mathematicians sometimes call the aesthetics of mathematics? The answer is a vigorous 'yes' indeed.

York Academy of Sciences 357 (1980) 249–259. For an historical account of the discovery read the foreword in Part One of this book and Mandelbrot's article *Fractals and the Rebirth of Iteration Theory* in *The Beauty of Fractals*, H.-O. Peitgen and P. H. Richter, Springer-Verlag, Heidelberg, 1986, pages 151–160.

14.1 From the Structural Dichotomy to the Potential Function

We learned in chapter 13 that for each parameter value of c in the iteration of $z \to z^2 + c$ there is a unique prisoner set P_c and a corresponding escape set E_c. Moreover, we also learned how to draw images of P_c and E_c. The equipotentials and field lines reveal the natural structure of E_c, leading to a deeper understanding of the dynamics of $z \to z^2 + c$ in the escape set and its boundary, the Julia set J_c. But so far we have only seen a glimpse of the infinite variety of Julia sets that can be explored.

Speaking in terms of a metaphor, we are confronted with an infinite picture book, each page of which holds the image of one particular Julia set J_c and where the page numbers are the complex parameters c belonging to the Julia sets. How many chapters does this infinite book have? And is there a principle of order? The answer to the latter query is 'yes' and follows from results already contained in the mathematical masterpieces of Julia and Fatou. The key is the structural dichotomy, which states that for any choice of c the associated Julia set J_c and the prisoner set P_c are both

- either one piece (mathematically: connected)
- or a dust (mathematically: totally disconnected).

We already discussed this dichotomy in the previous chapter. In fact, in the case of totally disconnected prisoner sets the study of their encirclements explained that the corresponding Julia sets can be understood as generalized Cantor sets. In other words, the infinite book of Julia sets can be organized in two chapters: the first for all connected Julia sets and the other for those that are totally disconnected. Around 1979 Mandelbrot had the idea of picturing this dichotomy within the set of all parameters c varying in the complex plane $\mathbf{C}$. This led directly to the *Mandelbrot set*

$$M = \{c \in \mathbf{C} \mid J_c \text{ is connected}\} .$$

He colored each point (pixel of a computer screen) in the plane of c-values black or white depending on whether the associated Julia set turned out to be one piece or a dust (see figure 14.2). The result was a black and white image (see insert in figure 14.3) which, given the graphics technology of that time, did not look very impressive or promising. But Mandelbrot realized that he had discovered one of the gems of mathematics and pushed further. Thus, he initiated a revitalization of a mathematical field which had been dormant for nearly 60 years.

The Mandelbrot Set — Dichotomy of Julia Sets

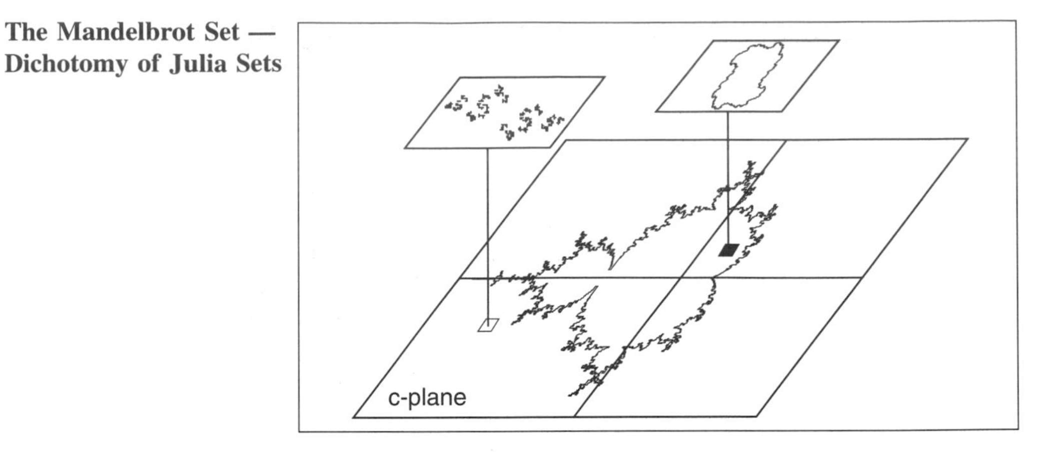

Figure 14.2 : Any point in the c-plane, interpreted as a parameter c for the iteration of $z \to z^2 + c$, corresponds to a Julia set. The point is colored black, if the corresponding Julia set is connected, and white if the set is disconnected. This is the essence of Mandelbrot's experiment from 1979.

Characterization by the Orbit of the Critical Point

But how did he actually let the computer make the decision whether a parameter c belongs to the Mandelbrot set or not? He was one of the few people at that time who knew the works of Julia and Fatou very well. In particular, he was aware of the fact that there is a tight interrelation between the dichotomy of Julia sets and the fate of the critical point. Let us recall this fact from section 13.7:

Fact. *The prisoner set P_c is connected if and only if the the critical orbit $0 \to c \to c^2 + c \to \cdots$ is bounded.*

This fact provides an alternative definition for the Mandelbrot set. In other words, Mandelbrot used

$$M = \left\{c \in \mathbf{C} \mid c \to c^2 + c \to \cdots \text{ remains bounded}\right\} \quad (14.1)$$

as the definition for M in his 1979 experiments.

Note that this definition is very similar to that of the prisoner set P_c (see page 375), written as

$$P_c = \left\{z_0 \in \mathbf{C} \mid z_0 \to z_0^2 + c \to \cdots \text{ remains bounded}\right\}.$$

However, while the Julia set is part of the plane of initial values z_0 whose orbits reside in the same complex plane, the Mandelbrot set is in the plane of parameter values c, and it is not appropriate to plot any orbits from the iteration of $z \to z^2 + c$ in this plane.

The First Clues About M

From our studies we already have a few clues about the Mandelbrot set. First of all, everything outside of a disk of radius 2 is not part of the Mandelbrot set, because, if $|c| > 2$, then the

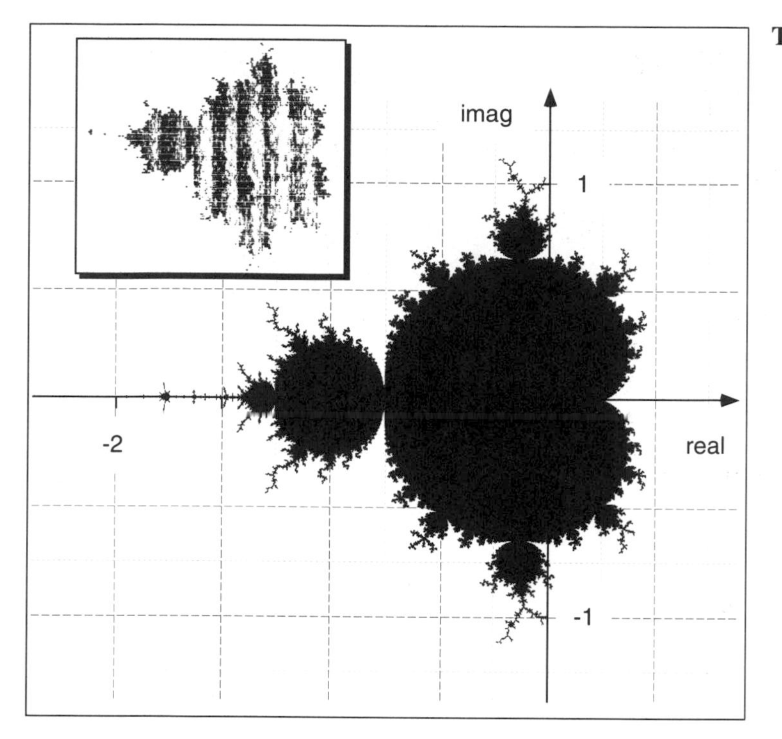

The Mandelbrot Set — Old and New Rendering

Figure 14.3 : The insert shows an original printout from Mandelbrot's experiment. We have produced the large Mandelbrot set using a modern laser printer and a more accurate mathematical algorithm.

critical point escapes to infinity,[3] and the Julia set is a Cantor set. Further, in chapter 13 we saw that for real parameters c with $-2 \leq c \leq 0.25$, the iteration of the critical point is bounded and the Julia set is connected. Thus, the interval $[-2, 0.25]$ on the real axis belongs to M, which is contained in a disk of radius 2 centered at the origin.[4]

Encirclement of M

Given a parameter c, how can we computationally decide whether the orbit of c is bounded or not, i.e., whether $c \in M$? Theoretically, this might require knowledge of the complete critical orbit, i.e., an infinite number of iterations. The problem is the same as computing whether an initial point z_0 is in the prisoner set or not. Therefore, we again look at encirclements M_k of M, which we may define analogous to the encirclements of prisoner

[3] See the technical section on page 378.

[4] In fact, the point $c = -2$ is the only point of the Mandelbrot set that has an absolute value equal to 2.

Figure 14.4 : A connected and a disconnected Julia set.

sets, namely as

$$M_k = \left\{ c \in \mathbf{C} \;\middle|\; \lim_{l \to \infty} \frac{\log_2 |z_l|}{2^l} \leq 2^k, \; z_0 = c \right\} \qquad (14.2)$$

where

$$z_{l+1} = z_l^2 + c, \quad l = 0, 1, 2, ...$$

Encirclements of M

In this technical section we derive the formula (14.2). Since M is contained in a disk of radius 2 around the origin we may use a disk of radius greater than or equal to 2 as a target set T. We set

$$R^{(-k)}(T) = \{ c \in \mathbf{C} \mid z_k \in T, \; z_0 = c \}, \; k = 0, 1, 2, ...$$

where $z_0, z_1, z_2, ...$ denotes the critical orbit, i.e.,

$$z_{k+1} = z_k^2 + c, \quad z_0 = c, \quad k = 0, 1, 2, ...$$

Let us first point out in this paragraph that $z_k \in T$, $k \geq 1$, implies that all the previous iterates $z_0, z_1, ..., z_{k-1}$ are also in T. Let $r \geq 2$ be the radius of the disk T and consider the case $|c| \geq r$ with initial point $z_0 = c$. Let z_m be any point of the orbit of z_0 with $|z_m| \geq |c|$. For the next point z_{m+1} of the orbit it follows that

$$\begin{aligned} |z_{m+1}| &= |z_m^2 + c| \geq |z_m^2| - |c| \\ &\geq |z_m|^2 - |z_m| = (|z_m| - 1)|z_m| \\ &\geq (r-1)|z_m| \geq |z_m| \ . \end{aligned}$$

Since already $|z_0|$ fulfills the assumption for $|z_m|$ we conclude

$$r \leq |c| = |z_0| \leq |z_1| \leq |z_2| \leq \cdots$$

In other words, the sequence $|z_0|, |z_1|, |z_2|, \ldots$ is monotonically increasing. Thus, if z_m is in the disk T, then also z_0 to z_{k-1} must be in T. Let us now consider the remaining case $|c| < r$ and assume that $|z_m| > r$, i.e. $z_m \notin T$. Then

$$\begin{aligned} |z_{m+1}| &= |z_m^2 + c| \geq |z_m^2| - |c| \\ &> r^2 - r = (r-1)r \\ &= (1+\varepsilon)r > r \ . \end{aligned}$$

By induction it follows that also all the following points in the orbit are not in T, i.e.

$$z_m, z_{m+1}, z_{m+2}, z_{m+3}, \ldots \notin T \ .$$

Thus, if $z_k \in T$, then also all iterates z_0 to z_{k-1} must be in T, which was to be shown.

In other words, the encirclement $R^{(-k)}(T)$ is given by precisely those parameter values c for which all of the first k iterations of $z_0 = c$ hit the target set T. Recall from the characterization in equation (14.1) that c belongs to the Mandelbrot set if the entire critical orbit is in the target set. In other words, the encirclement $R^{(-k)}(T)$ is an approximation of M which improves as the number k increases. Let us elaborate this important point.

We already pointed out above that M is contained in a disk of radius 2. This implies, that M is also contained in the target set T (which must contain the disk). Expressed in a formula,

$$M \subset T = R^{(0)}(T) \ .$$

Now $R^{(-1)}(T)$ is the set of all parameters c for which the critical value $z_0 = c$ is in T *and* the first iterate $z_1 = z_0^2 + c = c^2 + c$ is also in T. Thus, $R^{(-1)}(T)$ is a subset of $R^{(0)}(T)$. Moreover, M is contained in $R^{(-1)}(T)$:

$$M \subset R^{(-1)}(T) \subset R^{(0)}(T) \ .$$

Now $R^{(-2)}(T)$ is the set of all parameters c for which $z_0 = c$ is in T, $z_1 = z_0^2 + c$ is in T, *and* the second iterate $z_2 = z_1^2 + c$ is also in T. Thus, $R^{(-2)}(T)$ is a subset of $R^{(-1)}(T)$ and contains M:

$$M \subset R^{(-2)}(T) \subset R^{(-1)}(T) \subset R^{(0)}(T) \ .$$

This reasoning can go on producing better and better encirclements $R^{(-k)}(T)$ for $k = 3, 4, 5$ and so on. In the limit we obtain the Mandelbrot set itself,

$$\bigcap_{k=0}^{\infty} R^{(-k)}(T) = M \ . \tag{14.3}$$

There is an important difference between the encirclements of prisoner sets and those of the Mandelbrot set, which must be kept in mind when interpreting pictures of the Mandelbrot set. This is the fact that the encirclements $R^{(-k)}(T)$ of M are *not* iterated preimages of T with respect to some fixed transformation (i.e., $R^{(-k)}(T)$ is not an image of $R^{(-k-1)}(T)$).

The encirclements $R^{(-k)}(T)$ of the Mandelbrot set depend on the choice of the target set T. Besides disks of radius greater than or equal to 2 we could as well have chosen ellipses, squares, or any other shape as long as the property in eqn. (14.3) is guaranteed. In order to remove this ambiguity, we proceed along the same line as carried out for the prisoner sets. We use the disks

$$D^l = \left\{z \mid \log_2 |z| \le 2^l\right\}, \ l = 0, 1, 2, ...$$

as target sets T and define the kth encirclement of M as

$$M_k = \lim_{l \to \infty} R^{(k-l)}(D^l)$$

for any integer k. Let us put this rather abstract formula into a more accessible form. The parameter c is in $R^{(k-l)}(D^l)$, if

$$\log_2 |z_{l-k}| \le 2^l$$

or, after dividing by 2^{l-k},

$$\frac{\log_2 |z_{l-k}|}{2^{l-k}} \le 2^k \ .$$

In this case taking the limit $l \to \infty$ is the same as letting $l - k \to \infty$. Thus,

$$M_k = \left\{c \in \mathbf{C} \ \middle| \ \lim_{l \to \infty} \frac{\log_2 |z_l|}{2^l} \le 2^k, \ z_0 = c\right\} ,$$

where

$$z_{l+1} = z_l^2 + c, \quad z_0 = c \ .$$

This is now in the form suitable for computer implementation. Moreover, we clearly see the analogy to the encirclements $P_c^{(k)}$ of the prisoner set P_c (compare eqn. (13.3)).

For positive indices k, the encirclements M_k are almost identical to large disks. This is due to the fact that $z \longrightarrow z^2 + c$ acts like $z \longrightarrow z^2$ when $|z|$ is large. For example, M_3 is approximately a disk with radius $2^8 = 256$. In other words, we can accept a parameter c as an element of M_k if $k + 3$ iterations, started at $z_0 = c$, do not escape the disk of radius 256. The result is given in figure 14.5, which shows the boundaries of the encirclements M_0 through M_{-10}.

Encirclement of the Mandelbrot Set

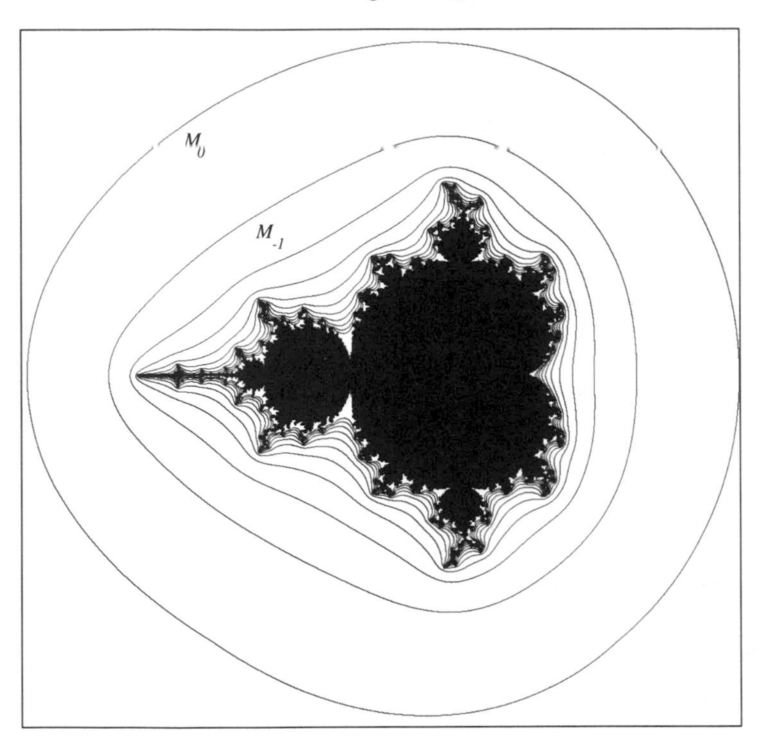

Figure 14.5 : The Mandelbrot set M and its approximation by encirclements M_0 through M_{-10}.

A Simple Algorithm for the Mandelbrot Set

For a given parameter c the fate of the critical point $z = 0$ must be determined by the algorithm. If $|c| > 2$, we already know that the orbit must escape, thus an algorithm can terminate immediately and return the result that the Julia set is disconnected. So let us assume that $|c| \leq 2$. In the most simple algorithm we would iterate $z \longrightarrow z^2 + c$, starting with $z_0 = c$, and check the points on the orbit. If a point from the orbit is outside a disk of radius

$$R = \max(2, |c|) = 2 ,$$

we are sure that the orbit must escape to infinity,[5] and again the

[5]See the derivation on page 376.

algorithm may terminate with the same result (i.e., c is not in M). Of course, some maximum number m of iterations must be prescribed to avoid infinite loops. But computing the parameters for which the iteration does not leave $T = \{c \in \mathbf{C} \mid |c| \leq 2\}$ within m steps is nothing other than computing $R^{(-m)}(T)$, which is an approximation of M. Instead of using 2 as the threshold radius of the target set, we propose using 256, which ensures that the resulting boundaries of encirclements can be interpreted as equipotentials of the Mandelbrot set.

```
k = 0
z = c
while (k < m)
    if (|z| > 256) then
       return (Julia set disconnected, c in M_{4-k})
    end if
    z = z*z + c
    k = k + 1
end while
return (result = c in M_{4-m})
```

The Mandelbrot Set is Connected

The concept of a potential has been the key tool in the mathematical analysis of the Mandelbrot set. A necessary prerequisite for that is the important fact that the Mandelbrot set is connected. This is known since 1982 through a paper of Douady and Hubbard,[6] in which they showed that the encirclement of the Mandelbrot set always generates domains which are bounded by circle-like curves. If the encirclement is properly manufactured it can be shown that the bounding curves are in fact equipotentials of the Mandelbrot set. This is the case for the encirclements

$$M_k = \left\{ c \in \mathbf{C} \;\middle|\; \lim_{l \to \infty} \frac{\log_2 |z_l|}{2^l} \leq 2^k,\ z_0 = c \right\} .$$

The boundaries of the sets M_k are equipotentials of the explicit potential function

$$p_M(c) = \lim_{l \to \infty} \frac{\log_2 |z_l|}{2^l} ,$$

where

$$z_l = z_{l-1}^2 + c, \quad z_0 = c .$$

From Field Lines to the Pinching Model

There is, however, a major difference to the encirclements of the prisoner sets. When computing those encirclements, we keep the parameter c fixed and work with only one fixed transformation $z \to z^2 + c$, while in the case of the Mandelbrot set, we keep the

[6] A. Douady, J. H. Hubbard, *Iteration des pôlynomes quadratiques complexes,* CRAS Paris 294 (1982) 123–126.

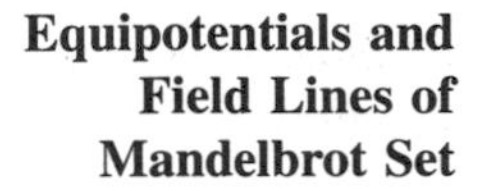

Equipotentials and Field Lines of Mandelbrot Set

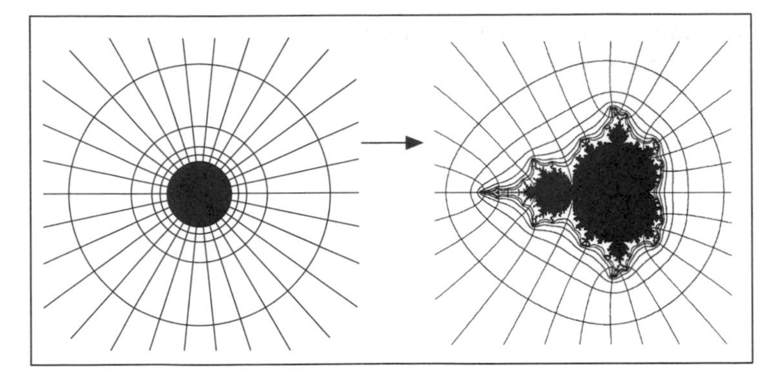

Figure 14.6 : The system of equipotentials and field lines provides a polar coordinate system for the complement of the Mandelbrot set.

initial point $z_0 = 0$ fixed, but change the value of c from pixel to pixel. This means that we cannot associate a dynamics with the encirclements.

In a way the system of equipotentials and field lines can be viewed as a particular polar coordinate system for the complement of the Mandelbrot set (see figure 14.6). More precisely, there is a one-to-one correspondence between the equipotentials and field lines of the unit disk and those of the Mandelbrot set. Each field line is then given by an angle α, where $0 \leq \alpha < 1$. In fact, this correspondence can be made in such a way that the binary expansion carries essentially the information about where the field line will land, as long as α is a rational number. For example, the point where the major period two bud is attached to the cardioid is a pinching point where two field lines land on the boundary of the Mandelbrot set and the angles are $\alpha = 1/3$ and $\alpha = 2/3$.

Details of this field line model yield an almost complete understanding of the Mandelbrot set as a pinching model.[7] Incidentally, it is known that each field line with a rational angle α lands on the Mandelbrot set, while for irrational angles this is still unknown in general. This is related to the fundamental problem of whether the Mandelbrot set is not only connected but also locally connected, which is one of a whole variety of unsolved research problems. A conjecture, also related to this problem, was recently proven positively by M. Shishikura, a young Japanese mathematician.[8] The conjecture was that the boundary of the Mandelbrot set has frac-

[7] See A. Douady, J. H. Hubbard, *Étude dynamique des pôlynomes complexes,* Publications Mathematiques d'Orsay 84-02, Université de Paris-Sud, 1984. See also H.-O. Peitgen and P. H. Richter, *The Beauty of Fractals*, Springer-Verlag, Heidelberg, 1986.

[8] M. Shishikura, *The Hausdorff dimension of the boundary of the Mandelbrot set and Julia sets,* SUNY Stony Brook, Institute for Mathematical Sciences, Preprint #1991/7.

tal dimension 2, which would somehow characterize the incredible complexity of the magnifications which we have seen.

14.2 The Mandelbrot Set — A Road Map for Julia Sets

The complexity of the Mandelbrot set is in an altogether different class compared to that of Julia sets. On the one hand, the Mandelbrot set has a solid interior without any structure, and on the other hand it is bordered by a very complex boundary with an infinity of different shapes. For a first impression of this variety, we provide a selection of images around the boundary (figure 14.7) and a zoom sequence (figure 14.8).

Journey Around The Mandelbrot Set

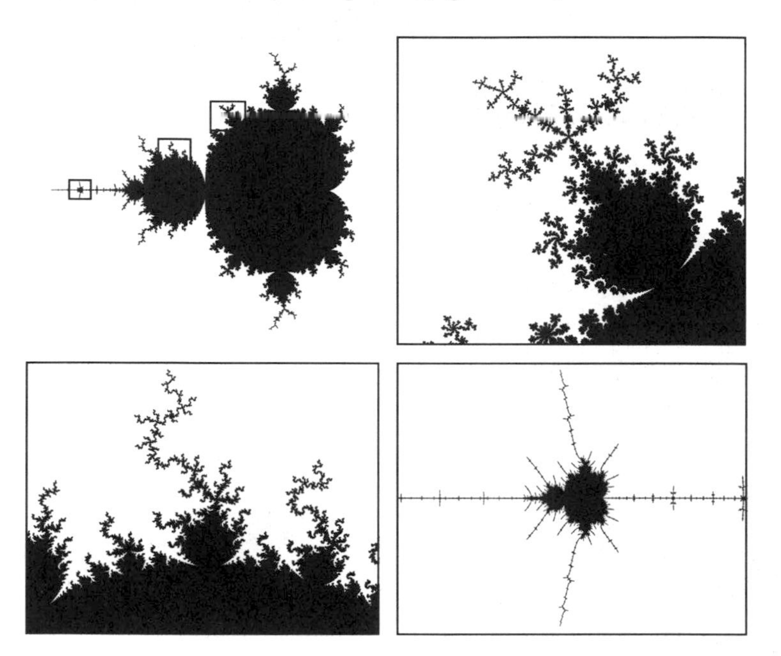

Figure 14.7 : Journey around the Mandelbrot set with locations of the individual images being marked on the initial one.

The Buds in the Mandelbrot Set

The first striking feature of the Mandelbrot set are its small buds which are lined up along the big, heart-shaped, central region. These buds have a meaning for the associated Julia sets. Let us first take a look at the main body of the set in the center. This heart-like set intersects the real axis in the interval from -0.75 to 0.25. We recall that the Julia set for $c = 0$ is a circle with an attractive fixed point at the origin. This fixed point is *super attractive*; the critical point is equal to the fixed point (see chapter 11). It is a fact that the parameters c on the line between -0.75 and 0.25 are precisely those real parameters for which one of the fixed points of $z \to z^2 + c$ is an attractor.[9] Therefore, it is no surprise that

[9]This c-interval $(-0.75, 0.25)$ corresponds to the a-interval $(1, 3)$ for $x \to ax(1 - x)$ (see chapter 1).

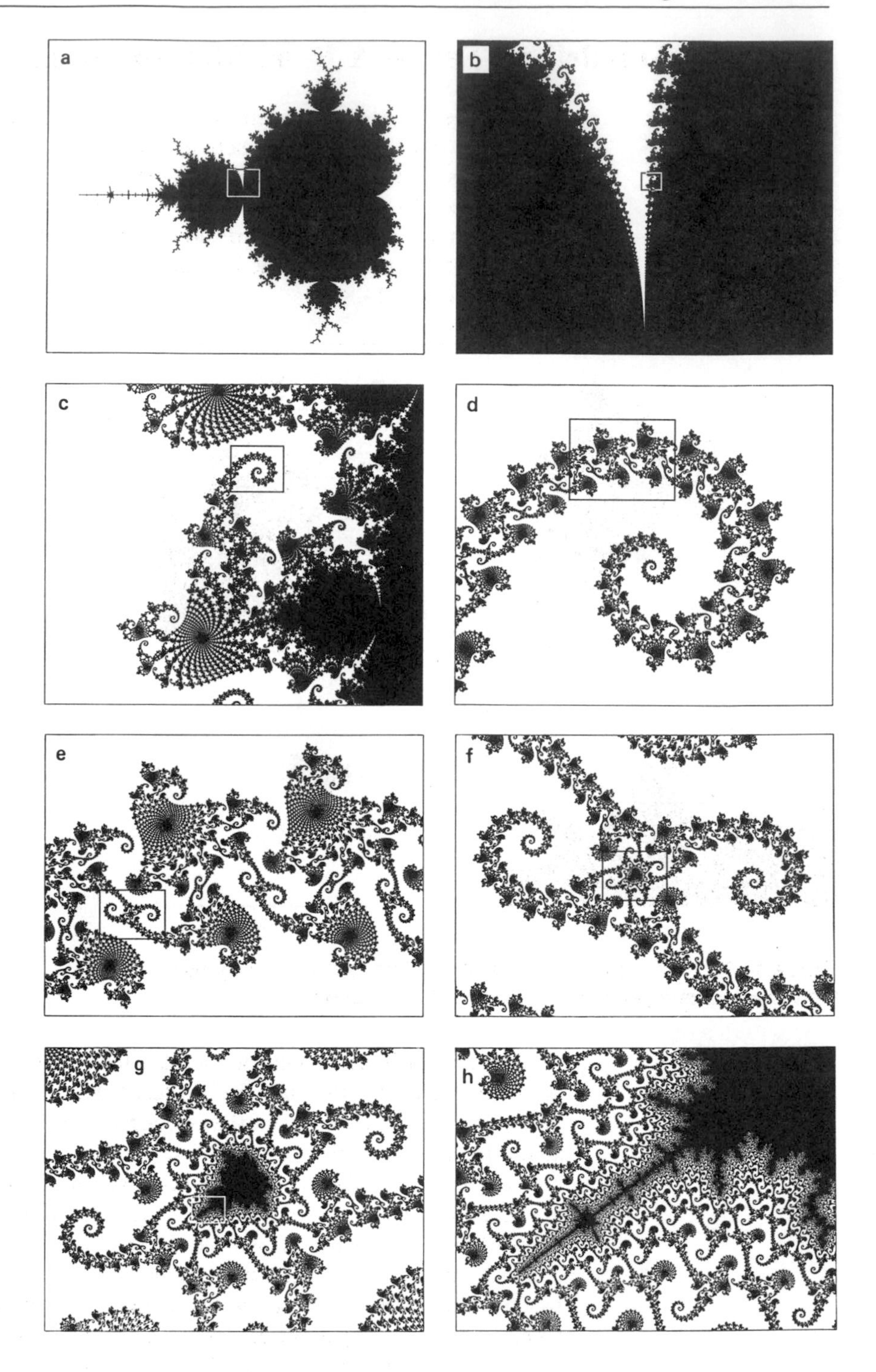

Figure 14.8 : Zoom into the Mandelbrot set.

the big heart-shaped region is the set of all (complex) parameters c for which one of the two fixed points of $z \to z^2 + c$ is attractive. Figure 14.9 shows two examples of Julia sets of this type (see also figure 13.15). Observe the position of the fixed points.

Basin of Attractive Fixed Points

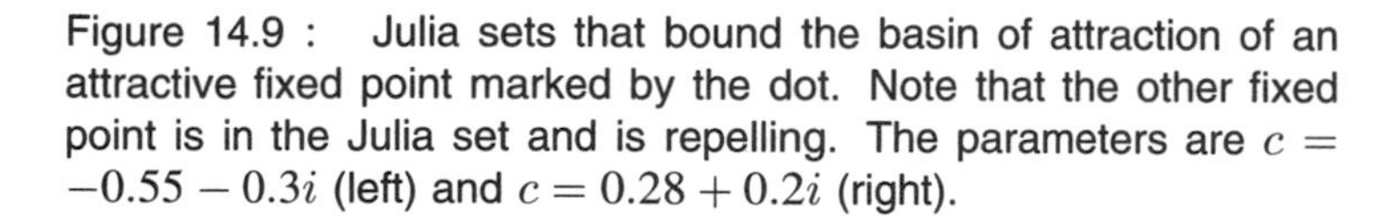

Figure 14.9 : Julia sets that bound the basin of attraction of an attractive fixed point marked by the dot. Note that the other fixed point is in the Julia set and is repelling. The parameters are $c = -0.55 - 0.3i$ (left) and $c = 0.28 + 0.2i$ (right).

The Real Parameters $-0.75 < c < 0.25$

The fixed points of the quadratic iteration $z \to z^2 + c$ are given by the formula

$$z_{1,2} = \frac{1 \pm \sqrt{1-4c}}{2} .$$

The derivative of $z^2 + c$ at $z_{1,2}$ is just twice that number. As c varies from -0.75 to 0.25, we have that $1 - 4c$ varies from 4 to 0, and that the root $\sqrt{1-4c}$ goes from 2 to 0. Thus, $1 - \sqrt{1-4c}$ is between -1 and 1, and smaller than 1 in absolute value which identifies an *attractive* fixed point. Thus, the fixed point $(1 - \sqrt{1-4c})/2$ is an attractor for c between -0.75 and 0.25.

The Boundary of the Heart-Like Central Region

We can determine an explicit formula for the outline of the 'heart' of the Mandelbrot set using the derivative criterion as follows. On the outline we have that the derivative of $z \to z^2 + c$ at one of the fixed points is equal to 1 in absolute value (the interior of the heart consists of parameters for which one of the fixed points has a derivative less than 1 in absolute value).

Assume that z is a fixed point of $z \to z^2 + c$, i.e., z solves

$$z^2 - z + c = 0 .$$

The derivative at z is given by $2z$ which we write in polar coordinates as

$$2z = re^{i\phi}$$

with $r \geq 0$ and $0 \leq \phi < 2\pi$. We combine the two equations and arrive at

$$\left(\frac{re^{i\phi}}{2}\right)^2 - \frac{re^{i\phi}}{2} + c = 0 .$$

This last equation is now solved for c:

$$c = \frac{1}{2}re^{i\phi} - \frac{1}{4}r^2 e^{2i\phi} .$$

Given an arbitrary number $re^{i\phi}$ this result specifies a parameter c such that the derivative of $z \to z^2 + c$ at one of the corresponding fixed points matches the given number. For example, in order to obtain a representation of the interior of the heart-like center of the Mandelbrot set we consider $r < 1$, while $r = 1$ yields its boundary curve. Writing $c = x + yi$, we can split the real and the imaginary components of the above equation (for $r = 1$):

$$\begin{aligned} x &= \frac{\cos\phi}{2} - \frac{\cos 2\phi}{4} \\ y &= \frac{\sin\phi}{2} - \frac{\sin 2\phi}{4} . \end{aligned} \tag{14.4}$$

These final equations readily produce a complex point for any given argument ϕ. Such a representation of a curve is called a *parametrization* (ϕ being the parameter of the curve in this case). Here are examples of points on the curve for 5 values of ϕ:

ϕ	x	y
0	0.25	0.0
$2\pi/5$	0.35676	0.32858
$2\pi/4$	0.25	0.5
$2\pi/3$	−0.12500	0.64952
π	−0.75	0.0

It turns out, that at the parameter values

$$\phi = \frac{2\pi}{k}, \quad k = 2, 3, 4, 5, 6, \ldots$$

one of the main buds of the Mandelbrot set is attached to the heart-shaped center. Moreover, the period of the attractive cycles that belong to these buds is given by the number k in $\phi = 2\pi/k$.

Let us make a final remark. The fixed points of $z \to z^2 + c$ are

$$z_{1,2} = \frac{1 \pm \sqrt{1-4c}}{2} .$$

and the derivatives at these fixed points are twice that, namely $1 \pm \sqrt{1-4c}$. From this representation it follows that if one derivative is inside the unit circle, then the other must be outside. Thus, if one fixed point is attracting, then the other must be repelling. Also, if one fixed point is indifferent, then the other must be repelling (except in the case where the fixed points are identical, i.e., for the case $c = 1/4$).

π and the Mandelbrot Set

Let us briefly interrupt our tour of the Mandelbrot set to present a splendid observation that relates the Mandelbrot set to the constant $\pi = 3.141592...$ It is a result of a small computer experiment, carried out by Dave Boll in 1991 and communicated by him using an electronic bulletin board called USENET which is read worldwide at most universities and schools. The following paragraphs are taken directly from his message.[10]

Boll's Observation

"I was writing a 'quick-n-dirty' program to verify that the 'neck' of the Mandelbrot set at $c = -0.75 + 0i$ is actually of zero thickness.[11] Accordingly, I was testing the number of iterations that points of the form $-0.75 + \varepsilon i$ (ε being a small number) went through before escaping. Here's a quick list for special values of ε:

ε	Iterations
0.1	33
0.01	315
0.001	3143
0.0001	31417
0.00001	314160
0.000001	3141593
0.0000001	31415928

Does the number of iterations strike you as a suspicious number? How about the product of the number of iterations with ε? It's π, to within $\pm\varepsilon$. My initial reaction was 'What the HELL is π doing here?'.

Adopting the motto 'When in doubt, keep going', I tried the same experiment at the 'butt' of the Mandelbrot set,[12] located at

[10] Slightly edited by the authors to adapt to the notation used in this book.

[11] This refers to the fact that the heart-shaped part of the Mandelbrot set and the major disk-shaped bud to the left touch in precisely one point, $c = -0.75 + 0i$.

[12] This refers to the cusp in the heart-shaped part of the Mandelbrot set.

$c = 0.25+0i$. I was now trying points of the form $c = 0.25+\varepsilon+0i$, with ε again a small number. Here are some more results for various values of ε:

ε	Iterations
0.1	8
0.01	30
0.001	97
0.0001	312
0.00001	991
0.000001	3140
0.0000001	9933
0.00000001	31414
0.000000001	99344
0.0000000001	314157
0.00000000001	993457
0.000000000001	3141625

Again, we get the same type of relationship, this time it is

$$\lim_{\varepsilon \to 0} N(\varepsilon) \cdot \sqrt{\varepsilon} = \pi$$

where $N(\varepsilon)$ is the number of iterations. [...] Has anyone seen this? What's going on?"

So far Boll's experiment. Initially, he did not get any responses, except for a couple saying in effect 'that's pretty strange'. A year later he re-posted his findings on the bulletin board,[13] and this time, there were some reactions with attempts for an explanation. In fact, it turned out that the second occurrence of π near $c = 1/4$ had already been noticed even long before 1991 in the context of intermittency. This can be illustrated by graphical iteration since the computations in the experiment near $c = 1/4$ involve only real numbers (see figure 14.10). Clearly there is a connection to the tangent bifurcation we considered in the study of intermittency in section 11.5. For that case we already demonstrated that $N(\varepsilon)\sqrt{\varepsilon}$ tends to a constant for the iteration of $g_a(x) = ax^2 \sin(\pi x)$ as a approaches the bifurcation point a^* ($\varepsilon = a^* - a$). However, the constant is about 2.36 while it is a surprise that for the system belonging to the Mandelbrot set we get π. In the following section we provide a heuristic explanation for this astonishing fact.[14]

[13] D. Boll, *Pi and the Mandelbrot set (again),* USENET article <1992Feb26.222630.36612@yuma.acns.colostate.edu>.

[14] It stems from J. Guckenheimer, P. Holmes, *Nonlinear Oscillations, Dynamical Systems, and Bifurcations of Vector Fields,* Springer-Verlag, New York, 1983 (see page 344) and is in the spirit of the paper by Y. Pomeau and P. Manneville, *Intermittent transition to turbulence in dissipative dynamical systems,* Commun. Math. Phys. 74 (1980) 189–197, which initiated the discussion of intermittency.

Tangent Bifurcation at $c = 1/4$

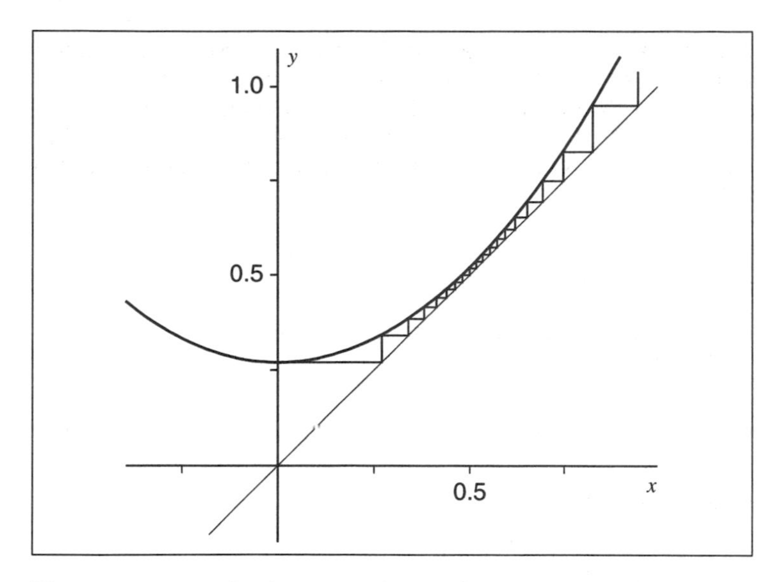

Figure 14.10 : In the second experiment reported by Boll the number $N(\varepsilon)$ of iterations of $x \to x^2 + 1/4 + \varepsilon$ starting at $x = 0$ is measured in relation to ε obtaining $N(\varepsilon)\sqrt{\varepsilon} \to \pi$ as $\varepsilon \to 0$.

Why is it π?

We consider the iteration

$$x_{k+1} = x_k^2 + \frac{1}{4} + \varepsilon, \quad x_0 = 0 \tag{14.5}$$

and interpret this as a sequence of Euler steps for a corresponding differential equation

$$x'(t) = f_\varepsilon(x(t)) .$$

Thus, we claim that the iteration can be written in the form

$$x_{k+1} = x_k + hf_\varepsilon(x_k) .$$

for some function f_ε with parameter ε and step size h.[15] In fact, using the step size $h = 1$, we can solve this equation for the function f_ε, obtaining

$$f_\varepsilon(x_k) = x_{k+1} - x_k = x_k^2 + \frac{1}{4} + \varepsilon - x_k .$$

We arrive at the differential equation

$$x'(t) = x^2(t) - x(t) + \frac{1}{4} + \varepsilon . \tag{14.6}$$

Starting a solution at the initial condition $x(0) = 0$, we now compute the time t that the solution of the differential equation spends until

[15]Then x_k approximates the solution $x(kh)$ for $k = 0, 1, 2, \ldots$

it arrives at $x(t) = 1$. Translated into the context of the discrete iteration in eqn. (14.5), the time t is the number of Euler steps until $x_k \geq 1$, which corresponds to the number of iterations in the computer experiment. As we let $\varepsilon \to 0$ we will see that $\sqrt{\varepsilon}t \to \pi$. We divide eqn. (14.6) by the right hand side and integrate from time 0 to time t.

$$\int_0^t \frac{x'(s)}{x(s)^2 - x(s) + \frac{1}{4} + \varepsilon} ds = \int_0^t ds \ .$$

Using the substitution rule we replace $x'(s)ds$ by dx and t by $x(t)$ on the left.

$$\int_0^{x(t)} \frac{dx}{x^2 - x + \frac{1}{4} + \varepsilon} = t \ . \qquad (14.7)$$

An antiderivative is given by

$$\frac{1}{\sqrt{\varepsilon}} \arctan\left(\frac{x}{\sqrt{\varepsilon}} - \frac{1}{2\sqrt{\varepsilon}}\right) .$$

Using this to evaluate the definite integral in eqn. (14.7) and multiplying by $\sqrt{\varepsilon}$ we obtain

$$\arctan\left(\frac{x(t)}{\sqrt{\varepsilon}} - \frac{1}{2\sqrt{\varepsilon}}\right) - \arctan\left(-\frac{1}{2\sqrt{\varepsilon}}\right) = \sqrt{\varepsilon}t \ .$$

We are interested in the time t when $x(t) = 1$. And we get

$$2 \arctan\left(\frac{1}{2\sqrt{\varepsilon}}\right) = \sqrt{\varepsilon}t \ .$$

Letting $\varepsilon \to 0$, the left hand side tends to π, and we see the result

$$\lim_{\varepsilon \to 0} \sqrt{\varepsilon}t = \pi \ .$$

Although this is not a rigorous proof for the observed phenomenon, it provides a supporting argument for it. To be rigorous, it must still be shown that passing from the differential equation to the corresponding discrete iteration using the Euler method in fact does not destroy the asymptotics that we have derived above. Concerning the other experiment for the point $c = -3/4$ a similar but more complicated approach can be taken.[16]

The Period-Two Disk

We now return to the discussion of the Mandelbrot set itself. At the left end of the heart-shaped region (at $c = -0.75$, where the number π can be seen in the iteration counts) there is a bud. It is a perfect disk of radius 0.25 centered at $c = -1$. Figure

[16] See G. Edgar, *Pi and the Mandelbrot set,* USENET article <1992Mar27.135743.28423@zaphod.mps.ohio-state.edu>.

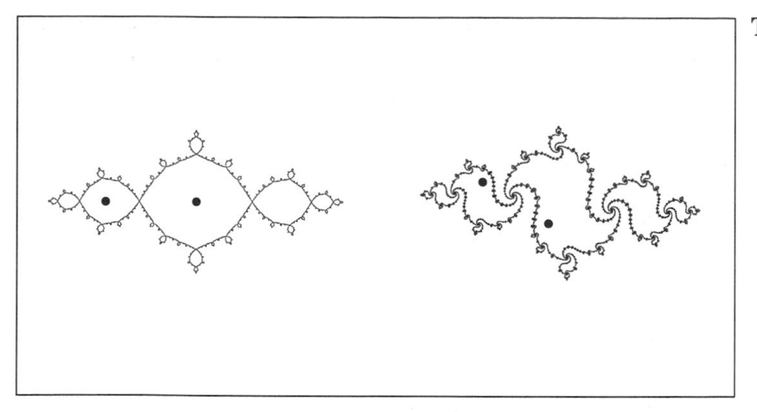

Two Julia Sets from the Big Bud of M

Figure 14.11 : Julia sets that bound the basin of attraction of an attractive cycle (marked by large dots) of period 2. Left $c = -1$ (the super attractive case), right $c = -0.83 + 0.16i$.

14.11 shows two Julia sets for parameters from this disk. For such parameters neither one of the two fixed points of $z \to z^2 + c$ can be attractive because c is outside of the heart-shaped center of M. Now what are the dynamics of the iteration within the prisoner set in this case? Let us check this with an experiment. For $c = -1$ we pick two initial points close to the fixed points

$$z_{1,2} = \frac{1 \pm \sqrt{5}}{2}$$

which are

$$\begin{aligned} z_1 &= +1.61803398... \\ z_2 &= -0.61803398... \end{aligned}$$

First, the iteration confirms that *neither* one of the fixed points is attractive.[17] The table 14.12 lists the first 16 iterations of two initial points (the fixed points, however, are *rounded* to two decimal places).

While the iteration of the first orbit leads to infinity, the iteration of the second orbit reveals the essential dynamics in the prisoner set. Table 14.13 lists another 18 iterations of this orbit. It is dominated by the attractive orbit of period 2:

$$0 \to -1 \to 0 \to \cdots$$

All initial values of the interior of the prisoner set are attracted by this orbit, and the Julia set is the boundary of this basin of attraction.

[17] Again the derivative criterion for attracting and repelling fixed points can be applied to verify our findings. The absolute values of the derivatives are $|2z_1|$ and $|2z_2|$. In fact, both of these numbers are greater than 1. Thus, both fixed point are repelling.

Dynamics Near the Fixed Points for $c = -1$

	Orbit 1		Orbit 2	
	x	y	x	y
z_0	1.62	0.00	−0.62	0.00
z_1	1.6244	0.00	−0.6156	0.00
z_2	1.63868	0.00	−0.62104	0.00
z_3	1.68523	0.00	−0.61431	0.00
z_4	1.84009	0.00	−0.62262	0.00
z_5	2.38593	0.00	−0.61235	0.00
z_6	4.69268	0.00	−0.62503	0.00
z_7	21.69268	0.00	−0.60933	0.00
z_8	440.89443	0.00	−0.62871	0.00
z_9	194386.8964	0.00	−0.60472	0.00
z_{10}		0.00	−0.63431	0.00
z_{11}		0.00	−0.59765	0.00
z_{12}		0.00	−0.64282	0.00
z_{13}		0.00	−0.58679	0.00
z_{14}		0.00	−0.65568	0.00
z_{15}		0.00	−0.57008	0.00
z_{16}		0.00	−0.54437	0.00

Table 14.12 : The iteration of two initial points for $z \to z^2 + c$, $c = -1$ started close to the fixed points $z_1 = 1.61803398...$ and $z_2 = -0.618033988...$ Both orbits diverge.

Dynamics Near the Fixed Points Continued

	Orbit 2 (continued)			Orbit 2 (continued)	
	x	y		x	y
z_{17}	−0.70367	0.00	z_{26}	−0.11015	0.00
z_{18}	−0.50485	0.00	z_{27}	−0.98787	0.00
z_{19}	−0.74512	0.00	z_{28}	−0.02412	0.00
z_{20}	−0.44479	0.00	z_{29}	−0.99942	0.00
z_{21}	−0.80216	0.00	z_{30}	−0.00116	0.00
z_{22}	−0.35654	0.00	z_{31}	−1.00000	0.00
z_{23}	−0.87288	0.00	z_{32}	0.00000	0.00
z_{24}	−0.23808	0.00	z_{33}	−1.00000	0.00
z_{25}	−0.94332	0.00	z_{34}	0.00000	0.00

Table 14.13 : Continuation of the computation of the second orbit of the last table 14.12. It converges to the periodic orbit which oscillates between 0 and −1.

In fact, for all parameters c from the interior of the 'period-two disk' of the Mandelbrot set we obtain an attractive orbit of period two, and the Julia set is the boundary of its basin of attraction. Note that in this disk of parameters, the value $c = -1$ is special.

Here the critical point coincides with one of the periodic points. Therefore, this point is called super attractive.

The Derivative Criterion for the Periodic Cycle

The derivative criterion for repelling and attracting fixed points (see page 399) can be generalized to periodic cycles. Let us take for example $c = -1$. Here the points

$$z_0 = 0 \text{ and } z_1 = -1$$

are periodic points for $z \to z^2 - 1$. Thus, they are *fixed points* of the transformation iterated *twice*, i.e., of

$$z \to (z^2 - 1)^2 - 1 = z^4 - 2z^2 \ .$$

Now the same criterion applies as in the case of fixed points for the quadratic transformation. If the absolute value of the derivative of this transformation at the fixed point exceeds 1, then the fixed point is a repeller. If the absolute value is less than 1, then the fixed point is an attractor. In our case the derivative is $4z^3 - 4z$. The evaluation of the derivative at the two fixed points gives

$$|4z_0^3 - 4z_0| = 0 < 1$$

and

$$|4z_1^3 - 4z_1| = |-4+4| = 0 < 1 \ .$$

Thus, the fixed points are attractive for $z \to z^4 - 2z^2$. This translates to the original situation as claimed. Namely that the periodic sequence of points $0, -1, 0, -1, \ldots$ is also attractive with respect to the original transformation $z \to z^2 - 1$.

The derivative criterion allows us also to compute the period-2 disk, i.e. the locus of all parameters c such that an orbit of period 2 is attractive. First we derive a quadratic equation for the periodic points with (minimal) period 2. By definition a point of period 2 must satisfy

$$(z^2 + c)^2 + c - z = 0 \ .$$

We note that also the two fixed points, given by $z^2 + c - z = 0$, solve this equation. Therefore, the polynomial $z^2 + c - z$ must be a factor of the polynomial $(z^2 + c)^2 + c - z$ of degree 4. In fact,

$$(z^2 + c)^2 + c - z = (z^2 + z + 1 + c) \cdot (z^2 + c - z) \ .$$

From this it follows that the solutions z_1 and z_2 of

$$z^2 + z + 1 + c = 0 \qquad (14.8)$$

are precisely the points of period 2. In other words, we have

$$\begin{aligned} z_1^2 + c &= z_2 \\ z_2^2 + c &= z_1 \end{aligned}$$

which, of course, can also be checked directly, which is, however, more cumbersome. We now compute the derivative of the twice iterated transformation $z \to (z^2 + c)^2 + c$ at the periodic point z_1 (using the chain rule),

$$2(z_1^2 + c)2z_1 = 4z_1z_2 \ .$$

(The derivative at z_2 is identical.) It follows from Vieta's law applied to the solutions z_1, z_2 of eqn. (14.8) that

$$z_1z_2 = 1 + c \ .$$

Thus, the above computed derivative is equal to $4(1+c)$, a quantity that is less than 1 in magnitude provided that

$$|1 + c| < \frac{1}{4} \ .$$

This inequality describes a disk of radius $1/4$ centered at $c = -1$, and gives us the result that precisely for parameters c in this disk the periodic points z_1 and z_2 are attractive.

Two Julia Sets from the Next Buds of M

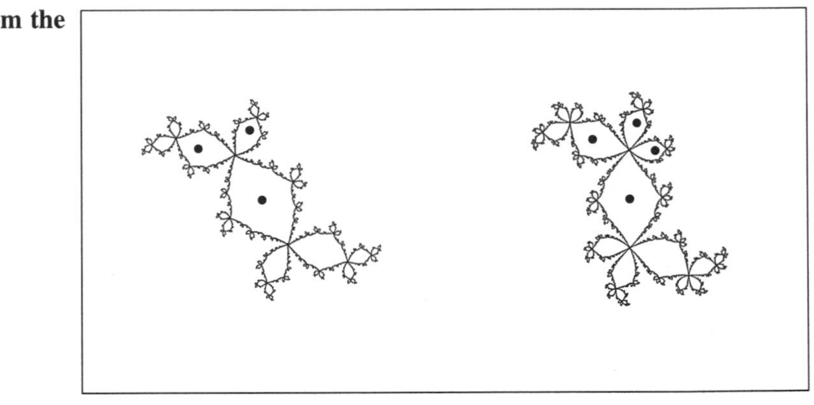

Figure 14.14 : Julia sets that bound the basin of attraction of an attractive cycle of period 3 and 4. Left $c = -0.13 + 0.76i$, right $c = 0.28 + 0.53i$.

The next big buds attached at the edge of the heart-shaped center of M correspond to period-three behavior; then there are buds which house parameters belonging to attractive cycles of period 4, and so on. Figure 14.14 shows Julia sets which bound basins of attraction for period-3 and period-4 cycles. Figure 14.15 presents an overview of the periodic behavior associated with the buds, or *atoms* as Mandelbrot calls them. Clearly they are strictly organized. Each bud carries on its boundary another complete set of smaller buds with corresponding sequences of periodic attractive

The Mandelbrot Set and its Atoms

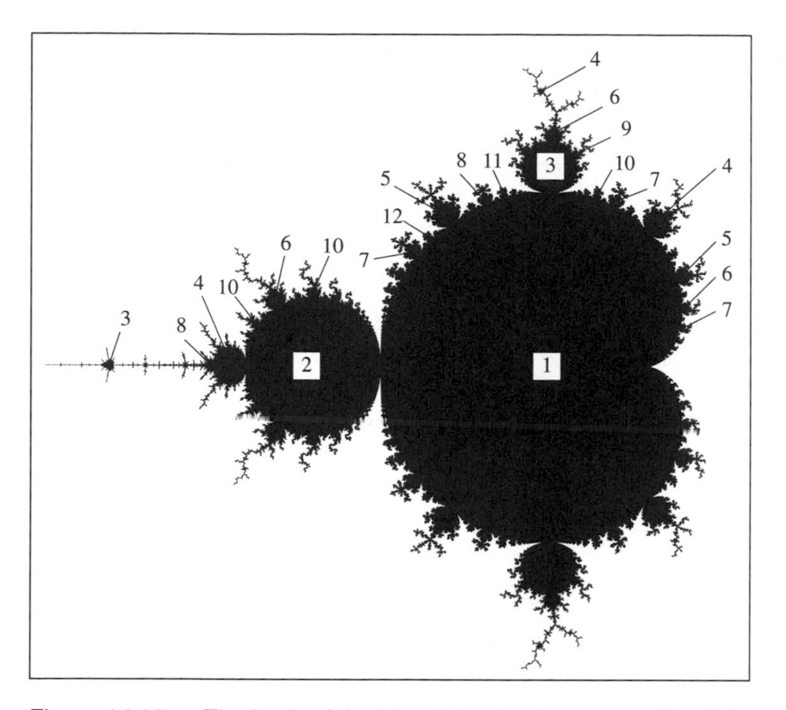

Figure 14.15 : The buds of the Mandelbrot set correspond to Julia sets that bound basins of attraction of periodic orbits. The numbers in the figure indicate the periods of these orbits.

cycles. Note that there is an amazing rule for the periods corresponding to the buds. Two given buds of periods p and q at the cardioid determine the period of the largest bud in between them as $p + q$. Similar rules are true for the buds on buds.

The Touching Points of Buds

Now let us take a look at the points of the cardioid again. The fixed point which has been attractive for parameters within the heart-shaped region looses this property right on the cardioid. Here the corresponding fixed point is said to be indifferent. In other words, the iteration $z \to z^2 + c$ is at the fixed point neither attractive nor repelling; it is related to a rotation (see eqn. (14.4)). Depending on whether that rotation is given by a rational or irrational number, the associated prisoner set is dramatically different. Moreover, when the rotation is irrational, several striking cases can be distinguished. In the following we will not make an attempt to discuss the classification to the extent it is known. Rather we will pick some particular cases to demonstrate the intricacies which evolve in understanding the prisoner sets. For the points where the buds are attached this rotation is given by an angle $\phi = 2\pi\alpha$, which is a rational multiple of 2π (i.e., α is a rational number). For example, we have $\alpha = 1/2$ at the point where the 'period-two disk' is attached and we have $\alpha = 1/3$ where the 'period-three bud' above the cardioid is attached. Figure 14.16 shows the Julia sets for these two examples.

Parabolic Fixed Points

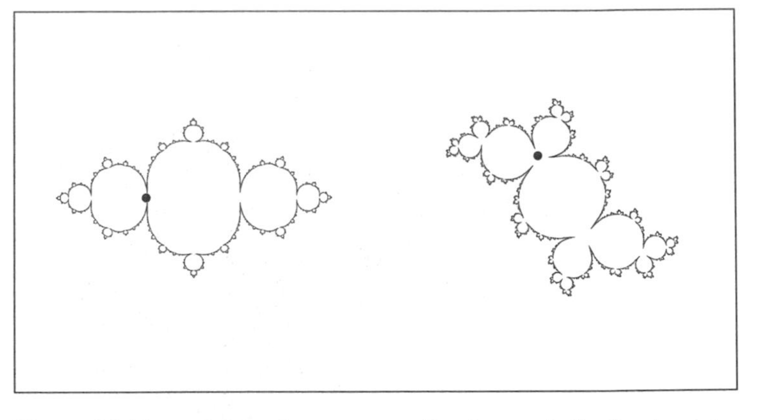

Figure 14.16 : Julia sets corresponding to parabolic fixed points. $c = 0.75 + i$ (left) and $c = -0.125 + 0.64925i$ (right), the point where the 'period-three bud' is attached to the heart-shaped center of M.

The Derivative Criterion for Indifferent Fixed Points

We have characterized a fixed point z of $z \to z^2 + c$ by its derivative $\lambda = 2z$. It is

- *attractive* if $|\lambda| < 1$ and
- *repelling* if $|\lambda| > 1$.

Now we turn to the intermediate case. We call the fixed point *indifferent* if the derivative $|\lambda|$ is equal to 1 in absolute value. In this case the derivative is determined by the argument $\phi = \arg \lambda$,

$$\lambda = \cos \phi + i \sin \phi$$

with $0 \leq \phi < 2\pi$. Setting as usual $\phi = 2\pi\alpha$, we call an indifferent fixed point

- *rationally indifferent* (or *parabolic*) if α is a rational number and
- *irrationally indifferent* if α is an irrational number.

The characterization of periodic orbits can be extended in the same way. In this case we check the derivative of the corresponding iterated map.

But what happens if the fixed point is indifferent and the rotation angle α (of the derivative) is irrational? This is really a complicated case and leads to Julia sets which are only slightly understood. For a moment let us stay with those cases which are

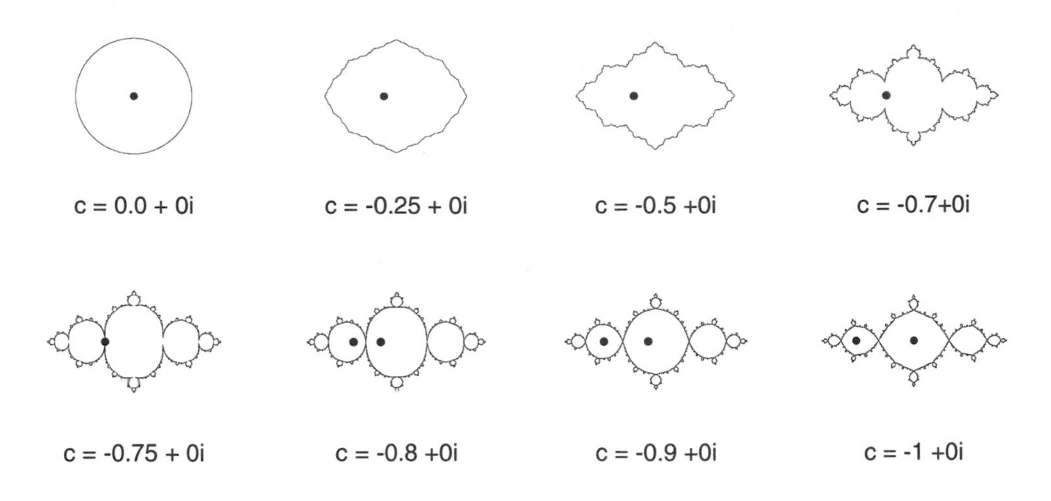

Figure 14.17 : Starting from the Julia set for $c = 0$ (the circle) we decrease the parameter to $c = -1$. The Julia set develops a pinching point for $c = -0.75$ and is the boundary of a period-2 attractor for the remaining plots.

more accessible. We know that any irrational number can be approximated by sequences of rational numbers. Some sequences approach a given irrational number faster than others. Some irrational numbers admit approximating sequences which converge very rapidly, others only admit sequences which converge rather slowly.[18] It turns out that these differences matter substantially for the character of a Julia set corresponding to an angle α. If α is an irrational number such that any approximating sequence of rational numbers converges very slowly (in some precise sense), then then the prisoner set is a so called *Siegel disk.*[19] It turns out that among all irrational numbers there is one which stands out as the one which is worst with respect to approximation by rational numbers. This number is the golden mean $\alpha = (\sqrt{5}+1)/2$. Figure 14.18 shows the corresponding, most prominent Siegel disk. Note that in this case the indifferent fixed point lies in the interior of the prisoner set, while it is known that for the rationally indifferent case it is on the boundary (i.e., part of the Julia set). Thus, if α_n, $n = 1, 2, \ldots$ is a sequence of rational numbers which approaches α, the golden mean, the associated Julia sets are fundamentally

[18]The technical section on page 443 explains in what sense the approximation of irrationals by rationals can be poor.

[19]This case is named after the German mathematician Carl Ludwig Siegel who established characteristics of that case in his paper *Iteration of analytic functions,* Ann. of Math. 43 (1942) 607–616. More precisely, the Siegel disk is only the disk-like component of the prisoner set which contains the fixed point. The other components are preimages of that disk.

A Siegel Disk

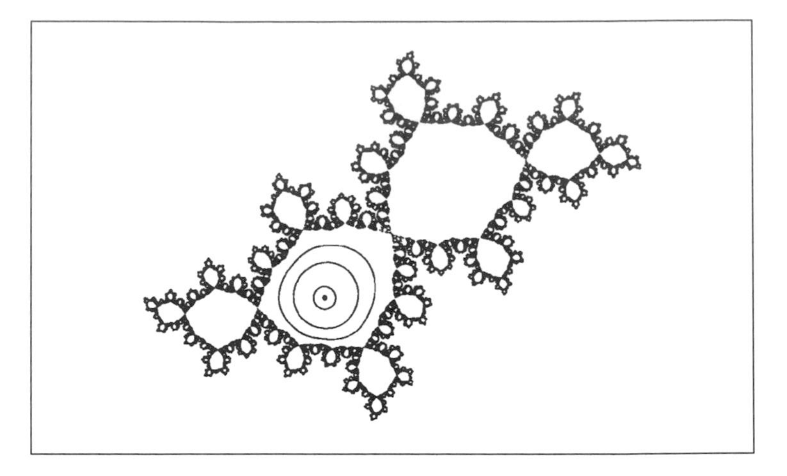

Figure 14.18 : The Julia sets that belongs to $c \approx -0.3905407802-0.5867879073i$ is a Siegel disk. With α being the golden mean, use $\phi = 2\pi\alpha \approx 3.883222077$ in eqn. (14.4) on page 430 to obtain the real and imaginary components of c. The dynamics near the fixed point is characterized by invariant curves on which the iteration acts like a rotation by the angle α.

different from the one corresponding to the limit α.

What we have looked at so far are the Julia sets that correspond to the attractive and the super attractive case, the rationally indifferent case and the Siegel disk case.[20] All these cases are characterized by the fact that the prisoner set has non-empty interior. Besides those prisoner sets which are totally disconnected (with parameter values c that are not in M) and those which have a non-empty interior, there are also boundary cases: prisoner sets with no interior point, but which, however, are still connected. These typically have many branches at all levels of details and are called *dendrites*. The most popular, certainly, is the dendrite for $c = i$. But also the prisoner set for $c = -2$, which is simply the real interval $[-2, 2]$ (which has no branches), is in this class. See the figures 14.19 and 14.24 for more examples. While these Julia sets are quite easy to compute and draw, this class of dendritic Julia sets also contains an infinity of monsters for which it probably is impossible to provide a valid computer graphic representation.

[20]For more general than quadratic iterations, given by rational functions $f(x) = p(x)/q(x)$, with p and q being polynomials, there is even a fifth different type of prisoner set with non-empty interior. The other four cases are: attractive case, super-attractive case, rationally indifferent case, and Siegel disk case. This is a result which partially goes back to Julia and Fatou and was completed by Dennis Sullivan (see *Quasiconformal homeomorphisms and dynamics I,* Annals of Math. 122 (1985) 401–418). This fifth type is a so-called *Herman ring*, named after M. R. Herman who constructed the first example (see M. Herman, *Exemples de fractions rationelles ayant une orbite dense sur la sph/'ere de Riemann,* Bull. Soc. Math. France 112 (1984) 93–142). See also A. F. Beardon, *Iteration of Rational Functions,* Springer-Verlag, New York, 1991.

Dendrites

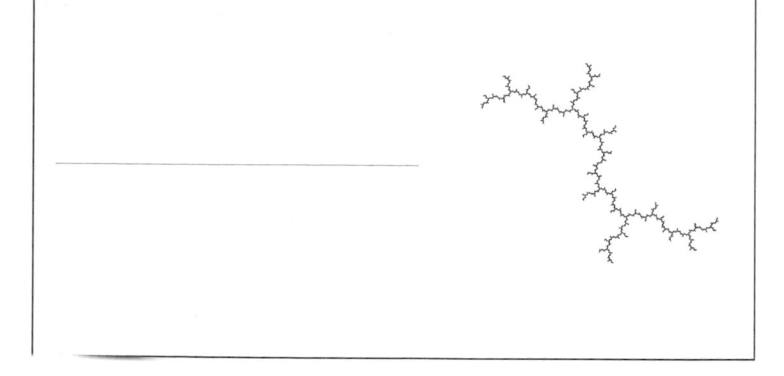

Figure 14.19 : Julia sets for $c = -2$ (left), a line segment, and $c = i$ (right), a more typical dendrite.

Irrational and Liouville Numbers

A number α is said to be *badly approximable* by rational numbers provided α satisfies the following number theoretical condition. There are $\varepsilon > 0$ and $\mu > 0$ such that

$$\left|\alpha - \frac{p}{q}\right| > \frac{\varepsilon}{q^{\mu}} \tag{14.9}$$

holds for all integers p and positive integers q. The famous result of Joseph Liouville dating from 1844 states that any (irrational) algebraic number[21] α of degree $n \geq 2$ is badly approximable where the constant μ in eqn. (14.9) is $\mu = n$. For example, the golden mean $\alpha = (\sqrt{5} - 1)/2$ is algebraic of degree 2 (it solves the equation $x^2 + x - 1 = 0$) and fulfills eqn. (14.9).

Liouville's estimate allows the explicit construction of *transcendental* numbers, i.e., irrational numbers which are not algebraic. Most real numbers are transcendental, but only relatively few are explicitly known. Indeed, algebraic numbers are countable, as Cantor argued by observing that algebraic equations with integer coefficients are countable.

For the explicit construction of a transcendental number, take for example, the number

$$\alpha = a_1 10^{-1!} + a_2 10^{-2!} + \cdots + a_k 10^{-k!} + \cdots , \tag{14.10}$$

where the coefficients a_i are arbitrary digits between 1 and 9 and $k!$ is 'k factorial' ($k! = k \cdot (k-1) \cdots 2 \cdot 1$). Thus, we may write the decimal expansion of α as

$$\alpha = 0.a_1 a_2 000 a_3 00000000000000000 a_4 000\ldots$$

[21] A number α is called *algebraic* of degree n provided it solves an equation of the form $b_0 + b_1 x + \cdots + b_n x^n = 0$ with integer coefficients b_i and where n is minimal.

The blocks of zeroes in the decimal expansion of α grow rapidly. Now we introduce α_k to be the rational number obtained by only considering terms up to (and including) the kth term in the definition of α in eqn. (14.10). Then we have that

$$|\alpha - \alpha_k| < \frac{10}{10^{(k+1)!}} .$$

Now let us assume that α is algebraic of some degree n. Setting

$$\alpha_k = \frac{p}{q} = \frac{p}{10^{k!}}$$

we would have from Liouville's result that there is $\varepsilon > 0$ and

$$|\alpha - \alpha_k| > \frac{\varepsilon}{10^{nk!}} .$$

Combining the two estimates for $|\alpha - \alpha_k|$ yields

$$\frac{\varepsilon}{10^{nk!}} < |\alpha - \alpha_k| < \frac{10}{10^{(k+1)!}}$$

or

$$\frac{(10^{k!})^{k+1}}{(10^{k!})^n} = (10^{k!})^{k+1-n} < \frac{10}{\varepsilon} .$$

However, the last inequality cannot be right; as k grows the left hand expression grows beyond all bounds. This means that α cannot be algebraic of degree n; it must be transcendental. This number is an example of a so called *Liouville number*. It does not satisfy the above number theoretical condition in eqn. (14.9), which characterizes poor approximation by rational numbers.

Liouville Monster

It is known that Julia sets which belong to fixed points around which $z \to z^2 + c$ is not linearizable are not locally connected.[22] An example is given by $z \to z^2 + c$ such that there is a fixed point which is irrationally indifferent and α is a Liouville number, i.e., it can be approximated well by rational numbers (see the technical section above). To get some flavor of the unimaginable complexity of such a Julia set, let us discuss this case from the viewpoint of field lines of the potential p_c as introduced in the previous chapter. The Julia set is not locally connected; and according to a result of Constantin Carathéodory, this implies that not all field lines land

[22]The transformation $f_c(z) = z^2 + c$ is called *linearizable* around a fixed point z_0 provided in a neighborhood of that fixed point the mapping is (analytically) equivalent to λz, where λ is the derivative of f_c at the point z_0. Examples are fixed points for which there is a Siegel disk, or attracting, or repelling fixed points, see A. F. Beardon, *Iteration of Rational Functions,* Springer-Verlag, New York, 1991, page 133. A set X is called *locally connected* provided every point p in X and every neighborhood U of p contains an open set V with $p \in V$ which is connected. For the theorem about the absent local connectedness see D. Sullivan, *Conformal dynamical systems,* Lecture Notes in Mathematics 1007 (1983) 725–752.

in a point of the Julia set.[23] However, it is known that this is not the rule; on the contrary, almost all field lines land in points of the Julia set. Douady and Hubbard have succeeded in showing that in particular all field lines corresponding to angles which are rational multiples of 2π must land in a point of the Julia set. It is beyond imagination to picture the complexity which such a strange condition must enforce: all rational field lines land but at least some irrational field lines do not! Note, that it is also not known whether Julia sets associated to Siegel disks (i.e. their boundaries) are locally connected in general. In other words, we begin to see how important the idea of the potential of prisoner sets is for understanding the structure of Julia sets. Moreover, concepts such as local connectedness are not just a phantasy of too theoretically minded mathematicians. They are decisive in understanding very physical properties of real objects if we accept Julia sets as real objects!

Do Field Lines Land?

It is not easy to imagine a structure of a Julia set so that it has field lines that do not land on the Julia set. In figure 14.20 we present a model (it is certainly not a Julia set) where the nature of this problem becomes apparent and understandable.

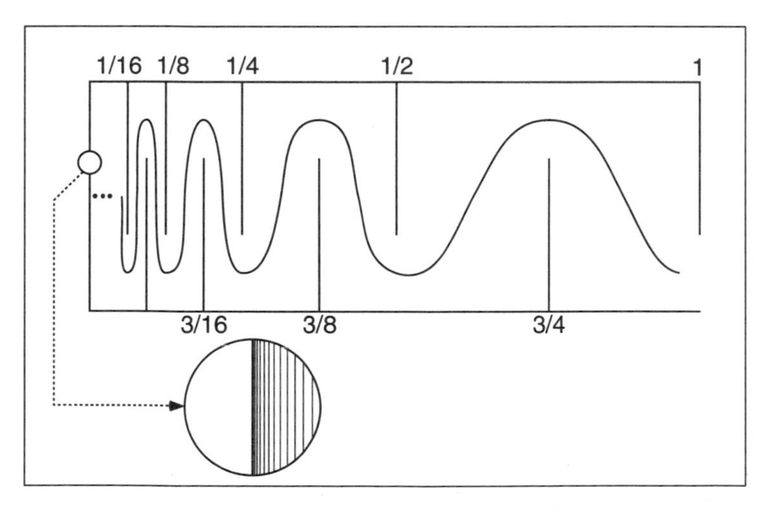

Figure 14.20 : For this model of a double-comb, there is a field line which oscillates infinitely often and which does not converge to a point at the left end of the comb. The double comb is not locally connected: in any small disk centered around the midpoint of the far left tooth, there is an infinite set of separate line segments piling up towards the left end.

[23] See C. Carathéodory, *Untersuchungen über die konformen Abbildungen von festen und veränderlichen Gebieten,* Math. Ann. 72 (1912).

Miniature M-Set I

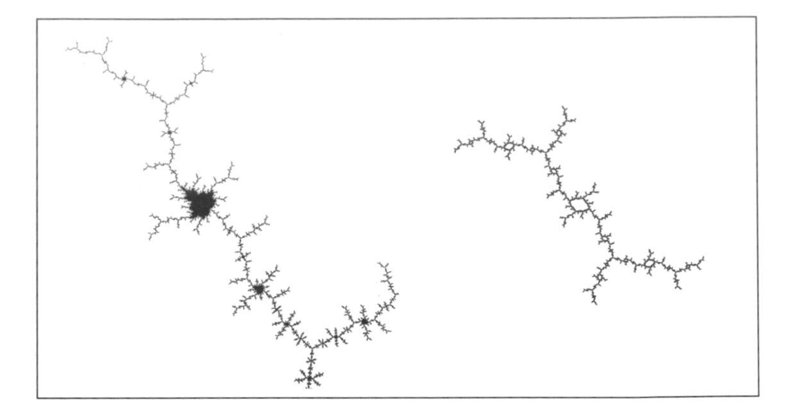

Figure 14.21 : Enlargement of a secondary Mandelbrot set in the upper region of the Mandelbrot set. The Julia set is for $c = 0.159789 + 1.03332i$.

Miniature M-Set II

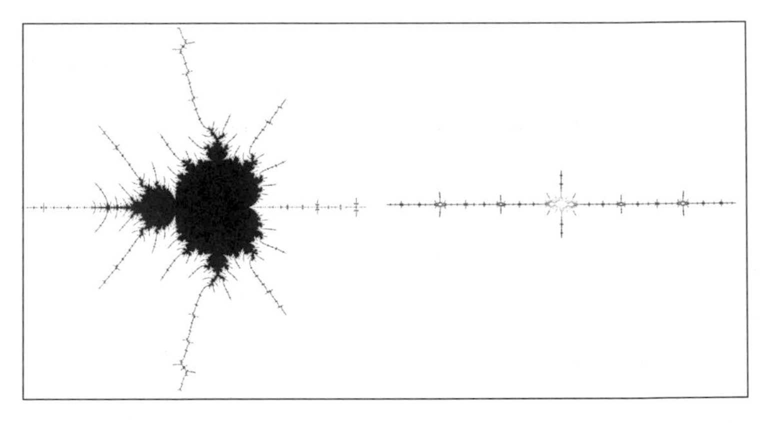

Figure 14.22 : Enlargement of a secondary Mandelbrot set in the left region of the Mandelbrot set (the tip). The Julia set is for $c = -1.77578$.

We show a double comb-like structure which is essentially given by two combs with infinitely many intertwined teeth positioned as shown in the figure. The combs are connected by an additional, longer tooth at the left end. A field line which is trying to get to this far left tooth must wiggle more and more to get around the infinitely many teeth and therefore will never arrive at the goal. But there must be a field line which oscillates infinitely often and which separates lines that land on the upper comb from those that land on the lower one. Pathologies like this were studied around 1912 by Carathéodory, who found that a compact, connected subset K of the plane has a potential with field lines that land everywhere on the boundary of K provided that the boundary is locally connected.[24]

[24]Another condition that must be satisfied is that the exterior of K is simply connected, i.e. must not contain any

Final-State Diagram and Mandelbrot Set

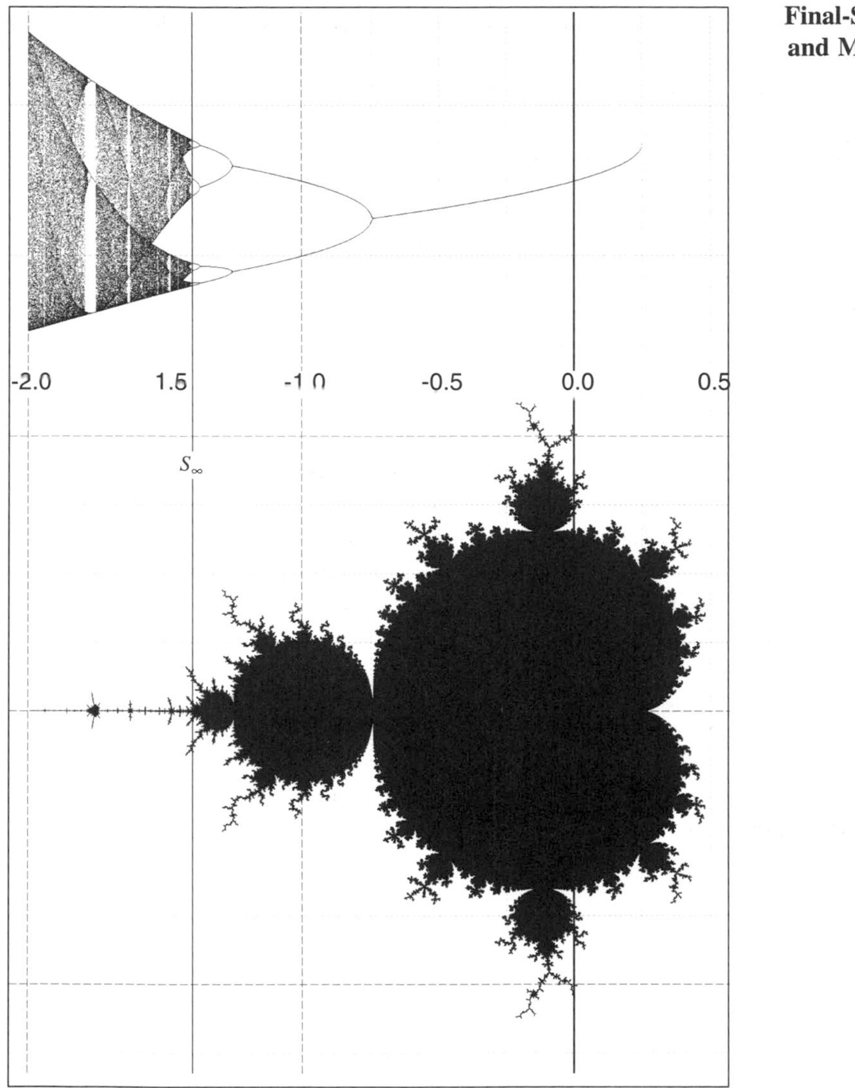

Figure 14.23 : The final-state diagram of $z \rightarrow z^2 + c$ in comparison to the Mandelbrot set.

This explains the problems in figure 14.20. Although the infinite double comb is connected, it is not locally connected.

'holes'. See C. Carathéodory, *Über die Begrenzung einfach zusammenhängender Gebiete,* Math. Ann. 73 (1913).

Secondary Mandelbrot Sets

Let us now briefly look at those prisoner sets which appear to be hybrids of Julia sets which are dendrites and those which bound prisoner sets with interior points. We find such cases for parameter values c from so called *secondary Mandelbrot sets*. These are small copies of M which, however, are subsets of M. In fact, one can find such small copies of M in the neighborhood of any point of the boundary of the Mandelbrot set. In other words, this is a self-similarity feature of M.[25] Figure 14.21 shows an enlargement of the upper region of M and a corresponding Julia set. The interior of the prisoner set is attracted by a period-3 cycle. The same is true for our other example of a secondary Mandelbrot set, which is located on the real axis and shown in figure 14.22.

This brings us back to the discussion of $z \rightarrow z^2 + c$ for real numbers z and c. In chapter 11 we used the final-state diagram as an ordering principle for the long term behavior of the quadratic iterator. Figure 14.23 shows the final-state diagram of $z \rightarrow z^2 + c$ in comparison to the Mandelbrot set. At real parameters from the heart-shaped region of M we see just one branch in the diagram, which identifies the attractive fixed point for the varying parameters c. Reading from right to left this branch splits, representing the two points forming period-2 cycles. This corresponds to the real parameters of the 'period-2 bud' in M. We observe that the complete period-doubling tree corresponds to a stack of smaller and smaller buds aligned at the real axis starting at the heart-shaped region of M. Now look at the periodic windows in the final-state diagram. They correspond to the secondary Mandelbrot sets at the tip of M.

[25] The exact relation of a copy with M is discussed in A. Douady, J. H. Hubbard, *On the dynamics of polynomial-like mappings,* Ann. Sci. Ecole Norm. Sup. 18 (1985) 287–344.

14.3 The Mandelbrot Set as a Table of Content

We saw in the last section that the Mandelbrot set is a visualization of the structural dichotomy of Julia sets for the quadratic iterator. It can also be used as a road map to all possible kinds of Julia sets. This road map is fantastically detailed, and this fascinating property of the Mandelbrot set — Mandelbrot already observed and documented it as early as 1980 — is the topic of this section.

Comparing the Boundary of M and Julia Sets

Let us look at the sequence of plots in figure 14.24. The upper portion of this figure shows a region along the boundary of the heart-shaped central region of the Mandelbrot set which is continuously enlarged (from left to right the magnification factor increases from 1 to about 6) and stretched out, so that the respective segment of the cardioid becomes a line segment. We see disk-like components out of which grow dendritic structures with 2, 3, ..., 6 major branches (from left to right). Now we choose as particular c-values the major branch points of these dendrites (see arrows). For each of these 5 c-values we compute the associated Julia set, displayed in the lower part of the figure.

We observe a striking resemblance in the structures and the combinatorics of the dendritic structures in the Mandelbrot set and the Julia sets. The geometry of these Julia sets is also visible in the Mandelbrot set when magnified appropriately at the corresponding parameter values. In this sense it is a visual *table of content* of the book of Julia sets (see also figures 14.25 and 14.26).

Figure 14.27 shows another more systematic example, which, however, still defies a rigorous mathematical understanding. The upper part shows the Mandelbrot set together with an enlargement around the c-value given by the cross hair cursor. Incidentally, the magnification factor is about 10^6, and the result is identical with part (f) in figure 14.8. The bottom part of figure 14.27 shows the Julia set for the c-value which is determined by the cross hair cursor and a blow up of a tiny portion of that Julia set around the point of the cursor in the Julia set. The magnification factor is about 10^5 here, and the resulting image is rotated by approximately $55°$ clockwise. The resulting similarity is striking and cannot be an accident. But if there is a systematic relation between the enlargements seen in the Mandelbrot set and enlargements of corresponding Julia sets, it cannot be easy to grasp in this case.

Note that the center of the upper magnification in figure 14.27 is a tiny copy of the Mandelbrot set, as figure 14.28 reveals, while the center of the lower left enlargement in figure 14.27 is definitely not a Mandelbrot set, as figure 14.29 shows. Thus, the similarity between the Mandelbrot set and the particular Julia set seen in figure 14.27 is a transitional one, i.e., it depends on the magnification factor.

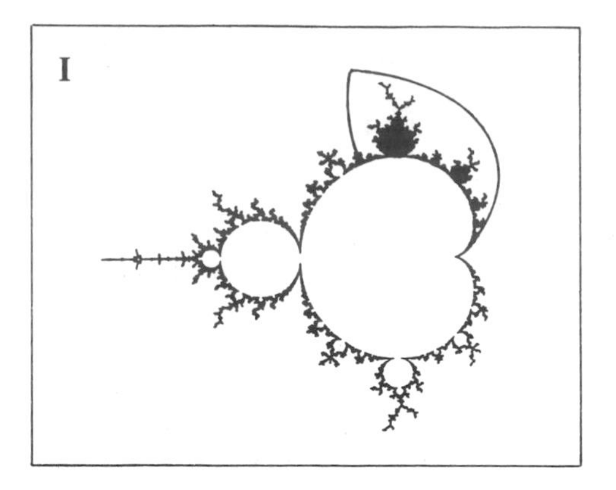
I

II
1
2

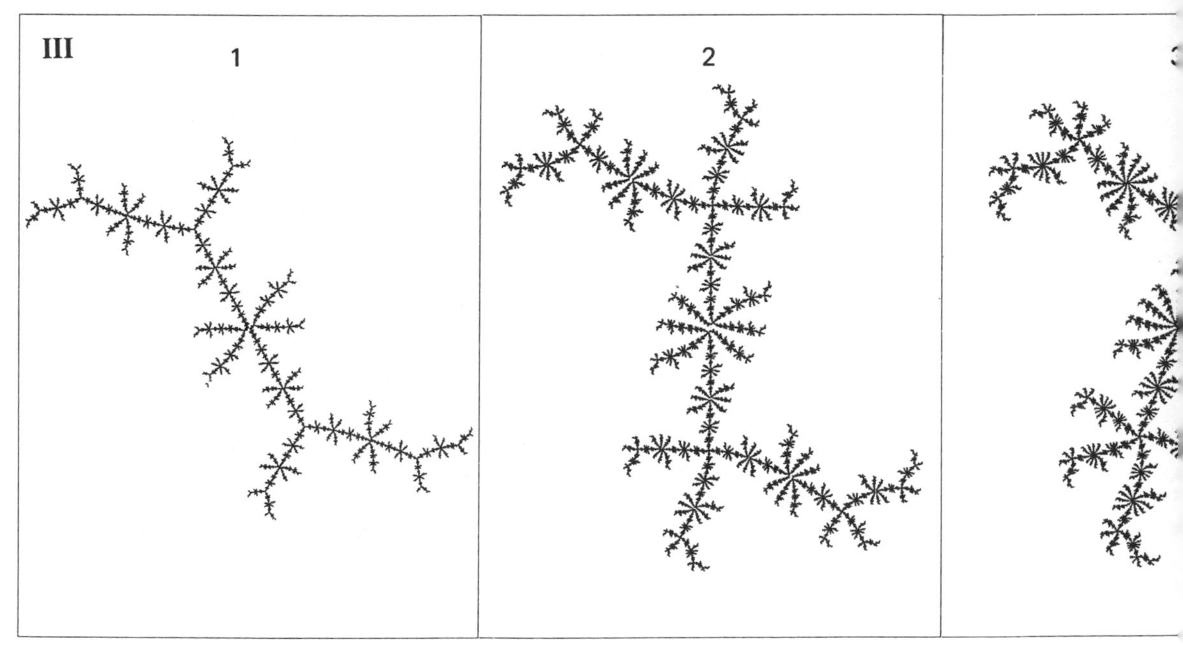
III
1
2

Figure 14.24 : Correspondence between Mandelbrot set and Julia sets. The section of the Mandelbrot set indicated in part I (top) is enlarged in part II (center). The magnification factor is not uniform; it increases from left to right in order to make the disk-like parts of the Mandelbrot set appear about the same size. The Julia sets computed for the parameter values indicated by the arrows are shown below.

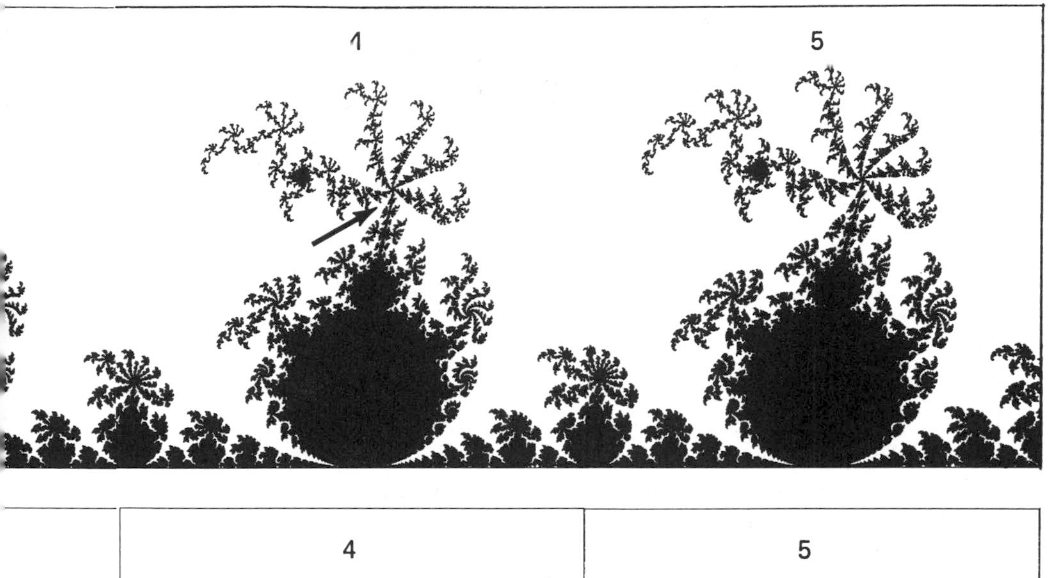

Zoom into the Mandelbrot Set

Figure 14.25 : In these 9 images a zoom into the boundary of the Mandelbrot set is shown. The final magnification is 300,000,000 fold.

For a large variety of Julia sets, however, the similarity is now understood quite clearly and with all mathematical rigor. This is a result of the beautiful work of the Chinese mathematician Tan Lei.[26] To discuss her results we need a few new tools. In particular, we have to extend our concept of self-similarity.

Magnification at a Particular Point

In order to discuss the similarity between Julia sets and the Mandelbrot set we have to magnify in a very particular fashion: we scale up around a given point, rotate by a certain angle, and repeat this procedure infinitely often. Recall that an image I (i.e., a subset of the plane) is *self-similar* if it is composed of parts which are small copies of the whole. Thus,

$$I = w_1(I) \cup w_2(I) \dots \cup w_n(I)$$

where the transformations w_i are similarity transformations. Turned the other way around, a suitable magnification (i.e., scaling and rotation) of the parts produces the complete set I. We center the magnification at one point, for example at $z_0 = 0$. This can be described as a scaling operation using a factor $\rho > 1$. Then ρI denotes all points of I scaled up by ρ (i.e., $\rho I = \{\rho z \mid z \in I\}$). Now assume that we have an object I which is invariant with

[26] Tan, Lei, *Similarity between Mandelbrot set and Julia sets,* Commun. Math. Phys. 134 (1990) 587–617.

Zoom into a Julia Set

Figure 14.26 : A parameter c is chosen from the center of the last image in the previous figure 14.25. We compute successive enlargements centered about this parameter c as for the Mandelbrot set in figure 14.25. Note how similar these Julia set sections are to the Mandelbrot set closeups. In the final images the objects are practically indistinguishable except for the scale and a rotation.

respect to scaling by ρ,

$$I = \rho I \ .$$

An object which satisfies this condition obviously must extend to infinity. The infinite stars shown in figure 14.30 are examples. But let us focus our attention on the neighborhood of the point z_0. We choose a disk $D_r(z_0)$ centered at z_0 with radius r as our area of interest. If there is a number $\varepsilon > 0$ such that

$$D_r(0) \cap I \ = \ D_r(0) \cap \rho I$$

holds for all radii $r < \varepsilon$, we say that the set I is *self-similar in 0.* The infinite and finite stars (see figure 14.30) are examples.[27] More generally, we may even allow that the scaling factor ρ is a *complex* number. In chapter 13 we showed that the multiplication by a complex number can be understood as scaling by $|\rho|$ and rotating by the angle $\phi = \arg \rho$.[28] In other words, the multiplication of

[27]A cover of a book that contains on it a picture of a hand holding that very book provides another example (see figure 3.11 in Part One of this book).

[28]Recall that $\rho = |\rho|(\cos\phi + i\sin\phi) = |\rho|e^{i\phi}$.

Figure 14.27 : Similarity of an enlargement of M and the closeup of the Julia set J_c at $c = -0.745429 + 0.113008i$.

Enlargement of Mandelbrot Set

Figure 14.28 : An enlargement centered at $c = -0.7454285 + 0.1130089i$. The width of the figure is 0.000006.

Enlargement of Julia Set

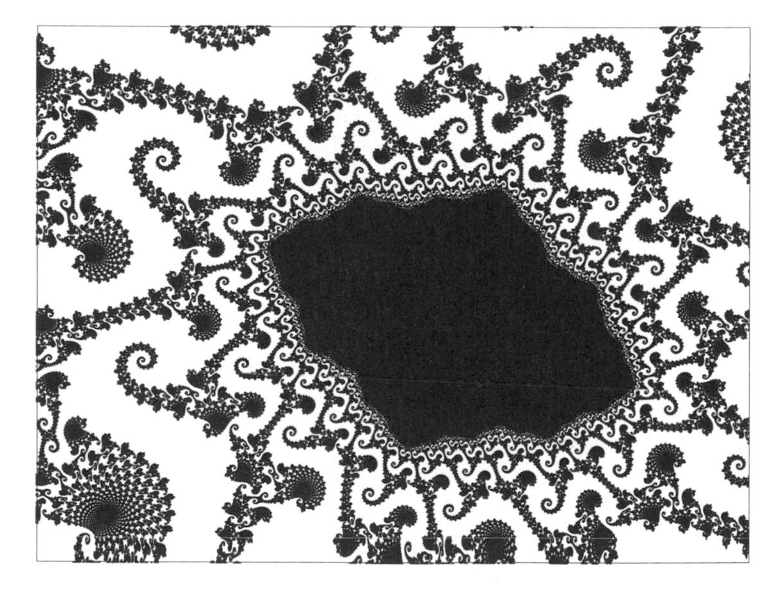

Figure 14.29 : An enlargement for the Julia set for $c = -0.7454285 + 0.1130089i$. The figure is centered at c and has width 0.000045.

Self-Similarity of Stars

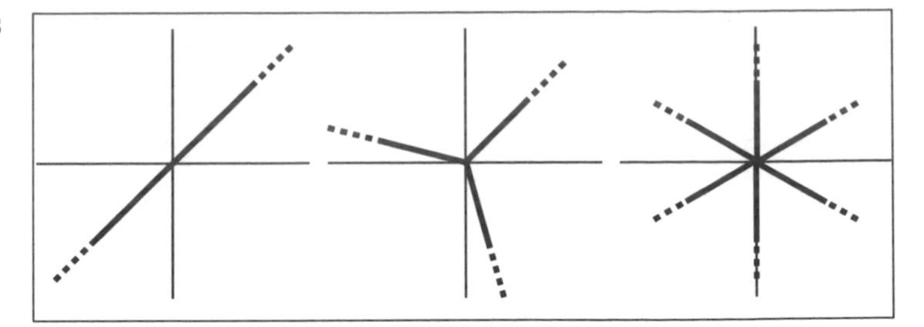

Figure 14.30 : The rays extend beyond all bounds. The stars are self-similar at the center.

an object (in the complex plane) by the complex number ρ with $|\rho| > 1$ can be interpreted as magnification in our sense. Look for example again at figure 14.30. The multiplication could involve a rotation by a multiple of 120 degrees (right) or 60 degrees (left). But let us go for more interesting examples.

Can you imagine an object which is self-similar in 0 under scaling with the golden mean $g = (\sqrt{5} - 1)/2$ and rotation by 90 degrees (i.e. $\rho = ge^{i\pi/2}$) and which is not a star like the objects in figure 14.30? You have already seen such an object! It is the golden spiral familiar from chapter 4. Figure 14.31 shows a small collection of logarithmic spirals. You will recall that logarithmic spirals are best described in polar coordinates (r, ϕ). For the points of a logarithmic spiral the equation

$$\log r = a\phi,$$

where r is the radius of the spiral at angle ϕ. From this notation we see that if we rotate by, for example, 90 degrees, we have to scale by $e^{a\pi/2}$ to obtain the same spiral. In other words, it is self-similar in 0 under the multiplication by $\rho = e^{a\pi/2}e^{i\pi/2}$.[29] Observe how the appearance of the spirals changes from almost circular to almost straight. In the latter case it is hard to see any rotation at all. The stretching is so large that it overrides the rotational effect in the visualization.

Logarithmic spirals are good models to describe some of the patterns we are going to discuss in the Mandelbrot set. Even compositions of spirals such as the double spiral from figure 14.32 and more complicated patterns can be found. But before we turn to those beautiful images, take a look at figure 14.33. This is indeed a fantastic spiral, don't you agree? Do you see the spiral? Which spiral? Is there a spiral at all? You might already begin to see that the world of spirals alone can be so confusing that an untrained eye can easily be fooled. Thus it is nice to have a bit of solid

[29]This result can be extended; the spiral is self-similar under multiplication with $e^{a\phi+i\phi}$ for all angles ϕ.

Different Scaling Values

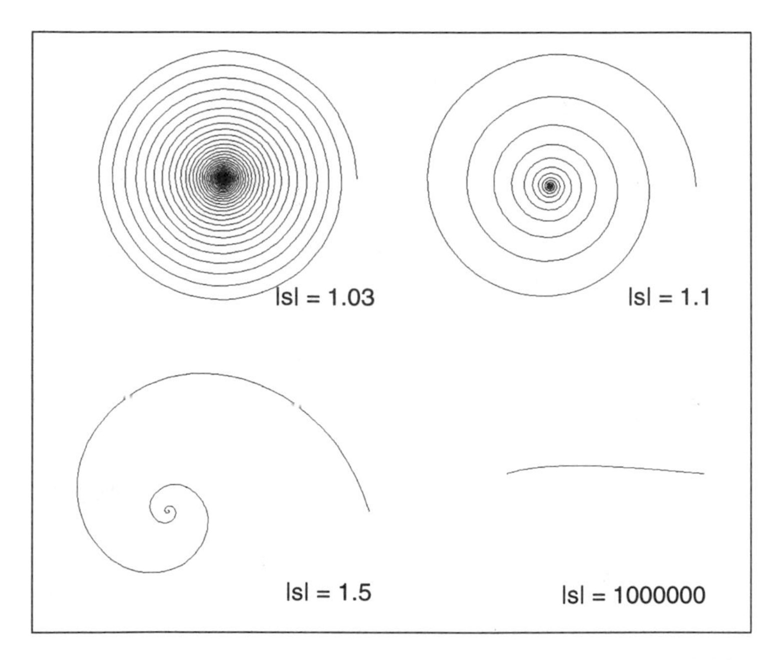

Figure 14.31 : The logarithmic spiral with $a = -0.019$, -0.061, -0.258 and -8.795 (from upper left to lower right) are invariant under rotating by 90 degrees (clockwise) and scaling by $|\rho| = 1.03$, 1.1, 1.5, and $1,000,000$.

Interwoven Spirals

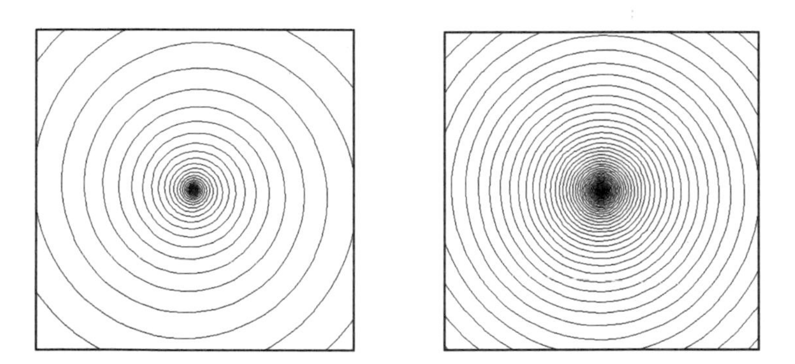

Figure 14.32 : One of the two spirals shown is a simple spiral, while the other is a double spiral, i.e., it is a composition of two interwoven logarithmic spirals. Can you decide at a first glance which is which? If not, just follow the spirals for one turn.

mathematics to hold onto when studying the wealth of structures in the Mandelbrot set.

We now return to the development of the description of the self-similarity features of the Mandelbrot and Julia sets. So far

Spiral or not Spiral?

Figure 14.33 : The artwork is by Nicholas Wade. Reproduced with the kind permission of the artist. From: Nicholas Wade, *The Art and Science of Visual Illusions,* Routledge & Kegan Paul, London, 1982.

we have discussed the term self-similarity of a set I in a point z_0 for $z_0 = 0$. In this case the similarity transformation is simply described by the multiplication by a complex number ρ. If $z_0 \neq 0$, the transformation consisting of a magnification (with possible rotation) around z_0 is technically more complicated. We first have to translate the object by $-z_0$ in order to move the center z_0 of the magnification to the origin. For this purpose let us introduce the notation

$$I - z_0 = \{z - z_0 \mid z \in I\}$$

Then we proceed as before; we magnify and rotate using multiplication by a complex number ρ. In other words, for the self-similarity property at a point z_0 we require that

$$D_r(0) \cap (I - z_0) = D_r(0) \cap \rho(I - z_0)$$

holds for all sufficiently small radii $r > 0$.[30]

[30]An alternative would be to use the similarity transformation $\theta(z) = \rho z + (1 - \rho)z_0$. This is a similarity

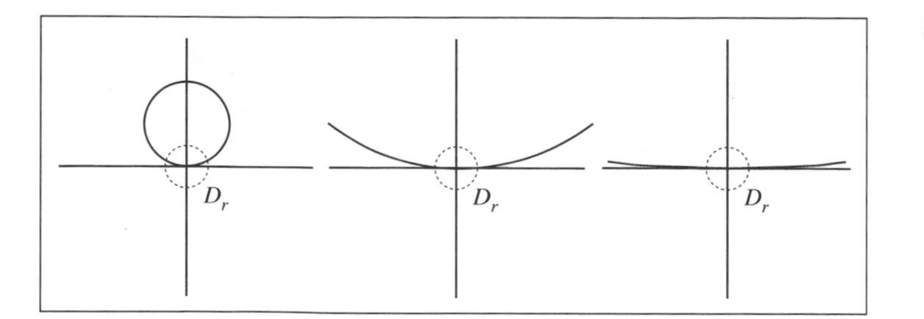

Figure 14.34 : The circle is asymptotically self-similar at any of its points. The given circle of radius 1 around the center $z = i$ (left) is scaled by a factor of $\rho = 5$ (center) and again (right). In the limit, a straight line is approached.

Self-Similarity of the Circle Segment

Asymptotic Self-Similarity at a Point

Let us return to the Mandelbrot set M and the Julia sets J_c. Unfortunately, we cannot postulate the self-similarity of M or a Julia set J_c at any point. The local structures of these objects are somewhat too complex for plain self-similarity, but not by much. The self-similarity at a point does not become apparent after a single magnification and rotation. The magnified and rotated copy is not identical to the original, not even in the small neighborhood of the center of magnification. However, when we repeat this rescaling procedure over and over, the resulting objects *converge* to a set which is self-similar at a point. Therefore, we call this property of the Mandelbrot set and of Julia sets *asymptotic self-similarity at a point*. Formally we can describe this as follows: we call I asymptotically self-similar at the point z_0 if there are

- a complex scaling factor ρ, called *multiplier*, with $|\rho| > 1$,
- a small radius $r > 0$,
- and a limit object L (a subset of the complex plane) which is self-similar at the origin

such that the relation

$$\lim_{n\to\infty} D_r(0) \cap \rho^n(I - z_0) = L \cap D_r(0)$$

holds.

Let us consider a simple example: a circle. Figure 14.34 shows the circle S, given by $|z - i| = 1$, and its scaled copies, using the scaling factor $\rho = 5$. Observe that at the intersection of $\rho^n S$ and the disk $D_r(0)$ about the origin, an arc from the (rescaled) circle, straightens more and more from stage to stage. It converges to the line segment $L = D_r \cap \{(x + yi \mid y = 0\}$ as n increases.

transformation which has z_0 as fixed point. Then we require that $D_r(z_0) \cap I = D_r(z_0) \cap \theta(I)$.

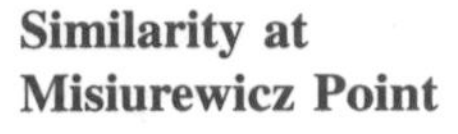

Similarity at Misiurewicz Point

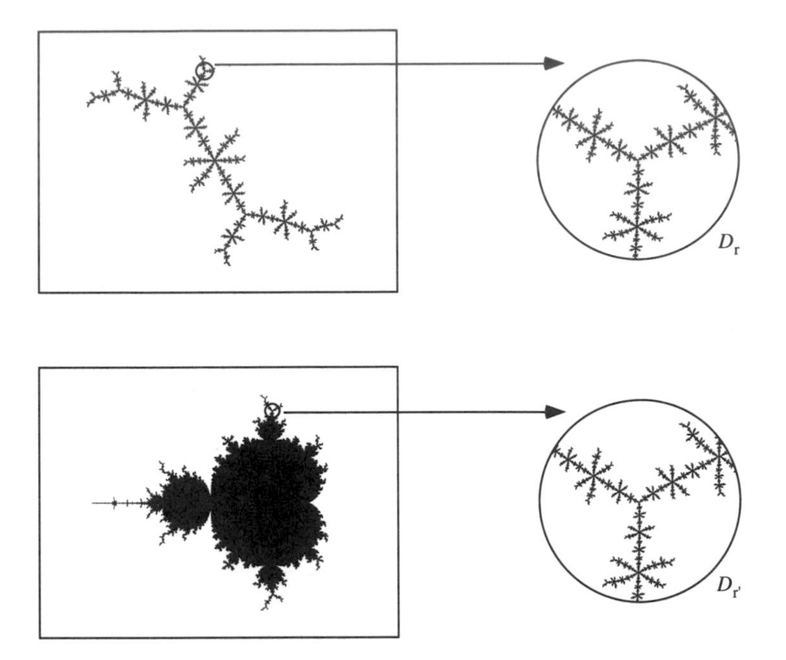

Figure 14.35 : The point $\gamma \approx -0.1011 + 0.9563i$ is a Misiurewicz point. This figure shows the Julia set for $c = \gamma$ (top) and the Mandelbrot set (bottom) with enlargements of small disks centered at the point $z = \gamma$ (upper right) and $c = \gamma$ (lower right). The proper rescaling operation, according to Tan Lei's theorem, is carried out only a couple of times at those points and the results can be compared on the right. Indeed, both structures are almost identical. Further rescaling operations would make the pictures even indistinguishable.

Now we can state Tan Lei's result:[31] Loosely speaking, if the parameter $c \in M$ is a particular type of point, called a Misiurewicz point[32] then the following is true.

- The Julia set J_c and the Mandelbrot set are both asymptotically self-similar in the point $z = c$ using the same multiplier ρ.
- The associated limit objects L_J and L_M are essentially the same; they differ only by some scaling and a rotation ($L_M = \lambda L_J$, where λ is a suitable complex number).

[31] Tan Lei, *Similarity between the Mandelbrot set and Julia sets,* Report Nr 211, Institut für Dynamische Systeme, Universität Bremen, June 1989, and, Commun. Math. Phys. 134 (1990) 587–617.

[32] A parameter c is a Misiurewicz point, if the critical point $z_0 = 0$ of $z \to z^2 + c$ is pre-periodic — but not periodic itself, i.e., the iteration starting at z_0 leads to a periodic cycle, which does not contain z_0 itself. Michal Misiurewicz is a Polish mathematician who became well known in the study of dynamics of one-dimensional maps like $x \to ax(1-x)$. Also he showed that the Lozi model of the Hénon system bears a strange attractor (see section 12.1).

Misiurewicz Point i

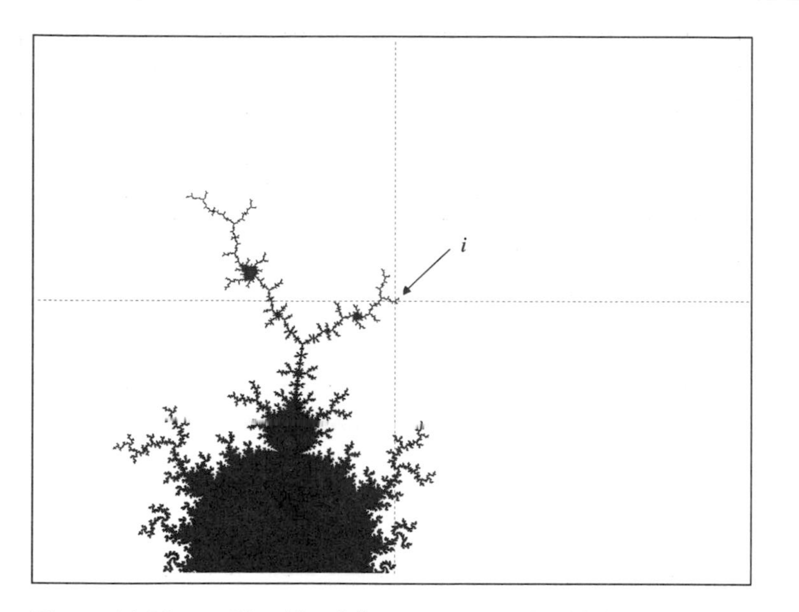

Figure 14.36 : The Mandelbrot set near the Misiurewicz point $c = i$.

Fast Spiral

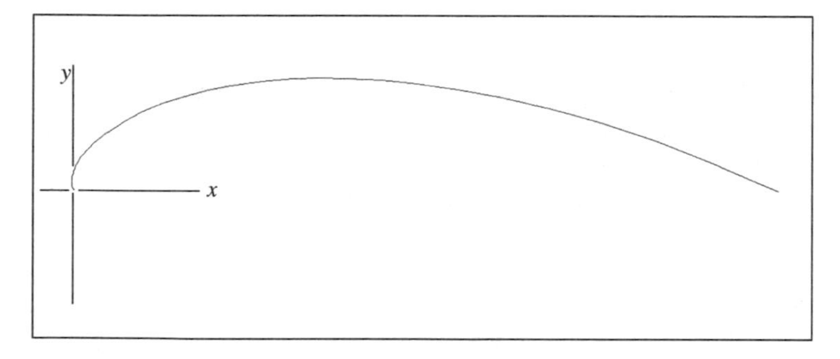

Figure 14.37 : The figure shows a spiral which is invariant under scaling by $r \approx 5.66$ and simultaneous rotation by $\phi = 45$ degrees.

The second fact is illustrated in figure 14.35 for the Misiurewicz point $c \approx -0.1011 + 0.9563i$.

Since Misiurewicz points are dense at the boundary of the Mandelbrot set,[33] Tan Lei's theorem is quite powerful. In a way, near all these points in the Mandelbrot set we can see the shape of the corresponding Julia sets. Or we could say the boundary of M is a visual *table of contents* for infinitely many Julia sets.

[33] This means that in the neighborhood of any point of the boundary, there exists a Misiurewicz point. See A. Douady, J. H. Hubbard, *Étude dynamique des pôlynomes complexes,* Publications Mathematiques d'Orsay 84-02, Université de Paris-Sud, 1984.

Closeup at $c = i$

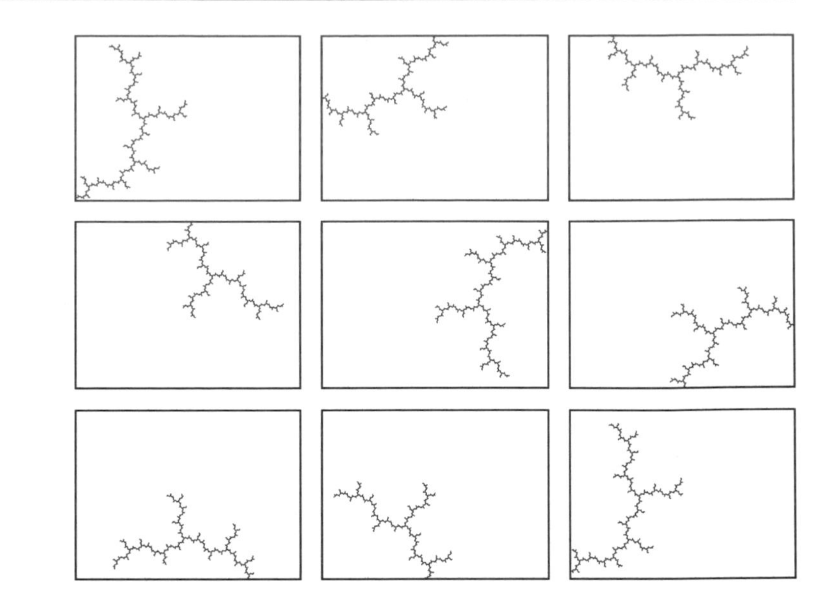

Figure 14.38 : Successive magnification by a scaling factor $4\sqrt{2}$ reveal a rotation by 45 degrees from one image to the next.

Is the Dendrite at $c = i$ a Spiral?

Let us close this section with two amazing observations which can be explained by means of Tan Lei's result. First, let us reconsider the Misiurewicz point at $c = i$ (see figure 14.36). The multiplier ρ for the asymptotic self-similarity of the Mandelbrot set and the Julia set for $c = i$ can be calculated; it is $\rho = 4 + 4i$. This means that the similarity transformation that yields the asymptotic self-similarity of the Mandelbrot set and the Julia set at the point $c = i$ is given by scaling factor of $|\rho| = \sqrt{32} \approx 5.66$ and a rotation by $\arg \rho = \phi = 5$ degrees.

Inspecting the dendrite near $c = i$ in figure 14.36 we seem to encounter a contradiction. On the one hand there appears to be an ordinary dendrite with branches. On the other Tan Lei's result tells us that we really should see a spiral because the limit model L is invariant under multiplication with ρ with $\arg \rho \neq 0$. But what does that mean graphically? Let us look in figure 14.37 at a logarithmic spiral which is self-similar at 0 under multiplication with the same complex factor ρ. Now we understand: it is very hard to see that this is really a spiral, the rotational effect is dominated by the stretching too much. Looking back at the number ρ, this should really be no surprise, a 45 degree turn stretches the spiral by a factor of $4\sqrt{2} \approx 5.66$, and thus, a full turn means stretching by the same factor raised to the eighth power, $(4\sqrt{2})^8 = 2^{20}$, i.e. about a million times!

Figure 14.39 : This sequence of enlargements zooms in on a small copy of the Mandelbrot set. Its diameter is about 10^{-8}. At the center of the spiral in the upper left, there is a Misiurewicz point.

Pollution With Small M-Sets

Now let us try to reveal the same spiraling structure that must be present at the tip of the dendrite at the parameter value $c = i$. In figure 14.38 we take successive magnifications centered at $c = i$, each one enlarged by a factor of $4\sqrt{2}$. However, we do not rotate! According to Tan Lei's result we should see that the resulting dendrite appears as if it were rotated by 45 degrees from step to step. Thus, after eight magnifications we should see a repetition; and indeed we do!

Finally we consider another amazing consequence from Tan Lei's result. As we have seen, close-ups of the Mandelbrot set near its boundary typically result in images which apparently contain small copies of the Mandelbrot set. In fact, it is known that there are infinitely many small copies of the Mandelbrot set populating the boundary region.[34] Figure 14.39 confirms that for a typical c-value.

The existence of small copies of the Mandelbrot set everywhere along its boundary seems to contradict Tan Lei's theorem, which postulates that near Misiurewicz points the Mandelbrot set and corresponding Julia sets look similar. Moreover, the similarity should become increasingly stronger when we magnify at a Misiurewicz

[34] See A. Douady, J. H. Hubbard, *Etudes dynamiques des polynômes complexes I, II,* Publ. Math. d'Orsay 84-02 (1984), 85-02 (1985).

Two Spirals

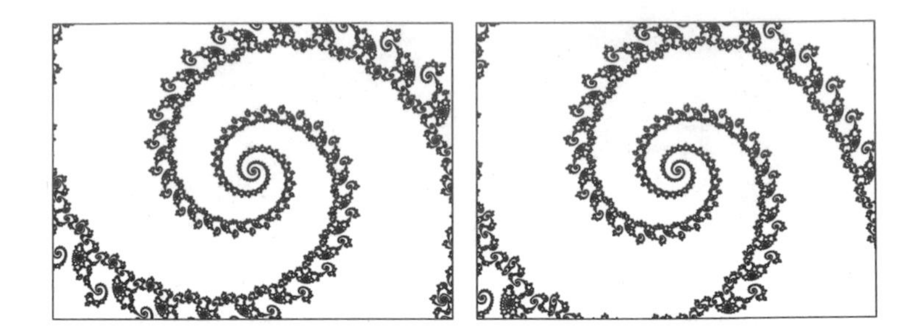

Figure 14.40 : An enlargement of the Julia set J_c with $c = -0.77568377 + 0.13646737i$, a Misiurewicz point, is shown on the left. The image is centered at c and has width 0.00036. Pictured on the right is an enlargement of the Mandelbrot set at the same Misiurewicz point (the width is 0.00048). The double spirals are almost identical, although the right one must contain infinitely many small copies of the Mandelbrot set, while the left one must not have any such copies in it (see figure 14.39.

point. On the other hand Julia sets cannot contain small copies of the Mandelbrot set; these would contradict the self-similarity of the Julia sets. Thus, continuous magnification of the Mandelbrot set at a Misiurewicz point will reveal copies of M without end, while magnification of the corresponding Julia set cannot produce these copies (see figure 14.40). Yet, according to Tan Lei's theorem, both structures should become indistinguishable as the magnification factor increases. How can we resolve this contradiction? The answer is surprising and is related to the meaning of asymptotic self-similarity at some point z_0.

Figure 14.41 shows a rather artificial but instructive example which exhibits the same 'contradiction'. We consider a straight line segment of length 1 placed with one end at the origin. At the other end we attach a square with side length $d_0 = 1$. Then we bisect the line segment. The left half interval is bisected again, and we repeat the procedure again and again. In other words, we construct the bisection points at

$$x_1 = \frac{1}{2},\ x_2 = \frac{1}{4},\ \ldots,\ x_n = \frac{1}{2^n},\ \ldots$$

At these bisection points we place squares T_n of side lengths

$$d_1 = \frac{1}{4},\ d_2 = \frac{1}{16},\ \ldots,\ d_n = \frac{1}{2^{2n}},\ \ldots$$

This is the set B. Now we successively magnify around 0, each time by a factor of $\rho = 2$, and watch what we see in the disk of radius 1. After each magnification the side length of the squares

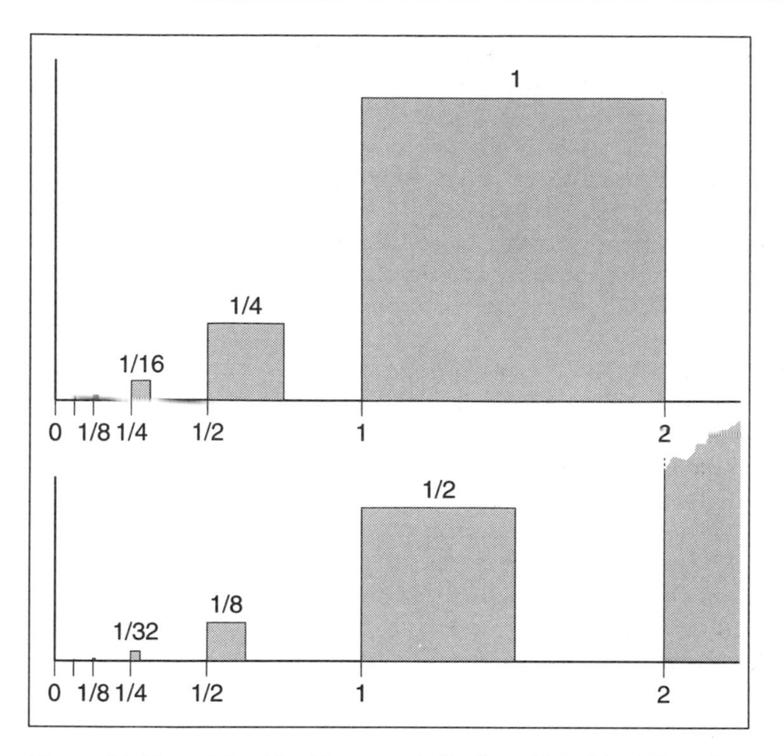

Figure 14.41 : This object is asymptotically self-similar at the origin 0, and the limit object is a straight line. The lower part is obtained from the upper part by a scaling by a factor of 2.

Asymptotic Self-Similarity

is doubled, but on the other hand they seem to move outward. Now look at a bisection point, say, $x = 1/2^n$. After the first magnification, we have the square T_{n+1} at this point, scaled by the factor $\rho = 2$. This square ρT_{n+1} has side length

$$2d_{n+1} = \frac{2}{2^{2(n+1)}} = \frac{1}{2^{2n+1}} .$$

After the second magnification we have the square $\rho^2 T_{n+2}$ at the point $x = 2^{-n}$, which has side length

$$2^2 d_{n+2} = \frac{2^2}{2^{2(n+2)}} = \frac{1}{2^{2n+2}} .$$

In general, after k such iterations we see the square $\rho^k T_{n+k}$ at the point $x = 2^{-n}$, which has side length

$$2^k d_{n+k} = \frac{2^k}{2^{2(n+k)}} = \frac{1}{2^{2n+k}} .$$

Thus in the limit for $k \to \infty$ these side lengths diminish and, moreover, this is true at all the bisection points. In other words, we have shown that our set B is asymptotically self-similar at 0 and the limit model L is just the straight line segment of length 1, which is quite a surprise when comparing B with L visually.

The reason for this effect lies obviously in the fact that the size of the squares in B decrease much more rapidly than the magnification can compensate for in successive magnifications. In other words, if the sizes of the squares were rather to go like 1/2, 1/4, 1/8, ..., for example, then each magnification would yield the same object in the disk of radius 1, i.e., in this case B would be even self-similar in 0.

Thus, we have learned by means of this construction that if we take a limit model which is self-similar at a point z_0 and pollute it with 'garbage' near z_0 making sure that the size of the 'garbage' decreases sufficiently rapidly close to z_0, then we obtain a structure which is still asymptotically self-similar in z_0! Since the 'garbage' can be arbitrary as long as it decreases in size sufficiently, that means that one limit model L stands for a whole class of such structures. We may replace some squares in our set B by triangles, others by hexagons, or whatever shape we like. As long as their sizes are chosen appropriately, the set is still asymptotically self-similar.

This explains why the upper left double spiral in figure 14.39 does not show any copies of the Mandelbrot set, and we will never see any such copies if we continue to magnify around the Misiurewicz point $c^* \approx -0.7756838 + 0.1364674i$ (the center of the double spiral). Yet there are infinitely many copies of the Mandelbrot set in the image. But we cannot see them because their sizes decrease so rapidly towards c^*. Or, to put it in another way, the two images in figure 14.40 are the same from any practical point of view, i.e., as an assembly of pixels. But for the underlying mathematical objects which are visualized here, there is a world of a difference.

It is interesting to turn the argument around: If a set B is asymptotically self-similar at a point with a limit object L, then whatever the pollution in B is, it must decrease in size sufficiently rapidly near that point. In other words, Tan Lei's result gives us an estimate for how small copies of the Mandelbrot set must be near the center c^* of the double spiral. The sizes of the small Mandelbrot set copies are beautifully balanced, they are just at the threshold between becoming submerged into invisibility and explosion upon successive magnification.

14.4 Program of the Chapter: The Mandelbrot Set

Computer programs that render the Mandelbrot set and produce enlargements have become very common. The majority of these programs simply scan a rectangular array of pixels. Then each pixel representing a point of the Mandelbrot set[35] is colored black. This simple approach is implemented by the program 'Mandelbrot Set Pixel Game' at the end of this section. Its major drawback is the fact that it is very slow, especially as an interpreted BASIC program. The program of the chapter takes a very different approach. It traces the boundaries of the encirclements M_k discussed in section 14.1. In other words, we follow the equipotential lines given by $p_M(c) = 2^k$ that surround the Mandelbrot set M and approximate its boundary. This is much faster than the usual scanning algorithm.

Screen Image of Program

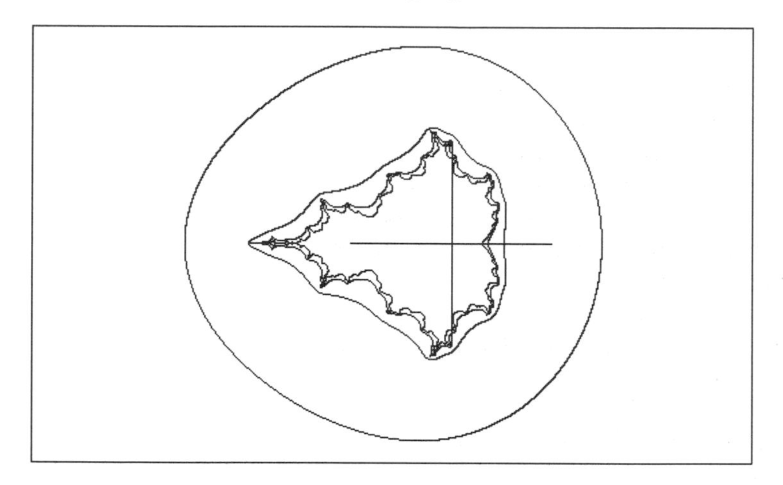

Figure 14.42 : Output of the program 'Mandelbrot' for levels $k = 0, -3, -6, -9$.

The basis of the algorithm is a triangulation of the complex plane. This is a subdivision of the plane into small triangles (see figure 14.43). From a given triangle (specified by three vertices z_1, z_2, and z_3) we can compute all three adjacent triangles by reflecting a vertex across the center of the opposite side. For example, if we reflect z_1, then

$$z_{1new} = z_2 + z_3 - z_1$$

is obtained, and z_{1new}, z_2, and z_3 are the vertices of an adjacent triangle of the triangulation. By appropriately repeating this procedure we can generate any triangle of the triangulation.

[35]More precisely, it is determined whether a point is part of a given encirclement M_k (e.g., M_{-20}) or not. However, in most programs the test used merely checks for $R^k(D^0)$.

Triangulation

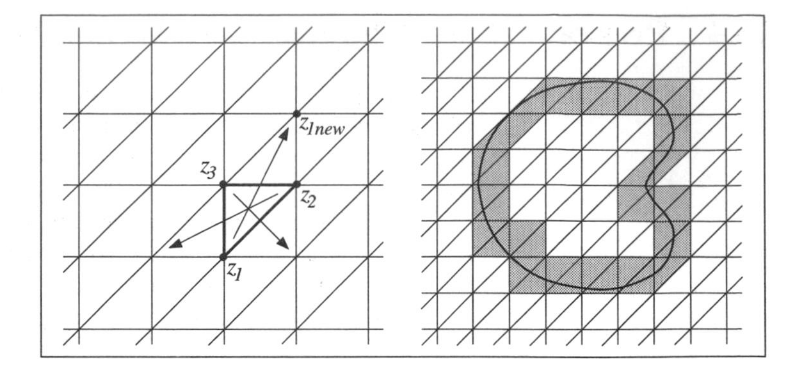

Figure 14.43 : The basis of the algorithm is a triangulation of the complex plane that can be obtained by reflecting triangle vertices. The boundary of M_k is traced by a chain of triangles which intersect this boundary.

The goal is to follow the boundary of M_k by pivoting through a chain of triangles which approximate the boundary. A triangle is called *completely labeled* if at least one of the three vertices lies in M_k (label 1) and another one is outside of M_k (label 0). Such a triangle must intersect the boundary of M_k. The algorithm generates a chain of completely labeled triangles tracing out the boundary of M_k. The approach is similar to the method of escaping from a planar maze. You can walk through the paths of the maze always touching the wall (which corresponds to the boundary of M_k) with your right hand. Either you return to the place where you started, or you will find an exit from the maze. In the same way we construct a chain of triangles which intersect the boundary of M_k having one vertex inside of M_k.

A triangle side is called *transversal* if one end point lies in M_k and the other is outside of M_k. Thus, a triangle is completely labeled if it has at least one transversal side. We can be more precise. Any triangle has either *no transversal sides*, or exactly *two transversal sides*.[36] Thus, a completely labeled triangle must have exactly two neighboring triangles[37] which are also completely labeled. Furthermore, we conclude that each completely labeled triangle is one in a chain of completely labeled triangles. From these observations we can derive an algorithm which computes such a chain. Assume that an initial completely labeled triangle is given. Choose one of the two transversal sides as the reflection side and then repeat the following steps:

[36] Essentially there are only four ways to 'assign labels' to the vertices of a triangle. The triangle has 0, 1, 2, or all 3 of its vertices in M_k. The corresponding numbers of transversal sides are 0, 2, 2, and 0.

[37] A triangle is a neighbor when it shares a common side.

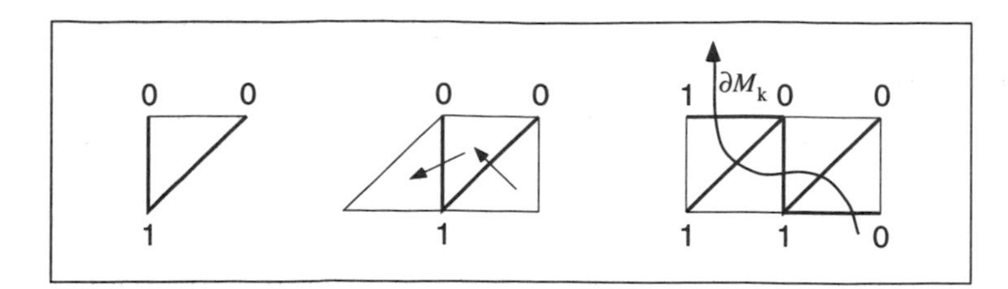

Completely Labeled Triangles

Figure 14.44 : A completely labeled triangle has exactly to transversal faces. Adjacent to each of these faces is another completely labeled triangle. This provides a chain of triangles that follows the boundary of M_k.

Step 1. Reflect the opposite vertex across the center of the reflection side. Now the reflected point and the reflection side form another completely labeled triangle of the triangulation that intersects the boundary of M_k.

Step 2. The new triangle intersects the boundary. It has two transversal sides, the reflection side used in the previous step and the next one. This second transversal side now becomes our new reflection side.

We repeat these steps until we return to the initial completely labeled triangle of the triangulation from which we started to trace the boundary of M_k.[38]

Let us now discuss some details of the program starting at label 500 of the routine 'Mandelbrot test'. The program determines whether a parameter c (`rec` is the real and `imc` the imaginary part) is inside or outside of the encirclement M_k. This test uses $R^{k-2}(D^2)$ as an approximation of M_k (compare with section 14.1). You can see this by looking at the `IF` statement which checks `re2 + cim2 > 256`. In other words, we use $D^2 = \{z \mid |z| < 16\}$ as the target set. If the program traces the boundary of M_k for the level $k = 2$, the result is the circle of radius 16, the boundary of the target set. For smaller values of k, we get curves closer to the Mandelbrot set M. Figure 14.42 shows the boundary of M_k for $k = 0, -3, -6$, and -9. Appending this routine to the program 'Mandelbrot' or 'Mandelbrot Set Pixel Game' completes these programs and makes them ready for execution.

[38] This type of curve following can be also generalized to curves in three or more dimensions. See H. O. Peitgen, M. Prüfer, *The Leray-Schauder continuation method is a constructive element in the numerical study of nonlinear eigenvalue and bifurcation problems,* in: *Functional Differential Equations and Approximation of Fixed Points,* H.-O. Peitgen, H.-O. Walther (eds.), Springer Lecture Notes, Berlin, 1979. See also E. Allgower, K. Georg, *Numerical Continuation Methods — An Introduction,* Springer-Verlag, New York, 1990. Also boundaries of solid objects in higher dimensions may be tracked by suitably adapted methods. See also C. Zahlten, *Piecewise linear approximation of isovalued surfaces,* in: *Advances in Scientific Visualization, Eurographics Seminar Series,* (F. H. Post, A. J. S. Hin (eds.), Springer-Verlag, Berlin, 1992.

BASIC Programm **Mandelbrot**
Title Equipotential lines for the Mandelbrot set

```
DIM x(3), y(3)
left = 20
w = 300
r = 2
s = 2*r/w
INPUT "level k (start, end, step):", kstart, kend, kstep

REM draw coordinate axes
LINE (-1/s+left+w/2, left+w/2) - (1/s+left+w/2, left+w/2)
LINE (left+w/2, -1/s+left+w/2) - (left+w/2, 1/s+left+w/2)

FOR k = kstart TO kend STEP kstep

    REM search boundary of level set
    FOR i = 0 TO 10000
        rec = s*i + recen : imc = imcen
        GOSUB 500
        IF outside = 1 GOTO 100
    NEXT i
    END

    REM place triangle on boundary
100 vin = 1 : vout = 2 : vnew = 3
    x(vin) = i-1 : x(vout) = i : x(vnew) = i
    y(vin) = 0 : y(vout) = 0 : y(vnew) = 1
    xin = x(vin) : yin = y(vin) : xout = x(vout) : yout = y(vout)
    PSET (x(vin)+left+w/2, -y(vin)+left+w/2)

    REM scan boundary of level set
200 rec = s*x(vnew) + recen : imc = s*y(vnew) + imcen
    GOSUB 500
    IF outside = 1 GOTO 300
        LINE - (x(vnew)+left+w/2, -y(vnew)+left+w/2)
        vref = vin : vin = vnew : vnew = vref
        GOTO 400
300     vref = vout: vout = vnew : vnew = vref
400 x(vnew) = x(vin) + x(vout) - x(vref)
    y(vnew) = y(vin) + y(vout) - y(vref)
    IF x(vin) <> xin OR y(vin) <> yin GOTO 200
    IF x(vout) <> xout OR y(vout) <> yout GOTO 200
NEXT k
END
```

BASIC Programm **Mandelbrot Test**
Title Testing for encirclement M-k of Mandelbrot set

```
REM Mandelbrot set test for level k
500 outside = 1
    re = rec : im = imc
    FOR j = 0 TO 2 - k
        re2 = re*re : im2 = im* im
        IF re2 + im2 > 256! THEN RETURN
        im = 2*re*im + imc
        re = re2 - im2 + rec
    NEXT j
    outside = 0
    RETURN
```

BASIC Programm **Mandelbrot Set Pixel Game**
Title The classical algorithm

```
left = 20 : w = 300
r = 2 : s = 2*r/w
recen = 0 : imcen = 0
INPUT "level k:", k
FOR y= 0 TO w
    FOR x = 0 TO w
        rec = s*(x-w/2) + recen : imc = s*(y-w/2) + imcen
        GOSUB 500
        IF outside = 0 THEN PSET(x+left, left+w-y)
    NEXT x
NEXT y
END
```

We now look at the implementation details of tracing the boundary of M_k using triangles. First, the program operator can enter the levels k that will be traced by the program. The input requires an initial level, a last level, and an increment, in that order. For example, if you enter $0, 0, 1$ you will obtain only M_0, if you enter $0, -4, -2$ you will obtain M_0, M_{-2}, and M_{-4}. For each value of k, the program finds a triangle that intersects the boundary of M_k. Starting at $c = 0$, which is a point of M,[39] we step along the positive real axis in increments of s until a point c is found that is not in M_k. At that point a transversal side is found on the real axis and a first completely labeled triangle can be easily constructed. The arrays x and y hold the coordinates of its three vertices. We set

[39]The routine 'Mandelbrot test' obtains the result `outside = 0`.

Screen Image of Program

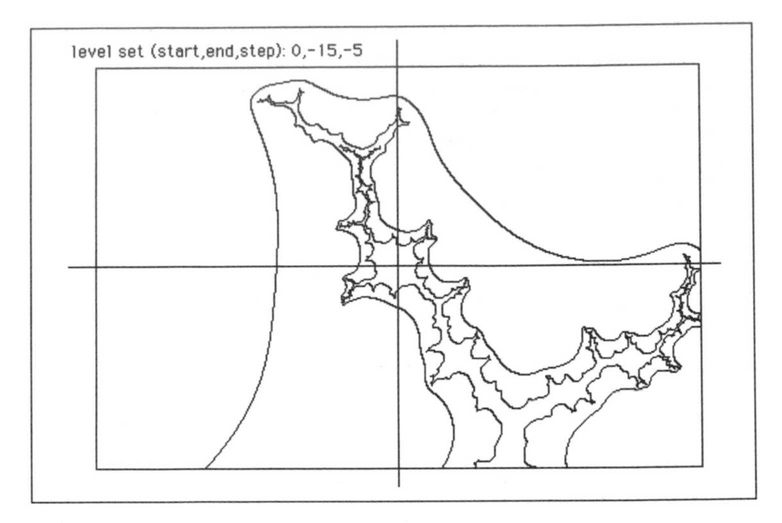

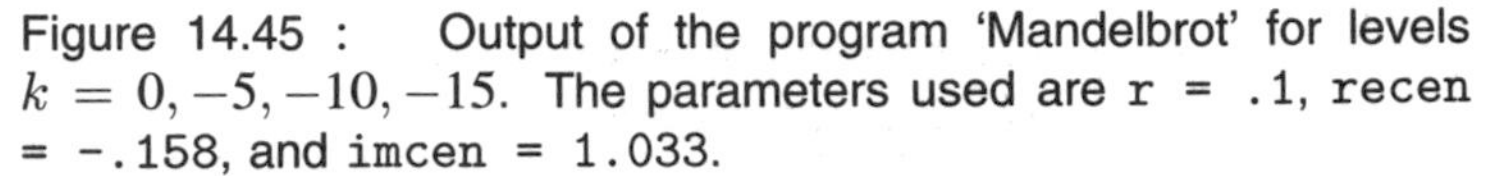
Figure 14.45 : Output of the program 'Mandelbrot' for levels $k = 0, -5, -10, -15$. The parameters used are `r = .1`, `recen = -.158`, and `imcen = 1.033`.

`vin = 1`, `vout = 2`, and `vnew = 3`, and define `x(vin) = i-1` and `y(vin) = 0` which represent the triangle vertex that is inside of M_k (the 'in-point'). `x(vout) = i` and `y(vout) = 0` represent the vertex that is outside of M_k (the 'out-point'). For the third vertex we set `x(vnew) = i` and `y(vnew) = 1` (the 'new-point'). Furthermore, we draw the in-point and store the coordinates of the in- and out-points for later comparison.

Now we start tracing the boundary. First, we test the new-point. If it is inside we draw a line to this point and rename it (i.e., the new-point gets the index `vin` and the old in-point gets the index `vref`). This point is now the subject of reflection across the center of the opposite triangle side given by the new in-point and the out-point. The reflected point becomes the new-point. On the other hand, if the new-point is outside, then it is not drawn. Furthermore, it gets the index `vout` and the old out-point gets the index `vref` and is reflected. This procedure is repeated until we get back to the initial completely labeled triangle. The stopping criterion is determined by comparing the in- and out-point with the corresponding saved values `xin`, `yin`, `xout`, and `yout`.

Finally let us discuss how you can change the focus of the algorithm. You might know how one produces enlargements for the Pixel Game algorithm. One selects a point of the complex plane that becomes the center of the image and a radius that specifies the size of a circle in the complex plane which corresponds in the drawing area to a width of `w`. This selection is done by setting the variables `recen`, `imcen` and `r` to appropriate values at the beginning of the program (for example, `r = 0.5`, `recen = -0.2`, and `imcen = 0.8`). The same can be done with the main program of the chapter. We must only prevent the algorithm from leaving the drawing region. For this purpose include the statements

```
IF x(new) < w/2 OR w/2 < x(new) THEN RETURN
IF y(new) < w/2 OR w/2 < y(new) THEN RETURN
```

just after the statement with label 500 of the routine 'Mandelbrot test'. Then this auxiliary routine will return the setting `outside = 1` for all points which are not part of the drawing area. With that you can do enlargements like the one shown in figure 14.45.[40] You only have to make sure that the center point (defined by `recen` and `imcen`) is inside of M_k.

[40]An alternative strategy would be to search the bounds of the drawing area for triangles that intersect the boundary of M_k and start the tracing with these triangles.

Bibliography

1. Books

[1] Abraham, R. H., Shaw, C. D., *Dynamics, The Geometry of Behavior,* Part One:*Periodic Behavior* (1982), Part Two: *Chaotic Behavior* (1983), Part Three: *Global Behavior* (1984), Aerial Press, Santa Cruz. Second edition Addison-Wesley, 1992.

[2] Allgower, E., Georg, K., *Numerical Continuation Methods — An Introduction,* Springer-Verlag, New York, 1990.

[3] Arnold, V. I., *Ordinary Differential Equations,* MIT Press, Cambridge, 1973.

[4] Avnir, D. (ed.), *The Fractal Approach to Heterogeneous Chemistry: Surfaces, Colloids, Polymers,* Wiley, Chichester, 1989.

[5] Banchoff, T. F., *Beyond the Third Dimension,* Scientific American Library, 1990.

[6] Barnsley, M., *Fractals Everywhere,* Academic Press, San Diego, 1988.

[7] Beardon, A. F., *Iteration of Rational Functions,* Springer-Verlag, New York, 1991.

[8] Becker K.-H., Dörfler, M., *Computergraphische Experimente mit Pascal*, Vieweg, Braunschweig, 1986.

[9] Beckmann, P., *A History of Pi,* Second Edition, The Golem Press, Boulder, 1971.

[10] Bélair, J., Dubuc, S., (eds.), *Fractal Geometry and Analysis,* Kluwer Academic Publishers, Dordrecht, Holland, 1991.

[11] Bondarenko, B., *Generalized Pascal Triangles and Pyramids, Their Fractals, Graphs and Applications,* Tashkent, Fan, 1990, in Russian.

[12] Borwein, J. M., Borwein, P. B., *Pi and the AGM — A Study in Analytic Number Theory,* Wiley, New York, 1987.

[13] Briggs, J., Peat, F. D., *Turbulent Mirror,* Harper & Row, New York, 1989.

[14] Bunde, A., Havlin, S. (eds.), *Fractals and Disordered Systems,* Springer-Verlag, Heidelberg, 1991.

[15] Campbell, D., Rose, H. (eds.), *Order in Chaos,* North-Holland, Amsterdam, 1983.

[16] Chaitin, G. J., *Algorithmic Information Theory,* Cambridge University Press, 1987.

[17] Cherbit, G. (ed.), *Fractals, Non-integral Dimensions and Applications,* John Wiley & Sons, Chichester, 1991.

[18] Collet, P., Eckmann, J.-P., *Iterated Maps on the Interval as Dynamical Systems,* Birkhäuser, Boston, 1980.

[19] Crilly, A. J., Earnshaw, R. A., Jones, H. (eds.), *Fractals and Chaos,* Springer-Verlag, New York, 1991.

[20] Cvitanović, P. (ed.), *Universality in Chaos,* Second Edition, Adam Hilger, New York, 1989.

[21] Devaney, R. L., *An Introduction to Chaotic Dynamical Systems, Second Edition,* Addison-Wesley, Redwood City, 1989.

[22] Devaney, R. L., *Chaos, Fractals, and Dynamics,* Addison-Wesley, Menlo Park, 1990.

[23] Durham, T., *Computing Horizons,* Addison-Wesley, Wokingham, 1988.

[24] Dynkin, E. B., Uspenski, W., *Mathematische Unterhaltungen II,* VEB Deutscher Verlag der Wissenschaften, Berlin, 1968.

[25] Edgar, G., *Measures, Topology and Fractal Geometry,* Springer-Verlag, New York, 1990.

[26] Engelking, R., *Dimension Theory,* North Holland, 1978.

[27] Escher, M. C., *The World of M. C. Escher,* H. N. Abrams, New York, 1971.

[28] Falconer, K., *The Geometry of Fractal Sets,* Cambridge University Press, Cambridge, 1985.

[29] Falconer, K.,*Fractal Geometry, Mathematical Foundations and Applications,* Wiley, New York, 1990.

[30] Family, F., Landau, D. P. (eds.), *Aggregation and Gelation,* North-Holland, Amsterdam, 1984.

[31] Family, F., Vicsek, T. (eds.), *Dynamics of Fractal Surfaces,* World Scientific, Singapore, 1991.

[32] Feder, J., *Fractals,* Plenum Press, New York 1988.

[33] Fleischmann, M., Tildesley, D. J., Ball, R. C., *Fractals in the Natural Sciences,* Princeton University Press, Princeton, 1989.

[34] Garfunkel, S., (Project Director), Steen, L. A. (Coordinating Editor) *For All Practical Purposes, Second Edition,* W. H. Freeman and Co., New York, 1988.

[35] GEO Wissen — Chaos und Kreativität, Gruner + Jahr, Hamburg, 1990.

[36] Gleick, J., *Chaos, Making a New Science,* Viking, New York, 1987.

[37] Golub, G. H., Loan, C. F. van, *Matrix Computations,* Second Edition, Johns Hopkins, Baltimore, 1989.

[38] Guckenheimer, J., Holmes, P., *Nonlinear Oscillations, Dynamical Systems, and Bifurcations of Vector Fields,* Springer-Verlag, New York, 1983.

[39] Guyon, E., Stanley, H. E., (eds.), *Fractal Forms,* Elsevier/North-Holland and Palais de la Découverte, 1991.

[40] Haken, H., *Advanced Synergetics,* Springer-Verlag, Heidelberg, 1983.

[41] Haldane, J. B. S., *On Being the Right Size,* 1928.

[42] Hall, R., *Illumination and Color in Computer Generated Imagery,* Springer-Verlag, New York, 1988.

[43] Hao, B. L., *Chaos II,* World Scientific, Singapore, 1990.

[44] Hausdorff, F., *Grundzüge der Mengenlehre,* Verlag von Veit & Comp., 1914.

[45] Hirsch, M. W., Smale, S., *Differential Equations, Dynamical Systems, and Linear Algebra,* Academic Press, New York, 1974.

[46] Hommes, C. H., *Chaotic Dynamics in Economic Models,* Wolters-Noordhoff, Groningen, 1991.

[47] Jackson, E. A., *Perspectives of Nonlinear Dynamics,* Volume 1 and 2, Cambridge University Press, Cambridge, 1991.

[48] Knuth, D. E., *The Art of Computer Programming, Volume 2, Seminumerical Algorithms,* Addison-Wesley, Reading, Massachusetts.

[49] Kuratowski, C., *Topologie II,* PWN, Warsaw, 1961.

[50] Lauwerier, H., *Fractals,* Aramith Uitgevers, Amsterdam, 1987.

[51] Lehmer, D. H., Proc. 2nd Symposium on Large Scale Digital Calculating Machinery, Harvard University Press, Cambridge, 1951.

[52] Leven, R. W., Koch, B.-P., Pompe, B., *Chaos in Dissipativen Systemen,* Vieweg, Braunschweig, 1989.

[53] Lindenmayer, A., Rozenberg, G., (eds.), *Automata, Languages, Development,* North-Holland, Amsterdam, 1975.

[54] Mandelbrot, B. B., *Fractals: Form, Chance, and Dimension,* W. H. Freeman and Co., San Francisco, 1977.

[55] Mandelbrot, B. B., *The Fractal Geometry of Nature,* W. H. Freeman and Co., New York, 1982.

[56] Marek, M., Schreiber, I., *Chaotic Behavior of Deterministic Dissipative Systems,* Cambridge University Press, Cambridge, 1991.

[57] McGuire, M., *An Eye for Fractals,* Addison-Wesley, Redwood City, 1991.

[58] Menger, K., *Dimensionstheorie,* Leipzig, 1928.

[59] Mey, J. de, *Bomen van Pythagoras,* Aramith Uitgevers, Amsterdam, 1985.

[60] Moon, F. C., *Chaotic Vibrations,* John Wiley & Sons, New York, 1987.

[61] Parchomenko, A. S., *Was ist eine Kurve,* VEB Verlag, 1957.

[62] Parker, T. S., Chua, L. O., *Practical Numerical Algorithms for Chaotic Systems,* Springer-Verlag, New York, 1989.

[63] Peitgen, H.-O., Richter, P. H., *The Beauty of Fractals,* Springer-Verlag, Heidelberg, 1986.

[64] Peitgen, H.-O., Saupe, D., (eds.), *The Science of Fractal Images,* Springer-Verlag, 1988.

[65] Peitgen, H.-O. (ed.), *Newton's Method and Dynamical Systems,* Kluver Academic Publishers, Dordrecht, 1989.

[66] Peitgen, H.-O., Jürgens, H., *Fraktale: Gezähmtes Chaos,* Carl Friedrich von Siemens Stiftung, München, 1990.

[67] Peitgen, H.-O., Jürgens, H., Saupe, D., *Fractals for the Classroom, Part One,* Springer-Verlag, New York, 1991.

[68] Peitgen, H.-O., Jürgens, H., Saupe, D., Maletsky, E., Perciante, T., Yunker, L., *Fractals for the Classroom, Strategic Activities, Volume One,* and *Volume Two,* Springer-Verlag, New York, 1991 and 1992.

[69] Peters, E., *Chaos and Order in the Capital Market,* John Wiley & Sons, New York, 1991.

[70] Press, W. H., Flannery, B. P., Teukolsky, S. A., Vetterling, W. T., *Numerical Recipes,* Cambridge University Press, Cambridge, 1986.

[71] Preston, K. Jr., Duff, M. J. B., *Modern Cellular Automata,* Plenum Press, New York, 1984.

[72] Prigogine, I., Stenger, I., *Order out of Chaos,* Bantam Books, New York, 1984.

[73] Prusinkiewicz, P., Lindenmayer, A., *The Algorithmic Beauty of Plants,* Springer-Verlag, New York, 1990.

[74] Rasband, S. N., *Chaotic Dynamics of Nonlinear Systems,* John Wiley & Sons, New York, 1990.

[75] Richardson, L. F., *Weather Prediction by Numerical Process,* Dover, New York, 1965.

[76] Ruelle, D., *Chaotic Evolution and Strange Attractors,* Cambridge University Press, Cambridge, 1989.

[77] Sagan, C., *Contact,* Pocket Books, Simon & Schuster, New York, 1985.

[78] Schröder, M., *Fractals, Chaos, Power Laws,* W. H. Freeman and Co., New York, 1991.

[79] Schuster, H. G., *Deterministic Chaos,* Physik-Verlag, Weinheim and VCH Publishers, New York, 1984.

[80] Sparrow, C., *The Lorenz Equations: Bifurcations, Chaos, and Strange Attractors,* Springer-Verlag, New York, 1982.

[81] Stauffer, D., *Introduction to Percolation Theory,* Taylor & Francis, London, 1985.

[82] Stauffer, D., Stanley, H. E., *From Newton to Mandelbrot,* Springer-Verlag, New York,1989.

[83] Stewart, I., *Does God Play Dice,* Penguin Books, 1989.

[84] Stewart, I., *Game, Set, and Math,* Basil Blackwell, Oxford, 1989.

[85] Thompson, D'Arcy, *On Growth an Form,* New Edition, Cambridge University Press, 1942.

[86] Toffoli, T., Margolus, N., *Cellular Automata Machines, A New Environment For Modelling,* MIT Press, Cambridge, Mass., 1987.

[87] Vicsek, T., *Fractal Growth Phenomena,* World Scientific, London, 1989.

[88] Wade, N., *The Art and Science of Visual Illusions,* Routledge & Kegan Paul, London,1982.

[89] Wall, C. R., *Selected Topics in Elementary Number Theory,* University of South Caroline Press, Columbia, 1974.

[90] Wegner, T., Peterson, M., *Fractal Creations,* Waite Group Press, Mill Valley, 1991.

[91] Weizenbaum, J., *Computer Power and Human Reason,* Penguin, 1984.

[92] West, B., *Fractal Physiology and Chaos in Medicine,* World Scientific, Singapore, 1990.

[93] Wolfram, S., Farmer, J. D., Toffoli, T., (eds.) *Cellular Automata: Proceedings of an Interdisciplinary Workshop,* in: Physica 10D, 1 and 2 (1984).

[94] Wolfram, S. (ed.), *Theory and Application of Cellular Automata,* World Scientific, Singapore, 1986.

[95] Zhang Shu-yu, *Bibliography on Chaos,* World Scientific, Singapore, 1991.

2. General Articles

[96] Barnsley, M. F., *Fractal Modelling of Real World Images,* in: The Science of Fractal Images, H.-O. Peitgen, D. Saupe (eds.), Springer-Verlag, New York, 1988.

[97] Cipra, B., A., *Computer-drawn pictures stalk the wild trajectory,* Science 241 (1988) 1162–1163.

[98] Davis, C., Knuth, D. E., *Number Representations and Dragon Curves,* Journal of Recreational Mathematics 3 (1970) 66–81 and 133–149.

[99] Dewdney, A. K., *Computer Recreations: A computer microscope zooms in for a look at the most complex object in mathematics,* Scientific American (August 1985) 16–25.

[100] Dewdney, A. K., *Computer Recreations: Beauty and profundity: the Mandelbrot set and a flock of its cousins called Julia sets,* Scientific American (November 1987) 140–144.

[101] Douady, A., *Julia sets and the Mandelbrot set,* in: The Beauty of Fractals, H.-O. Peitgen, P. H. Richter, Springer-Verlag, 1986.

[102] Dyson, F., *Characterizing Irregularity,* Science 200 (1978) 677–678.

[103] Gilbert, W. J., *Fractal geometry derived from complex bases,* Math. Intelligencer 4 (1982) 78–86.

[104] Hofstadter, D. R., *Strange attractors : Mathematical patterns delicately poised between order and chaos,* Scientific American 245 (May 1982) 16–29.

[105] Mandelbrot, B. B., *How long is the coast of Britain? Statistical self-similarity and fractional dimension,* Science 155 (1967) 636–638.

[106] Peitgen, H.-O., Richter, P. H., *Die unendliche Reise,* Geo 6 (Juni 1984) 100–124.

[107] Peitgen, H.-O., Haeseler, F. v., Saupe, D., *Cayley's problem and Julia sets,* Mathematical Intelligencer 6.2 (1984) 11–20.

[108] Peitgen, H.-O., Jürgens, H., Saupe, D., *The language of fractals,* Scientific American (August 1990) 40–47.

[109] Peitgen, H.-O., Jürgens, H., *Fraktale: Computerexperimente (ent)zaubern komplexe Strukturen,* in: *Ordnung und Chaos in der unbelebten und belebten Natur,* Verhandlungen der Gesellschaft Deutscher Naturforscher und Ärzte, 115. Versammlung, Wissenschaftliche Verlagsgesellschaft, Stuttgart, 1989.

[110] Peitgen, H.-O., Jürgens, H., Saupe, D., Zahlten, C., *Fractals — An Animated Discussion,* Video film, W. H. Freeman and Co., 1990. Also appeared in German as *Fraktale in Filmen und Gesprächen,* Spektrum Videothek, Heidelberg, 1990. Also appeared in Italian as *I Frattali,* Spektrum Videothek edizione italiana, 1991.

[111] Ruelle, D., *Strange Attractors,* Math. Intelligencer 2 (1980) 126–137.

[112] Ruelle, D., *Chaotic Evolution and Strange Attractors,* Cambridge University Press, Cambridge, 1989.

[113] Stewart, I., *Order within the chaos game?* Dynamics Newsletter 3, no. 2, 3, May 1989, 4–9.

[114] Sved, M. *Divisibility — With Visibility,* Mathematical Intelligencer 10, 2 (1988) 56–64.

[115] Voss, R., *Fractals in Nature,* in: *The Science of Fractal Images,* H.-O. Peitgen , D. Saupe (eds.), Springer-Verlag, New York, 1988.

[116] Wolfram, S., *Geometry of binomial coefficients,* Amer. Math. Month. 91 (1984) 566–571.

3. Research Articles

[117] Abraham, R., *Simulation of cascades by video feedback,* in: "Structural Stability, the Theory of Catastrophes, and Applications in the Sciences", P. Hilton (ed.), Lecture Notes in Mathematics vol. 525, 1976, 10–14, Springer-Verlag, Berlin.

[118] Aharony, A., *Fractal growth,* in: *Fractals and Disordered Systems,* A. Bunde, S. Havlin (eds.), Springer-Verlag, Heidelberg, 1991.

[119] Bak, P., *The devil's staircase,* Phys. Today 39 (1986) 38–45.

[120] Bandt, C., *Self-similar sets I. Topological Markov chains and mixed self-similar sets,* Math. Nachr. 142 (1989) 107–123.

[121] Bandt, C., *Self-similar sets III. Construction with sofic systems,* Monatsh. Math. 108 (1989) 89–102.

[122] Banks, J., Brooks, J., Cairns, G., Davis, G., Stacey, P., *On Devaney's definition of chaos,* American Math. Monthly 99.4 (1992) 332–334.

[123] Barnsley, M. F., Demko, S., *Iterated function systems and the global construction of fractals,* The Proceedings of the Royal Society of London A399 (1985) 243–275

[124] Barnsley, M. F., Ervin, V., Hardin, D., Lancaster, J., *Solution of an inverse problem for fractals and other sets,* Proceedings of the National Academy of Sciences 83 (1986) 1975–1977.

[125] Barnsley, M. F., Elton, J. H., Hardin, D. P., *Recurrent iterated function systems,* Constructive Approximation 5 (1989) 3–31.

[126] Bedford, T., *Dynamics and dimension for fractal recurrent sets,* J. London Math. Soc. 33 (1986) 89–100.

[127] Benedicks, M., Carleson, L., *The dynamics of the Hénon map,* Annals of Mathematics 133,1 (1991) 73–169.

[128] Benettin, G. L., Galgani,L., Giorgilli, A., Strelcyn, J.-M., *Lyapunov characteristic exponents for smooth dynamical systems and for Hamiltonian systems; a method for computing all of them. Part 1: Theory, Part 2: Numerical application,* Meccanica 15, 9 (1980) 21.

[129] Berger, M., *Encoding images through transition probablities,* Math. Comp. Modelling 11 (1988) 575–577.

[130] Berger, M., *Images generated by orbits of 2D-Markoc chains,* Chance 2 (1989) 18–28.

[131] Berry, M. V., *Regular and irregular motion,* in: Jorna S. (ed.), Topics in Nonlinear Dynamics, Amer. Inst. of Phys. Conf. Proceed. 46 (1978) 16–120.

[132] Blanchard, P., *Complex analytic dynamics on the Riemann sphere,* Bull. Amer. Math. Soc. 11 (1984) 85–141.

[133] Borwein, J. M., Borwein, P. B., Bailey, D. H., *Ramanujan, modular equations, and approximations to π, or how to compute one billion digits of π,* American Mathematical Monthly 96 (1989) 201–219.

[134] Brent, R. P., *Fast multiple-precision evaluation of elementary functions,* Journal Assoc. Comput. Mach. 23 (1976) 242–251.

[135] Brolin, H., *Invariant sets under iteration of rational functions,* Arkiv f. Mat. 6 (1965) 103–144.

[136] Cantor, G., *Über unendliche, lineare Punktmannigfaltigkeiten V,* Mathematische Annalen 21 (1883) 545–591.

[137] Carpenter, L., *Computer rendering of fractal curves and surfaces,* Computer Graphics (1980) 109ff.

[138] Caswell, W. E., Yorke, J. A., *Invisible errors in dimension calculations: geometric and systematic effects,* in: *Dimensions and Entropies in Chaotic Systems,* G. Mayer-Kress (ed.), Springer-Verlag, Berlin, 1986 and 1989, p. 123–136.

[139] Cayley, A., *The Newton-Fourier Imaginary Problem,* American Journal of Mathematics 2 (1879) p. 97.

[140] Charkovsky, A. N., *Coexistence of cycles of continuous maps on the line,* Ukr. Mat. J. 16 (1964) 61–71 (in Russian).

[141] Corless, R. M., *Continued fractions and chaos,* The American Math. Monthly 99, 3 (1992) 203–215.

[142] Corless, R. M., Frank, G. W., Monroe, J. G., *Chaos and continued fractions,* Physica D46 (1990) 241–253.

[143] Cremer, H., *Über die Iteration rationaler Funktionen,* Jahresberichte der Deutschen Mathematiker Vereinigung 33 (1925) 185–210.

[144] Crutchfield, J., *Space-time dynamics in video feedback,* Physica 10D (1984) 229–245.

[145] Dekking, F. M., *Recurrent Sets,* Advances in Mathematics 44, 1 (1982) 78–104.

[146] Derrida, B., Gervois, A., Pomeau, Y., *Universal metric properties of bifurcations of endomorphisms,* J. Phys. A: Math. Gen. 12, 3 (1979) 269–296.

[147] Devaney, R., Nitecki, Z., *Shift Automorphism in the Hénon Mapping,* Comm. Math. Phys. 67 (1979) 137–146.

[148] Douady, A., Hubbard, J. H., *Iteration des pôlynomes quadratiques complexes,* CRAS Paris 294 (1982) 123–126.

[149] Douady, A., Hubbard, J. H., *Étude dynamique des pôlynomes complexes,* Publications Mathematiques d'Orsay 84-02, Université de Paris-Sud, 1984.

[150] Douady, A., Hubbard, J. H., *On the dynamics of polynomial-like mappings,* Ann. Sci. Ecole Norm. Sup. 18 (1985) 287–344.

[151] Dress, A. W. M., Gerhardt, M., Jaeger, N. I., Plath, P. J, Schuster, H., *Some proposals concerning the mathematical modelling of oscillating heterogeneous catalytic reactions on metal surfaces,* in: L. Rensing, N. I. Jaeger (eds.), Temporal Order, Springer-Verlag, Berlin, 1984.

[152] Dubuc, S., Elqortobi, A., *Approximations of fractal sets,* Journal of Computational and Applied Mathematics 29 (1990) 79–89.

[153] Eckmann, J.-P., Ruelle, D., *Ergodic theory of chaos and strange attractors,* Reviews of Modern Physics 57, 3 (1985) 617–656.

[154] Eckmann, J.-P., Kamphorst, S. O., Ruelle, D., Ciliberto, S., *Liapunov exponents from time series,* Phys. Rev. 34A (1986) 4971–4979.

[155] Elton, J., *An ergodic theorem for iterated maps,* Journal of Ergodic Theory and Dynamical Systems 7 (1987) 481–488.

[156] Faraday, M., *On a peculiar class of acoustical figures, and on certain forms assumed by groups of particles upon vibrating elastic surfaces,* Phil. Trans. Roy. Soc. London 121 (1831) 299–340.

[157] Farmer, D., *Chaotic attractors of an infinite-dimensional system,* Physica 4D (1982) 366–393.

[158] Farmer, J. D., Ott, E., Yorke, J. A., *The dimension of chaotic attractors,* Physica 7D (1983) 153–180.

[159] Fatou, P., *Sur les équations fonctionelles,* Bull. Soc. Math. Fr. 47 (1919) 161–271, 48 (1920) 33–94, 208–314.

[160] Feigenbaum, M. J., *Universality in complex discrete dynamical systems,* in: Los Alamos Theoretical Division Annual Report (1977) 98–102.

[161] Feigenbaum, M. J., *Quantitative universality for a class of nonlinear transformations,* J. Stat. Phys. 19 (1978) 25–52.

[162] Feigenbaum, M. J., *Universal behavior in nonlinear systems,* Physica 7D (1983) 16–39. Also in: Campbell, D., Rose, H. (eds.), *Order in Chaos,* North-Holland, Amsterdam, 1983.

[163] Feit, S. D., *Characteristic exponents and strange attractors,* Comm. Math. Phys. 61 (1978) 249–260.

[164] Fine, N. J., *Binomial coefficients modulo a prime number,* Amer. Math. Monthly 54 (1947) 589.

[165] Fisher, Y., Boss, R. D., Jacobs, E. W., *Fractal Image Compression,* to appear in: *Data Compression,* J. Storer (ed.), Kluwer Academic Publishers, Norwell, MA.

[166] Fournier, A., Fussell, D., Carpenter, L., *Computer rendering of stochastic models,* Comm. of the ACM 25 (1982) 371–384.

[167] Franceschini, V., *A Feigenbaum sequence of bifurcations in the Lorenz model,* Jour. Stat. Phys. 22 (1980) 397–406.

[168] Fraser, A. M., Swinney, H. L., *Independent coordinates for strange attractors from mutual information,* Phys. Rev. A 33 (1986) 1034–1040.

[169] Frederickson, P., Kaplan, J. L., Yorke, S. D., Yorke, J. A., *The Liapunov dimension of strange attractors,* Journal of Differential Equations 49 (1983) 185–207.

[170] Geist, K., Parlitz, U., Lauterborn, W., *Comparison of Different Methods for Computing Lyapunov Exponents,* Progress of Theoretical Physics 83,5 (1990) 875–893.

[171] Goodman, G. S., *A probabilist looks at the chaos game,* in: *Fractals in the Fundamental and Applied Sciences,* H.-O. Peitgen, J. M. Henriques, L. F. Peneda (eds.), North-Holland, Amsterdam, 1991.

[172] Grassberger, P., *On the fractal dimension of the Hénon attractor,* Physics Letters 97A (1983) 224–226.

[173] Grassberger, P., Procaccia, I., *Measuring the strangeness of strange attractors,* Physica 9D (1983) 189–208.

[174] Grebogi, C., Ott, E., Yorke, J. A., *Crises, sudden changes in chaotic attractors, and transient chaos,* Physica 7D (1983) 181–200.

[175] Grebogi, C., Ott, E., Yorke, J. A., *Attractors of an N-torus: quasiperiodicity versus chaos,* Physica 15D (1985) 354.

[176] Grebogi, C., Ott, E., Yorke, J. A., *Critical exponents of chaotic transients in nonlinear dynamical systems,* Physical Review Letters 37, 11 (1986) 1284–1287.

[177] Grebogi, C., Ott, E., Yorke, J. A., *Chaos, strange attractors, and fractal basin boundaries in nonlinear dynamics,* Science 238 (1987) 632–638.

[178] Großman, S., Thomae, S., *Invariant distributions and stationary correlation functions of one-dimensional discrete processes,* Z. Naturforsch. 32 (1977) 1353–1363.

[179] Haeseler, F. v., Peitgen, H.-O., Skordev, G., *Pascal's triangle, dynamical systems and attractors,* to appear in Ergodic Theory and Dynamical Systems.

[180] Haeseler, F. v., Peitgen, H.-O., Skordev, G., *On the fractal structure of limit sets of cellular automata and attractors of dynamical systems,* to appear.

[181] Hart, J. C., DeFanti, T., *Efficient anti-aliased rendering of 3D-linear fractals,* Computer Graphics 25, 4 (1991) 289–296.

[182] Hart, J. C., Sandin, D. J., Kauffman, L. H., *Ray tracing deterministic 3-D fractals,* Computer Graphics 23, 3 (1989) 91–100.

[183] Hénon, M., *A two-dimensional mapping with a strange attractor,* Comm. Math. Phys. 50 (1976) 69–77.

[184] Hentschel, H. G. E., Procaccia, I., *The infinite number of generalized dimensions of fractals and strange attractors,* Physica 8D (1983) 435–444.

[185] Hepting, D., Prusinkiewicz, P., Saupe, D., *Rendering methods for iterated function systems,* in: *Fractals in the Fundamental and Applied Sciences,* H.-O. Peitgen, J. M. Henriques, L. F. Peneda (eds.), North-Holland, Amsterdam, 1991.

[186] Hilbert, D., *Über die stetige Abbildung einer Linie auf ein Flächenstück,* Mathematische Annalen 38 (1891) 459–460.

[187] Holte, J., *A recurrence relation approach to fractal dimension in Pascal's triangle,* ICM-90.

[188] Hutchinson, J., *Fractals and self-similarity,* Indiana University Journal of Mathematics 30 (1981) 713–747.

[189] Jacquin, A. E., *Image coding based on a fractal theory of iterated contractive image transformations,* to appear in: IEEE Transactions on Signal Processing, 1992.

[190] Judd, K., Mees, A. I. *Estimating dimensions with confidence,* International Journal of Bifurcation and Chaos 1,2 (1991) 467–470.

[191] Julia, G., *Mémoire sur l'iteration des fonctions rationnelles,* Journal de Math. Pure et Appl. 8 (1918) 47–245.

[192] Jürgens, H., *3D-rendering of fractal landscapes,* in: *Fractal Geometry and Computer Graphics,* J. L. Encarnacao, H.-O. Peitgen, G. Sakas, G. Englert (eds.), Springer-Verlag, Heidelberg, 1992.

[193] Kaplan, J. L., Yorke, J. A., *Chaotic behavior of multidimensional difference equations,* in: *Functional Differential Equations and Approximation of Fixed Points,* H.-O. Peitgen, H. O. Walther (eds.), Springer-Verlag, Heidelberg, 1979.

[194] Kawaguchi, Y., *A morphological study of the form of nature,* Computer Graphics 16,3 (1982).

[195] Koch, H. von, *Sur une courbe continue sans tangente, obtenue par une construction géometrique élémentaire,* Arkiv för Matematik 1 (1904) 681–704.

[196] Koch, H. von, *Une méthode géométrique élémentaire pour l'étude de certaines questions de la théorie des courbes planes,* Acta Mathematica 30 (1906) 145-174.

[197] Kummer, E. E., *Über Ergänzungssätze zu den allgemeinen Reziprozitätsgesetzen,* Journal für die reine und angewandte Mathematik 44 (1852) 93–146.

[198] Lauterborn, W., *Acoustic turbulence,* in: *Frontiers in Physical Acoustics,* D. Sette (ed.), North-Holland, Amsterdam, 1986, pp. 123–144.

[199] Lauterborn, W., Holzfuss, J., *Acoustic chaos,* International Journal of Bifurcation and Chaos 1, 1 (1991) 13–26.

[200] Li, T.-Y., Yorke, J. A., *Period three implies chaos,* American Mathematical Monthly 82 (1975) 985–992.

[201] Lindenmayer, A., *Mathematical models for cellular interaction in development, Parts I and II,* Journal of Theoretical Biology 18 (1968) 280–315.

[202] Lorenz, E. N., *Deterministic non-periodic flow,* J. Atmos. Sci. 20 (1963) 130–141.

[203] Lorenz, E. N., *The local structure of a chaotic attractor in four dimensions,* Physica 13D (1984) 90–104.

[204] Lovejoy, S., Mandelbrot, B. B., *Fractal properties of rain, and a fractal model,* Tellus 37A (1985) 209–232.

[205] Lozi, R., *Un attracteur étrange (?) du type attracteur de Hénon,* J. Phys. (Paris) 39 (Coll. C5) (1978) 9–10.

[206] Mandelbrot, B. B., Ness, J. W. van, *Fractional Brownian motion, fractional noises and applications,* SIAM Review 10,4 (1968) 422–437.

[207] Mandelbrot, B. B., *Fractal aspects of the iteration of* $z \mapsto \lambda z(1-z)$ *for complex* λ *and* z*,* Annals NY Acad. Sciences 357 (1980) 249–259.

[208] Mandelbrot, B. B., *Comment on computer rendering of fractal stochastic models,* Comm. of the ACM 25,8 (1982) 581–583.

[209] Mandelbrot, B. B., *Self-affine fractals and fractal dimension,* Physica Scripta 32 (1985) 257–260.

[210] Mandelbrot, B. B., *On the dynamics of iterated maps V: conjecture that the boundary of the M-set has fractal dimension equal to 2,* in: Chaos, Fractals and Dynamics, Fischer and Smith (eds.), Marcel Dekker, 1985.

[211] Mandelbrot, B. B., *An introduction to multifractal distribution functions,* in: *Fluctuations and Pattern Formation,* H. E. Stanley and N. Ostrowsky (eds.), Kluwer Academic, Dordrecht, 1988.

[212] Mañé, R., *On the dimension of the compact invariant set of certain nonlinear maps,* in: *Dynamical Systems and Turbulence, Warwick 1980,* Lecture Notes in Mathematics 898, Springer-Verlag (1981) 230–242.

[213] Marotto, F. R., *Chaotic behavior in the Hénon mapping,* Comm. Math. Phys. 68 (1979) 187–194.

[214] Matsushita, M., *Experimental Observation of Aggregations,* in: *The Fractal Approach to Heterogeneous Chemistry: Surfaces, Colloids, Polymers,* D. Avnir (ed.), Wiley, Chichester 1989.

[215] Mauldin, R. D., Williams, S. C., *Hausdorff dimension in graph directed constructions,* Trans. Amer. Math. Soc. 309 (1988) 811–829.

[216] May, R. M., *Simple mathematical models with very complicated dynamics,* Nature 261 (1976) 459–467.

[217] Menger, K., *Allgemeine Räume und charakteristische Räume, Zweite Mitteilung: Über umfassenste n-dimensionale Mengen,* Proc. Acad. Amsterdam 29 (1926) 1125–1128.

[218] Misiurewicz, M., *Strange Attractors for the Lozi Mappings,* in Nonlinear Dynamics, R. H. G. Helleman (ed.), Annals of the New York Academy of Sciences 357 (1980) 348–358.

[219] Mitchison, G. J., Wilcox, M., *Rule governing cell division in Anabaena,* Nature 239 (1972) 110–111.

[220] Mullin, T., *Chaos in physical systems,* in: *Fractals and Chaos,* Crilly, A. J., Earnshaw, R. A., Jones, H. (eds.), Springer-Verlag, New York, 1991.

[221] Musgrave, K., Kolb, C., Mace, R., *The synthesis and the rendering of eroded fractal terrain,* Computer Graphics 24 (1988).

[222] Norton, V. A., *Generation and display of geometric fractals in 3-D,* Computer Graphics 16, 3 (1982) 61–67.

[223] Norton, V. A., *Julia sets in the quaternions,* Computers and Graphics 13, 2 (1989) 267–278.

[224] Olsen, L. F., Degn, H., *Chaos in biological systems,* Quarterly Review of Biophysics 18 (1985) 165–225.

[225] Packard, N. H., Crutchfield, J. P., Farmer, J. D., Shaw, R. S., *Geometry from a time series,* Phys. Rev. Lett. 45 (1980) 712–716.

[226] Peano, G., *Sur une courbe, qui remplit toute une aire plane,* Mathematische Annalen 36 (1890) 157–160.

[227] Peitgen, H. O., Prüfer, M., *The Leray-Schauder continuation method is a constructive element in the numerical study of nonlinear eigenvalue and bifurcation problems,* in: *Functional Differential Equations and Approximation of Fixed Points,* H.-O. Peitgen, H.-O. Walther (eds.), Springer Lecture Notes, Berlin, 1979.

[228] Pietronero, L., Evertsz, C., Siebesma, A. P., *Fractal and multifractal structures in kinetic critical phenomena,* in: *Stochastic Processes in Physics and Engineering,* S. Albeverio, P. Blanchard, M. Hazewinkel, L. Streit (eds.), D. Reidel Publishing Company (1988) 253–278. (1988) 405–409.

[229] Pomeau, Y., Manneville, P., *Intermittent transition to turbulence in dissipative dynamical systems,* Commun. Math. Phys. 74 (1980) 189–197.

[230] Prusinkiewicz, P., *Graphical applications of L-systems,* Proc. of Graphics Interface 1986 – Vision Interface (1986) 247–253.

[231] Prusinkiewicz, P., Hanan, J., *Applications of L-systems to computer imagery,* in: "Graph Grammars and their Application to Computer Science; Third International Workshop", H. Ehrig, M. Nagl, A. Rosenfeld and G. Rozenberg (eds.), (Springer-Verlag, New York, 1988).

[232] Prusinkiewicz, P., Lindenmayer, A., Hanan, J., *Developmental models of herbaceous plants for computer imagery purposes,* Computer Graphics 22, 4 (1988) 141–150.

[233] Prusinkiewicz, P., Hammel, M., *Automata, languages, and iterated function systems,* in: *Fractals Modeling in 3-D Computer Graphics and Imaging,* ACM SIGGRAPH '91 Course Notes C14 (J. C. Hart, K. Musgrave, eds.), 1991.

[234] Rayleigh, Lord, *On convective currents in a horizontal layer of fluid when the higher temperature is on the under side,* Phil. Mag. 32 (1916) 529–546.

[235] Reuter, L. Hodges, *Rendering and magnification of fractals using iterated function systems,* Ph. D. thesis, School of Mathematics, Georgia Institute of Technology (1987).

[236] Richardson, R. L., *The problem of contiguity: an appendix of statistics of deadly quarrels,* General Systems Yearbook 6 (1961) 139–187.

[237] Rössler, O. E., *An equation for continuous chaos,* Phys. Lett. 57A (1976) 397–398.

[238] Ruelle, F., Takens, F., *On the nature of turbulence,* Comm. Math. Phys. 20 (1971) 167–192, 23 (1971) 343–344.

[239] Russell, D. A., Hanson, J. D., Ott, E., *Dimension of strange attractors,* Phys. Rev. Lett. 45 (1980) 1175–1178.

[240] Salamin, E., *Computation of π Using Arithmetic-Geometric Mean,* Mathematics of Computation 30, 135 (1976) 565–570.

[241] Saltzman, B., *Finite amplitude free convection as an initial value problem — I,* J. Atmos. Sci. 19 (1962) 329–341.

[242] Sano, M., Sawada, Y., *Measurement of the Lyapunov spectrum from a chaotic time series,* Phys. Rev. Lett. 55 (1985) 1082.

[243] Saupe, D., *Efficient computation of Julia sets and their fractal dimension,* Physica D28 (1987) 358–370.

[244] Saupe, D., *Discrete versus continuous Newton«s method : A case study,* Acta Appl. Math. 13 (1988) 59–80.

[245] Saupe, D., *Point evalutions of multi-variable random fractals,* in: *Visualisierung in Mathematik und Naturwissenschaften - Bremer Computergraphiktage 1988,* H. Jrgens, D. Saupe (eds.), Springer-Verlag, Heidelberg, 1989.

[246] Sernetz, M., Gelléri, B., Hofman, F., *The Organism as a Bioreactor, Interpretation of the Reduction Law of Metabolism in terms of Heterogeneous Catalysis and Fractal Structure,* Journal Theoretical Biology 117 (1985) 209–230.

[247] Siegel, C. L., *Iteration of analytic functions,* Ann. of Math. 43 (1942) 607–616.

[248] Sierpinski, W., *Sur une courbe cantorienne dont tout point est un point de ramification,* C. R. Acad. Paris 160 (1915) 302.

[249] Sierpinski, W., *Sur une courbe cantorienne qui contient une image biunivoquet et continue detoute courbe donnée,* C. R. Acad. Paris 162 (1916) 629–632.

[250] Simó, C., *On the Hénon-Pomeau attractor,* Journal of Statistical Physics 21,4 (1979) 465–494.

[251] Shanks, D., Wrench, J. W. Jr., *Calculation of π to 100,000 Decimals,* Mathematics of Computation 16, 77 (1962) 76–99.

[252] Shaw, R., *Strange attractors, chaotic behavior, and information flow,* Z. Naturforsch. 36a (1981) 80–112.

[253] Shishikura, M., *The Hausdorff dimension of the boundary of the Mandelbrot set and Julia sets,* SUNY Stony Brook, Institute for Mathematical Sciences, Preprint #1991/7.

[254] Shonkwiller, R., *An image algorithm for computing the Hausdorff distance efficiently in linear time,* Info. Proc. Lett. 30 (1989) 87–89.

[255] Smith, A. R., *Plants, fractals, and formal languages,* Computer Graphics 18, 3 (1984) 1–10.

[256] Stanley, H. E., Meakin, P., *Multifractal phenomena in physics and chemistry,* Nature 335 (1988) 405–409.

[257] Stefan, P., *A theorem of Šarkovski on the existence of periodic orbits of continuous endomorphisms of the real line,* Comm. Math. Phys. 54 (1977) 237–248.

[258] Stevens, R. J., Lehar, A. F., Preston, F. H., *Manipulation and presentation of multidimensional image data using the Peano scan,* IEEE Transactions on Pattern Analysis and Machine Intelligence 5 (1983) 520–526.

[259] Sullivan, D., *Quasiconformal homeomorphisms and dynamics I,* Ann. Math. 122 (1985) 401–418.

[260] Sved, M., Pitman, J., *Divisibility of binomial coefficients by prime powers, a geometrical approach,* Ars Combinatoria 26A (1988) 197–222.

[261] Takens, F., *Detecting strange attractors in turbulence,* in: *Dynamical Systems and Turbulence, Warwick 1980,* Lecture Notes in Mathematics 898, Springer-Verlag (1981) 366–381.

[262] Tan Lei, *Similarity between the Mandelbrot set and Julia sets,* Report Nr 211, Institut für Dynamische Systeme, Universität Bremen, June 1989, and, Commun. Math. Phys. 134 (1990) 587–617.

[263] Tél, T., *Transient chaos,* to be published in: *Directions in Chaos III,* Hao B.-L. (ed.), World Scientific Publishing Company, Singapore.

[264] Thompson, J. M. T., Stewart, H. B., *Nonlinear Dynamics and Chaos,* Wiley, Chichester, 1986.

[265] Velho, L., de Miranda Gomes, J., *Digital halftoning with space-filling curves,* Computer Graphics 25,4 (1991) 81–90.

[266] Voss, R. F., *Random fractal forgeries,* in : Fundamental Algorithms for Computer Graphics, R. A. Earnshaw (ed.), (Springer-Verlag, Berlin, 1985) 805–835.

[267] Voss, R. F., Tomkiewicz, M., *Computer Simulation of Dendritic Electrodeposition,* Journal Electrochemical Society 132, 2 (1985) 371–375.

[268] Vrscay, E. R., *Iterated function systems: Theory, applications and the inverse problem,* in: Proceedings of the NATO Advanced Study Institute on Fractal Geometry, July 1989. Kluwer Academic Publishers, 1991.

[269] Williams, R. F., *Compositions of contractions,* Bol.Soc. Brasil. Mat. 2 (1971) 55–59.

[270] Willson, S., *Cellular automata can generate fractals,* Discrete Appl. Math. 8 (1984) 91–99.

[271] Witten, I. H., Neal, M., *Using Peano curves for bilevel display of continuous tone images,* IEEE Computer Graphics and Applications, May 1982, 47–52.

[272] Witten, T. A., Sander, L. M., Phys. Rev. Lett. 47 (1981) 1400–1403 and Phys. Rev. B27 (1983) 5686–5697.

[273] Wolf, A. Swift, J. B., Swinney, H. L., Vastano, J. A., *Determining Lyapunov exponents from a time series,* Physica 16D (1985) 285–317.

[274] Yorke, J. A., Yorke, E. D., *Metastable chaos: the transition to sustained chaotic behavior in the Lorenz model,* J. Stat. Phys. 21 (1979) 263–277.

[275] Young, L.-S., *Dimension, entropy, and Lyapunov exponents,* Ergod. Th. & Dynam. Sys. 2 (1982) 109.

[276] Zahlten, C., *Piecewise linear approximation of isovalued surfaces,* in: *Advances in Scientific Visualization, Eurographics Seminar Series,* (F. H. Post, A. J. S. Hin (eds.), Springer-Verlag, Berlin, 1992.

Index

Bold entries refer to this volume, and the entries in regular type indicate page numbers from *Part One* of *Fractals for the Classroom.*